d. Atomic Units

Quantity	Atomic Units	SI Units
Length (Bohr radius)	$a_o = 1$	0.529177×10^{-10} m
Mass	$m_e = 1$	9.1095×10^{-31} kg
Time	$a_o/\alpha c = 1$	2.418884×10^{-17} s
Bohr velocity	$\alpha c = 1$	2.18769×10^6 m s^{-1}
Energy	$m_e c^2 \alpha^2 = 1$	27.2116 eV
Charge	$e = 1$	1.60219×10^{-19} C
Frequency	$4\pi R_\infty c = 1$	4.134137×10^{-16} Hz

e. Conversions (see also Table 1.1)

$$1 \text{ eV} = 1.60219 \times 10^{-19} \text{ J}$$
$$1 \text{ eV/particle} = 2.30602 \times 10^4 \text{ cal mol}^{-1}$$
$$1/e = 6.241460 \times 10^{12} \text{ charge } \mu\text{C}^{-1}$$
$$1 \text{ Pa} = 7.5 \times 10^{-3} \text{ Torr} = 10 \text{ } \mu\text{bar}$$
$$N = \rho N_0/A \text{ atoms cm}^{-3}$$
$$2(2 \ln 2)^{1/2} = 2.35482$$
$$1 \text{ Å} = 10^{-10} \text{ m} = 0.1 \text{ nm}$$
$$1 \text{ tesla} = 10^4 \text{ gauss}$$

f. SI Prefixes

K	M	G	T	P	E
kilo	mega	giga	tera	peta	exa
10^3	10^6	10^9	10^{12}	10^{15}	10^{18}
m	μ	n	p	f	a
milli	micro	nano	pico	femto	atto
10^{-3}	10^{-6}	10^{-9}	10^{-12}	10^{-15}	10^{18}

g. Radiation Units (see also Table 2.6)

Absorbed dose	1 Gy (gray)	= 100 rad
Absorbed dose rate	1 Gy s^{-1}	= 3.6×10^5 rad h^{-1}
Dose equivalent	1 Sv (sievert)	= 100 rem
Dose rate	1 Sv s^{-1}	= 3.6×10^5 rem h^{-1}
Activity	1 Bq (becquerel)	= 2.7×10^{-11} Ci

Symbols are listed on back endpapers.

Ion Beams for Materials Analysis

Ion Beams for
Materials Analysis

Edited by

J. R. Bird
*ANSTO Lucas Heights Research Laboratories
Menai, Australia*

and

J. S. Williams
*Microelectronics and Materials Technology Centre, RMIT
Melbourne, Australia*

ACADEMIC PRESS
Harcourt Brace Jovanovich, Publishers
Sydney   San Diego   New York   Berkeley
Boston   London   Tokyo   Toronto

ACADEMIC PRESS AUSTRALIA
30–52 Smidmore Street, Marrickville, NSW 2204

United States edition published by
ACADEMIC PRESS INC.
1250 Sixth Avenue
San Diego, California 92101–4311

United Kingdom edition published by
ACADEMIC PRESS, INC. (LONDON) LTD.
24/28 Oval Road, London NW1 7DX

Printed in Australia

National Library of Australia Cataloguing-in-Publication Data

Ion beams for materials analysis.

    Bibliography.
    Includes index.
    ISBN 0 12 099740 1.

    1. Ion bombardment—Industrial applications. 2.
Materials—Analysis. I. Bird, J. R. (John Roger)
II. Williams, J. S. (James Stanislaus).

539.7′3

Library of Congress Catalog Card Number: 86-72997

Contents

Contributors

The numbers in parentheses indicate the pages on which the authors' contributions begin.

J. E. E. Baglin (103), IBM Almaden Research Laboratories, San Jose, CA 95120–6099, USA.

J. R. Bird (3, 149, 515, 551, 581, 607), ANSTO Lucas Heights Research Laboratories, Private Mailbag 1, Menai, NSW 2234, Australia.

R. A. Brown (607), School of Physics, University of Melbourne, Parkville, Vic. 3052, Australia.

E. Clayton (209), ANSTO Lucas Heights Research Laboratories, Private Mailbag 1, Menai, NSW 2234, Australia.

D. D. Cohen (209, 607), Australian Institute of Nuclear Science and Engineering, Private Mailbag 1, Menai, NSW 2234, Australia.

R. G. Elliman (261), Microelectronics and Materials Technology Centre, Royal Melbourne Institute of Technology, Melbourne, Vic. 3000, Australia.

L. C. Feldman (413), AT & T Bell Laboratories, Murray Hill, NJ 07974-2070, USA.

M. J. Kenny (47), CSIRO Division of Applied Physics, Private Mailbag 7, Menai, NSW 2234, Australia.

B. V. King (335), Department of Physics, University of Newcastle, Shortland, NSW 2308, Australia

G. J. F. Legge (443), MARC School of Physics, University of Melbourne, Parkville, Vic. 3052, Australia.

R. J. MacDonald (335, 373), Department of Physics, University of Newcastle, Shortland, NSW 2308, Australia.

D. J. O'Connor (373), Department of Physics, University of Newcastle, Shortland, NSW 2308, Australia.

L. S. Wielunski (581), CSIRO Division of Applied Physics, Private Mailbag 7, Menai, NSW 2234, Australia.

J. S. Williams (3, 103, 261, 515, 551, 581, 607), Microelectronics and Materials Technology Centre, Royal Melbourne Institute of Technology, Melbourne, Vic. 3000, Australia.

Preface

The properties of ion beams and their usefulness for the study of materials remain very much a mystery to many people, even in the general scientific community. The growth in applications of low energy ion beam techniques such as ion implantation and secondary ion mass spectrometry has changed this situation somewhat but higher energy techniques are still regarded as a specialist interest pursued in large laboratories able to afford expensive facilities. This is in spite of the development of purpose-designed accelerators for sample analysis, radioisotope dating and materials modification. The purpose of this book is to provide answers to two questions:

i.   when should I use ion beam techniques rather than other more familiar methods? and

ii. what do I need to know to make effective use of ion beam techniques?

HISTORICAL DEVELOPMENT

Following the discovery of "positive rays" by Goldstein 100 years ago, measurements of their mass provided vital evidence for the atomic theory of the elements. The mass spectrometer was at first a research tool but eventually became widely used as an analytical tool. Other properties of ions, which were discovered during the first half of the twentieth century, such as Rutherford scattering (Geiger and Marsden, 1913), channeling (Stark, 1912), ion induced X-ray emission (Chadwick, 1912) and nuclear reactions (Rutherford, 1919), had to wait until the 1950s, or even the 1970s, before being put to serious use in sample analysis.

Ion induced activation analysis was demonstrated by Seaborg and Livingood (1938) but was outpaced for many years by neutron activation analysis and did not start to grow appreciably until 1955. The first use of Rutherford scattering was reported by Rubin (1950) in a paper entitled "Chemical Analysis by Proton Scattering". Likewise, Rubin *et al.* (1957) reported the first use of ion–ion and ion–gamma reactions for the "Chemical Analysis of Surfaces by Nuclear Methods". These and other demonstration experiments paved the way for the introduction of ion beam analysis during the 1960s and a spectacular growth in the last two decades which has been supported by a number of factors:

- the wide variety of types of information that can be obtained from ion beam analysis, including composition, spatial distributions and atomic structure;
- the special features such as speed, sensitivity, versatility and non-destructive capability;
- the growth in interest in surface properties, atomic structure and trace elements in many rapidly advancing areas of science and technology; and
- the increasing availability of accelerators which are no longer required for nuclear research as well as the design of special accelerators for both low energy and high energy ion beam analysis.

A chart of analytical techniques ordered according to incident ion parameters and type of observed radiation is shown in Fig. 1 and the acronyms used here and throughout the book are defined in Table 1. Ion beam techniques fall naturally into two categories:

i. low energy techniques (less than 50 keV and usually less than 10 keV incident ions); and
ii. high energy techniques (more than 100 keV and usually more than 500 keV incident ions).

Secondary ion mass spectrometry (SIMS), low energy ion scattering (LEIS) and other related methods have become prominent since the early 1970s with the availability of commercial instruments and have shown great sensitivity for characterising the surface layer of atoms of a sample; the quantitative interpretation of results is an area of active physics interest.

BEAM		VIS UV	X	GAMMA	ELECTRON	ION	NEUTRON
				PRODUCT			
	VIS	OES , AAS ,RS				RIMS	
ELECTRO-	UV				UPS	RIMS	
MAGNETIC	X		XRF/D		XPS		
RADIATION	G			PNA,GAA	GAA		PNA
ELECTRON	keV	SIPS			SEM,TEM,AES		
ION	keV	SIPS			AES	SIMS,LEIS	
	MeV		PIXE	PIGME	PAA	RBS ,RBS -C	PNA
				PAA		RFS ,ERA,PNA	
						PNA	
NEUTRON	meV			NAA,PNA	NAA	PNA	ND,NT
	MeV			NAA	NAA	PNA	NT

Fig. 1 Analytical techniques ordered according to incident and observed radiation.

TABLE 1
List of acronyms

AA	Activation Analysis
AAS	Atomic Absorption Spectrometry
ADC	Analog to Digital Converter
AE	Auger Electrons
AES	Auger Electron Spectrometry
AMP	Amplifier
AMS	Accelerator Mass Spectrometry
AXIL	PIXE spectrum processing program
BATTY	PIXE spectrum processing program
BGO	Bismuth Germanate detector
BMDP	BioMedical Data Package
BS	BackScattering
BSE	Backscattered Electrons
CAMAC	Computer Automated Modular Acquisition Control
CFD	Constant Fraction Discriminator
CH	Channeling
CPWBA	Coulomb corrected, Plane Wave Born Approximation
CRO	Cathode-Ray Oscilloscope
CS	Charge Sensitive
CURFIT	PIXE background fitting program
DAC	Digital-to-Analog Converter
DC	Direct Current
DISC	Discriminator
ECPSSR	Energy loss and Coulomb corrected, Perturbed Stationary State Relativistic theory
EDS	Energy Dispersive Spectrometer
EELS	Electron Energy Loss Spectroscopy
EDX	Energy Dispersive X-ray analysis
EPR	Electron Paramagnetic Resonance
EIXE	Electron Induced X-ray Spectroscopy
EMP	Electron MicroProbe
ERA	Elastic Recoil Analysis
ESA	Electrostatic Analyser
ESCA	Electron Spectrometry for Chemical Analysis
ESR	Electron Spin Resonance
EVOLVE	Monte Carlo sputter profiling program
EXAFS	Extended X-ray Absorption Fine Structure
FAS	Flame Atomic Spectrometry
FNAA	Fast Neutron Activation Analysis
FIM	Field Ion Microscope
FS	Forward Scattering
FTIR	Fourier Transform Infra-Red spectroscopy
FWHM	Full Width at Half Maximum
FWTM	Full Width at Tenth Maximum
Ge(Li)	Lithium drifted Germanium Detector
HEX	PIXE spectrum processing program
HIXE	Heavy ion Induced X-ray Emission
HV	High Voltage
IBA	Ion Beam Analysis

TABLE 1 (*cont.*)

ICISS	Impact Collision Ion Surface Scattering
ICP	Inductively Coupled Plasma
IEEE	Institute for Electrical and Electronic Engineers
IG	Intrinsic Germanium detector
II	Ion Implantation
IIXE	Ion Induced X-ray Emission
IMP	Ion MicroProbe
IMS	Ion MicroScope
IS	Ion Scattering
ISO	International Standards Organisation
ISS	Ion Scattering Spectrometry
LAMMA	Laser Microprobe Mass Analyser
LEED	Low Energy Electron Diffraction
LEIS	Low Energy Ion Scattering
LEPS	Low Energy Photon Spectrometer
LLD	Low Level Discriminator
LMIS	Liquid Metal Ion Source
LMP	Laser MicroProbe
LSS	Lindhard-Scharff-Schiott theory
MBE	Molecular Beam Epitaxy
MCA	MultiChannel Analyser
MDL	Minimum Detection Limit
MEIS	Medium Energy Ion Scattering
MS	Mass Spectrometry
MS-ICP	Mass Spectrometry-Inductively Coupled Plasma source
MS-SS	Mass Spectrometry-Spark Source
NA	Not Applicable
NAA	Neutron Activation Analysis
NBS	National Bureau of Standards (USA)
NCRP	National Committee on Radiation Protection
ND	Neutron Diffraction
NIM	Nuclear Instrumentation Modules
NMR	Nuclear Magnetic Resonance
NRA	Nuclear Reaction Analysis
NRA, GG	Nuclear Reaction Analysis, Gamma–Gamma
NRA, GI	Nuclear Reaction Analysis, Gamma–Ion
NRA, GN	Nuclear Reaction Analysis, Gamma–Neutron
NRA, IG	Nuclear Reaction Analysis, Ion-Gamma
NRA, II	Nuclear Reaction Analysis, Ion–Ion
NRA, IN	Nuclear Reaction Analysis, Ion–Neutron
NRA, N	Nuclear Reaction Analysis, Neutron–Neutron
NRA, NG	Nuclear Reaction Analysis, Neutron–Gamma
NRA, NI	Nuclear Reaction Analysis, Neutron–Ion
NT	Neutron Transmission
OES	Optical Emission Spectrometry
OS	Optical Spectrometry
PAA	Particle Induced Activation Analysis
PE	PhotoElectron
PHA	Pulse Height Analyser
PIGE	Proton Induced Gamma-ray Emission

TABLE 1 (*cont.*)

PIGME	Particle Induced Gamma-ray Emission
PIXE	Particle Induced X-ray Analysis
PIXRF	Proton Induced X-ray induced Fluorescence
PM	PhotoMultiplier
PMP	Proton MicroProbe
PNA	Prompt Nuclear Analysis
PSS	Perturbed Stationary State theory
PWBA	Plane Wave Born Approximation
QMA	Quadrupole Mass Analyser
RACE	PIXE yield calculation
RBS	Rutherford Backscattering Spectrometry
RBS-C	Rutherford Backscattering Spectrometry with Channeling
RF	Radio Frequency
RFS	Rutherford Forward Scattering
RHEED	Reflection High Energy Electron Diffraction
RIMS	Resonance Ionisation Mass Spectroscopy
RMS	Root Mean Square
RS	Raman Spectroscopy
RUMP	Ion interaction simulation program
SB	Surface Barrier detector
SCA	Single Channel Analyser
SD	Standard Deviation
SE	Secondary Electrons
SEM	Scanning Electron Microscope
SEMP	Scanning Electron MicroProbe (b = bright field; d = dark field)
Si(Li)	Lithium drifted Silicon Detector
SIMP	Scanning Ion MicroProbe
SIMS	Secondary Ion Mass Spectrometry
SIPS	Secondary Ion Photon Spectrometry
SLS	Strained Layer Superlattice
SNMS	Secondary Neutral Mass Spectrometry
SP	Surface Peak
SPMP	Scanning Proton MicroProbe
SPSS	Statistical Package for the Social Sciences
STEM	Scanning Transmission Electron Microscope
STIM	Scanning Transmission Ion Microscope
STM	Scanning Tunneling Microscopy
TAC	Time-to-Amplitude Converter
TEM	Transmission Electron Microscope
TLA	Thin Layer Activation
TRIM	Ion Interaction Simulation Program
TRYDYN22	Monte Carlo sputter profiling program
TZM	Ti, Zr, Mo alloy
UHV	Ultra High Vacuum
UPS	Ultraviolet Photoelectron Spectrometry
US	UnScattered
WD, WDS	Wavelength Dispersive Spectrometer
XPS	X-ray Photoelectron Spectrometry
XRD	X-Ray Diffraction
XRF	X-Ray Fluorescence

High energy ion scattering (RBS) is the simplest technique to apply and has become a routine tool, together with channeling (RBS-C), for many applications including the characterisation of semiconductor and other thin film and crystalline materials which have assumed such enormous importance in modern technology.

Ion induced X-rays (PIXE) were overshadowed for a long period by X-ray fluorescence and other related techniques but their use for sample analysis has experienced an explosive growth since 1970 because of greater sensitivity, particularly in such areas of renewed significance as pollution studies and microprobe analysis.

Ion induced activation started earlier than other ion techniques and continues to be used in spite of the power of the better known neutron activation analysis. Different sensitivities for many elements and new uses for thin layer activation in wear and corrosion create a growing interest in this technique.

Prompt nuclear reaction analysis (NRA) is the most complex field, offering many different capabilities in surface and depth analysis and profiling; applications have developed rapidly since the mid-1960s but there is still a considerable amount of work needed to fully establish its capabilities.

SCOPE OF BOOK

The use of ion beams for materials analysis involves many different ion–atom interaction processes which previously have largely been considered in separate reviews and texts. A list of books and conference proceedings is given in Table 2. This book is divided into three parts, the first of which treats all ion beam techniques and their applications in such diverse fields as materials science, thin film and semiconductor technology, surface science, geology, biology, medicine, environmental science, archaeology and so on.

For readers not familiar with ion interactions and ion beam techniques, these topics are discussed in general terms in Chapters 1 and 2 respectively and simple examples of their use for composition analysis and depth profiling using elastic scattering of high energy ions can be found in Chapter 3. A detailed description of individual low and high energy techniques is provided by the expert reviews in Chapters 3 to 9. The use of the techniques in the important field of microprobe analysis (with beam spots of the order of 1 μm or less) is treated in Chapter 10 which, together with Chapter 11, includes a critical appraisal of the attributes and limitations of ion beam methods in comparison with other instrumental techniques for composition, spatial and atomic structure

TABLE 2
General references to ion beam analysis

Andersen, H.H., Bottiger, J. and Knudsen, H. (eds). (1980). "Ion Beam Analysis", *Nucl. Instrum. Methods*, **168**.

Benninghoven, A., Evans, C.A., Powell, R.A. and Storms, H.A. (eds). (1979). "Secondary Ion Mass Spectrometry II", Springer-Verlag, Berlin.

Benninghoven, A., Giber, J., Laszlo, J. and Werner, H.W. (eds). (1982). "Secondary Ion Mass Spectrometry III", Springer-Verlag, Berlin.

Biersack, J.P. and Wittmaack, K. (eds). (1986). "Ion Beam Analysis 7", *Nucl. Instrum. Methods*, **B15**.

Bird, J.R. and Clark, G.J. (eds). (1981). "Ion Beam Analysis 5", *Nucl. Instrum. Methods*, **191**.

Chu, W.-K., Mayer, J.W. and Nicolet, M.-A. (1978). "Backscattering Spectrometry", Academic Press, N.Y.

Deconninck, G. (1978). "Introduction to Radioanalytical Physics", Elsevier, Amsterdam.

Demortier, G. (ed.). (1982). "Chemical Analysis: Applications Using Charged Particle Accelerators", *Nucl. Instrum. Methods*, **197**.

Duggan, J.L. and Morgan, I.L. (eds). (1987). "Application of Accelerators in Research and Industry '86", *Nucl. Instrum. Methods*, **B24/25**.

Duggan, J.L., Morgan, I.L. and Martin, J.A. (eds). (1985). "Application of Accelerators in Research and Industry '84", *Nucl. Instrum. Methods*, **B10/11**.

Feldman, L.C. and Mayer, J.W. (1986). "Fundamentals of Surface and Thin Film Analysis", North Holland, N.Y.

Feldman, L.C., Mayer, J.W. and Picraux, S.T. (1982). "Materials Analysis by Ion Channeling", Academic Press, N.Y.

Grime, G.W. and Watt, F. (eds). (1988). "1st Nucl. Microprobe Technol. and Applications". *Nucl. Instrum. Methods*, **B30**.

Lanford, W.A., Tsong, I.S.T. and Williams, P. (eds). (1983). "Ion Beam Analysis 6", *Nucl. Instrum. Methods*, **218**.

Martin, B. (ed.). (1984). "Particle Induced X-ray Emission and its Analytical Applications", *Nucl. Instrum. Methods*, **B3**.

Mayer, J.W. and Rimini, E. (eds). (1977). "Ion Beam Handbook for Materials Analysis", Academic Press, N.Y.

Mayer, J.W. and Ziegler, J.F. (eds). (1974). "Ion Beam Surface Layer Analysis", Elsevier, Lausanne.

Morgan, D.V. (ed.). (1973). "Channeling", Wiley & Sons, London.

Van Rinsvelt, H., Bauman, S., Nelson, J.W. and Winchester, J.W. (eds). (1987). "Particle Induced X-ray Emission and its Analytical Applications", *Nucl. Instrum. Methods*, **B22**.

Watt, F. and Grime, G.W. (eds). (1987). "Principles and Applications of High-Energy Ion Microbeams", Adam Hilger, Bristol.

Wolicki, E.A., Butler, J.W. and Treado, P.W. (eds). (1978). "3rd Int. Conf. on Ion Beam Analysis", *Nucl. Instrum. Methods*, **149**.

Ziegler, J.F. (ed.). (1975). "New Uses of Ion Accelerators", Plenum Press, N.Y.

analysis. Chapter 11 includes lists of examples to be found elsewhere in the book, according to field of application, with an assessment of the usefulness of IBA in each field.

Part 2 of the book is devoted to *reference information*, starting with mathematical and statistical treatments which are common to many techniques (Chapter 12). Chapter 13 contains information on materials properties and processing.

Part 3, Chapter 14, gives *tabulations* of ion beam interaction and related data.

REFERENCES

Chadwick, J. (1912). *Phil. Mag.* **24**, 594.
Geiger, H. and Marsden, E. (1913). *Phil. Mag.* **25**, 604.
Rubin, S. (1950). *Phys. Rev.* **78**, 83.
Rubin, S., Passell, T.O. and Bailey, L.E. (1957). *Anal. Chem.* **29**, 736.
Rutherford, E.J. (1919). *Nature* **103**, 415.
Seaborg, G.T. and Livingood, J.J. (1938). *J. Am. Chem. Soc.* **60**, 1784.
Stark, J. (1912). *Phys. Z.* **13**, 973.

Section 1
Ion Beam Analysis

1

Concepts and Principles of Ion Beam Analysis

J.S. WILLIAMS

Microelectronics and Materials Technology Centre, RMIT
Melbourne, Australia

J.R. BIRD

ANSTO Lucas Heights Research Laboratories
Menai, Australia

ION BEAMS FOR
MATERIALS ANALYSIS
ISBN 0 12 099740 1

1.1 ION–ATOM INTERACTIONS

Ion–atom interactions are controlled by many parameters including ion velocity and energy, ion and atom size, atomic number and mass, and the distance of closest approach. Useful parameters are defined by Equations (1.1) to (1.5), Table 1.1. Important processes which differ from those of the everyday world include:

- the extremely small size of the objects involved;
- the large number of such objects;
- the extremely high velocities involved; and
- the nature of the forces

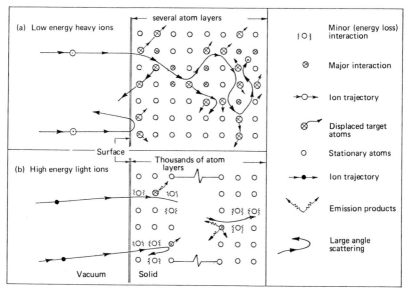

Fig. 1.1 Schematic illustration of the major and minor interactions occurring during ion penetration of a sample for (a) low energy heavy ions, and (b) high energy light ions.

As an ion approaches and interacts with sample atoms we must first consider minor interactions which control the slowing down of an energetic ion and perturb its trajectory. Ions also undergo major interactions such as large angle scattering, atomic displacements, sputtering, inner shell ionisation or nuclear reactions. Emission products from major interactions yield information on the number, type, distribution and structural arrangement of atoms present in the sample; their detection is the basis of ion beam analysis applications. Fig. 1.1 illustrates the differences between interactions of low energy heavy ions which penetrate only a few atom layers undergoing many major interactions and high energy light ions which penetrate thousands of atom layers with only occasional major interactions.

1.1.1 Basic Concepts

a. Dimensions

Ion–atom interactions involve sizes and distances covering many orders of magnitude as illustrated in Fig. 1.2. Values of atomic dimensions are given in Chapter 14.8. A single atom has a radius, Equation (1.3), of the order of 10^{-10} m (1 Å) and is composed of a central nucleus which has a radius of the order of 10^{-14} m and a number of orbiting electrons. The innermost electrons orbit in the K-shell which has a mean radius of the order of 10^{-12} m. Isolated atoms are electrostatically neutral and can easily approach each other until their outermost electrons experience a Coulomb interaction. In a gas, atoms scatter from each other in Brownian motion whereas, in a solid, interactions lead to the formation of bonds which hold atoms together with average spacings of the order of 10^{-9} to 10^{-10} m. For example, 1 mm^3 of Si contains approximately 5 $\times$ 10^{19} atoms. However, atoms are not stationary but undergo thermal vibrations to an extent dependent on the sample temperature. Atomic vibrations which, at room temperature, are typically one-tenth of the atomic spacing in amplitude are significant when using ion beams to investigate atomic structure (Section 1.4). Bonded atoms share outer electrons and, on an atomic scale, a solid consists of a large amount of space occupied by orbital electrons amongst a network of minute atomic nuclei. The question of interest to us is 'What happens to a foreign atom or ion which approaches this complex system?'

The particle/wave duality must be taken into account in the study of particles of atomic dimensions by consideration of the de Broglie wavelengths, Equation (1.1). At 10 eV or higher, a proton has a wavelength which is much less than atomic spacings and it can easily pen-

TABLE 1.1
General equations

	Equations	Units	
Dimensions			
Wavelength of Particle	$\lambda = 2.872 \times 10^{-11} (ME)^{-\frac{1}{2}}$	m	(1.1)
Radius of Nuclei	$R = 1.2 \times 10^{-15} A^{\frac{1}{3}}$	m	(1.2)
Radius of Atomic Shell n	$R_n = 5.292 \times 10^{-11} n^2/Z'_n$	m	(1.3)
Screening Radius	$a = 4.69 \times 10^{-13} (Z_1^{2/3} + Z_2^{2/3})^{-\frac{1}{2}}$	m	(1.4a)
	$a = 4.69 \times 10^{-13} (Z_1^{0.23} + Z_2^{0.23})$	m	(1.4b)
Distance of Closest Approach	$D_0 = 1.440 \times 10^{-9} Z_1 Z_2 (M_1 + M_2)/M_2 E_1$	m	(1.5)
Velocities			
Velocity of V volt ions	$v = 1.389 \times 10^4 \sqrt{E/M}$	m/s	(1.6)
Time of Flight	$t = 71.99 \sqrt{M/E}$	μs/m	(1.7)
Velocity of n-shell Electrons	$v_e = 2.188 \times 10^6 Z'_n/n$	m/s	(1.8)
Interactions			
Coulomb Barrier	$E_b = Z_1 Z_2 (M_1 + M_2)/M_2 (A_1^{1/3} + A_2^{1/3})$	MeV	(1.9)
Coulomb Force	$F = Z_1 Z_2 e^2/r^2 = -Z_1 e \nabla\Phi$		(1.10)
Screened Coulomb Potential	$\Phi_s = (Z_1 Z_2 e^2/r) F(r/a)$		(1.11)
Molière Potential	$F(r/a) = 0.35 e^{-0.3(r/a)} + 0.55\, e^{-1.2(r/a)}$ $+ 0.1 e^{-6(r/a)}$		(1.12)

Lindhard Potential

$$\Phi_L = Z_1 Z_2 e^2 \left(1/r - 1/[r^2 + c^2 a^2]^{\frac{1}{2}}\right) \qquad (1.13)$$
where $c \sim 3$ is a fitting parameter

Angular Momentum

$$L = M_1 v D_c \qquad (1.14)$$

Equilibrium Charge

$$\bar{q} = Z_1(1 - \exp[-0.741 v_R]) \qquad (1.15)$$

for solids $(Z_1 > 3)$

$$v_R = (v_1 \hbar / e^2)^{0.903}/Z_1^{0.41}$$

Yields

Interaction Probability

$$P = \sigma \, dx \qquad (1.16)$$

Number of Interactions

$$N = N_1 \sigma \, dx \qquad (1.17)$$

Mean Free Path

$$p = 1/\sigma \text{ atom cm}^{-2} = m/\sigma \text{ g cm}^{-2} \qquad (1.18)$$

Attenuation

$$N_t = N_1 \exp(-\mu t) \qquad (1.19)$$

Product Yield

$$Y_3 = eN_3 (E_3, \theta) = eN_1 (d\sigma_3/d\theta) \, d\Omega \, dx \qquad (1.20)$$

(E in eV, M in amu, m in g.atom^{-1}, dx in atom cm^{-2}, σ in cm^2 atom^{-1}, μ in cm^{-1}, t in cm, Ω in sr, e = fractional efficiency) etrate between the atoms of a solid. Heavier particles have wavelengths which are inversely proportional to the square root of the mass, Equation (1.1), and the nucleus of an energetic heavy ion penetrates amongst sample atoms quite easily. However, the heavy ions used in ion beam experiments are usually only partially ionised and still have many orbital electrons surrounding the nucleus and there are frequent interactions between electrons of ions and atoms.

Some examples are included in Fig. 1.2 of the depths to which energetic ions can penetrate solid samples of various atomic masses. However, the physical distance travelled by an ion through a sample is not a relevant parameter in ion–atom interactions. The probability of an interaction depends on the number of atoms that an ion passes rather than the actual distance travelled. The important parameter is the areal density (atoms cm^{-2}). At low energies, the number of atoms available for a particular type of interaction is often expressed as a fraction of a monolayer which is the order of 10^{15} atoms cm^{-2}. A distance scale in monolayers is shown at the right of Fig. 1.2.

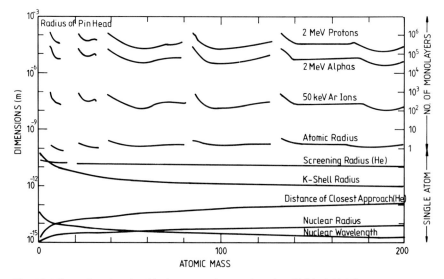

Fig. 1.2 Dimensions involved in ion–atom interactions (see Table 1.1). The upper curves show approximate ranges of various ions as a function of sample atomic mass.

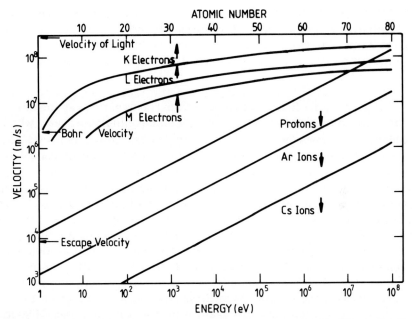

Fig. 1.3 Typical velocities involved in ion–atom interactions (see Table 1.1) as a function of sample atomic number (upper curves) or ion energy (lower curves). The escape velocity is the magnitude for spacecraft escape from the earth (arrows indicate scale to be used).

b. Velocities

The velocities of protons, Ar and Cs ions are shown in Fig. 1.3 as a function of ion energy, Equation (1.6). At very high energies (> 100 MeV) protons become relativistic with velocities asymptotic to the velocity of light. Heavy ion velocities become relativistic at much higher energies. From ion velocities we can calculate their time-of-flight, Equation (1.7), which for a proton is approximately 72.3 μs m^{-1} at 1 eV. The velocity of orbital electrons, Equation (1.8), as a function of atomic number (Z) are included in Fig. 1.3. The inner or K-shell electrons have the highest velocity and are relativistic for the atoms of heavy elements. For high Z elements, the successively filled outer shells of electrons have velocities of 10^6 to 10^7 ms^{-1}.

c. Forces

Electromagnetic and nuclear forces are the basis of keV and MeV ion–atom interactions and hence of methods for determining the elemental and/or isotopic composition of a sample. The most important force in ion–atom interactions is the Coulomb or electrostatic force between two charges (Z_1e and Z_2e) when they are separated by a distance r, Equation (1.10). Two processes modify this force — charge exchange and screening.

i. *Charge Exchange* (loss or gain of electrons) takes place as an ion moves amongst the atoms of a sample and it adopts an equilibrium charge which depends on the velocity. At very low velocities, the ion is neutralised during its first interaction with the electron distribution of surface atoms, and ensuing collisions are neutral atom–atom collisions. For heavy ions at velocities less than the velocity of orbital electrons, quasi-molecular electron orbits may be temporarily established between the ion and a target atom. An ion or atom leaving the surface after an interaction also adopts a neutral or ionised state depending on its velocity and the local Coulomb field strength. These effects are discussed in more detail in Chapter 8. At MeV energies, light ions such as alpha particles are fully stripped immediately they enter a solid sample and only the positive nuclei penetrate the sample. After these have been slowed down to low energies they pick up electrons again from the sample atoms. Heavier ions, with velocities greater than those of orbiting electrons in sample atoms reach an equilibrium charge state after travelling through sufficient thickness of sample (typically 5 to 50 μg cm^{-2}). The mean charge ($\bar{q}$) is given approximately by Equation (1.15) (Betz, 1983). However, this is the mean value of a broad, approximately Gaussian charge state distribution which, being velocity dependent, changes as the ion slows down.

ii. *Screening* occurs because outer electrons do not feel the full electrostatic potential of the positively charged nucleus. This effect can be approximated by the use of an effective charge (Z') in place of the atomic number (Z) to describe the influence of the nuclear charge. Typical values of effective charge are ($Z - 0.3$) and ($Z - 4.15$) for considering the effective potential experienced by electrons in the K and L shells. Screening also affects the interaction between an incident ion and sample atoms. The degree of interpenetration of the electron clouds of ion and atom determines the extent to which the field of the positively charged nucleus of each is balanced by the field of intervening electrons. A screened Coulomb potential can be used to include these effects, Equation (1.11), and two forms of the screening distance (a) are given in Equation (1.4). If the screening function (F) is zero, we have isolated atoms at large separations. If F is approximately unity, i.e. at small separations, the full Coulomb potential operates between the positive charges of the nuclei of both ion and atom.

There are several possible choices for interatomic potentials which include screening. The most accurate and useful potentials for ion beam analysis are based on the Thomas-Fermi model. The most widely used approximations are the Molière potential, Equation (1.12), and the Lindhard potential, Equation (1.13), for large impact parameters. The Molière potential is useful over wide energy and mass regimes but other forms of the screening function are often employed to match theory and experiment, for example in calculations of energy loss and range (Biersack and Ziegler, 1982; O'Connor and Biersack, 1986). The Lindhard potential is well suited for describing channeling phenomena (Chapters 6 and 9). Finally, we must consider nuclear forces. If the nuclei of ion and atom approach so closely that their separation is less than the sum of the nuclear radii, a potential well is created. The attractive nuclear force confines the component nucleons of both ion and atom, at least temporarily, within a 'compound nucleus'. This is the regime of nuclear reactions.

d. Distance of Closest Approach

The distance of closest approach (D_c) between atoms and nuclei is important in determining the nature of ion–atom interactions. It is useful to consider its dependence on initial velocity and the impact parameter (D_1), which is the initial separation of ion and atom centres perpendicular to the ion trajectory. These parameters are shown in Fig. 1.4a in relation to the hyperbolic trajectory of an ion subject to Coulomb repulsion. For a large impact parameter, the ion undergoes small angle scattering and the target atom recoils as in an elastic billiard ball collision. For a head-on collision (zero impact parameter, Fig. 1.4b), the distance of closest approach has its minimum value (D_0) which is given by Equation (1.5). If the distance of closest approach is less than the sum of atomic or nuclear radii, excitations can take place and classical collision mechanics must be replaced by a quantum mechanical treatment of the collision process.

The *Coulomb Barrier* is the energy which an ion must have for the distance of closest approach to be such that two nuclei just touch, Equation (1.9). This energy represents a qualitative lower limit to ion energy at which nuclear reactions readily take place. It is not a strict limit because quantum-mechanical tunnelling through the potential barrier can lead to nuclear reactions but with a low probability. For example, the Coulomb barriers for proton or He ion irradiation of Al are approximately 3 or 6 MeV respectively whereas some nuclear reactions can be

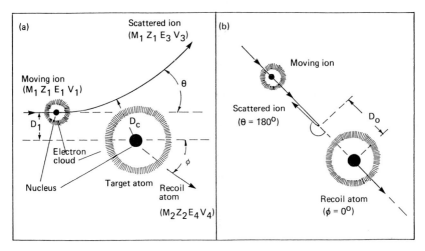

Fig. 1.4 Ion paths for different impact parameters (D_i) showing the distance of closest approach (a) D_c for large impact parameters and (b) D_0 for zero impact parameter.

initiated using energies from 0.5 to 2 MeV. The probability for nuclear reactions increases rapidly at energies above the Coulomb barrier.

e. Angular Momentum, Spin and Parity

An ion deflected by an atomic potential has an angular momentum given by Equation (1.14) where v and D_c are the velocity and separation at the point of closest approach. In addition, atoms and nuclei have internal angular momentum associated with orbiting electrons or nucleons. Spin and parity are also important in describing stable or excited states of atoms and nuclei. Excitation involves changes in these parameters according to selection rules and conservation laws which control the probabilities of excitation during ion–atom interactions and subsequent de-excitation. This topic is taken up as necessary in chapters on individual techniques but a knowledge of measured probabilities is the main requirement for ion beam analysis.

f. Interactions

In any ion–atom interaction, if no excitation of the target atom or nucleus occurs, the interaction is called an elastic collision. However, there is an energy transfer from projectile to target which is determined by the laws of conservation of energy and momentum (Section 1.1.1*i*).

The energy transfer becomes negligibly small for the very large impact parameters which are involved in forward (low angle) scattering. If ion and atom come close enough together for either atomic or nuclear excitation to occur, an inelastic collision is involved. Such events are given a variety of names depending on the type of excitation. An energy deficit is introduced into the equations of motion and a corresponding amount of energy becomes available for emission of interaction products such as X-rays or γ-rays. With two main types of force (Coulomb and nuclear) and two types of collision (elastic and inelastic), there are four processes to consider (Fig. 1.5):

- elastic atomic collisions;
- inelastic atomic collisions with atomic excitation;
- elastic nuclear collisions; and
- inelastic nuclear collisions with nuclear excitation.

i. *Elastic Atomic Collisions* are of primary importance at very low energies (typically less than 10 keV for ions such as Ar^+) for which ion

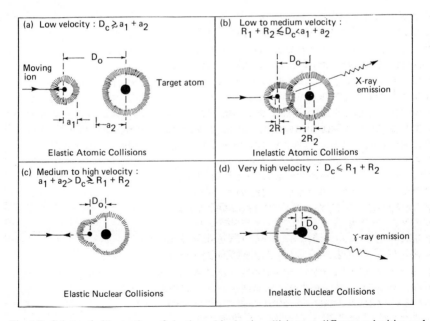

Fig. 1.5 Schematic illustration of elastic and inelastic collisions at different velocities and distances of closest approach for head-on collisions ($D_1 = 0$) (symbols defined as shown and as in Fig. 1.4).

and atom scarcely penetrate before the ion is scattered and the atom recoils (Fig. 1.5a). The distance of closest approach is greater than the sum of the screening radii of ion and atom ($a_1 + a_2$) and the interaction involves the screened Coulomb potential. Approximate energy limits below which screening is important are given in Table 1.2 for various ion and atom types.

ii. *Inelastic Atomic Collisions* can occur for an energy characteristic of atomic energy levels. As the ion velocity increases, the probability for atomic excitation, including ionisation, increases — at first for outer shell electrons and then for inner shell electrons. At the same time, the probability for hard-sphere elastic collisions decreases as a proportion of the total collision probability. The increasing probability of excitation leads to an increased emission of photons and electrons (Fig. 1.5b). These processes are most probable when the incident ion velocity is close to that of orbital electrons in specific shells. In general, protons must have energies of many keV to MeV and heavier ions must have higher energies in order to maximise the probability for inner shell excitation. Although inelastic collisions have a high probability at such energies, their effect on the incident ion is limited to a slow loss in energy and small changes in direction.

iii. *Elastic Nuclear Collisions* (Coulomb or Rutherford Scattering) are important if the distance of closest approach of the nuclei of ion and sample atom is of the order of the sum of the screening radii, when the unscreened Coulomb potential accurately describes the interaction process. For small impact parameters, such as the head-on collision in Fig. 1.5c, the incident ion undergoes a major change in direction while the nucleus of the target atom receives considerable recoil energy.

TABLE 1.2
Energy limits for elastic ion–atom collisions (eV)

TARGET	C	Si	Cu	Ag	Au
ION					
H	0.414	1.163	2.93	5.46	10.75
He	1.087	2.68	6.30	11.45	22.2
Ar	46.8	68	107	160	26.6
Kr	200	254	338	453	676
Xe	498	585	722	903	1250

iv. *Inelastic Nuclear Collisions* are least common since they require a distance of closest approach smaller than the sum of the nuclear radii of ion and atom ($R_1 + R_2$) (Fig. 1.5d). Excitation of the target nucleus can be produced through the effects of either the Coulomb force or the nuclear force. Coulomb interactions are only possible when the incident ion energy exceeds that of the first excited state of the nucleus (often 100 keV or greater). The energy lost by the ion, over and above the kinematic energy transfer, appears as excitation of the target nucleus and is eventually given up by photon emission. Interactions involving the nuclear force include excitation or the transfer of one or more nucleons from one nucleus to the other followed by the emission of high energy particles and photons and often involves the formation of new isotopes.

g. Cross-sections

The probability of a particular type of interaction is quantified by the interaction cross-section (σ). The cross-section is the effective area presented to the incoming ions at each interaction centre (atom or nucleus) and has the units cm^2 per atom or barns where 1 barn $= 10^{-24}$ cm^2 per atom. The probability that an incoming particle will interact within a given sample thickness is the product of the cross-section and the areal atom density (dx in atoms cm^{-2}), Equation (1.16). The number of interactions is given by Equation (1.17). For example, the elastic atomic scattering of 30 keV Kr ions by Cu atoms gives the effective interaction

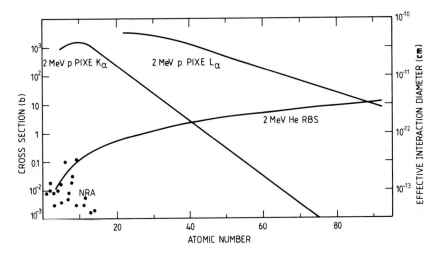

Fig. 1.6 Typical ion-interaction cross-sections.

diameter of each scattering centre as approximately 0.4Å. Since this is of the order of one atom spacing, most Kr ions interact with the first atom that they meet. By contrast, the elastic nuclear scattering of 2 MeV He ions is much less frequent and corresponds to an effective diameter of 10^{-4} Å for Cu nuclei. Nuclear reactions are much less frequent again. Some typical cross-sections for high energy ion–atom interactions are shown in Fig. 1.6.

The Mean Free Path (p) of an ion is the average distance between collisions and can be calculated from Equation (1.18). High cross-sections imply small mean free paths and, in cases such as scattering of low energy ions (e.g. < 10 keV), the mean free path is so small that there is a significant attenuation of the incident beam flux by removal of ions from the forward directed beam through major interactions with early atomic layers of a sample. The beam flux is then no longer constant with depth (x) but is given by Equation (1.19) where μ is the attenuation coefficient. This equation also applies to photons and neutrons which are attenuated when passing through material but do not undergo energy loss. Attenuation of ions is negligible at MeV energies since the cross-sections for major interactions are quite small.

h. Angular Distributions

Equations (1.16) and (1.17) define the total cross-section for all inter-actions of a specific type, irrespective of the geometry and efficiency with which the interaction products may be observed. If we replace σ in these equations by $d\sigma_3/d\theta$ we obtain a relation for the probability of an interaction leading to scattering or emission of radiation in direction θ, Equation (1.20). The probability of emission usually varies with angle and this angular distribution must be taken into account when selecting suitable conditions for sample analysis.

i. Kinematics

The energies of ion and atom after an interaction are governed by the laws of conservation of energy and of momentum (both parallel and perpendicular to the incident ion direction). The resulting kinematic relations, Equations (1.21) to (1.30), are summarised in Highlight 1.1 for elastic scattering and nuclear reactions with symbols defined in Fig. 1.7c. In most cases, the positive sign is used in Equation (1.21a or b) but there are two special cases which are important in IBA. Firstly, the energy release (Q), which is positive for most useful nuclear reactions, is sometimes negative. If so, the incident energy must exceed the threshold value, Equation (1.26), before an interaction can occur. For energies above the threshold but below E_{lim}, Equation (1.28), conservation of forward momentum requires that product particles can only be observed within a forward cone whose angle is given by Equation (1.27) and their energies are double-valued at each angle (both signs in Equation (1.21a)). This is illustrated in Fig. 1.7a which shows the ratio of product energy to total energy as a function of angle for the $^{7}\text{Li(p,n)}^{7}\text{Be}$ reaction (Chapter 4). For $E_1 = E_{\text{lim}}$, $\theta_{\text{max}} = 90°$ and above this energy normal kinematics (positive sign only) apply.

Secondly, if the incident ion is heavier than the target atom, a similar situation applies in scattering which is then only possible within a cone defined by Equation (1.30). The scattering energies at a specific angle are again double-valued (both signs in Equation (1.21b)). However, the recoil energy is single-valued, Equation (1.22b), and can exceed the scattered energy — a fact that is exploited in Elastic Recoil Analysis (ERA, Chapter 3). Fig. 1.7b shows the ratios E_3/E_1 and E_4/E_1 as a function of θ and ϕ respectively when $M_1 = 2M_2$. The high E_3 group corresponds to large impact parameter collisions in which the incident ion transfers very little energy to the target atom. This is similar to Rutherford Forward Scattering (RFS) ($M_1 < M_2$) which has application in the Proton Microprobe (PMP) (Chapter 10). The low energy group in Fig. 1.7b corresponds to head-on collisions when the heavy incident ion follows through with enough energy to satisfy the conservation of momentum. Since both energy and cross-section depend on angle, the choice of experimental geometry is important in most IBA applications.

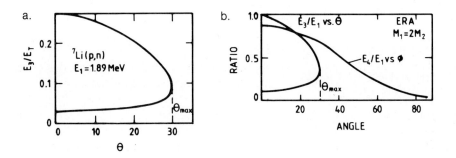

Fig. 1.7 (a) E_3/E_T versus θ for the $^{7}\text{Li(p,n)}$ ^{7}Be reaction; (b) E_3/E_1 versus θ and E_4/E_1 versus ϕ for Elastic Recoil Analysis.

HIGHLIGHT 1.1

KINEMATIC RELATIONS

$$\mu = M_2/M_1$$
$$E_T = E_1 + Q = E_3 + E_4$$
$$W = (M_2M_3E_T + M_1M_3Q)/M_1M_4E_1$$
$$X = M_1M_3E_1/E_T(M_1 + M_2)(M_3 + M_4)$$
$$Y = M_2M_4(1 + M_1Q/M_2E_T)/(M_1 + M_2)(M_3 + M_4)$$
$$K = [(\cos\theta \pm (\mu^2 - \sin^2\theta)^{1/2})/(1 + \mu)]^2$$

---

a. NUCLEAR REACTIONS

Normal Kinematics $(M_2 > M_1; Y \geq X)$

  Use positive sign only unless $Y > X$

PRODUCT ION	$E_3 = XE_T[\cos\theta \pm Y/(X\text{-}\sin^2\theta)^{\frac{1}{2}}]^2$	(1.21a)
RECOIL ATOM	$E_4 = E_1 + Q - E_3$	(1.22a)
RECOIL ANGLE	$\sin\theta - (M_3E_3/M_4E_4)^{\frac{1}{2}}\sin\theta$	(1.23a)

Centre of Mass Conversion

PRODUCT ION $\quad \dfrac{\sigma(\Theta)}{\sigma(\theta)} = \dfrac{\sin\theta\, d\theta}{\sin\Theta\, d\Theta} = [XY(Y/X\text{-}\sin^2\theta)]^{\frac{1}{2}}E_T/E_3$

(1.24a)

RECOIL ATOM $\quad \dfrac{\sigma(\Phi)}{\sigma(\phi)} = \dfrac{\sin\phi\, d\phi}{\sin\Phi\, d\Phi} = [XY(W - \sin^2\phi)]^{\frac{1}{2}}E_T/E_3$

(1.25a)

Threshold Reactions $(Q < 0)$ Use both signs in Equation (1.21a)

THRESHOLD ENERGY $\quad E_t = -Q(M_1 + M_2)/M_2$ $\qquad$ (1.26)

$\qquad\qquad\qquad\qquad$ and for $E_t < E_1 < E_{\lim}$

CONE ANGLE

$$\theta_{\max} = \sin^{-1}\{M_2M_4[E_1 + Q(1 + M_1/M_2)]/M_1M_3E_1\}^{\frac{1}{2}} \quad (1.27)$$

ENERGY LIMIT $\quad E_{\lim} = M_4Q(M_1 + M_2)/(M_1M_2 - M_2M_4) \qquad$ (1.28)

PRODUCT ENERGY
VERSUS ANGLE

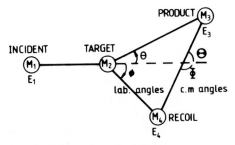

Fig. 1.7 (c) Parameters used in kinematics calculations.

b. **ELASTIC SCATTERING**

$(M_3 = M_1; M_4 = M_2; Q = 0)$

Use positive sign only unless $M_1 > M_2$

$$E_3 = E_1\{(\cos\theta \pm [\mu^2 - \sin^2\theta]^{1/2})/(1 + \mu)\}^2 = KE_1 \qquad (1.21b)$$

$$E_4 = 4 M_1 M_2 E_1 \cos^2\phi/(M_1 + M_2)^2 = (1 - K)E_1 \qquad (1.22b)$$

$$\sin\phi = [K/\mu(1 - K)]^{\frac{1}{2}}\sin\theta$$

$$\frac{\sigma(\Theta)}{\sigma(\theta)} = \frac{\sin\theta\,d\theta}{\sin\Theta\,d\Theta} = M_2(\mu^2 - \sin^2\theta)^{\frac{1}{2}}/K(1 + \mu)^2 \qquad (1.24b)$$

$$\frac{\sigma(\Phi)}{\sigma(\phi)} = \frac{\sin\phi\,d\phi}{\sin\Phi\,d\Phi} = 1/4\cos\phi \qquad (1.25b)$$

Elastic Recoil $(M_2 > M_1)$ Use both signs in Equation (1.21b)

PRODUCT AND RECOIL
ENERGY VERSUS ANGLE

ENERGY LIMIT $E_{4max} = 4M_1 M_2 E_1/(M_1 + M_2)^2$; $\phi = 0°$ (1.29)

ANGLE LIMIT $\theta_{max} = \sin^{-1}\mu$; $\phi_{max} = \pi/2$ (1.30)

1.1.2. Minor Interactions

In this section we will consider energy loss and multiple scattering during sample penetration by medium to high energy ions. These processes involve a multitude of low energy transfer ion–atom collisions and, although each collision usually has only a minor effect on the ion energy and direction, they are very important in bringing to the methods of ion beam analysis a unique capability for deriving depth information (Section 1.3). Another consequence of minor interactions is secondary electron emission which also accompanies ion irradiation of a sample. This process is not useful for sample analysis but can be used as an imaging tool in the high energy proton microprobe (Chapter 10).

Useful parameters, which depend on energy loss processes, include the rate of energy loss, stopping power or stopping cross-section, range and path length, straggling and multiple scattering. The last two arise from the statistical distribution of the number and type of collisions which each ion undergoes and cause a parallel monoenergetic incident beam to become increasingly distributed in energy and direction as it penetrates a sample. Energy loss equations are summarised in Table 1.3.

a. Energy Loss

At non-relativistic energies, energy loss can be conveniently divided into two independent processes, namely electronic stopping (inelastic ion–atom collisions) and nuclear stopping (elastic ion–atom collisions). The relative magnitude of the nuclear and electronic terms depends on the ion energy and also on the particular target–projectile combination (Chu,

TABLE 1.3
Energy loss equations

RATE OF ENERGY LOSS (Stopping cross-section, $eV cm^2 atom^{-1}$ or stopping power, $eV cm^2 g^{-1}$)

Components

$$dE/dx = (dE/dx)_n + (dE/dx)_e + (dE/dx)_r \tag{1.31}$$

Thomas-Fermi Energy

$$E_{TF} = 4.68 \times 10^{-11} M_2/(M_1 + M_2)Z_1Z_2(Z_1^{2/3} + Z_2^{2/3})^{\frac{1}{2}} \tag{1.32}$$

Lindhard-Scharff Region

$$dE/dx = -(8\pi a_o/\hbar)\, \xi Z_1 Z_2 v_1/(Z_1^{2/3} + Z_2^{2/3}) \tag{1.33}$$

Lindhard-Scharff Limit

$$v_1 = Z_1^{2/3} e^2/\hbar \tag{1.34}$$

Bethe Region

$$dE/dx = -4\pi(N_0 e^4 Z_1^2 Z_2/m_o v_1^2 A_2)L \tag{1.35}$$

$$L = ln(2m_o v_1^2/I) + \text{Correction terms} \tag{1.36}$$

1980; Kumakhov and Komarov, 1981). For example, for a Si target, the nuclear and electronic energy loss components are equal in magnitude at 10 keV for B ions, 150 keV for P ions, 750 keV for As ions and 1.8 MeV for Sb ions.

The variation of rate of energy loss over a wide range of incident energies is shown in Fig. 1.8 for a Si sample. Nuclear stopping is important at energies low enough for the distance of closest approach to be greater than the screening radius; i.e. for energies less than the Thomas-Fermi energy, Equation (1.32). The contribution from nuclear stopping is evident in the low energy peak for the energy loss of As ions in Fig. 1.8. The corresponding effect for protons is not apparent since nuclear stopping is only about 16% of the total at 1 keV and much less at higher energies. Nuclear stopping is only important near the end of the range of high energy light ions. For energies above the Thomas-Fermi energy, electronic energy loss dominates and the rate of energy loss, Equation (1.33), is proportional to the ion velocity (Lindhard and Scharff, 1961). For ion velocities approaching and exceeding the limit given by Equation (1.34), charge-exchange effects become important and the rate of energy loss decreases. This is known as the Bethe region which can be described with reasonable accuracy by the Bethe relation, Equations (1.35) and (1.36). For much higher velocities (approaching the velocity of light) relativistic processes such as bremmstrahlung and

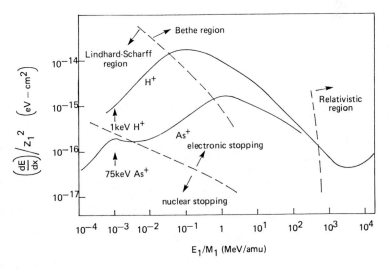

Fig. 1.8 Dependence of energy loss on ion energy (reduced units) for H^+ and As^+ illustrating different energy loss regimes.

Cerenkov radiation introduce a third term into Equation (1.31) and cause the rate of energy loss to slowly increase.

For analysis applications using low energy (< 10 keV) heavy ions, nuclear stopping is the dominant energy loss mechanism. In the Lindhard-Scharff region both nuclear and electronic processes contribute but this region (e.g. 10 to 100 keV light ions) is not often used for IBA. With MeV light ions, for which the Bethe region of electronic energy loss dominates, extensive data tabulations are available and semi-empirical relations are also used to obtain values of the rate of energy loss (Chapter 14.1).

The wide range of impact parameters and atomic excitation parameters involved in energy loss events leads to a statistical distribution in both number and size of such events for a specific depth of penetration. This is known as energy straggling. The stopping cross-section and stopping power are thus mean values and the energy distribution of ions broadens as a function of distance travelled through the sample (Fig. 1.9). For short distances, the distribution is an asymmetric Vavilov distribution (Deconninck and Fouilhe, 1975). For large distances where energy loss is greater than a few per cent of the energy, the distributions become approximately Gaussian. The distribution in depths at which ions reach a specific energy (lower than the incident energy but greater than zero) can be obtained from the straggling distribution. However in most cases,

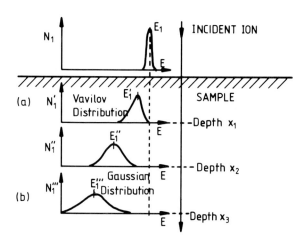

Fig. 1.9. Energy loss straggling versus depth: (a) Vavilov distributions at short distances; and (b) Gaussian distributions at large distances.

it is sufficient to use the relation: $\Delta x = \Delta E/(dE/dx)$ to convert energy-loss straggling into an equivalent path-length straggling provided that the mean energy loss is relatively small. Details of ion straggling are given in Chapter 14.2.

In MeV ion beam analysis, it is of interest to consider the sample thickness for which there is a sufficient drop in ion energy to change the probability for a major interaction (e.g. large-angle scattering, X-ray emission or nuclear reaction) by a significant amount. A thin sample (with negligible energy loss) can be taken as typically of the order of 100 μg cm^{-2} for MeV protons and 10 μg cm^{-2} for similar energy He ions. Corresponding thick samples (thick enough to stop an ion) are of the order of 10 mg cm^{-2} and 1 mg cm^{-2} respectively.

b. Range and Path Length

The total path length of an ion can be calculated by integration of the energy loss equation from zero to the incident energy. This is not especially useful since frequent collisions cause a departure from linearity of the ion trajectory and also give rise to a statistical distribution in path lengths. The projection of the path length (R_t) on the original ion direction is known as the projected range (R_p) and is the quantity that is most useful in practice (Fig. 1.10). Fluctuations in the shape of ion trajectories affect the ratio of projected range to path length and introduce distributions in both the projected range and lateral displacement (at right angles to the initial ion direction). For high energy ions (Fig. 1.10a), approximately Gaussian shaped profiles are observed for the longitudinal and lateral distributions having standard deviations, ΔR_p and ΔR_l respectively. For low energy ions a significant fraction can be scattered out of the surface so that the range distribution of ions remaining within the sample is both asymmetric and truncated (Fig. 1.10b). The average depth penetrated by an ion for a given value of energy loss can be determined by taking the difference between tabulated ranges for the initial and final energy. Details of ion ranges are given in Chapter 14.1.

c. Multiple Scattering and Lateral Spread

The gradual increase in the diameter of an ion beam as it passes into a sample is termed the lateral spread while the associated increasing distribution in direction of ions relative to the initial direction is referred to as the angular spread. These can be estimated from theories of multiple scattering (Sigmund and Winterbon, 1974; Marwick and

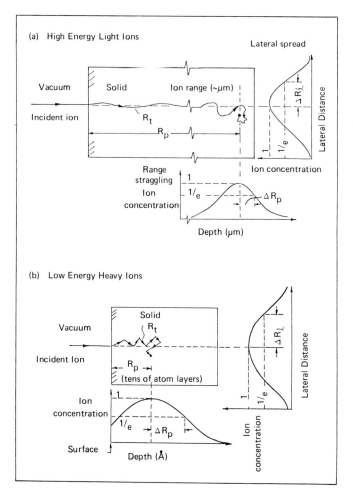

Fig. 1.10 Schematic illustration of the path length (R_t), range (R_p), range straggling (ΔR_p) and lateral spread (ΔR_l) for (a) high energy light ions, and (b) low energy heavy ions.

Sigmund, 1975) as shown in Chapter 14.2. Path length fluctuations also apply to a scattered or product ion as it travels out of the sample. Energy loss and path length parameters must therefore be estimated for both ingoing and outgoing paths. Angular spread and lateral spread will increase the path length and hence energy loss fluctuations especially if either path is not normal to the sample surface. These processes are limiting factors for depth determination and resolution but can be adequately taken into account (Chapter 12.2).

1.1.3 Major Interactions

Major interactions are those characterised by emission products which can be used to derive information on material composition and structure. For low energy ions, observable interaction products include scattered ions, sputtered ions and atoms, displaced atoms and de-excitation products such as photons and electrons (Fig. 1.11a). Scattered or sputtered ions and to a lesser extent photon emission are the most useful products for materials analysis. In the high energy regime, the major processes are inner shell ionisation (X-ray emission), elastic scattering and nuclear reactions (Fig. 1.11b). These interactions occur only occasionally but are very important in materials analysis. Sputtering is negligible at MeV energies and atomic displacements are rare and of limited significance. Low energy techniques predominantly use medium to high atomic weight ions ($M > 10$) whereas high energy techniques are based largely on the use of light ions ($M < 5$). However, there are important exceptions to this trend and the variety of ions and energies used is increasing as new methods and applications are developed.

a. Low Energy Interactions

i. *Low Energy Ion Scattering (LEIS)* uses elastic scattering of ions at energies less than 10 keV. The energy of the scattered ions is dependent on the atomic mass of the target atoms through Equation (1.21b). However, the shape of the energy spectrum depends on the neutralisation

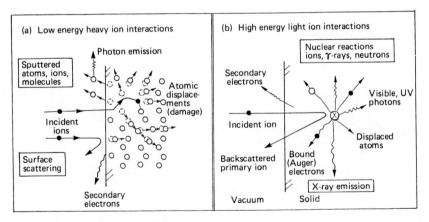

Fig. 1.11 Schematic illustration of the interactions exploited in ion beam analysis methods using (a) low energy, and (b) high energy ions.

probability, the scattering cross-section and the probability that double or multiple scattering events are detected. Discrete peaks are observed corresponding to each atomic species present in the surface layer. Double and multiple scattering events are important, especially at the lowest energies and for ions incident at an oblique angle to the surface. The final ion energy after multiple collisions is given by the product of terms having the form of Equation (1.21b) and can exceed the single scattering energy for the same angle and scattering mass.

Medium Energy Ion Scattering (MEIS) involves incident ions with low neutralisation probability and sufficient energy to penetrate many atomic layers. LEIS and MEIS techniques are widely employed for the study of atomic constituents in surface layers (Chapter 8) and the surface arrangement of atoms (Section 1.4).

ii. *Ion Beam Modification of Materials* during low energy ion irradiation, particularly with heavy ions, can substantially modify the structure and composition of the surface layers of a sample. There are four main processes involved (Fig. 1.12):

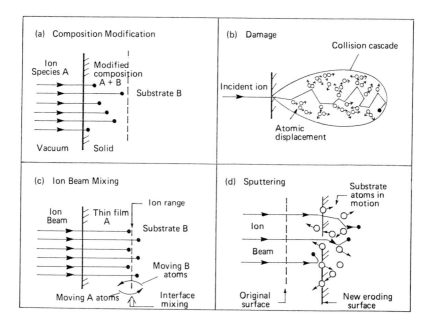

Fig. 1.12 Schematic illustration of four ion beam modification processes.

- ion implantation — the introduction of a new atomic species;
- radiation damage — the displacement of sample atoms;
- ion beam mixing — the promotion of diffusion and migration of atomic species; and
- sputtering — the ejection of surface atoms.

The first three of these processes are not useful for IBA but they must be taken into account in the interpretation of results. Fortunately, in most analysis situations, modification can be limited or controlled. Ion beam analysis techniques are very useful for the study of ion beam modification processes.

The near-surface composition of a sample can be substantially modified by ion implantation (Fig. 1.12a) and this is now widely used for changing materials properties (Williams, 1986). For example, ion implantation is a standard method of controllably introducing dopant atoms into semiconductors in the fabrication of integrated circuits. High dose implantation is also used to produce new surface alloys and compounds for modifying the mechanical (wear, hardness and friction), chemical (catalysis and corrosion), optical and electrical properties of metals, insulators and semiconductors.

When ions lose energy in nuclear collisions with target atoms (particularly for low energy heavy ions), many atoms are displaced from their normal locations. Target atoms recoiling from these collisions can themselves carry enough energy to cause additional displacements — sometimes producing a collision cascade which affects many atoms at a distance from the original ion path (Fig. 1.12b). The impact of a single heavy ion such as a 50 keV Ar ion, can lead to the displacement of tens or even hundreds of atoms in a volume surrounding the ion trajectory. The number of displaced atoms (N_d) can be estimated from the modified Kinchin-Pease formula (Sigmund, 1969):

$$N_d = 0.42 \, \gamma_n(E)/E_d \qquad (1.37)$$

where $\gamma_n(E)$ is the energy deposited during nuclear collisions and E_d is the atomic displacement threshold energy (approximately 25 eV). Not all radiation damage is stable since defects can be mobile at room temperature (particularly in metals). Nevertheless, radiation damage during ion beam analysis can directly influence the results of structure determinations (Section 1.4).

Ion irradiation can also promote diffusion through both collisional effects and increases in local temperature in the irradiated region (Fig. 1.12c). Ion beam mixing of atomic constituents is a process which can be usefully exploited for the development of new materials (Williams, 1986) but it can also change the target composition during ion beam analysis, particularly when high doses of heavy ions are employed (sputter profiling).

iii. *Sputtering* accompanies collision cascades which cause target atoms to be ejected (Fig. 1.12d). For example, between 1 and 10 atoms may be ejected for each 10 keV Ar ion incident on a sample surface. The number is termed the sputter coefficient (Y_s) which is proportional to the energy deposited close to the surface in displacement collisions and inversely proportional to the surface binding energy (Chapter 7). Sputtering is an important method for the controlled removal of surface layers from a solid. It is the basis of sputter profiling in which a number of techniques may be used to measure the elemental composition of either the sputtered atoms or of the freshly exposed layer of surface atoms. Chemical information can sometimes be obtained by examining sputtered molecules.

Sputtering produces neutral atoms and positive or negative ions. The detection of sputtered ions is the basis of Secondary Ion Mass Spectrometry (SIMS). A laser or electron beam can be used to selectively ionise neutral atoms of one isotope which is the basis of Sputtered Neutral Mass Spectrometry (SNMS). The deexcitation of sputtered ions leads to the emission of photons which is the basis of Sputter Ion Photon Spectrometry (SIPS). The sputtered surface can also be analysed by electron or photon beam irradiation, e.g. by Auger Electron Spectrometry (AES), X-ray Photoelectron Spectrometry (XPS) or Ultra-violet Photoelectron Spectrometry (UPS). All these methods give information on the elemental composition of a sample as well as some information on chemical structure. Sputter profiling and related techniques, which are a versatile and rapidly expanding field of sample analysis, are discussed in Chapter 7.

b. High Energy Interactions

i. *Rutherford or Coulomb Scattering* was the basis of Rutherford's discovery of atomic nuclei and is exploited in Rutherford Backscattering Spectrometry (RBS), a widely used technique for sample analysis. The kinematics of RBS are similar to those of LEIS analysis. RBS employs deeply penetrating light ions (e.g. MeV H^+ or He^+) and the Rutherford cross-section is very much smaller than that of elastic atomic collisions which characterise LEIS. The kinematic relation, Equation (1.2lb), relates the energy of scattered ions to the mass of sample atoms. Individual isotopes of light elements can be resolved with MeV He^+ ions but, for heavy target elements, isotopes (and in many cases adjacent elements) cannot usually be resolved (Chapter 3). Energy loss by incident and scattered ions, when the scattering occurs at some depth below the surface, gives rise to a continuous backscattered energy spectrum up to a maximum energy dictated by the kinematic relation for the heaviest atomic mass in the target. Depth profiling is possible with RBS and is discussed in Section 1.3 and Chapter 3. Finally, RBS is commonly used to study defects and the structure of crystalline materials by combining it with the channeling technique (RBS-C) (Section 1.4 and Chapters 6 and 9).

ii. *Elastic Recoil Analysis (ERA)* uses recoiling target atoms which can be detected if they receive sufficient energy in ion-atom collisions and originate close enough to the surface. A high energy heavy ion beam is used at a small angle relative to the plane of the sample surface. It makes possible the depth profiling of light atoms as discussed in Chapter 3.

iii. *Inner Shell Ionisation* arises from ion collisions with small impact parameters which remove electrons from an inner shell of sample atoms. The resulting excitation energy may be released either by emission of electrons or X-rays with energies characteristic of a particular element. Electron transition and X-ray emission probabilities determine the relative strengths of the many X-ray lines observed. The probability of exciting particular shells is a maximum when the ion and electron velocities are similar and Proton Induced X-ray Emission (PIXE) at 2 to 3 MeV provides optimum conditions for the determination of many medium atomic number elements. Other types of ions are sometimes used to favour excitation of specific elements or to confine analysis to thinner layers. Additional processes, such as self-absorption in the surface layer of the sample and self-excitation are also important (Chapter 5).

iv. *Nuclear Reactions* of many kinds can be used for sample analysis. Perhaps surprisingly, the change from atomic to nuclear interactions is quite subtle. In elastic scattering, a departure from the Rutherford cross-section, Equation (3.4), Chapter 3, is evidence of the influence of the nuclear force. This can also give rise to peaks (resonances) in the scattering cross-section. On the other hand, a purely Coulomb inter-action can lead to Coulomb excitation of the nucleus of a target atom. In most nuclear reactions the product radiation and recoil nucleus are quite different to the incident ion and target nucleus and there is an energy release of many MeV. The products include ions, neutrons and gamma-rays. An entire nuclear chemistry is available for sample analysis and an important feature of Nuclear Reaction Analysis (NRA) is that each isotope of an element undergoes different nuclear reactions. Isotopic analysis is therefore possible with little or no interference, especially for light elements ($Z > 20$). Proton or Particle Induced Gamma-Ray Emission (PIGE or PIGME) is particularly important in analysis of light to intermediate atomic number elements. Particle Induced Activation Analysis (PAA) uses the fact that some reactions produce a final nucleus which is radioactive and can be exploited for analysis of a different set of isotopes. NRA is discussed in Chapter 4.

c. Analytical Regimes

The ion energy and mass regimes which are most commonly used for low, medium and high energy analytical techniques are shown in Figure 1.13. For techniques based on sputtering (SIMS, SNMS, SIPS and IMP) O, Ar or Cs ions are most often used at energies around 5 keV. LEIS most commonly employs noble gas ions (or other selected species) at energies up to 10 keV. High energy techniques predominantly employ light ions (up to mass 4) at MeV energies. However, heavier ions are useful for ERA and for some PIXE and NRA measurements of light isotopes. At intermediate energies (10–200 keV), MEIS is the only technique which has an important role but it is not as widely used as LEIS or RBS.

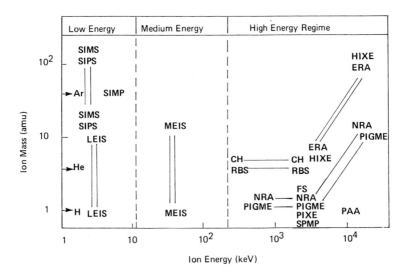

Fig. 1.13 Ion beam analysis regimes for different ion energies and masses.

1.2 DETERMINATION OF SAMPLE COMPOSITION

The procedure for quantitative determination of a specific isotope or element involves the following steps:

- selection of the most appropriate technique, irradiation conditions and a detection system which will distinguish radiation from the isotope(s) or element(s) of interest and from interfering interactions or background;
- measurement of the intensity of that radiation as a number of particles or photons emitted for a measured quantity of incident ions;
- determination of any corrections necessary for the effects of detector performance, interfering interactions and background; and
- conversion of the net yield or radiation to a concentration using a known cross-section or by comparison with the yield from a sample of known composition.

At low energies, only a few atom layers can be probed and the yield of interaction products is a measure of the fraction of surface atoms of a specific element. However, quantitative analysis of the surface composition is difficult because of the number of factors which influence the cross-sections and comparison with appropriate reference samples is usually necessary. At high energies, multiple events are unusual and cross-sections at specific values of E_1 and θ can be used to calculate absolute isotope or element concentrations. The most common experiment method is to measure the energy of each particle or photon recorded by the detector and to display the results as an energy spectrum (number of events versus energy). For a thin sample, the integrated number of events, $N_3(\theta)$, in a spectrum peak or other energy interval is used in Equation (1.20) to derive the number of atoms cm^{-2} of the isotope or element responsible for those events. For thick samples, Equation (1.20) must be integrated over incident energy or suitable approximations used as described in Chapter 12.1. The ability to use measured cross-sections to derive absolute atom concentrations is an important feature of high energy IBA. The simplest case is RBS and the principles just described are illustrated by examples in Chapter 3.2.

1.2.1 Sensitivity and Detection Limits

Cross-section values give some idea of relative sensitivities which can be expected in IBA (Fig. 1.6) but thick sample yields (radiation intensity per unit ion dose and atom concentration) are often more useful. However, the experimentally observed yields depend on angle of observation, solid angle, filters and detector efficiency and other factors (Chapter 12). Since these are all specific to a particular experimental setup, so also are the sensitivity and detection limits achieved. For the observed yield (Y) to be significant it must exceed the uncertainty (B) in the detector background by a factor k (i.e. $Y > kB$) which is used to define three types of limit:

- Critical level for which $k = 1.64$; the only statement that can be made is that the product is detected or not;
- Detection Limit for which $k = 3$; a qualitative estimate can be given of the concentration; and
- Quantitative Limit for which $k = 10$; the concentration can be determined with an uncertainty of 10%.

If the background uncertainty is obtained from the number of background counts (N_b) then $B = N_b^{1/2}$. A general comparison of methods of composition determination and sensitivities is given in Chapter 11.

1.2.2 Microprobes

A finely focused beam of incident ions can be used to irradiate a spot 1 μm or less in diameter and to scan the surface to obtain two dimensional information on sample composition. Three approaches have been developed for exploiting this concept:

i. Ion Microprobe (IMP) — the use of a low energy microbeam (1 to 5 μm diameter) rastered across the surface with synchronised measurements to obtain a spatial distribution of analysed species; depth information from time dependence is an additional capability;

ii. Ion Microscope (IMS) or Ion Microanalyser — a large area low energy sputtering beam is used together with imaging optics for sputtered ions to obtain a spatial distribution of analysed species; only secondary ions can be exploited; depth information from time dependence is an additional capability; and

iii. Proton or Nuclear Microprobe (PMP) — a high energy proton or heavier ion microbeam (1 to 5 μm diameter) is used to analyse a small spot or scanned across the surface with synchronised measurements to obtain multi-element spatial distributions; if RBS, ERA or NRA is used, depth information is also obtained.

Low and high energy microprobes are a major use of IBA with a number of commercial systems being available (Chapter 10).

1.3 DEPTH PROFILING

1.3.1 Energy Loss Profiling

Energy loss by incident ions before they interact with a sample atom provides the basis of a powerful tool for the non-destructive measurement of depth profiles of invidivual elements or isotopes. If the interaction product is also an ion then further energy loss occurs as it travels out of the sample towards the detector. The parameters involved in energy loss profiling are illustrated in Fig. 1.14. An incident ion which interacts at depth t (normal to the surface) has a path length l_1 where $t = l_1 \sin \alpha$. The energy of the incident ion at the point of interaction is therefore:

$$E_1' = E_1 - S_1 l_1 \tag{1.38}$$

if a constant stopping power (S_1) is applicable over the whole of the entrance path. The energy of product particles immediately following the interaction (E_3') is given by Equation (1.21). If these are ions (as in the case of scattering or an ion–ion nuclear reaction) the product path length is l_3 where $t = l_3 \sin \beta$ and the energy on emergence from the sample is:

$$E_3'' = E_3' - S_3 l_3 \tag{1.39}$$

where S_3 is the (constant) stopping power on the exit path.

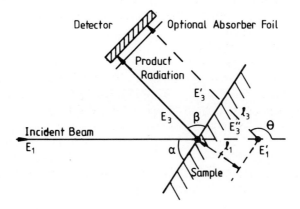

Fig. 1.14 Geometry and parameters for concentration determination and depth profiling.

The way in which these processes can be used to obtain depth information is shown schematically in Fig. 1.15. Within the sample, the line AB is drawn to show the drop in ion energy, Equation (1.38), during initial penetration of the sample with an energy scale increasing from left to right and depth increasing from top to bottom. Energy loss resulting from scattering interactions is shown by an appropriate gap between the points A and C or B and D. The size of the gap depends on the kinematic relation (1.21a). Each scattered ion locus (e.g. DE) shows the further drop in energy which occurs while the scattered ion emerges from the sample, Equation (1.39). These locii show how the depth of interaction dictates the final ion energy in the scattered ion spectrum (a) which is a plot of counts versus observed energy. Two forms of dependence of cross-section on incident ion energy are shown schematically (b) below the sample and two concentration profiles are shown to the left (c). The product of these two parameters determines the height of the observed spectrum at each energy. If either the cross-section or concentration varies as indicated by the dashed curves then a spectrum such as that dashed in Fig. 1.15a is observed. Examples of the use of these principles are presented in Chapter 3.2.

In the case of a nuclear reaction, there is usually an energy gain due to a positive value of Q and so the product radiation has a high energy illustrated by the distance between the points A and F or between B and G. Energy loss by a product ion while emerging from the sample (e.g. locus GH) has a different slope to that for scattered ions if the stopping power is different. The reaction spectrum (d) thus has a different dependence of energy on depth to that in (a). If the product radiation does not lose

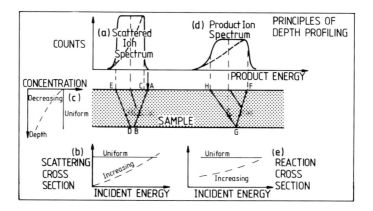

Fig. 1.15 Schematic illustration of depth profiling principles (see text).

energy in escaping from the sample (such as neutrons or gamma-rays), GH will be vertical and the product energy spectrum is dependent only on incident ion energy loss. Once again, the effect of a varying cross-section or concentration profile (as shown in Fig. 1.15e) produces the dashed spectrum in (d).

Using high ion energies, three approaches to depth profiling are available:

- interpretation of energy spectra of product radiation as depth profiles;
- use of a narrow resonance peak in the cross-section to probe the concentration as a function of depth; and
- interpretation of yield curves (yield versus incident energy) as concentration profiles.

These methods are unique features of ion beam analysis and further details are given in Chapter 12.2.

1.3.2 Sputter Profiling

Sputtering removes surface atoms at a rate which depends on the ion type, energy, current, angle of incidence and the sample composition (Chapter 7). The sputter rate also depends on the chemical species present at the surface of the sample including the presence of O or other reactive species. If an interaction of any kind is determined as a function of time during sputtering, or as a function of accumulated dose of primary ions for alternated periods of sputtering and surface analysis using a second ion or electron beam, a depth profile is obtained. The depth (x) is obtained from the sputtering coefficient, Y_s, a

$$x = Y_s \, \Phi \, \sin \alpha / N_s \qquad (1.40)$$

where Φ is the ion dose per cm^2, α is the angle between incident beam direction and the surface plane and N_s is the atomic number density of the substrate atoms. Typical current densities for sputtering are 1 to 10 μA cm^{-2} which give sputter rates of the order of 1 to 10 $nm \, s^{-1}$. Absolute depths can most readily be calibrated by measurement of the total depth of the sputtered crater at the end of a profile determination. The depth resolution is primarily determined by the depth range over which the interaction signal is averaged. This is typically 1 to 2 Å for low energy analysis techniques (e.g. SIMS or LEIS alternated with sputtering) and 1 to 2 nm for AES. The ability to control the sputtering rate together with excellent depth resolution (which, in principle, is independent of depth) make this attractive for determining depth profiles (Chapter 7).

1.3.3 Surface Removal

a. Wedge Scanning

Any ion beam analysis method can be applied with a beam of small diameter which is scanned across the surface of a sample. If the surface has been removed at a shallow angle, or is a cross-section through the sample, the surface scan gives information on the depth distribution of the elements or isotopes determined. The depth resolution is given by:

$$\Delta t = d \tan \alpha \qquad (1.41)$$

where d is the beam diameter and α is the wedge angle. An important advantage of wedge scanning is that the depth resolution is the same at all depths. Although destructive, it is a very useful technique.

b. Surface Dissolution and Abrasion

Successive application of IBA and chemical etching, electrolytic dissolution or oxidation, mechanical stripping or polishing to remove thin layers from the surface of the sample (Whitton, 1973) also result in a depth profile. Care must be taken to ensure uniformity in the thickness removed across all the area to be studied by IBA. Constant depth resolution is obtained to much larger depths than can normally be probed but the depth scale must be determined from weight loss or by other independent techniques. Chemical and electrolytic processes can have different removal rates for different areas of a polycrystalline sample, especially in the region of grain boundaries. Etch rates may depnd on other factors such as the presence of radiation damage and are also specific in their action to particular chemical species. Mechanical methods, on the other hand, are generally applicable to all sample materials although deformation and smearing may cause problems. Anodisation, followed by mechanical or chemical removal of the oxide is the most reproducible technique because of the precise relation between anodisation voltage and layer thickness. However, it is only suitable for a relatively small number of materials, e.g. Al, Si, Mo, Ta and W.

1.3.4 Thickness Determination

An additional straightforward use of ion beam techniques is in the determination of the thickness of foils, thin films on a solid substrate or of the depth to which atoms have been implanted in a solid sample. The principles involved are the same as those described for the determination of the depth scale (Section 1.3.1 and Chapter 12.2) and many such examples are given in later chapters.

1.4 STRUCTURE DETERMINATION

Diffraction techniques, using electrons, X-rays or neutrons, are vital tools for the study of the atomic structure of crystalline solids. Ion diffraction is only observed at very low energies (of the order of a few eV or less) for which the de Broglie wavelength is comparable to atomic spacings in crystals. In such cases, angular variation in ion scattering yields from crystalline solids show a number of diffraction maxima (Heiland and Taglauer, 1983). However, at higher energies which are of prime interest for materials analysis applications, the de Broglie wavelength is much less than atomic dimensions (Fig. 1.2) and the ion interacts with individual atoms. Nevertheless, the nature and frequency of ion–atom interactions can be dramatically influenced by the regular atomic arrangements in crystals and can be employed to provide important structure information.

1.4.1 Basic Concepts Used in Structure Determination

Three basic concepts are exploited for the study of the atomic structure of a sample using ion beams:

- Shadowing
- Correlated (Multiple) Collisions; and
- Crystal Transparency.

Shadow cones and correlated collisions are important for very low energy ions (say less than 10 keV). At high energy, transparency is important in the phenomena of channeling and blocking. Depth and composition analysis (Sections 1.2 and 1.3) are also used in conjunction with structure determination making ion beam analysis a very versatile tool for materials characterisation. The nomenclature used to describe crystal structure is described in Chapter 14.6.

TABLE 1.4
Structure equations

Shadow Cone Radius		
$$R_c = 2(Z_1 Z_2 e^2 d/E)^{\frac{1}{2}}$$	m	(1.42)
Channeling Critical Angle		
Low energy		
$$\psi_c \cong \left\{ \frac{a}{d}(2Z_1 Z_2 e^2/Ed)^{\frac{1}{2}} \right\}^{\frac{1}{2}}$$	rad	(1.43)
High energy		
$$\psi_c \cong (2Z_1 Z_2 e^2/Ed)^{\frac{1}{2}}$$	rad	(1.44)
Minimum yield		
$$\chi_{min} = \pi N d \rho^2$$		(1.45)
Substitutional fraction		
$$S = \chi_I/(1 - \chi_s)$$		(1.46)

a. Shadow Cones

A shadow cone arises from an inverse relationship between impact parameter and scattering angle; for a large impact parameter the scattering angle is small and vice versa. This can be seen by drawing the trajectories of initially parallel ions with different impact parameters incident on a single atom (Fig. 1.16a). A paraboloidal space 'behind' the atom is not accessible to any ions from a strictly parallel incident beam and this space is described as the shadow cone. The cone angle (half-angle at a distance equal to the atomic spacing in a solid sample) depends on the incident energy and is given by Equation (1.42). If another atom lies within the shadow cone, it cannot interact with the incident beam. However, if the incident direction is changed, the shadow cone will swing around like an 'inverse' searchlight and a specific second row atom may be 'visible' for some angles and not for others. The observed angular dependence of the number of interaction products can thus be used to study atomic structure — especially to determine the location of specific types of surface or near-surface atoms. Such techniques are discussed in Chapters 8 and 9.

b. Correlated Collisions

Ions which are incident at a glancing angle to the sample surface may experience multiple small-angle (large impact parameter) collisions with successive surface atoms. Fig. 1.16b shows the resultant reflection of ions from a crystalline surface. For low ion energies, typically less than a few keV, reflection dominates for angles up to 15° whereas, for high energy

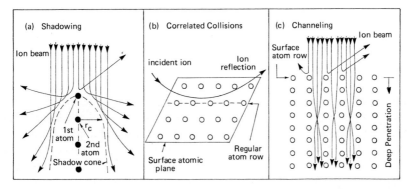

Fig. 1.16 Schematic illustration of ion paths demonstrating three concepts involved in structure determination by ion beam analysis: (a) shadowing, (b) correlated collisions, and (c) channeling.

ions (e.g. 2 MeV He$^+$ ions) incident angles must be less than about 1° before significant reflection is obtained. The observation of multiple scattering peaks in scattered (reflected) energy spectra is normally confined to low energy ions. The geometrical arrangement of atoms (with respect to the incident beam direction) can strongly influence the yield and energy spectrum of reflected ions. For example, multiple scattering is enhanced when ions are incident at a grazing angle along a low index crystallographic direction where the atoms are close-packed. Conversely, atomic steps and ledges and missing or additional atoms on the surface can markedly decrease the multiple scattering yield. Such structure-sensitive effects provide important information about the surface arrangement of atoms, including arrangements which differ from that of the bulk crystal. Another type of correlated collision sequence occurs when a low energy heavy ion enters a crystal along a low index direction. A head-on collision with the first atom in a row produces a collision sequence of recoiling target atoms along a low index (close-packed) crystal direction. Such processes are important in sputter profiling as described in Section 1.3.2. For high energy light ions, collision recoil sequences have extremely low probabilities.

c. Channeling and Blocking

Crystals have a degree of transparency to ions incident along a low index direction and this leads to enhanced penetration which is termed channeling. However, channeling is more than a simple crystal transparency effect. Ions which enter a channel at a large enough impact parameter, such that they are not scattered out of the channel by the surface atom row, undergo a sequence of correlated large impact parameter collisions and are steered along the channel by the confining potential of the bounding atom rows or planes (Fig. 1.16c). Channeled ions therefore undergo an oscillatory trajectory as illustrated pictorially in Fig. 1.17. Channeling occurs when ions are incident within a certain critical angle from a crystal axis or plane. For low energy ions (e.g. 10 keV Ne^+), critical angles can be up to 10° for major axes whereas high energy light ions (e.g. 2 MeV He^+) exhibit very small ($< 1°$) critical angles, Equations (1.43 and 1.44). For channeled ions, the correlation of successive large impact parameter collisions reduces the probability of interactions which require a small impact parameter. Channeled ions therefore penetrate further into a crystal with less energy loss and fewer major interactions than occur in the non-channeled (random) case. This reduced interaction probability for channeled ions is exploited to investigate the structure of crystals, surfaces and interfaces as discussed briefly below and in more detail in Chapters 6 and 9.

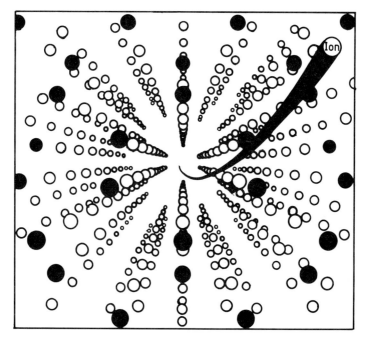

Fig. 1.17 A schematic representation of the path of a channeled ion through a diamond cubic lattice.

An analogous process, called blocking, can influence scattered ions or reaction products which travel towards a detector along a major crystal axis or plane. When an ion has a major interaction with a lattice atom, the scattered or reaction products are blocked from travelling along low index directions by the presence of neighbouring atoms and their shadow cones in such directions. Both channeling and blocking techniques are used for locating the lattice position of foreign atoms within a crystal lattice, (Section 1.4.3). Low energy ions (< 10 keV) are particularly sensitive to the surface structure whereas high energy analysis facilitates subsurface (or bulk-like) structure determination.

1.4.2. Dependence of Angular Interaction Yield on Crystal Structure

Some typical low energy angular yield effects are illustrated in Fig. 1.18. Fig. 1.18a shows the LEIS yield for reflection from the surface layer of atoms as a function of azimuthal angle γ for a forward scattering geometry where θ (the scattering angle) is 2 to 30°. Maxima in the multiple scattering yield (arrowed) arise when the incident beam is aligned along close-packed surface atomic rows. Fig. 1.18b illustrates the LEIS interaction yield dependence from first and second atom rows when the target is tilted about a shadowing direction or the detector is tilted about a blocking direction. The yield from the first (surface) row of atoms is essentially constant for small variations of tilt angle (i.e. the surface atoms are always 'visible' regardless of geometry). However, the yield from the second row varies with angle. The lowest yield occurs when the shadow cone axis intersects second row atoms and maximum yield occurs when second row atoms are completely 'visible' outside the shadow cone. For low energy heavy ions (e.g. 10 keV Ne^+ ions) the angular width of the dip at half height can be typically several degrees. For heavy ions it is given approximately by the critical angle for channeling, Equation (1.43). The shape and angular position of the angular yield curve for LEIS can be used to precisely locate the relative positions and spacings of atoms (either of the host crystal or adsorbed impurities) within the first few atomic rows. Finally, we show the angular dependence of sputtering yield in Fig. 1.18c. As the detector is moved to vary the exit angle, the sputtered yield changes. When the detector is aligned with close packed directions in the bulk lattice, the yield is maximum as a result of focused collision sequences along such directions.

For high energy (MeV) light ions, Fig. 1.19 illustrates the interaction yield variations in crystals as a function of incident angle and penetration

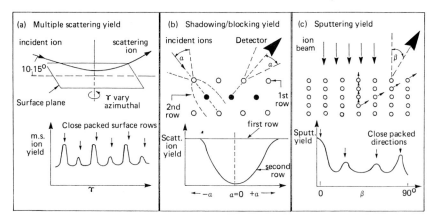

Fig. 1.18 Low energy ion paths and angular yields from a crystalline sample showing the effects of (a) multiple scattering, (b) shadowing/blocking, and (c) sputtering.

depth. The incident beam is initially channeled along a low index crystallographic direction ($\theta = 0°$). As the crystal is tilted about this direction, the interaction yield from the surface rows of atoms is independent of tilt angle. The angular yield from a depth increment Δx_1 just below the surface exhibits the expected minimum at $\theta = 0°$. Significantly away from zero degrees, the yield returns to the average yield (normalised value of 1). The full width at half height of the channeling dip ($\psi_{1/2}$) is proportional to and of the same magnitude as the critical angle for channeling (ψ_c) which is of the order of 1° for 2 MeV He$^+$ channeled along low index directions in crystals. For fast light ions, the critical angle depends on the lattice spacing, the ion mass and energy and target atomic number according to Equation (1.44). Exact relations between the angular half width and critical angle are given in Chapter 6.

Interestingly, at incident angles just greater than the critical angle, the normalised interaction yield is greater than one, indicating a greater than normal interaction probability (Chapter 6). The minimum yield (called χ_{min}) at $\theta = 0°$ is non-zero since not all the ions are channeled at the depth increment Δx_1. The random (i.e. non-channeled) component of the penetrating beam can be approximated by the fraction of ions which is scattered from the surface rows of atoms through angles greater than the critical angle for channeling. The major interactions of the random component of the beam with lattice atoms at a particular depth thus accounts for the non-zero yield. Consequently, χ_{min} can be represented by an effective cross-section, given by an interaction area around the atoms in the surface plane according to Equation (1.45). Typically, χ_{min} is of the

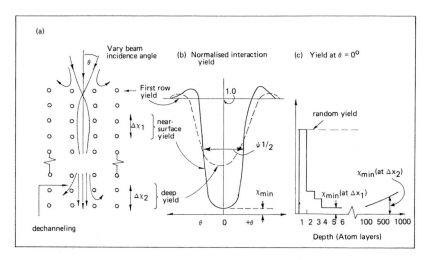

Fig. 1.19 High energy angular yield effects in crystals (a) channeling trajectories, (b) channeling yield as a function of angle, and (c) minimum channeling yield as a function of depth (atom rows).

order of 2% for channeling down low index axes in perfect crystals. The variation in interaction yield with penetration depth is illustrated in Fig. 1.19c for a beam aligned with a channeling direction ($\theta = 0°$). This idealised representation indicates the high non-channeled yield from the surface atomic layer and the rapid decrease in interaction yield from subsequent sub-surface layers as a result of shadowing. The minimum yield, χ_{min}, occurs below the first few atom layers; at greater penetration depths (e.g. Δx_2), dechanneling (the gradual reduction in the channeled component) causes an increase in the random component of the beam and hence an increase in yield.

1.4.3 Analysis of Bulk Defects and Atom Location

For LEIS, the signal is received only from the first few atom layers and the principles described in the previous section can be used for determining the positions of surface atoms (Chapter 8). High energy ion scattering can also be used for surface atom location by monitoring angular variations in the surface peak yield (Chapter 9). In addition, the greater penetration depth of high energy light ions allows the study of bulk defects and the determination of impurity atom locations in crystals (Chapter 6). We will only briefly introduce the concepts in this section. Atoms which are displaced significantly from atom rows at some depth in

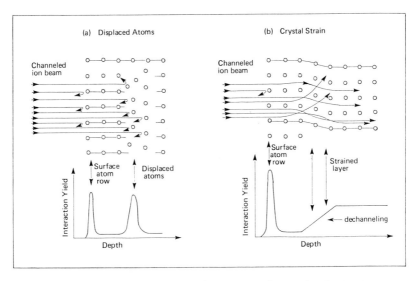

Fig. 1.20 Principles involved in defect studies by channeling: (a) for displaced atoms, and (b) for crystal strain.

a crystal (Fig. 1.20a) give rise to an enhancement in the large angle scattering or other interaction yield under channeling conditions. This produces an additional peak in the energy spectrum of interaction products at an energy corresponding to the depth of the disordered layer. The area (yield) associated with this peak is proportional to the number of displaced atoms. Alternatively, small angular distortions of the lattice rows or strain in the crystal lattice around defects (Fig. 1.20b) will cause enhanced dechanneling (small angle scattering) as channeled ions pass through this region of the crystal. The dechanneling process causes a measurable increase in the interaction yield beginning at an energy corresponding to the depth of the defects. The magnitude of the yield increase and its dependence on the incident beam energy and angle can be used to provide a semi-quantitative analysis of the bulk (or sub-surface) defects in crystals.

If the ion beam is incident along various channeling or random directions, differences will occur in the major interaction yield (large angle scattering, nuclear reaction or X-ray emission) between incident ions and foreign atoms which are preferentially located with respect to the host lattice positions. Such differences in the interaction yield arise from the shadowing effect along low index rows. Fig. 1.21 indicates how particular locations of foreign atoms can influence the interaction yield in different incident ion directions. Shadowed atoms give low yield

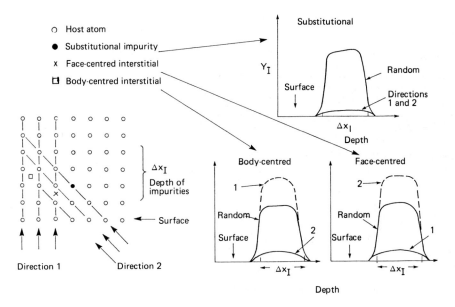

Fig. 1.21 Principles involved in foreign atom location by channeling.

whereas atoms lying in channel centres give high yield. Yield differences can therefore be used to give qualitative information on foreign atom location. It is also possible to quantify the fraction of atoms occupying specific sites under favourable conditions by comparing the impurity yields under various channeling and random alignment conditions. For substitutional impurities, the substitutional fraction S is given by Equation (1.46). For precise atom location, however, it is necessary to measure changes in yield when the incident beam is scanned in angle about low index directions (Chapter 6).

REFERENCES

Betz, H.D. (1983). *In* 'Condensed Matter', (Datz, S., ed.). Academic Press, Orlando, 2.

Biersack, J.P. and Ziegler, J.F. (1982). *In* 'Electrophysics', Springer, Berlin, Vol. **10**, 122.

Chu, W.-K. (1980). *In* 'Atomic Physics, Accelerators', (Richards, P., ed.). Academic Press, New York, 25.

Deconninck, G. and Fouilhe, Y. (1975). *In* 'Ion Beam Surface Layer Analysis', (Meyer, O., Kappeler, F. and Linker, G., eds.). Plenum Press, New York, Vol. **1**, 97.

Heiland, W. and Taglauer, E. (1983). *In* 'Condensed Matter', (Datz, S., ed.). Academic Press, Orlando, 238.

Kumakhov, M.A. and Komarov, F.F. (1981). 'Energy loss and Ion Ranges in Solids', Gordon and Breach, New York, 58.

Lindhard, J. and Scharff, M. (1961). *Phys. Rev.* **124**, 128.
Marwick, A.D. and Sigmund, P. (1975). *Nucl. Instrum. Methods* **126**, 317.
O'Connor, D.J. and Biersack, J.P. (1986). *Nucl. Instrum. Methods* **B15**, 14.
Sigmund, P. (1969). *Phys. Rev.* **84**, 383.
Sigmund, P. and Winterbon, K.B. (1974). *Nucl. Instrum. Methods* **119**, 541.
Whitton, J.L. (1973). *In* 'Channeling, Theory, Observation and Applications' (Morgan, D.V., ed.) Wiley and sons, New York, 225.
Williams, J.S. (1986). *Rep. Progr. Phys.* **49**, 1.

2

Techniques and Equipment

M.J. Kenny

Division of Applied Physics, CSIRO
Menai, Australia

ION BEAMS FOR
MATERIALS ANALYSIS
ISBN 0 12 099740 1

2.1 INTRODUCTION

The major components of an ion beam analysis facility are shown schematically in Fig. 2.1 and typical beam requirements are summarised in Table 2.1. Accelerators producing ion beams with energies in excess of 200 keV are usually classed as 'high energy'. Such facilities require at least 100 m² of specially serviced laboratory space. The key components such as accelerator, beam handling apparatus and target chamber are physically separate, being connected by evacuated beam-transport tubes. PIXE, ERA, RBS, RBS-C and NRA are performed in dedicated target chambers. Radiation shielding may be required, particularly for NRA. Typically the cost of a multi-technique high energy ion beam facility is considerably in excess of $US1M at 1987 prices. Single technique facilities cost less but both cost and space allocation would significantly exceed those of a commercial 200 keV transmission electron microscope.

The intermediate (30–200 keV) energy range is not widely used for ion beam analysis, but ion implanters are sometimes used to provide an ion beam for analytical and imaging functions (Chapter 10). Low energy accelerators used for SIMS, SIPS and LEIS are typically < 30 KeV and the entire apparatus (accelerator, beam analysis and detectors) are often

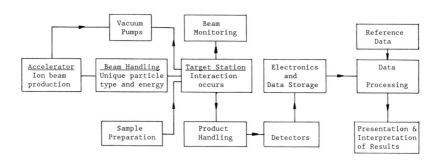

Fig. 2.1 Major components of an IBA facility.

TABLE 2.1
Typical beam requirements for ion beam analysis

Technique	Typical Ion Beam	Typical Energy Range
Sputter profiling (SIMS, SIPS, AES)	O, Ar, Cs	< 3 keV
LEIS	He, Ne, Ar	< 10 keV
MEIS	H, He	10–200 keV
PIXE	H	2.0–3.0 MeV
RBS/Ion Channelling Elastic Recoil	H, He	1.0–3.0 MeV
NRA	^{1}H, ^{2}H, ^{3}He, ^{4}He N, O, F	0.5–10 MeV
ERA	He, O, Cl	

integrated into a single vacuum chamber of comparable size to an electron microscope. The facility will thus fit into a standard laboratory and the cost is comparable with that of standard (< 100 keV) scanning electron microscopes having X-ray analytical capabilities. A schematic layout of a typical low energy system is illustrated in Fig. 2.2 for ion microprobe and ion microscope applications. Focussing, scanning and imaging functions are described in Chapter 10.

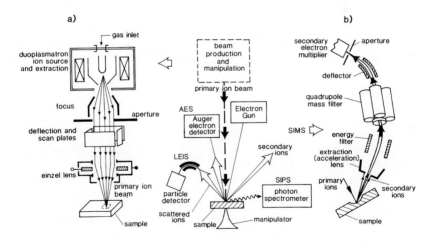

Fig. 2.2 Schematic layout of a low energy IBA chamber.

2.2 ION BEAM PRODUCTION

2.2.1 Ion Sources

Both positive and negative ion beams are used in analysis and requirements can vary from 10^{-10} A to 10^{-4} A with the nanoampere region being the most useful. Many different beam species are used and consequently no one type of ion source is suitable for all needs. Ion sources and accelerator types used in IBA are described in Table 2.2.

Ion source operation is based on one of a number of processes: electrical discharge in a gas or vapour, electron impact in gas or vapour, surface ionisation, surface sputtering, or field emission (Valyi, 1977). The most common sources for high energy accelerators are the radiofrequency (RF) source (Moak *et al.*, 1951) and the duoplasmatron source (Moak *et al.*, 1959). These are also useful for low energy accelerators as are liquid metal ion source (LMIS), surface ionisation and electron impact sources (e.g. colutron). An RF source is ideal for the production of ions of H, He and most gases with currents of tens of μA which are adequate for most ion beam analysis work. A duoplasmatron source gives

TABLE 2.2
Ion sources and accelerators for IBA

	Ion Source	Current	Ion Beams	Techniques	Accelerator Type
Low Energy	Radio frequency	1 mA	H, He, N, O	LEIS	Single gap
	Duoplasmatron	10 mA	H, He, N, O		
	Colutron				
	Penning	5 mA	C, N, Ne, Kr		
	Caesium sputter		Most solid	SIMS	
	Freeman	10 mA	Most solid		
	Electron impact				
	LMIS		Ga, In, Au, Bi		
High Energy	Positive ions				
	Radio-frequency	1 mA	H, He, N, O	RBS, PIXE, NRA	Single ended
	Duoplasmatron	10 mA	H, He, N, O		Electrostatic
	Negative ions				
	Duoplasmatron (off-axis extraction)	100 μA	H, O	RBS, PIXE, NRA	Tandem with −ve ion injection and stripping at terminal to give +ve ions
	Penning	2 mA	H, ^{2}H		
	Sputter-source		Most		
	RF with charge exchange	100 μA	H, He, N, O,		

higher currents of several mA. Both duoplasmatron and RF ion sources produce predominantly positive ions and the ion beam extracted from the source will usually contain a mixture of different atomic and molecular species and charge states. For example a hydrogen beam could have H^+, H^{2+} and H^{3+} components and a nitrogen beam N^+, N_2^+ and N^{++} components.

Atoms of alkali metals (e.g. Cs) can be ionised at the surface of a metal such as W to produce beams with good intensity and low angular divergence. An alternative method of obtaining beams from solid materials is to heat the substance to produce a vapour which may then be ionised by electron impact. Such a source is ideal for LEIS which does not require high beam currents. A high brightness LMIS produces beams (e.g. from Ga, In, Au or Bi) with an emitting area less than 30 nm and an energy spread less than 10 eV (Gnaser and Rudenauer, 1983). Sputter ion sources (Middleton, 1977) provide a means of obtaining beams from small quantities of solid material. A beam of inert gas (Ne, Ar, or Kr) or of an alkali metal (Cs) is used to sputter a solid sample of the element whose beam is required. Secondary ions of either positive or negative charge state are then extracted, focussed and mass analysed.

Negative ion extraction from duoplasmatrons for use in tandem accelerators is possible in the case of H and O beams. For He and some other ions direct negative extraction is not possible. Positive ions must be passed through a charge exchanger of Li vapour (or other alkali metal vapour) to obtain the negative ions needed. N does not form a negative ion, but NH_2^- ions can be produced and accelerated to the terminal where dissociation and charge exchange occur, producing $^{14}N^+$ ions. Sputter ion sources can also be used to obtain negative ion beams. The energy spread and stability of the extracted ion beam and its angular divergence influence the ability to transport the beam to a spot on the target. In the case of ion microbeams, which are an important application of both low and high energy ion beams, the smallest spot size and beam intensity that can be delivered to a sample are limited by ion source performance (Chapter 10).

2.2.2 Accelerators

Accelerators with voltages up to 50 keV are constructed using a high voltage power supply with single gap acceleration between ion source and ground potential. This type of accelerator has advantages of simplicity and low cost. The optical properties of the beam are controlled by extraction electrode shape and an einzel lens after extraction. An einzel lens consists of three cylindrical elements, the first and last being grounded and the centre element being at several kV. The electric fields are such that a diverging beam is defocussed in the first gap and focussed in the second, with a net forcussing effect. The focal length is controlled by electrode voltage. The attainable voltage is restricted by the insulator and the width of the extraction gap which typically is 2–10 mm. The achievement of higher energies is a major area of technology (England, 1974). Quite a number of accelerators capable of ion energies of several MeV were developed for fundamental nuclear physics research but are now being redeployed for ion beam analysis (Table 2.2). However in recent years, new accelerators are being installed for ion beam analysis.

In the Van de Graaff accelerator, charge is transferred by a moving belt to an equipotential surface (terminal) which builds up in voltage. By suitable choice of components and insulating gas, voltages of millions of volts can be obtained. An ion source in the terminal and an accelerating tube to ground complete the accelerator, as shown in Fig. 2.3a. The

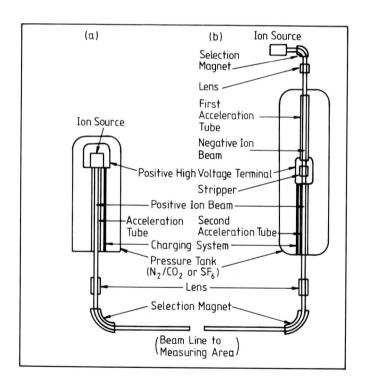

Fig. 2.3a. Single stage accelerator. b. Tandem accelerator.

Pelletron is a similar accelerator in which the belt is replaced by a charging chain of alternate insulating and conducting components. This leads to more uniform transport of charge and better voltage stability. The Pelletron also incorporates modular tube sections using UHV technology.

The single ended accelerator is the most popular form but has the disadvantage that the ion source is located within the terminal and the pressure vessel, making access for maintenance and modification slow and difficult. A development which overcomes this problem and at the same time allows for higher energies is the tandem or two stage accelerator. The ion source is located externally and negative ions are accelerated from ground to the terminal, stripped of their negative charge and accelerated through a second tube to ground to give an energy of $(n + 1)V$ eV where V is the terminal voltage and n the positive charge state after stripping (Fig. 2.3b). The Tandetron is a tandem accelerator in which the high voltage is obtained by means of a voltage doubler type power supply fed by an RF signal. There are no moving parts and the power supply is very stable. Single-ended and tandem accelerators have been built in both horizontal and vertical forms which have important differences in operational convenience. In a vertical accelerator, a 90° magnet is required to bend the ion beam into a horizontal direction since it is most convenient to establish numerous experimental stations in a horizontal layout.

Cyclotrons give high energy (≥ 10 MeV), high current (≥ 100 μA) beams which are only used in ion beam analysis for specific cases of activation analysis and ERA. Other forms of accelerator such as Linacs and Synchrotrons are rarely used for ion beam analysis.

2.3 BEAM SELECTION AND CONTROL

Once ions have been extracted from an accelerator, those of the desired ion species and charge state must be separated from unwanted components of the beam. The beam may then require focussing and/or angular collimation to provide a well defined size. For high energy beams a number of beam line components are usually required between ion source and target station as shown schematically in Fig. 2.4 and summarised along with useful procedures in Highlight 2.1. For low energy ion beam applications, much smaller component dimensions are involved and beam manipulation and optics are much simpler (Fig 2.2 and Section 2.3.2b).

An ion moves in a circular path when subjected to a magnetic or electric field, the radius depending on the mass M (amu), charge q (electronic

HIGHLIGHT 2.1
HIGH ENERGY BEAM TRANSPORT

Components

A typical high energy beam transport system consists of a selection of the following components (Fig. 2.4):

   i. analyser and switch magnets with deflection angles between 15° and 90°; electrostatic analysers to separate ions such as D_2^+, He^+ and O^{2+} which are transmitted by the same magnetic field;
  ii. magnetic or electrostatic quadrupole or einzel lenses;
 iii. electrostatic steering or scanning in X and Y directions; electrostatic deflection can also be used for on-demand pulsing to remove the beam from the sample while a detector signal is processed;
  iv. single or double (dogleg) deflection by approximately 10° to remove neutral beam particles which can constitute up to several per cent of the beam (especially if the vacuum is poor); this should be done as close as possible to the target (inset, Fig. 2.4);
   v. apertures or X, Y slits to define the size of the beam at the entrance and exit of the analyser magnet and before the sample; edges should not be thinner than the ion range (inset, Fig. 2.4);
       Additional cooled apertures may be needed to prevent beam heating of sensitive beam line components such as gaskets;
  vi. Vycor (quartz) viewers which can be inserted to intercept the beam and give a visual indication of size and shape;
 vii. beam profile monitors which give a remote indication of beam shape and intensity in X and Y directions; and
viii. beam stops to interrupt the beam at appropriate points along the beam line — preferably remote controlled.

Materials

   i. Slits, apertures, beam stops and any other surfaces receiving direct beam must be able to dissipate its power; water-cooled copper should be used for beam currents above 5 to 10 μA; graphite, Cu, brass, Ta or W can be used at lower currents but neutrons are emitted by Cu for protons above 2.17 MeV unless it is plated, for example with Au;
  ii. Al vacuum lines and chambers, neoprene gaskets (e.g. of ISO or Dependex design) can be used with HV systems; stainless steel components and metal gaskets are needed for UHV; other vacuum equipment is discussed in Section 2.5.1.

Alignment

All beam components must be accurately aligned and rigidly mounted. The following procedure is typical for aligning components between deflecting units:

  i. using a temporary beam line, locate the beam centre at the beginning and end using quartz viewers; the end spot should be on the axis of rotation when using a goniometer;

 ii. position cross wires at the beam centres and align a laser beam or viewing telescope to pass through the cross-wires;

iii. instal each unit centred on the optical beam; and

iv. fine adjustment may be necessary with the ion beam on to allow for the effects of stray magnetic fields and possible differences between magnetic and optical paths. The earth's magnetic field causes a deflection of 3 MeV protons by $\sim$ 1 mm per metre.

---

charges) and energy E_1 (eV) of the ion. The radius of curvature R (m) for a magnetic field B (T) is given by

$$R = (2ME_1)^{1/2}/qB \qquad (2.1)$$

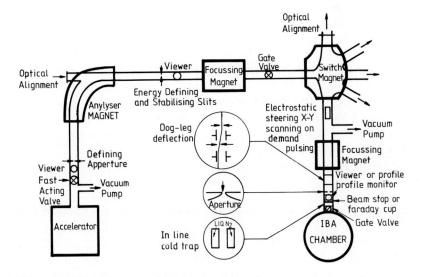

Fig. 2.4 Schematic diagram of high energy beam line components.

With entrance and exit path lengths of the order of 2 m, a field of about 1 T is necessary to deflect 3 MeV ions of mass 20 through 45°. For an electric field E (V m^{-1}), the radius is given by:

$$R = 2E_1/Eq \qquad (2.2)$$

If the entrance and exit points are defined by slits or apertures, only ions with a specific value of $(2ME_1/q^2)$ will pass through a magnetic field and in the electrostatic case $(2E_1/q)$ is the determining value. Thus magnetic deflection is mass selective for constant energy and charge state, whereas electrostatic deflection is energy and charge state selective.

An important principle is that of constant beam emittance at constant beam energy. The beam as it leaves the accelerator will have displacement coordinates and directions in two planes at right angles to each other. If all these quantities are zero, then the beam is travelling along the axis. If the angles are zero, but the displacements are finite, the beam is parallel. The optical properties of a beam are best represented by plots showing intensity distributions as a function of displacement and angle in each of the two planes. This takes the form of an ellipse whose area is called 'the phase space', or emittance, which remains constant unless the beam is accelerated, decelerated or collimated. As the beam is subjected to focussing or deflection by a magnetic or electric field, the shape of the beam will change, but the emittance remains constant. Beam which is lost on apertures or slits causes a reduction in the emittance.

2.3.1 Beam Selection

A velocity selector (e.g. Wien Filter incorporating electric and magnetic fields at right angles) is often used in low energy systems and in accelerators after the ion source and prior to acceleration. The fields are adjusted so that the desired beam component passes through with zero net deflection. Other components are deflected away from the central path. The equation for zero deflection is:

$$\sqrt{(2E_1/M)} = E/B \qquad (2.3)$$

Alternatively a deflection of typically 30° in either an electric or magnetic field may be used to select a component of specific M, E or q.

An analyser magnet is frequently used for energy stabilisation. Current signals from a set of beam defining slits (typically 1 mm wide) at the image point give a feed-back signal which can be used to raise or lower the accelerator voltage so that the beam passes symmetrically

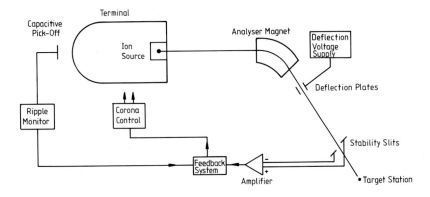

Fig. 2.5 Beam stabilisation schematic.

through the slits. This principle can be used to stabilise the terminal voltage to ~ 1 kV. A stabiliser has been developed (Amsel *et al.*, 1982) which incorporates fast and slow responses to allow for both transients and long term drifts. It also allows fine scale energy scanning, thus enabling resonance energies to be located quickly. A schematic diagram of this type of stabiliser is shown in Fig. 2.5. It will produce a beam stability of the order of 100 eV on a belt-charged acelerator. Chain charged or DC power supply type (Tandetron) accelerators can achieve this stability without additional stabiliser equipment. A generating voltmeter installed in the pressure vessel of an accelerator can also be used to measure and control the terminal voltage. A small charge, dependent on the terminal voltage, is converted to a voltage signal and used for determining and controlling terminal voltage and hence beam energy.

Either the analysing magnet itself or a second (switch) magnet or electrostatic deflector may be used to route the beam to any one of a number of beam lines and target stations. Field reversal enables either left or right deflection and doubles the number of available ports. In the case of a vertical accelerator, a rotating 90° analyser magnet can fulfill both analysis and switch roles.

2.3.2 Beam Transport and Focussing

a. High Energy Beams

Most target stations for high energy accelerators are located several metres or more from the accelerator and the position, size and divergence

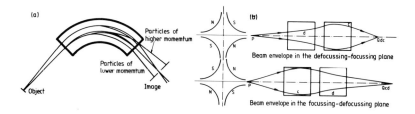

Fig. 2.6a. Typical ion paths through an analyser magnet. b. Beam focussing with quadrupole lenses with the shape of the pole pieces shown at the left and ion paths in X- and Y-planes at right.

of the beam must be controlled from the ion source to the target station. A small beam spot (area ~ 1 mm²) is commonly used for PIXE, RBS and NRA measurements, but a larger area may be desirable for samples which are sensitive to heating. Alternatively, electrostatic scanning of the beam over larger areas can achieve the same result. Beam sizes below 0.1 mm² require sophisticated techniques which are described in Chapter 10. Small angular divergence is particularly important for channeling studies and as a starting point for microbeams. The principles, which are well established for electron beam focussing, also apply to ion beams (Septier, 1967; Valyi, 1977).

Fig. 2.6a shows a typical set of paths through an analyser magnet. The shape of the edges of the pole pieces and hence of the fringing magnetic fields can be designed to produce a focussing action (Glavish, 1981). Additional focussing is best done with a set of quadrupole lenses as shown in Fig. 2.6b. A single quadrupole will focus a beam in one plane and at the same time defocus it in the plane at right angles, giving either a disc, horizontal line or vertical line, but the X- and Y- emittances remain constant. Combinations (doublets or triplets) are used to provide focussing in both planes. Magnetic quadrupoles are normally used at high energies, but electrostatic lenses have the advantage that the focal length is independent of ion mass.

b. Low Energy Beams

Low and high energy beam transport involve the same principles but low energy practice is affected by such factors as:

- the small size of the equipment;
- the predominant requirement for heavy ion beams (e.g. O_2^+, Ar^+ or Cs^+, etc.);

- the frequent use of small diameter beams($< 10~\mu$m) for micro-probe analysis:
- the greater likelihood of significant neutral beam components; and
- the use of UHV for control of surface conditions.

A choice of ions is desirable and the duoplasmatron or electron impact sources are often used for gas ions. A hot cathode may be necessary in the duoplasmatron for non-reactive gases. Typical current densities are 10 to 100 μA cm^{-2} at 5 keV with an energy spread of 30 eV for the duoplasmatron and 2 eV for electron impact. A surface ionisation source is needed for Cs$^+$ beams and the LMIS may be used for high spatial resolution. Negative ions are useful at times to minimise the charging of insulating samples. Occasionally it may be desirable to use a neutral primary beam which can be obtained by placing the sample in the straight through position of an analyser which deflects charged ions. Low energy ions (~ 1 keV) give low sputter rates and good depth resolution (low ion beam mixing) whereas higher energies (~ 1 keV) give a high sputter rate and smaller beam diameter.

Mass analysis of the primary beam is an advantage especially if trace elements are to be studied since these may be obscured by impurity components of the primary beam. Good mass resolution may be desirable and the field should be reversible for use of positive or negative ion beams. A small laboratory magnet or a Wien filter are suitable but with the latter a small deflector is needed to remove neutral ions.

Many experiments can be done with a beam diameter of ~ 1 mm but a microprobe requires additional focus and sweep systems (Fig. 1.2a). Electrostatic (einzel) lenses are often used and a condenser plus objective allow independent adjustment of focus and beam intensity. An electrostatic deflection system can be used in conjunction with synchronised electronic gating of the secondary ion signal to avoid measuring secondaries from the wall of the sputter pit (Section 2.6.6). Near normal incidence of the primary ion beam is advantageous when using reactive gas ions but at 45° or greater a higher sputter rate and better depth resolution is obtained. The requirements for achieving small beam diameters are discussed by Liebl (1975), Benninghoven *et al.* (1982) and Chapter 10.

2.4 TARGET CHAMBERS

Analysis chambers may be quite simple when devoted to one kind of measurement using little more than a basic target holder and a detection system to monitor the interaction yield. More complex systems can provide a choice of techniques, sophisticated target manipulation or automated sample changing. Many low energy systems have the entire ion beam apparatus, including ion source, acceleration and focussing elements, targets and detectors in a single vacuum chamber. However, it is more usual to have the beam production and manipulation components separated from the target and detection system, at least by a vacuum valve which facilitates target changing without letting the entire system up to atmospheric pressure (Fig. 2.2). Surface techniques may require in situ sample cleaning (e.g. a 1 keV ion gun for sputtering or heating) and additional apparatus for examining surfaces (e.g. AES and LEED).

A schematic diagram of a high energy target chamber is shown in Fig. 2.7. Optional features include:

- sample translation (X, Y, Z);
- sample rotation (ϕ, γ, ω);

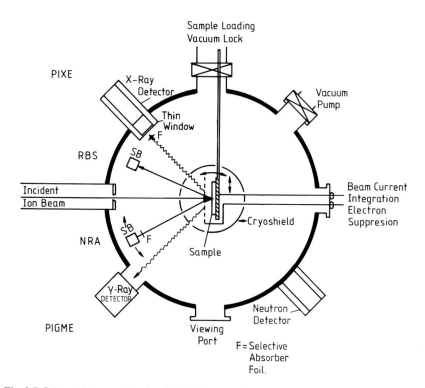

Fig. 2.7 Schematic target chamber for high energy IBA.

- vacuum lock sample loading;
- arrangement for current integration;
- collimators, secondary electron suppressors, anti-scattering baffles;
- anti-contaminant cryoshield;
- filters, absorbers;
- viewing port for positioning the beam on target;
- target cooling (or heating);
- Ion detectors (optionally movable) located inside the chamber and externally adjustable; and
- Photon and neutron detectors mounted inside or outside the chamber

Most of these features are discussed in more detail in later sections and in Chapters devoted to individual techniques, additional facilities may be included for associated non-ion beam studies.

2.4.1 Sample Manipulators

For a single measurement on a single sample the only manipulation involved is to locate the sample in a fixed geometry relative to the beam and detector(s). Measurements at a number of positions across the sample require translation in one or two directions. A number of multi-sample arrangements have been reported in the literature. These include linear drive (Duerden *et al.*, 1984), carousel (Hall *et al.*, 1984), and forked stick (Fallavier *et al.*, 1986). Combined rotation and linear drive is illustrated in Fig. 2.8a. For translation, reproducibility of < 0.25 mm is desirable and in microbeam measurements micron accuracy is essential (see Chapter 10). Manipulator requirements for various types of IBA are given in Table 2.3. For RBS and channelling, angular alignment of the sample to an accuracy < 0.1° is required, necessitating at least two angular rotations. Ideally rotation about three axes is necessary as shown in Fig. 2.8b (Scott *et al.*, 1985) and Chapter 6. Moving components must be designed to maintain HV or UHV during manipulation. The latter requires the use of magnetically coupled drives, bellows or carefully designed seals. The use of motors inside the vacuum is not recommended unless they are specifically designed for vacuum work as they are likely to degas. When a large number of samples is to be analysed, transfer to and from the chamber is best accomplished through an airlock.

a.

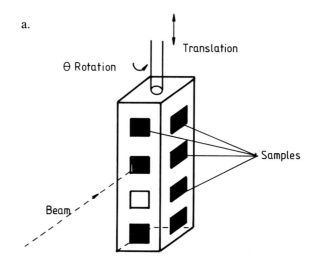

b.

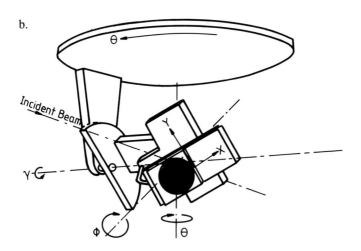

Fig. 2.8a. Schematic illustration of a linear drive multi-sample holder. b. Goniometer for channelling showing X, Y, θ, γ and ϕ motions relative to the incident ion beam.

TABLE 2.3
Target manipulator requirements

Technique	Minimum Manipulator	Preferred Setup
RBS	Fixed holder (no degree of freedom)	θ, X, Y (multi-sample)
RBS-C	Two orthogonal rotations	θ, ϕ, (γ), X, Y
PIXE	Fixed holder	X, Y (multi-sample)
NRA	Fixed holder	θ, X, Y
ERA	Two translations	θ, X, Y
SIMS, SIPS	Fixed holder (UHV)	θ, X, Y
LEIS	Two rotations (UHV)	θ, ϕ, X, Y
IMP, PMP	Precision X–Y	θ, ϕ, X, Y

2.4.2 Sample Mounting, Handling and Temperature Control

Samples should be loaded carefully to minimise contamination and the risk of damage. Gloves may be worn to reduce transfer of grease etc. to the surface. Water vapour may be removed by prepumping or by preheating, as long as it does not damage the sample. At the completion of a run it is desirable to let the target chamber up to an atmosphere of dry N_2 rather than air. Samples which are prone to absorb moisture should be stored in a dessicator. Special care is required with biological samples which may need freeze drying. More discussion on biological samples is given in Chapter 10.

In many situations the ion beam may cause sample damage either by radiation or heating. For high energy analysis, beam currents can be typically 1–100 nA at 2.5 MeV. This represents a power dissipation of up to 0.25 W, but the power is all dissipated in the top few microns of the sample and if the beam spot is focussed to 1 mm^2, then the power density is 25 kw cm^{-3}. If the target is metallic, then this level of power dissipation can be handled by the use of a support cooled by water, freon, oil or forced air. Water cooling may create problems with beam current integration because of low impedance. Insulating samples may reach surface temperatures of several hundred degrees. A description of methods to calculate and control of temperature in thin samples during ion beam analysis has been given by Cahill *et al.* (1986).

For crystalline materials, radiation damage can occur at ion doses in excess of 10^{13} ions cm^{-2} — equivalent to a few seconds irradiation with a 100 nA beam. In some cases it may be practical to defocus or scan the beam to many mm^2 in order to reduce the power density. In low energy analysis, even though the actual power dissipation is low, so is the depth of penetration. Surface erosion (sputtering) of the first few atomic layers is then the most important form of damage. Radiation damage is very dependent on technique and sample type and further discussion is included in Chapters 3, 6, and 10.

2.4.3 Beam Current Integration and Sample Charging

Measurement of beam current is essential for determination of ion dose for each measurement and to obtain absolute yields. Currents from 10 pA (6 x 10^8 ions s^{-1}) to 10 μA (6 x 10^{13} ions s^{-1}) may have to be measured to an accuracy of 1% or better. The relations between charge (Q), current (I), time (t) and number of ions or dose (N_1 ions cm^{-2}) are:

$$Q = It; N_1 = 6 \times 10^{18} Q \qquad (2.4)$$

Integration of the charge reading on the target is the most practical method of monitoring beam dose, but there are alternate methods. One of the best methods is to use a wire rotating through the beam (Roth *et al.*, 1974, Magee *et al.*, 1986) which scatters a fraction of the beam to a surface barrier detector which can be calibrated aginst a beam current measured by a Faraday cup. This method relies on the shape of the beam remaining constant during the rotation periods of the wire. The emission of associated particles, X-rays or gamma-rays can also be used as a beam monitor. It is also possible to periodically deflect the beam into an offset Faraday cup for a small fraction of the time.

Some of the experimental techniques which can be used to ensure accurate integration include:

- elimination of secondary electron effects by use of a suppression electrode;
- removal of unwanted species such as neutrals or multiply charged particles by use of suitable analysers;
- high insulation between sample and ground;
- elimination of charging followed by breakdown in the case of insulating samples;
- calibration of current integrator using a standard current source; and
- rapid fluctuations in beam intensity should be minimised.

Secondary electrons can be a major source of error. The problem is conventionally overcome by use of a Faraday cup as shown in Fig. 2.9.

The length to diameter ratio should be at least 3 to 1 to reduce the solid angle for electron emission through the cup entrance aperture. Further reduction of secondary electron losses is obtained either by placing a small ring in front of the cup and biasing it negative by a few hundred volts or by applying a small magnetic field in the region of the entrance to the cup to spiral electrons back into the cup. The base of the cup can be a rotating vane which intercepts the beam periodically and allows it to pass to the sample for the remainder of the time. Faraday cup design has been discussed in detail by Hemment (1980).

In IBA it is usually not practical to create an enclosed Faraday cup around the samples. Therefore either a negatively biased ring is placed in front of the target or the whole target chamber is electrically insulated and so acts as a Faraday cup. In the former method, care must be taken to ensure that the ion beam does not strike the negatively biased grid, thus producing secondary electrons which could reach the target. Insulating the entire target chamber may not be practical in complex chamber arrangements. Also at low currents ($\sim$ nA) it is difficult to isolate the chamber from capacitive pick up from the surrounding environment or leakage by conduction across insulated joints caused by humidity.

The presence of various charge stages and neutral particles in the beam leads to incorrect integration and they should be eliminated by magnetic or electrostatic beam deflection prior to the target chamber (Fig. 2.4.) Maintaining linearity of 1% or better in current integration is difficult to achieve. Calibration using a known current source should be checked regularly when such accuracy is required, even at the start and

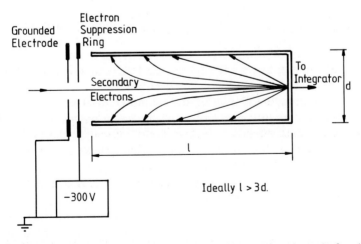

Fig. 2.9 Faraday cup for beam current integration with negative electrode for electron suppression.

finish of a series of measurements. More detailed descriptions of current integration have been given by England (1974) and Hemment (1980).

An insulating sample may charge and discharge in various ways leading to integration errors. The breakdown may be to ground rather than through the integrator or the discharge spike may be too rapid for the integrator to handle. The buildup of positive charge during ion irradiation of an insulating sample can cause a reduction of ion energy at the surface or beam deflection. Furthermore it can lead to background X-rays emitted by electrons accelerated towards the sample surface. A number of methods are available for minimising or removing these problems (Goclowski *et al.*, 1983).

i. Electrons from a hot filament near the sample are attracted to the surface as soon as ion irradiation causes the potential to rise. If the filament is hot enough, the electron current is self-regulating and the potential stays below 1 kV. This method is satisfactory except for very high resistance materials such as pure silica which may not completely discharge. An alternative is to use an electron gun with the current set to just compensate the ion current being used. For direct ion current integration, the target and electron flood apparatus need to be enclosed within a Faraday cup.

ii. Coating the surface of insulating samples with a thin layer of conducting material such as C is standard procedure in electron microscopy. This method can also be applied in IBA but the surface layer will cause energy loss by the incident ions and in some applications this may be unsatisfactory.

iii. Proximity foils can be placed on the sample surface surrounding the area to be analysed to provide a good conduction path and hence minimum charge build-up. Alternatively, it may be possible to deposit a conducting grid structure on the surface if analysis of several regions is desirable.

iv. An entrance foil placed immediately in front of the sample releases secondary electrons as the beam passes through. These provide an effective electron flood source for the target. However, energy loss in the foil and difficulty with charge integration must be accepted with this method.

v. Poor vacuum (~ 1 Pa) or a partial atmosphere of H or air ~ 100 Pa) can give similar results.

vi. Mingay and Barnard (1978) observed that a transverse magnetic field was as successful as an electron flood and resulted in lower probability of target contamination.

2.4.4 Detector Arrangements

Table 2.4 summarises the typical detector types and arrangements used for the various ion beam analysis techniques. Figs. 2.2 and 2.7 show typical detector layouts for low and high energy techniques. Specific details of detector performance are given in Section 2.6. In NRA and ERA, foils are used to absorb scattered ions. Plastics such as Mylar and Kapton are often used but these can charge up and the consequent sparking introduces noise into the electronics. Metal foils can also be used provided that they are free of inhomogeneities which degrade the detector resolution. Annealed foils suffer from recrystallisation so that partial channeling can lead to transmission of a fraction of the scattered

TABLE 2.4
Typical detector types and arrangements for various ion beam techniques

IBA	Product	Detector	Configuration	Vacuum
LEIS	Scattered Ions	Channeltron	Vacuum, movable advantageous Energy measurement requires electrostatic/magnetic analyser	10 nPa
SIMS	Secondary Ions	Channeltron	Vacuum, fixed geometry Low mass resolution with ESA, QMA High mass resolution with Sector Field Analyser	preferred < 1 μPa
SIPS	Optical photons	Spectro-photometer	External to chamber. Fixed geometry. High wavelength resolution.	"
PIXE	X-rays	Si (Li) IG	Vacuum or external. Filters Thin window. Liquid N cooling	< 1 mPa
RBS	Ions	Surf.barrier[a]	Vacuum, movable geometry Small and simple arrangement	
RBS-C	Ions	"		< 100 μPa
ERA	Ions	"	Glancing angle geometry for improved depth resolution	
NRA	Ions	"		
PIGME	Gamma-rays	Ge (Li) NaI	External with window, cryostat High Resolution, Low efficiency Poor Resolution, high efficiency	"
NRA	Neutrons	BF$_3$ Li glass Scintillator[b]	External, low efficiency Detection only Broad resolution by unfolding	

[a] Improved resolution can be obtained by use of magnetic or electrostatic spectrometer at expense of count rate. Time of flight sometimes useful.
[b] Time of flight can give energy resolution at expense of count rate.

ions. Beam contamination such as a D^+ component in a H_2^+ beam may also be transmitted by the foils and give a high background.

2.4.5 External Beams

High energy protons have sufficient range (typically 10 to 100 μm) to be able to pass through a thin window into the atmosphere. The range of a 3 MeV proton in air is about 140 mm. This offers special advantages in the following situations:

- samples which might change in composition or decompose in vacuum;
- extremely close geometry for X-ray or gamma-ray detectors which may be required for low yield situations;
- samples too large for target chamber; and
- samples which require special cooling.

The window is usually a very thin foil (typically 5 to 10 μm) (see, for example, Deconninck and Bodart, 1978) made from Cu, Fe, Al, Ni or plastic. Alternatively the beam may pass through a number of differentially pumped stages and, finally, a pin hole to air. Fig. 2.10 shows an external beam arrangement using a fine mesh made of graphite with about 130 holes each 0.3 mm diameter which acts as a support and a heat sink (Anttila *et al.*, 1985). The ion energy is reduced in the foil and also in the

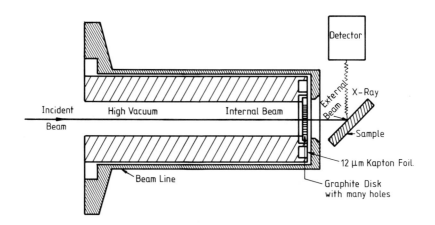

Fig. 2.10 External beam arrangement for large area irradiation of a sample outside the accelerator vacuum system.

air and allowance must be made for this when estimating yields. Energy loss and energy spread may be reduced by using He rather than air.

There is always a risk of vacuum failure with a thin window. It is important, therefore, to incorporate a fast acting gate valve which will seal off the window from the accelerator vacuum in the event of failure. Foil life is a function of thickness, diameter, material and power dissipation but it can be thousands of hours. However, restrictions are placed on the use of ions heavier than H^+ as an external beam due to the large heat dissipation in the foil (Raisanen, 1984). For example, using a cooled 8 μm Kapton foil, the lifetime was found to be only 10 minutes for a beam current of 50 nA mm^{-2} of 2.4 MeV α-particles. A review of suitable foil materials is given by Raisanen and Anttila (1982). They recommend Be, Al or aluminised Mylar. If Be is used the incident energy must be kept below the neutron threshold of 1.8 MeV.

Direct measurement of the beam current in an external beam is made difficult by high secondary electron production in the gas. The dose can be determined by monitoring the X-ray or RBS yield from the window or a rotating wire in the beam. External beams are mostly used for PIXE and PIGME analysis. A novel external beam RBS facility has been developed by Doyle (1983). The backscattered particles repenetrate the exit foil and are detected inside the vacuum chamber, thus minimising the path length of ions in air. An external beam PIXE system with carousel type multi-sample holder has been constructed by Hall *et al.* (1984).

2.5 ADDITIONAL EQUIPMENT CONSIDERATIONS

2.5.1 Vacuum Requirements

Typical vacuum requirements for low and high energy analysis techniques are included in Table 2.4 and useful procedures are summarised in Highlight 2.2. To eliminate contamination of the surface most low energy techniques use UHV equipment (target chamber pressures below 1 μPa) and in situ ion or electron beam cleaning. Most high energy methods only require vacuum in the range 1 mPa to 100 μPa. At 10 mPa, the hydrocarbon content of the residual gas causes a build-up of contamination at the sample surface. With 1 μPa partial pressure of hydrocarbon vapour, the deposition rate may exceed 10^{13} atoms cm^{-2}s^{-1} depending on the beam current and type of hydrocarbon (Bottiger *et al.*, 1973; Möller *et al.*, 1981).

HIGHLIGHT 2.2
USEFUL VACUUM PROCEDURES

- Apertures and slits present a high pumping impedance, increasing pump out time and leading to poor vacuum on one side unless a bypass is used. If a pressure differential is needed while running, the bypass can be closed off.
- Arrangements should be made to pump hidden volumes such as those behind screw threads or locating devices.
- Do not subject plastics to direct beam or heat which could cause degassing.
- UHV components need to be baked at > 100° to remove volatiles. Plastics or metals such as Zn, Bi,brass or bronze should not be used in UHV systems because of the release of vapours during baking. Al degasses more than stainless steel unless it is highly polished.
- Samples should be loaded/unloaded through a vacuum lock, especially when using UHV; if this is not possible porous samples, such as pressed powders, and other samples which outgas considerably should be prepumped in a separate chamber.
- Dry N_2 or Ar should be used when opening a vacuum system; cold shields can tolerate dry N_2 for short periods without icing up but the system should be pumped down again even if it is only to be left for ~ 5 minutes.
- Samples should be loaded/unloaded through a vacuum lock, especially when using UHV; if this is not possible porous samples, vapour into the target chamber.
- Roughing pump exhausts should be piped away from working areas to avoid exposure of personnel to vapours.

The target chamber can be operated at a different pressure (p_2) to that of the accelerator beam lines (p_1) by the use of differential pumping techniques. A pumping impedance (Z) is used which usually consists of a relatively small diameter tube (~ 3 mm) to separate the two regions so that:

$$p_1 - p_2 = Zg \qquad (2.5)$$

and

$$Z = (0.4l/d^3 + 0.53/d^2)\rho^{1/2} \qquad (2.6)$$

where g is the rate of gas flow, l and d are the length and diameter of the tube and ρ is the gas density. The target chamber pumps must also remove any gas evolved from the target and chamber so that a high speed pump or clean conditions or both are needed to achieve UHV near the target. Gases and vapours can be removed from the beam lines and chamber by the use of liquid N_2 or He cooled traps. These may be placed:

- as a side arm to a beam line or analysis chamber;
- as a cooled sleeve through which the beam passes; or
- as a cooled surface within a chamber.

The unwanted components are frozen to the cold surface and are inactive until the surface becomes warm and this should be done periodically under controlled conditions (e.g. when no ion beam is in use) to purge the impurities out of the system. Bottiger *et al.* (1973) have shown that a cold shield at 25 K surrounding the target can reduce C build-up by a factor of 40. Contamination of the target arises from beam-induced cracking of oil or grease present within the vacuum or by vapours emitted by a sample. A method of measuring the extent of carbon build-up by means of the reaction $^{12}C(p, \gamma)^{13}N$ has been reported by Rudolph *et al.* (1981). The characteristics of various types of pumping system are summarised in Table 2.5

TABLE 2.5
Vacuum pumping systems

Pump type	Operating Range	Fore-Pump Required	Major Use	Comments
Rotary	Atm–1 Pa	No	Backing or pump from atmosphere	Exhaust should be vented externally Cold trap desirable
Cyro-sorb	Atm–1 Pa	No	Clean pump down	Bake out after extended use to drive off trapped gas. Requires liquid N
Vac-ion Ti-sublim	0.1 Pa–1 μPa	No	Small HV/UHV	Not suitable for noble gases
Diffusion pump	1.0 Pa–1 μPa	Yes	Ion source and beam line	Liquid N trap and baffle to minimise backstreaming Water cooling necessary
Turbo-pump	10 Pa–0.1 μPa	Yes	Ion source, beam linc, chamber	Mesh over entrance to stop solid components dropping into turbines
Cryo-pump	0.1 Pa–1 nPa	No	UHV-chamber beam line	Needs periodic bakeout Compressor vibrates

2.5.2 Radiation Safety and Shielding

Accelerators with maximum ion energies below 50 keV do not present a significant radiation hazard. Soft X-ray generation in the region of the ion source and acceleration gap is easily absorbed within metal vacuum chambers. Lead glass windows can be used where direct viewing of the ion source region or extraction electrode is required. However, all high energy accelerators produce some X- and gamma-radiation and, in some cases, neutrons. It is essential that accelerator installations have adequate shielding to protect personnel and to satisfy all regulations made by appropriate authorities. Gamma-ray and X-ray yields increase exponentially with the particle energy and linearly with the target current. Table 2.6 lists current limits set by the National Council on Radiation Protection (NCRP, 1971) for radiation levels. Local regulations may be

TABLE 2.6
Dose limit recommendations NCRP (1971)

Region	Dose mSv/year	
Occupational exposure limits		
Whole body, gonads, eye lens, red bone marrow	50	
Skin	150	
Hands	750	
Forearms	300	
Other organs, tissues and organ systems	150	
Fertile women (with respect to fetus)	5	(gestation period)
Public or occasionally exposed individuals		
Individual or occasional	5	
Students	1	
Population dose limits		
Genetic	1.7	
Somatic	1.7	
Emergency lifesaving		
Individual (older than 45 years if possible)	1000	
Hands and forearms	2000	additional
Emergency dose limits — less urgent		
Individual		
Hands and forearms	1000	total
Family of radioactive patients		
Individual (under 45 years)	5	
(over 45 years)	50	

more stringent. Interlocks must be installed to prevent entry to any area with a radiation level above maximum permissible levels. If entry for short periods is permissible, exposure should be measured to ensure that it is kept within the prescribed limits.

X-rays are predominantly produced in the region of the acceleration column and ion source. The pressure vessel attenuates this radiation but, at high beam currents, radiation levels of hundreds of microsieverts per hour may be produced. For a beam energy below 3 MeV and beam currents of protons or He ions of less than 1 μA, neutrons are not a serious radiation problem unless light elements such as ^{3}H, ^{7}Li or ^{9}Be are used as targets. However, many metals (including Cu) have neutron thresholds below 3 MeV and currents in excess of a few μA on slits and apertures can produce appreciable neutron levels. If deuterons are accelerated, they produce significant neutron radiation, possibly many times the maximum permissible level and the area must be appropriately shielded. If neutron radiation levels are high, 20–100 cm of heavy concrete or borated paraffin shielding would be required. In ion beam analysis, most of the X-rays produced in the target are absorbed in the walls of the target chamber and radiation levels outside are usually negligible. Gamma-ray shielding (e.g. Pb) is necessary if high energy ions or high beam currents are used.

Radiation can be produced by "ion-induced" radioactivity in accelerator components or targets and can remain long after the ion beam is turned off, depending on the half-life of the particular isotope. For example a ^{7}Li target used with a proton beam above 1.88 MeV will undergo (p,n) reactions producing ^{7}Be which decays by positron emission with a half-life of 53 days. Alpha- and beta-radiation are absorbed by small amounts (mm) of shielding so any hazard usually only arises when handling samples after irradiation. The higher the beam energy and dose the greater the activity produced. Beam line components which are subjected to extended periods of irradiation may be radioactive and should be monitored before being handled. In order to minimise both 'prompt' and 'residual' radiation, slits and apertures should be constructed from materials such as graphite or Ta which normally have low radiation yields.

2.5.3 Accelerator Energy Calibration

A typical analyser magnet with 1 mm entrance and exit slits can be used to define the energy with a resolution of 1 keV. Accurately known reaction energies (such as resonances or thresholds) are used to calibrate the measured magnetic field in terms of beam energy. For example a thin

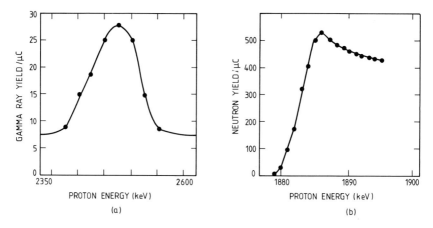

Fig. 2.11 Resonance and threshold techniques for accelerator calibration. (a) Yield from 1043 kev γ-ray in 2374 keV resonance. (b) Neutron yield at O° from ^{7}Li(p, n)^7 Be reaction.

Al target (≈ 0.1 μm, corresponding to an energy loss of ≈ 2 keV) gives a sharp peak in the yield of the 1013 keV gamma-ray at a proton energy of 2374 keV (Fig. 2.11a). Another suitable reaction is ^{24}Mg(p, p', γ)^{24}Mg where a resonance at 2330 keV produces a very high back angle yield in the 1368 keV gamma-ray. The reaction ^{7}Li (p, n) ^{7}Be has a threshold at 1881 keV and Fig. 2.11b shows the 0° neutron yield as a function of beam energy for a 6 keV thick Li target. Table 2.7 gives a list of some (p, n) thresholds and resonance energies for calibration purposes.

It is also possible to calibrate the analyser magnet against an alpha-particle source such as ^{204}Po which emits 4.88 MeV alpha-particles. If the accelerator is equipped with a precision generating voltmeter, this can also be calibrated in a similar way. At the energies used in most ion beam analysis work, relativistic corrections for increase in mass are not necessary. Calibrations should be checked regularly, particularly if there has been major maintenance which could in any way change the alignment of the accelerator and analyser magnet. For example, a small shift in accelerator position may lead to a change in the direction of the ion beam as it passes through the defining slits and enters the magnetic field and hence a change in energy calibration. Even asymmetric erosion of the ion source canal may cause a similar effect. For the same reason, beam steering in the plane of the ion path should not be used prior to the analyser magnet.

Table 2.7
Some useful reactions for accelerator calibration

Reaction	Ion Energy (keV)	Type
$^{19}F(p, \alpha)^{16}O$	340.04 870.20	Resonance
$^{27}Al(p, \gamma)^{28}Si$	991.87	Resonance
$^{3}H(p, n)^{3}He$	1019	Threshold
$^{13}C(p, \gamma)^{14}N$	1747.6	Resonance
$^{16}O(d, n)^{17}F$	1829.2	Threshold
$^{7}Li(p, n)^{7}Be$	1880.6	''
$^{9}Be(p, n)^{9}B$	2057	''
$^{24}Mg(p, p\,\gamma)^{24}Mg$	2330	Resonance (1368 keV γ-ray)
$^{27}Al(p, \gamma)^{28}Si$	2374	Resonance (1013 keV γ-ray)
$^{27}Al(p, \gamma)^{28}Si$	2403	Resonance (843 keV γ-ray)
$^{13}C(p, n)^{13}N$	3235.7	Threshold
$^{31}P(\alpha, p)^{35}S$	3655	Resonance
$^{19}F(p, n)^{19}Ne$	4234.3	Threshold
$^{60}Ni(p, n)^{60}Cu$	7023.6	Threshold

Additional calibration energies are given by Marion and Young (1968) and by Helmer *et al.* (1979).

2.6 DETECTION OF INTERACTION PRODUCTS

Ion beam analysis measurements are made by counting the number of emitted interaction products (particles or photons) and/or by measuring their energy. The latter can be done by the detector itself or by placing a detector after an electrostatic, magnetic or time-of-flight analyser. Common detector arrangements for IBA techniques were summarised in Table 2.4 and typical detector types and parameters are listed in Table 2.8.

2.6.1 Detection Methods and Principles

Detectors exploit the interaction between the emitted product radiation and detector material. Four types of material are commonly used:

Table 2.8
Detector types and parameters for IBA

Radiation	Energy	Device	Detection Process	Resolution	Bias
Photons					
Visible, UV	< 100 eV	Channeltron	PE generation and ionisation		
X-rays	1–100 keV	Si(Li) IG	Electron-hole pairs	140 eV at 5 keV	1 kV
γ-rays	.05–10 MeV	IG Ge(Li) NaI BGO	PE, Compton, pair production electron-hole production	200 eV at 6 keV 2 keV at 1.1 Mev 70 keV at 1.1 MeV	3 kV 3–5 kV 0.5 kV
Particles					
Neutrals		SB	Electron-hole pair production	12 keV	50 V
Ions	< 100 eV	ESA	Deflection in field	1% or 300 eV at 30 keV	
	1–30 keV	QMA		M~ unity	
	0–300 keV	SB	Electron-hole		
	> 300 keV	ESA,Mag	Deflection in field	12 keV at 4 MeV 1–5 keV	50 V
Neutrons	Thermal	BF_3, ^3He	NRA		
	> 100 keV	Liquid Plastic	Elastic Recoil	Requires time-of-flight	
Electrons		Channeltron	Ionisation		

i. Alkali Metals — as coating of electrodes which emit one or more electrons for every particle or photon striking the surface. Electron multipliers and photomultipliers use many electrodes to achieve a gain of 10^8 or more and the amplified signal is detected electronically at the anode. The most common detector for low energy ions is a microchannel plate detector such as the Channeltron (Evans, 1965). Photomultipliers are used to detect optical emission or the fluorescence of scintillators.

ii. Gas — used in an ionisation chamber in which electrons and ions are accelerated to opposite electrodes and give rise to an output pulse proportional to the incident ion energy; in the Geiger counter an avalanche generates a fixed height pulse irrespective of incident radiation energy.

iii. Scintillator — a crystal, plastic or liquid which fluoresces emitting low energy photons which are detected by a photomultiplier.

iv. Semiconductors — such as reverse-biased p-n junctions which operate as solid state ionisation chambers.

Semiconductor detectors are of special interest since they are the most widely used type in high energy IBA. Electron-hole pairs produced by ionising radiation are separated by an electric field and attracted towards opposite electrodes. The efficiency of collection depends on the extent of recombination or trapping caused by impurities and defects. For thicknesses up to the order of 1 mm, a Si surface barrier detector (SB) can be used. The reverse bias voltage sweeps free charge carriers out of the depletion region whose thickness depends on the bias. Fig. 2.12 is a nomogram relating specific resistance, capacitance, bias voltage and particle energy which may be stopped within the depletion region. The dotted line shows that 13 000 Ω-cm, n-type Si requires a bias voltage of 65 V to fully stop a 23 MeV α-particle, a 6 MeV proton or a 2.5 MeV electron. The energies are defined by the dashed line drawn from the point where the dotted line cuts the specific capacitance scale. To obtain greater sensitive depths, the semiconductor (usually Si or Ge) must be of high purity. Alternatively, Li can be diffused through the semiconductor at an elevated temperature so that it will compensate any electron traps present. The use of semiconductors is the topic of Highlight 2.3.

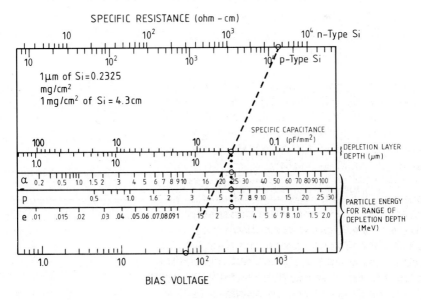

Fig. 2.12 Nomograph linking bias, depth and energy scale for ion detection with a surface barrier detector.

HIGHLIGHT 2.3
USE OF SEMICONDUCTOR DETECTORS

Surface Barrier Detectors

- SB detectors are sensitive to light and should be kept dark when bias is applied.
- Only ruggedised SB detectors can be operated at atmospheric pressure; they can also be baked during use in UHV chambers.
- Do not touch the electrode of SB detectors or the windows of detector cryostats.

Cryostat Detectors

- Si(Li) and IG detectors can be stored at room temperature but do not let biased detectors warm up; Ge(Li) detectors must be kept cooled.
- Si(Li) detectors should be stored with a vacuum outside the thin window to avoid gas penetration through any pinholes in the window; any condensation or contamination on the window or detector attenuates low energy X-rays so check the efficiency periodically.
- He can diffuse through Be windows and should not be used as the filling gas when opening target chambers with Si(Li) detectors.
- The window on a Si(Li) detector may not be thick enough to stop scattered protons which will generate a high count rate unless an additional filter is used.
- Li segregation can reduce the active volume and hence the efficiency of Si(Li) and Ge(Li) detectors; redrifting will usually restore the efficiency but sometimes with some deterioration in resolution.
- Mount detectors so that they are free of vibration (e.g. from vacuum pumps) which can increase the electronic noise.
- Use an aperture close to the detector to accurately define the solid angle and minimise edge effects.

a. Photon Detection

Visible and ultra-violet light from ion interactions (e.g. SIPS) can be directed and focussed by mirrors, lenses and diffraction gratings and detected with a photomultiplier. The wavelength of photons can be measured with a dispersive device, e.g. grating or prism. Dispersive detectors use a diffraction grating to concentrate a specific wavelength at a well defined position where a photosensitive detector is used to measure the intensity. A full energy spectrum is obtained by measuring the intensity as a function of angle. This can be a time consuming process. Monochromators are also used with SIPS and occasionally with PIXE techniques.

High energy photons undergo a number of types of interaction with detector material:

i. *Photoelectron Production.* An atomic electron is given the full energy of the photon less that required for ionisation. This is the process used in photomultipliers, but it can also be used at high energies to determine photon energies from a measurement of the photoelectron energy. The probability of photoelectron production depends on the fourth power of the atomic number and is most probable at energies below about 200 keV.

ii. *Compton Scattering.* The photon behaves like a particle and elastic scattering from an electron follows Equation (1.21b). The maximum energy that can be imparted to the electron is for 180° scattering when the energy of the gamma-ray is reduced to approximately 0.2 MeV. This value is almost independent of the initial energy of the gamma-ray. The probability of Compton scattering depends linearly on atomic number and decreases with increasing gamma-ray energy. At 1 to 2 MeV it is the most probable interaction process. The scattered photon may undergo another scattering event within the detector and be further reduced in energy.

iii. *Pair production.* A high energy photon can convert into an electron-positron pair in the presence of a Coulomb field. The rest mass of the pair (1.022 MeV) comes from the photon energy so that there is a threshold at this energy for pair production. Above the threshold energy the excess energy goes into kinetic energy of the electron and the positron. The probability of pair production depends on the square of the atomic number and rises sharply with energy above threshold. Associated with pair production is the reverse process in which the positron can annihilate with a free electron to produce two photons each having an energy of 0.511 MeV. These photons may also interact within a large detector volume, but sometimes one or both can escape.

The three photon interaction mechanisms transfer energy from the photon to one or more electrons which cause ionisation of the detector material. Resulting fluorescence (low energy photon emission) can be detected by a photomultiplier which produces a current pulse proportional to the original photon energy. This is the basis of detectors such as NaI or BGO. Semiconductor detectors can also be used for photon detection by collecting electron-hole pairs produced by energetic electrons from photon interactions.

Photon intensities are attenuated as they pass through a detector material. At low energies the attenuation is rapid so that a thin (1 mm or less) detector can detect photons at close to 100% efficiency. At energies above a few keV, increased thickness is required to achieve this efficiency. For MeV energies the probability of detection will be less than 100% even if the detector is several cm thick. Some high energy gamma-ray detectors are at least 20–40 cm thick. The use of high Z material improves detector efficiency. Fig. 2.13 shows the probability of each of the

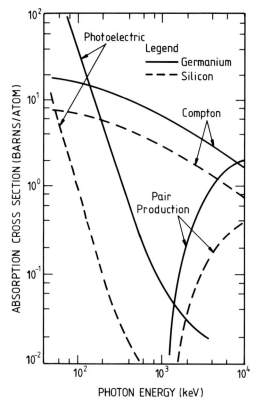

Fig. 2.13 Cross-sections for three types of photon interactions in Si and Ge (by permission of EG&G Inc.).

three interaction processes as a function of photon energy for Ge and Si. Efficiency and resolution both vary as a function of energy and a reliable calibration of detector performance is essential.

b. Particle Detection

Positive or negative ions and neutral atoms all lose energy by ionisation as they penetrate detector material creating electron-hole pairs. The electron current is collected and used as a measure of the energy deposited. Gas or semiconductor detectors are most commonly used for high energy ion detection while electron multipliers are used at low energies (Section 2.6.5) and to detect electrons. Neutrons do not directly produce ionisation, but undergo collisions with H atoms and the recoil protons then lose energy by ionisation. Neutrons may also be captured by light elements such as Li or B when heavy ions are given off which produce ionisation (Section 2.6.7).

2.6.2 X-ray Detection

a. Si(Li) Detectors

A relatively small Si crystal doped with Li (typical area 20–80 mm^2, depth 1–3 mm) is used for X-rays in the range 1–60 keV. A detailed description of these detectors can be found in Woldseth (1973) or in manufacturers' catalogues. A bias of about 1 kV is necessary for a large depletion depth ($\sim$ 9 mm). Energy resolution is typically 150 eV at 4 keV. The detector is operated at liquid nitrogen temperature to stop Li diffusion under bias and to improve depletion depth. It is therefore contained in an evacuated cryostat. A bare Si(Li) detector can be used within the target chamber if it is necessary to measure X-rays below 1 keV (e.g. for elements below Na). Otherwise the detector housing has a thin window (8–25 μm thick) enabling X-rays with energies in excess of 1 keV to pass through. If the target chamber is to be raised to atmospheric pressure it should be done carefully so as to minimise the risk of rupture. The window lifetime can be several years. At energies above 25 keV, attenuation by the window is negligible and there is an increasing probability of X-rays passing through the detector without being detected. Fig. 5.9 shows an efficiency curve for a thin window Si(Li) detector as a function of energy. When a windowless detector is being used, an ion deflector is necessary to prevent scattered ions from reaching the Si(Li) (Musket, 1986).

b. Intrinsic Germanium (IG) Detectors

A high purity Ge crystal can detect X-rays or gamma-rays in the energy range 3 keV to 1 MeV. These detectors are operated at liquid N temperature and have a thin entrance window ($\sim$ 125 μm thick) which absorbs characteristic X-rays from Ar or lighter elements. The detector area can be as large as 1000 mm^2 with a thickness from 1 to 10 mm. An energy resolution of about 200 eV is obtained at 6 keV and 500 eV at 122 keV. Some manufacturers refer to these as Low Energy Photon Spectrometers (LEPS). They are valuable for obtaining higher efficiency than Si(Li) detectors for 25–200 keV photons.

2.6.3 Gamma-ray Detection

In detecting gamma-rays below 100 keV an IG detector can be used. Above this energy, there is a choice available between high efficiency, low resolution performance obtainable with NaI or BGO detectors, or low efficiency, high resolution from a Ge(Li) detector. Because of the higher atomic number BGO scintillators have a higher efficiency than NaI for gamma-ray detection. For detectors with the same volume, an improvement of 20% at 100 keV and a factor of 4.5 at 1 MeV is obtained in the peak efficiency. However, the pulse height resolution is not as good as for a NaI detector. A Ge(Li) detector may have an efficiency of 15% to 30% of that of a 75 × 75 mm NaI detector, but the resolutions at 1330 keV are about 2 keV and 70 keV respectively. Most gamma-rays used for sample analysis have energies below 1.5 MeV and it is advantageous to use a high resolution Ge(Li) or IG detector to obtain accurate energy values and resolution of gamma-rays from different nuclear reactions. Unwanted low energy gamma-rays can be removed by the use of absorbers (e.g. 1.5 mm Pb).

The operating principle of Ge(Li) gamma-ray detectors has been described by Ewan and Tavendale (1963) and in catalogues produced by various manufacturers. They have a Ge crystal (volume 10–100 cm^3), usually in a coaxial form. Lithium is drifted inwards to obtain a region of intrinsic Ge, surrounded by an outer n$^+$ layer and, at the same time, a p-type core is maintained at the centre. A bias voltage (up to 5 kV) is placed between the core and the outside. The detector is operated at liquid N$_2$ temperature in order to stop field–induced Li diffusion. The cryostat contains a window typically (500–1000 μm Be) through which the gamma-rays enter the crystal. These detectors can operate in the energy range from 100 keV up to many MeV.

Because of the three possible interaction processes and the variation of their probabilities as a function of gamma-ray energy, the pulse height spectrum from a gamma-ray detector is complex. Fig. 2.14a shows a typical response function for the 1173 and 1333 keV gamma-rays from ^{60}Co using a Ge(Li) detector. Important features of the response function for gamma-ray energies below 1.5 MeV include:

i. the full energy peak arising from photoelectron production in the detector;

ii. the continuum below the peak caused by multiple Compton scattering events;

iii. the rise in the continuum at an energy of approximately (E − 200) keV which is the maximum energy deposited in the detector by a single Compton scattering event; and

iv. the broad peak at approximately 200 keV which arises from large angle scattering of gamma-rays before they reach the detector.

If the gamma-ray energy is sufficiently high for pair production to be significant, the two 0.511 MeV annihilation quanta must be considered. If these interact within the crystal by the photoelectric effect, then all the gamma-ray energy is deposited in the crystal and the observed pulse height reflects the gamma-ray energy. If one or both escape from the detector, then energy is lost. A spectrum of high energy gamma-rays contains three peaks known as full energy, single escape and double escape peaks, together with an underlying Compton continuum. Fig. 2.14b shows the high energy end of the gamma-ray spectrum obtained from the reaction ^{19}F(p, γ)^{16}O using a large volume ($\sim$ 100 cm^3, 18%) Ge(Li) detector. The relative sizes of the three peaks vary with gamma-ray energy and this variation is shown in Fig. 2.15 for the detector used to obtain the above spectrum.

Peaks from the 6917 and 7115 keV gamma-rays are also visible in Fig. 2.14b, but are Doppler broadened because they are emitted before the product ^{16}O nucleus comes to rest after the recoil caused by α-particle emission. The observed gamma-ray energy in such a case is given by:

$$E_\gamma = E_0(1 \pm \beta \cos \theta) \qquad (2.7)$$

where θ is the angle between the directions of recoil motion and gamma emission, E_0 is the transition energy and β is the ratio of the velocity of nuclear recoil to the velocity of light. There is an additional recoil associated with the gamma emission process itself, but this is insignificant compared with nuclear reaction recoil and only gives rise to a shift rather than broadening of the gamma-ray energy. Recoil broadening for (p, α) reactions is of the order 50–100 keV, being greatest when the

a.

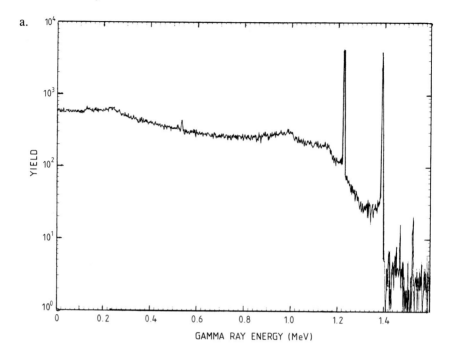

b.

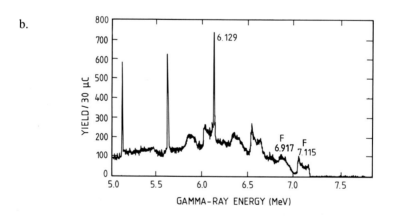

Fig. 2.14a. Response function of Ge(Li) detector showing a full energy peak and Compton continuum from 1173 and 1333 keV γ-rays. b. High energy gamma-ray spectrum from $^{19}F(p,\alpha)^{16}O$ showing the full energy, single and double escape peaks superimposed on Compton continuum for three γ-rays, two of which are subject to Doppler broadening.

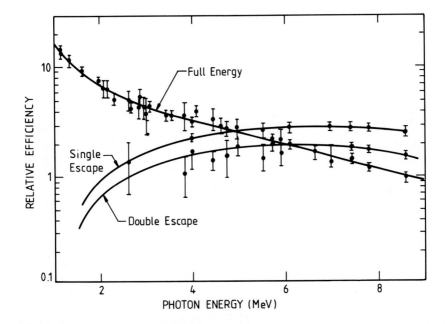

Fig. 2.15 Efficiency curve for large volume Ge(Li) detector.

product particle is heavier than the incident ion. It is zero if the lifetime of the emitting state is sufficiently long ($> 10^{-12}$ s) for the recoil nucleus to come to rest before gamma-ray emission occurs.

The pulse height spectrum from a NaI detector has the same general features but because of the large volume usually used (typically > 300 cm^3), the escape probability by a 0.511 MeV quantum is low. The spectrum therefore contains a large (and relatively broad) photopeak with a small single escape peak and an underlying continuum. The differences in resolution and efficiency for NaI and Ge(Li) detectors are seen in Fig. 2.16 where spectra from ^{60}Co decay are presented for each detector. Further information on the use of gamma-ray detectors is given in Section 4.3.1.

Dramatic reductions in continuum and escape peak count rates can be achieved by using an anti-coincidence spectrometer or pair spectrometer (Damjantschitsch *et al.*, 1983). For example, a Ge(Li) high resolution detector may be surrounded by high efficiency NaI detectors which respond to scattered or escape quanta. In the pair spectrometer mode an event is only recorded if a coincidence is observed between a double escape event in the Ge(Li) detector and two annihilation quanta in the NaI detectors. Because of the need for shielding, pair spec-

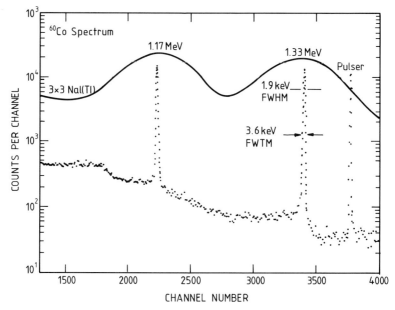

Fig. 2.16 Spectrum comparison for Ge(Li) and NaI detectors (by permission of EG&G Inc.).

trometers are quite large and have even lower efficiency than single detectors. The resulting low count rate is offset to some extent by the fact that a greater proportion of the counts appear in the peaks. In the anti-coincidence mode, gamma-rays which are Compton scattered in the central detector may be subsequently detected in the outer detectors and the event vetoed. The resulting spectrum contains only photoelectron events plus a small background from false coincidences.

2.6.4 High Energy Ion/Atom Detection and Analysis

SB detectors are most suitable for detection of ions with energy above about 500 keV. They are relatively small, typically 10 to 20 mm in diameter and a few mm thick. They are intrinsic Si devices with gold electrodes, operating in the reverse bias mode. The particle must come to rest totally within the depletion region if the current pulse is to be proportional to the ion energy. A typical bias voltage is 50 V and the best resolution is 8 to 12 keV for a 5 MeV α-particle. Fig. 2.17 shows the pulse height spectrum from a ^{241}Am source obtained using a 25 mm^2, 100 μm detector with 11.2 keV resolution. The resolution of a SB detector is less for heavy ions (Z > 6) because recombination occurs along the densely

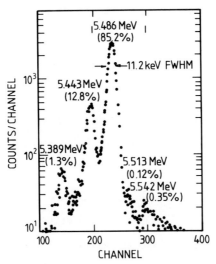

Fig. 2.17 Surface barrier detector pulse height spectrum for α-particles from ^{241}Am source (by permission of EG&G Inc.).

ionised path of such ions. Thin detectors may be used to measure the rate of energy loss of a particle. This technique may be used to distinguish between particles of different mass, but the same energy.

Detector resolution is adversely affected by increased capacitance and reverse leakage current and by incomplete charge collection. Increased capacitance caused for example by user inserted cables increases the noise contribution from the preamplifier and also causes deterioration in signal risetime. The effect of external capacitance on resolution is shown in Fig. 2.18. The best resolution is obtained by restricting the active area of the detector and optimising the sensitive depth. Stray capacitance from chamber leads, and so on, must also be minimised. Reverse leakage current increases the detector noise and can be reduced by cooling the detector to about $-10°C$, improving the resolution by about 2 keV. For low count rates, it is tempting to use a large detector. However, this will have a large capacitance which will lead to poor pulse height resolution. It is better to use a small detector mounted close to the sample.

SB detectors are made in various forms for different applications. The most common form is a circular type with rear mount and positive bias. The gold contact at the front surface is very sensitive and must not be touched. 'Ruggedised' detectors are less prone to damage and have a front surface which may be cleaned and they are less light sensitive than other detectors. Annular detectors are also useful when ions are to be

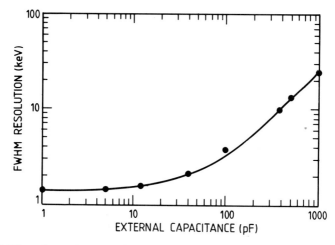

Fig. 2.18 Effect of capacitance on detector resolution.

detected as close as possible to 180°. It is important to note that SB detectors do not discriminate between different mass, charge state or neutral particles. SB detectors are subject to radiation damage. For example deterioration in resolution is observed if the integrated MeV-He$^+$ dose exceeds about 10^{18} ions cm^{-2} (for heavier ions this dose is lower) or the detector is operated in a fast neutron environment where the integrated dose exceeds 10^9 n cm^{-2}.

2.6.5 Magnetic and Electrostatic Analysers for High Ion Energies

Magnetic analysers are used for improved energy resolution in high energy IBA. A single SB or gas ionisation detector after the analyser receives ions whose radius of curvature, Equation (2.1), is a measure of ME/q^2. If M and q are known, the energy is determined with a resolution of 1 to 5 keV if a sufficiently large analyser is used. A position sensitive detector, or a bank of small SB detectors, can be used to determine both energy and radius of curvature for a range of ion energies (Hirvonen and Hubler, 1976). There are a number of disadvantages:

- Analysers must be large (typical radius of curvature of the ion path 50 to 100 cm) in order to achieve good resolution; this means that the detector is at a considerable distance from the sample and the count rate is correspondingly low.
- Only a small acceptance angle (< 2 msr) can be used, even with a

double focussing magnet, in order to limit the energy spread arising from variation in interaction angle.

- They are difficult and expensive to build and are not often available with small accelerators used for ion beam analysis.
- They cover only a very limited part of the energy spectrum.
- The low count rate means that a high dose is required on the sample to obtain good statistics and this can mean radiation damage to the sample.
- The increase in energy straggling with depth of ion penetration into the sample negates the high instrumental resolution after only quite moderate depths ($\sim$ 100 nm).

Electrostatic analysers are sometimes used for energy analysis independent of ion mass, Equation (2.2), but with similar limitations. It is also difficult to achieve uniform high electrostatic field gradients, requiring precision machining of curved electrostatic plates. Double focussing analysers can be used to increase the acceptance solid angle especially at low ion energies (Feuerstein *et al.*, 1976). Recent developments (O'Connor, 1987) have led to a significant reduction in the size and cost of electrostatic analysers for higher energies.

Time of flight techniques can be used to determine the velocity of high energy ions. A start pulse is obtained from a beam pulsing or chopping unit or from a channeltron positioned to detect secondary electrons produced when the ion passes through a very thin Cu foil. The stop pulse is obtained from an SB detector or another channeltron receiving the scattered particles.

2.6.6 Low Energy Ion/Atom Detection

SIMS equipment typically consists of three segments — extraction and transfer optics, mass analyser and ion detector. Sputtered ions have a broad range of energies and are emitted over a range of angles. The observed count rate therefore depends on the efficiency of the secondary ion extraction optics and the transmission of the mass analyser. The overall efficiency can be as low as 10^{-3} and may vary with changes in sputter conditions such as sweeping of the beam or sample charging.

a. Secondary Ion Extraction

A suitably shaped electrode is held at a positive or negative voltage for negative or positive ion extraction. The electrode shape may be designed to focus a major fraction of the sputtered ions into the mass analyser but this leads to relatively low mass resolution because of the ion energy spread. Alternatively, emphasis can be placed on high resolution which is necessary for molecular ions and low background. An energy filter is usually used for this purpose and several types are available.

- Deceleration — a high pass filter which does not transmit low energy ions;
- Acceleration — a broad band filter which reduces the relative energy spread by adding a large constant energy to each ion;
- Combination — a narrow band filter; for example, selection of ions with energies a little above the main energy peak reduces the number of ions accepted by approximately a factor of two while reducing the intensity of unwanted ions by a factor of 100; and
- Electrostatic Analyser — an alternative narrow band filter usually having low solid angle and transmission.

The extraction and transfer optics (filters) are designed to match the acceptance of the mass analyser, to avoid scattering of unwanted ions within the analyser and to achieve relatively constant transmission and resolution (Wittmaack, 1982).

An open geometry is desirable in the region of the extraction electrode to allow easy access for auxiliary equipment such as an electron gun, a heater or ion cleaning gun, AES or other additional analytical units. Open geometry also helps to minimise memory effects which can arise from resputtering of material deposited by electrodes that are too close to the sample.

b. Mass Analyser

An RF quadrupole mass analyser (QMA) (Dawson, 1976) is often used in SIMS. It transmits one mass at a time and the transmission, including extraction and transfer optics, can be up to 10%. QMA systems have been reviewed by Wittmaack (1982). The mass resolution is typically 1 in 300 or even up to 1 in 1000. A low resolution double-focussing magnetic sector analyser can also be used to achieve good transmission. High resolution (1 in 10 000) is needed for separation of molecular ions and large magnetic sectors are needed for this purpose. They can be used to study molecular species present at the sample surface or to resolve these from atomic species under study. Hydrides are particularly prevalent and interfere in the detection of many atomic species.

c. Detector

An electron multiplier such as the Channeltron is usually used for low energy ion detection. Because it will also detect an energetic neutral atom, the detector should be placed off the QMA axis or a small deflection should be introduced to allow separation of ions from neutrals. A low noise multiplier is needed to achieve a high dynamic range ($\sim 10^6$). If major elements are also to be measured the count rate may reach at least 10^7 Hz and it may be necessary to use a Faraday cup for such measurements.

d. Ion Microanalyser

High quality secondary ion optics must be used in the ion microanalyser to achieve a spatial resolution of < 0.5 μm and good detection limits (~ 10 ng g^{-1}). Secondary ions from a large area at the sample surface (> 1 mm diameter) are transmitted to the detector with their spatial distribution maintained. The best known example is the Cameca IMS (Ruberol *et al*, 1979; Furman and Evans, 1982) which uses a combination of good extraction and transfer optics, a double magnetic sector analyser (including an electrostatic mirror to produce a magnified image at the detector) for a selected mass/energy species. Both event counting and visual image detectors are used (Odom *et al.*, 1983).

2.6.7 Neutron Detection

Neutrons are detected indirectly through scattering or nuclear reactions. When a neutron enters hydrogeneous material, protons recoil with an energy E_4 which is related to the incident neutron energy E_1 by Equation (1.2lb). There is thus a continuous distribution of recoil proton energies for each neutron energy, depending on the recoil angle, up to a maximum when the scattering angle is $180°$. Deconvolution of proton recoil spectra is possible to obtain limited energy information. Certain types of liquid and plastic scintillator are useful for this purpose and are described in manufacturers' catalogues. These scintillators also detect gamma-rays through electron interactions. In some liquid scintillators the rise and decay time of the pulse is dependent on whether the pulse originates from an electron (gamma-ray interaction) or a proton (neutron interaction). Pulse shape discrimination circuitry (Brooks, 1959) can be used to reject gamma-ray induced events.

Alternate methods of neutron detection involve the high capture cross-section reactions $^6Li(n, t)^4He$ and $^{10}B(n, \alpha)^7Li$. A BF_3 counter, a 'sandwich' semiconductor detector (consisting of two surface barrier detectors either side of a Li containing layer) or a Li glass scintillator exploit these reactions for neutron counting.

Neutron energies can be measured by time-of-flight. A pulsed beam is required and can be obtained by pulsing the ion source typically for the order of 1 ns every 1 μs. A signal is obtained as the beam passes through a capacitive or inductive pickoff and this signal is related to the time at which the neutrons leave the sample. When the neutrons reach the detector, a second timing signal is generated. A 1 MeV neutron travels one metre in $\sim$ 72 ns and a 1 ns uncertainty leads to a 3% error in energy. Much of the timing uncertainty is in the width of the beam pulse and the thickness of the detector. The maximum repetition rate is determined by the flight times involved and the need to avoid overlap from consecutive beam pulses. For example, if the maximum flight time is 200 ns, a repetition rate of 5 MHz could be tolerated.

2.6.8. Detector Calibration

Detectors must be calibrated under the conditions used in experimental measurements. Important factors are solid angle, edge effects, absorbing

TABLE 2.9
Useful gamma-ray calibration sources

Isotope	Half-life	E (keV)	Photons/100 decays	
^{57}Co	271.6 ± 0.5 d	122.06	85.6	± 0.2
		136.47	10.8	± 0.2
^{152}Eu	13.6 ± 0.3 y	121.8	137.5	± 2.2
		244.3	35.8	± 0.6
		366.3	128.3	± 1.2
^{133}Ba	10.7 ± 0.1 y	276.40	11.4	± 0.2
		302.84	29.3	± 0.6
		356.00	100	
		384.84	14.8	± 0.3
^{137}Cs	30.0 ± 0.5 y	662.65	85.1	± 0.5
^{54}Mn	312 ± 0.2 d	834.8	93.4	± 0.7
^{88}Y	106.6 d	898.0	2.93	± 0.16
			7.33	± 0.30
^{60}Co	1925.7 ± 0.9 d	1173.2	99.86	± 0.02
		1332.5	99.98	± 0.003
^{22}Na	951.7 ± 2.7 d	1274.5	99.95	± 0.07
^{24}Na	15.0 h	2753.92	13.02	± 0.35

TABLE 2.10
Gamma-ray branching ratios

Reaction	E_p(keV)	Gamma-rays E_γ	I_γ	Reference
^{23}Na(p, γ)^{24}Mg	1395	1370	2.7	Boydell and Sargood (1975a)
		4120	3.0	
		4240	26	
		5240	46	
		7620	10	
^{27}Al(p, γ)^{28}Si	992	1522	3.0	Scott (1975)
		1779	94.1	
		2839	6.3	
		4744	11.5	
		6020	7.8	
		7925	5.2	
		10764	72.4	
	1262	1779	43	Gibson *et al.* (1968)
		4617	6	
		6878	32	
	1520	1779	76	Meyer *et al.* (1975)
		6878	18	
		7800	2	
	1699	1779	100	Antoufiev *et al.* (1964)
		2600	10	
		3300	8	
		5750	14	
		7420	14	
		11410	100	
^{31}P(p, γ)^{32}S	1557	2230	12	Boydell and Sargood (1975b)
		4280	40	
		4700	9.2	
		5410	12	
		6670	14	

materials, bias voltages, operating temperature, linearity and stability of associated electronics. Sources which emit protons of known energy and intensity are available to calibrate X-ray and gamma-ray detectors. Tables 5.5 and 2.9 list useful calibration sources. The lists are by no means exhaustive and other sources are available. A review of the calibration of gamma-ray detectors has been given by Helmer *et al.* (1979). It is difficult to obtain gamma-ray sources with absolute intensities known for energies above 2.0 MeV. However nuclear reactions such as ^{27}Al(p, γ)^{28}Si and ^{24}Mg(p, γ)^{25}Al have resonances for which the branching ratios are well known (Rofke *et al.*, 1966; Meyer *et al.*, 1975). These values are summarised in Table 2.10 Using a thin target and a

TABLE 2.11
α-Particle calibration sources for surface barrier detectors

Source	α-Particle Energy (MeV)	Half Life
^{210}Po	5.305	138.4 d
^{230}Th	4.682	8.0×10^4 y
	4.610	
^{231}Pa	7.36	3.25×10^4 y
	6.80	
	5.71	
	5.01	
	4.94	
	4.72	
^{234}U	4.768	2.5×10^5 y
	4.717	
^{241}Am	5.476	433 y
	5.433	
	5.378	

particular resonance, the ratio of the efficiency at high energy to that at low energy can be obtained from the observed intensities of cascade gamma-rays, and used to extend the efficiency calibration to high energies. For SB detectors, assuming that they are thick enough to stop the particle, pulse height calibration may be obtained from α-particle sources and threshold nuclear reactions. Some suitable sources are listed in Table 2.11.

2.7 ELECTRONICS AND DATA PROCESSING

2.7.1 General Considerations

The processing of detector signals depends on the measurement being made which may involve counting, energy spectra, time-of-flight or mass, angle or wavelength scans. Some typical equipment configurations are shown schematically in Fig. 2.19. Table 2.12 summarises the performance of individual components. Such components are often manufactured to conform to standard module specifications such as NIM (Nuclear Instrumentation Module) and CAMAC (Computer Automated Modular Acquisition Control). CAMAC equipment is specifically designed for computer interfacing and can incorporate NIM modules which are available to serve all the functions outlined in Table 2.12.

Signal processing involves two types of pulse — logic and analog.

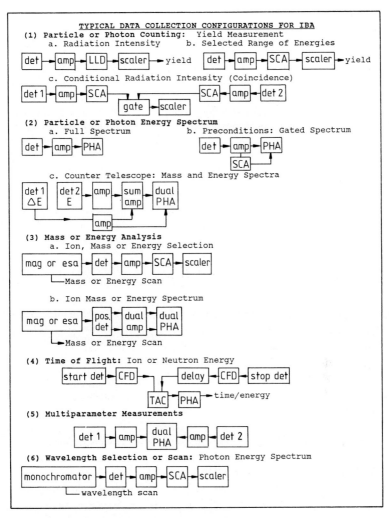

Fig. 2.19 Typical data collection configurations for IBA.

Logic signals have a fixed amplitude and duration (e.g. + 5 V, 500 μs or − 800 mV, 25 ns for NIM equipment). Analog signals have an amplitude which is proportional to parameters such as energy or time of flight. The shape of analog pulses is determined by the time constant (chosen to suit detector type) used in the preamplifier and main amplifier as well as other operations (Table 2.13). Short time constants (0.5 to 1.0 μs) give good time resolution but poor pulse height resolution. Long time

TABLE 2.12
Electronics modules used for IBA

Component	Use	Comment
Bias Supply	Detector Bias 5 kV Ge(Li), 3 kV Si(Li) 100 V Surface Barrier	Should be stable Low current drain
Preamplifier (preamp)	Collect charge from detector and produce small current pulse	High stability, low noise for good resolution Located near detector to reduce cap. pickup
Main Amplifier (amp)	Takes signal from preamp Shapes and amplifies 0–10 V	Linear, Stable, Shaping constants, Pole zero Baseline restore
Analog to digital converter (ADC)	Converts amplifier signal to digital pulse train for storage	Conversion gain 512–4096 channels Stable and linear
Lower level discriminator (LLD)	Provides output for all pulses above certain height	Indicates number of events above threshold
Single Channel Analyser (SCA)	Selects signals in a chosen pulse height range	Records number of events corresponding to a particular energy.
Multi Channel Analyser (MCA or PHA)	Provides full pulse height analysis	Incorporates ADC. Commonly computer interfaced
Scaler	Counts events, e.g. SCA output	Operated by computer or for preset time.
Coincidence or gate	Logic pulse output when pre-conditions are met	Signal timing important
Time to amplitude converter (TAC)	Measures time interval between two events	Stable and linear Output to MCA
Time pickoff	Start or stop signal for TAC	Minimal time jitter

constants (2.0 to 4.0 μs) give poor time resolution and good pulse height resolution. Amplifiers with low noise should be used and precautions taken against extraneous noise sources (Highlight 2.4). Signals may be transmitted from one module to another by means of a coaxial cable with an impedance matched to the output of the module (typically 50 Ω for analog signals and 10 Ω for logic signals). It is customary to keep modules close to one another, but it may be necessary to transmit signals from a preamplifier to a main amplifier or from an amplifier to a PHA over a distance of many metres and this is acceptable if impedances are matched. Particle or photon counting is illustrated in part (1) of Fig. 2.19.

HIGHLIGHT 2.4
PULSE HEIGHT RESOLUTION

To achieve the optimum pulse height resolution from detectors it is necessary to eliminate sources of noise which can affect the electronics.

- 33 kHz from turbo pump operation can be picked up through earth loops;
- Mains frequency pickup can be minimised by having a separate mains supply for sensitive electronics and power supplies with a solid copper strap to an earth location in the incoming mains; all motors, printers and other general purpose equipment should be connected to a different mains supply.
- SB detector power supplies can introduce noise and it is best to use a battery power supply; this is also true for the power supply for a flood gun for discharging insulating samples.
- Beam lines act as aerials to pickup RF signals (radar, radio, TV) and detector earth connections should not be linked to beam lines.
- Some power supply modules radiate signals which can increase the noise in an amplifier placed in the adjacent bin slot.
- A plastic absorber foil used in ERA or NRA can become charged unless it is aluminised; the charging and discharging induces random signals in the detectors.
- Light sensitive detectors may be affected by fluorescing targets.
- The choice of time constants for shaping the detector signals is vital for achieving maximum resolution. The use of very short or narrow band-pass time constants can filter out many noise frequencies but it also limits the time during which the signal is being integrated so that a smaller pulse is obtained with poorer resolution;
- Detector signals cannot be transmitted over long cables until after a preamp which provides an impedance match to the cable. Matching is important at both ends of the cables to avoid reflections.
- Magnetic fields are as important in noise radiation and pickup as electric fields so that electrostatic shielding is not sufficient for full noise rejection.

TABLE 2.13
Detector electronics for IBA

Detector	Bias Volts	Pre-Amp	Amplifier Time Constants μs	PHA Channels
NaI	1000	IS	1.0–2.0	512/1024
Ge(Li)	5000	CS	0.5–2.0	2048/4096
Si(Li)	1000	CS	0.5–1.5	1024/2048
Si	50	CS	1.0–2.0	512/1024
Liquid Scintillator	300	IS	2.0–3.0	512/1024
Channeltron	3500	IS	2.0–3.0	NA

IS-current sensitive, CS-charge sensitive

2.7.2 Pulse Height Analysis

Pulse height analysis is illustrated by part 2 of Fig.2.19. The detector signal is fed to the preamplifier if necessary through an insulated vacuum seal. The collected charge produces an analog voltage pulse (typically 0 to 50 mV). By keeping the preamplifier as close as possible to the detector, the capacitance and noise are minimised. Typically a capacitance of 10 pF will give a 2 keV FWHM spectrum broadening with an SB detector. If the capacitance increases to 100 pF, the broadening becomes 5 keV and at 1000 pF it is 25 keV. The effects of capacitance on resolution are shown in Fig 2.18. The preamplifier may be mounted in vacuum to reduce capacitance; however, an adequate heat sink must be provided and the unit must not contain components which will degas in the vacuum.

The main amplifier shapes (and inverts if necessary) the signal and amplifies it to give outputs in the range 0 to 10 V. The amplitude of this

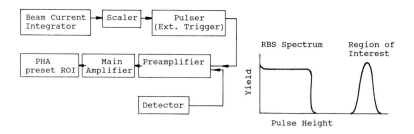

Fig. 2.20 Electronics (left) for external pulser measurement of deadtime with schematic spectrum (right) showing the pulser peak as well as an IBA spectrum (in this case RBS).

signal is a measure of the energy deposited in the detector. Amplifiers must be designed to be linear and to be stable to better than 0.005% °C^{-1}. They must be capable of handling the peak count rates expected in a particular measurement. Generally, with time constants of the order of microseconds, amplifiers can handle count rates up to 10 kHz. As the count rate increases there is an increasing probability that the signal from one event overlaps the next event and the signal from the latter pulse will be incorrectly recorded. Most modern amplifiers have a 'pole-zero' adjustment which allows either overswing or underswing to be minimised for the operating count rate. Many amplifiers incorporate baseline restorers so that even at high count rates, the baseline returns to zero after an event has occurred.

Sometimes it is advantageous to use a 'pile-up' reject system to sense when a signal is being processed and to send out a logic pulse which turns off the front end of the electronics until the processing is finished. The pile-up reject option can be built into an amplifier. Alternatively a constant fraction discriminator may be used to sense the arrival of a pulse and then trigger a fast beam deflection mechanism which will turn off the beam until the electronics is ready to process another signal.

In order to record the full spectrum of pulse heights, it is necessary to use a Pulse Height Analyser (PHA). The analog signal is first changed to a digital pulse train with an analog to digital converter (ADC). This may be either a plug-in module or an integral part of the PHA. The PHA then sorts out the events according to initial pulse height and produces a spectrum showing the number of events with each pulse height. A conversion range between 2^8 (256) and 2^{13} (8192) channels is chosen depending on the energy range of the pulses and the energy resolution of the detector system. Usually, about four channels correspond to the system energy resolution. Events are accumulated for a specified ion dose as determined by a beam current integrator. The signal processing time of a PHA depends on the pulse height and is typically 20 to 40 μs. A busy signal is generated which can be used to gate off the current integrator during the busy period for each pulse. Some PHA's incorporate an input buffer to store incoming events while one signal is being processed.

One method of ensuring that live time is accurately determined is shown in Fig. 2.20. The output from the beam current integrator is used to trigger a pulser signal which is fed into the preamplifier and through the electronics so that a pulser peak appears in a convenient part of the spectrum. The PHA is set to record a predetermined number of events in that peak. If the count rate is high and events are lost, then the run time is automatically extended until the number of events corresponding to the required charge have been recorded.

The standard mode of data collection is single parameter mode. A single spectrum is accumulated e.g. the energy spectrum of X-rays from a sample irradiated with protons. If two detectors are used (e.g. when backscattering spectra are to be measured at a glancing angle of 100° and a back angle of 170°), either two ADCs are used or a multiplexer combines the spectra by putting a voltage pedestal on pulses from the second detector. The lower half of the displayed spectrum contains the events from the first detector and the upper half contains those from the second. Some multiplexers can handle up to four individual inputs.

It is sometimes desirable to measure two dependent parameters such as the energy of scattered and recoil particles or energy and time of flight. This requires the use of two ADCs and it may be necessary to insert delays of several microseconds into one channel in order to ensure that the signals are processed in coincidence. The results constitute a three dimensional array, with yields as a function of each of the two parameters. If the count rates in each channel are low, long running periods will be required to obtain reasonable statistical accuracy. A schematic diagram of a dual parameter analysis facility is shown in part 5 of Fig. 2.19. Time intervals (in the range ns to ms) can be measured accurately using a time-to-amplitude converter (TAC) in conjunction with a PHA. The output pulse height from the TAC is directly proportional to the elapsed time between start and stop pulse.

2.7.3 Automated Operation

A relatively inexpensive microprocessor can handle the following:

- position each sample;
- start and stop data collection;
- monitor target current;
- monitor elapsed time and electronics live time and make necessary corrections for losses;
- monitor vacuum;
- transfer data to bulk storage at conclusion of run;
- carry out simple processing of data;
- change to next sample and commence next run;
- derive element identification composition or profile, and
- detect faults and apply corrections, or give warnings or
- close down the experiment if necessary.

Commercially available SIMS and sputter Auger systems normally provide these capabilities. Other low energy systems (e.g. LEIS) and many high energy facilities have partially automated systems. Descriptions of automated high energy analysis capabilities are given by Duerden *et al.* (1984), Giles and Peisach (1976) and Norman *et al.* (1984). Computer control may also be extended to the accelerator to monitor and control target current, beam energy, stability, component temperature, vacuum, radiation and so on.

REFERENCES

Amsel, G., d'Artemare, E., and Girard, E. (1982). *Nucl. Instrum. Methods* **205**, 5.
Antoufiev, Y.P., Darwish, D.A.E., Badaway, D.E., El-Nadi, L.M., Sorokin, P.V. (1964) *Nucl. Phys.* **56**, 401.
Anttila, A., Raisanen, J. and Lappalainen, R. (1985). *Nucl. Instrum. Methods* **B12**, 245.
Benninghoven, A., Giber, J., Laszlo, J., Riedel, M. and Werner, H.W. (eds). (1982). 'Secondary Ion Mass Spectrometry SIMS III', Springer-Verlag, Berlin.
Bottiger, J., Davies, J.A., Lozi, J. and Whitton J.L (1973). *Nucl. Instrum. Methods* **109**, 579.
Boydell, S.G. and Sargood, D.G. (1975a) *Aust. J. Phys.* **28**, 369.
Boydell, S.G. and Sargood, D.G. (1975b) *Aust. J. Phys.* **28**, 383.
Brooks, F.P. (1959) *Nucl. Instrum. Methods* **4**, 151.
Cahill, T., McColm, D.W., Kusko, B.H. (1986) *Nucl. Instrum. Methods* **B14**, 38.
Damjantschitsch, H., Weiser, M., Heusser, G., Kalbitzer, S. and Mannsperger, H. (1983) *Nucl. Instrum. Methods* **218**, 129.
Dawson, P.H. (1976) *Int. J. Mass Spectrom. and Ion Analysis* **21**, 317.
Deconninck, G. and Bodart, F. (1978) *Nucl. Instrum. Methods* **149**, 609.
Doyle, B.L. (1983) *Nucl. Instrum. Methods* **218**, 29.
Duerden, P., Bird, J.R., Clayton, E.J., Cohen, D.D. and Leach, B.F. (1984) *Nucl. Instrum. Methods* **B3**, 419.
England J.B.A. (1974) *In* 'Techniques in Nuclear Structure Physics', Macmillan, London.
Evans, D.S. (1965) *Rev. Sci. Instr.* **36**, 375.
Ewan, G.T. and Tavendale, A.J. (1963) *Nucl. Instrum. Methods* **25**, 185.
Fallavier, M., Chartoire, M.Y. and Thomas, J.P. (1986) *Nucl. Instrum. Methods* **B15**, 712.
Feuerstein, A., Grahmann, H., Kalbitzer, S. and Oetzman, H. (1976) *In* 'Ion Beam Surface Layer Analysis' (Meyer, O., Linker, G., and Kappeler, F., eds), Plenum Press, New York, 471.
Furman, B.K. and Evans Jr., C.A. (1982) *In* 'Secondary Ion Mass Spectrometry SIMS III' (Benninghoven, A., Giber, J., Laszlo, J., Riedel, M., and Werner, H.W., eds), Springer-Verlag, Berlin, 88.
Gibson, E.F., Battleson, K., McDaniels, D.K. (1968) *Phys. Rev.* **172**, 1004.
Giles, I.S. and Peisach, M. (1976) *J. Radioanal. Chem.* **32**, 105.
Glavish, H. (1981) *Nucl. Instrum. Methods* **189**, 43.
Gnaser, H. and Rudenauer, F.G. (1983) *Nucl. Instrum. Methods* **218**, 303.
Goclowski, M., Jaskola, M., Zemlo, L. (1983) *Nucl. Instrum. Methods* **204**, 553.
Hall, G.S., Roach, N., Simmons, U., Cong, H., Lee, M-L., Cummings, E. (1984) *J. Radioanal. and Nucl. Chem.* **82**, 329.
Helmer, R.G., Van Assche, P.H.M. and Van Der Leun, C. (1979) *Atomic and Nucl. Data Tables* **24**, 39.
Hemment, P.F. (1980) *Rad. Eff.* **44**, 31.
Hirvonen, J.K. and Hubler, G.K. (1976) *In* 'Ion Beam Surface Layer Analysis', (Meyer, O., Linker, G., and Kappeler, F., (eds) Plenum Press, New York, 457.
Liebl, H. (1975) *NBS Spec. Publ.* **427**, 1.

Magee, C.W., McFarlane, S.H. and Hewitt, L.R. (1986) *Nucl. Instrum. Methods* **B15**, 707.

Marion, J.B., and Young, F.C. (1968) *In* 'Nuclear Reaction Analysis', North-Holland, Amsterdam.

Meyer, M.A., Venter, I. and Reitmann, D. (1975) *Nucl. Phys.* **A250**, 235.

Middleton, R. (1977) *Nucl. Instrum. Methods* **141**, 373.

Mingay, D.W. and Barnard, E. (1978) *Nucl. Instrum. Methods* **157**, 537.

Moak, C.D., Reese, H.T. and Good, W.M. (1951) *Nucleonics* **9**, 18.

Moak, C.D., Banta, H.E., Thurston, J.N., Johnson, J., King, R.F. (1959) *Rev. Sci. Instr.* **30**, 694.

Möller, W., Pfeiffer, Th. and Schluckebier, M. (1981) *Nucl. Instrum. Methods* **182/183**, 297.

Muskett, R.G. (1986) *Nucl. Instrum. Methods* **B15**, 735.

NCRP 39 (1971). 'Basic Radiation Protection Criteria', National Council on Radiation Protection and Measurements, Washington D.C.

Norman, L., Jensen, B., Bauman, S.E., Houmere, P.D., Nelson, T.J.W. (1984) *Nucl. Instrum. Methods* **B3**, 122.

O'Connor, D.J. (1987). J. Phys. E. Sci. Instrum. **20**, 437.

Odom, R.W., Furman, B.K., Evans Jr, C.A. Bryson, C.E., Petersen, W.A., Kelly, M.A. and Wayne, D.H. (1983) *Anal. Chem.* **55**, 574.

Raisanen J. and Anttila, A, (1982) *Nucl. Instrum. Methods* **196**, 489.

Raisanen, J. (1984) *Nucl. Instrum. Methods* **B3**, 220.

Rofke, H., Anyas-Weiss, N. and Litherland, A.E. (1966) *Phys. Lett.* **27B**, 368.

Roth, J., Behrisch, R. and Scherzer, B.M.U. (1974) *J. Nucl. Mat.* **53**, 147.

Ruberol, J.M., Lepareur, M., Autier, B. and Gourgoutt, J.M. (1979) *In* 'VIIIth Int. Conf. on X-Ray optics and Micro-analysis' (Beaman, D.R. ed.), Pendell Publ. Co., Midland, 322.

Rudolph, W., Bauer, C., Gippner, P. and Hohmuth, K. (1981). *Nucl. Instrum. Methods* **191**, 373.

Scott, H.L. (1975) *Nucl. Instrum. Methods* **131**, 517.

Scott, M.D., Kenny M.J. and Janky, S. (1985) *Nucl. Instrum. Methods* **B12**, 181.

Septier, A. (1967) *In* 'Focussing of Charged Particles', Vol. I and II, Academic Press, New York.

Valyi, L. (1977) *In* 'Atom and Ion Sources', Wiley, London.

Wittmaack, K. (1982) *Vacuum* **32**, 65.

Woldseth, R. (1973) *In* 'X-Ray Energy Spectroscopy', Kevex, Burlingame.

3
High Energy Ion Scattering Spectrometry

J.E.E. BAGLIN
IBM Almaden Research Laboratories
California, USA

J.S. WILLIAMS
Microelectronics and Materials Technology Centre
RMIT, Melbourne, Australia

ION BEAMS FOR
MATERIALS ANALYSIS
ISBN 0 12 099740 1

Note

In ion scattering texts the usual notation for incident and scattered ion energies employs the subscripts 0 and 1 respectively. Subscripts 1 and 3 are used here for consistency with the more general notation applicable for nuclear reactions. Subscripts 2 and 4 refer to the target atom and recoil atom respectively.

3.1 INTRODUCTION

Rutherford backscattering spectrometry (RBS) is the simplest and most widely used of the ion beam analysis methods and the basic concepts are introduced in Chapter 1. It involves measurement of the number and energy distribution of energetic ions (usually MeV light ions such as He^+) backscattered from atoms within the near-surface region of solid targets. From such measurements it is possible to determine, with some limitations, both the atomic mass and concentration of elemental target constituents as a function of depth below the surface. RBS is ideally suited to determining concentration profiles of trace elements which are heavier than the major constituents of the substrate. Its sensitivity for light masses is extremely poor, but forward scattering geometries can then be used to advantage. If the incident ions are heavier than the light masses of interest, the latter can recoil in forward directions with high energies and can be detected with high sensitivity. This group of ion scattering techniques is thus a powerful and versatile component of IBA methods.

    Fig. 3.1 shows a typical backscattering geometry. A mono-energetic beam of ions of energy, E_1, and mass, M_1, is incident on a solid target. Ion penetration into the target and scattering of the ions by target nuclei produce information on target mass, concentration and depth. Firstly, mass information is given by the kinematic scattering Equations (3.1) and (3.2) in Table 3.1, where the kinematic factor, K, is a function of M_1, M_2 and θ. A tabulation of kinematic factors and backscattering energies is given in Chapter 14.3 for 2 MeV He^+ ions backscattered off each element in the periodic table for four typical backscattering geometries. The concentration of target constituents can be obtained from the

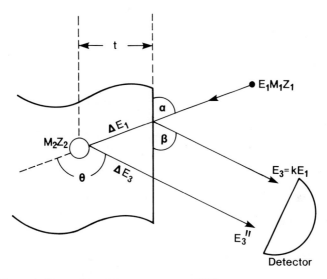

Fig. 3.1 Schematic illustration of the parameters of RBS.

Rutherford scattering cross-section, σ, which is a function of Z_1, Z_2, θ and E_1, Equation (3.4). The number of scattered particles measured by a detector (ΔA) can be converted to the concentration of a particular element in the target (N_M) by Equation (3.3). Finally, the incident particles lose energy in penetrating the solid both along the incident path (energy loss ΔE_1) prior to scattering and along the exit path (energy loss ΔE_3) following scattering, Equation (3.5), where $\Delta E_{1,3}$ is the total energy loss term. Energy loss can be obtained from a stopping cross-section, ε, or rate of energy loss, $S = dE/dx$ (Equations (3.6) to (3.8)). Thus, energy differences in RBS energy spectra can be used to identify scattering depths and hence provide a depth profile of target constituents.

3.2 BASIC CONCEPTS OF ION SCATTERING

The application of the Equations in Table 3.1 to the determination of mass and concentration profiles is developed in this section using examples which employ 2 MeV He$^+$ ions, a species commonly used for RBS analysis.

3.2.1 Mass Determination in Thin Films

Fig. 3.2 illustrates the RBS energy spectrum which would result from scattering of 2 MeV He$^+$ ions through $\theta = 165°$ by a thin Si film which results in negligible energy loss ($\Delta E_{1,3} = 0$ in Equation (3.5)) of the He$^+$

TABLE 3.1
Basic equations of RBS

Relationship	Equation	
Backscattered energy, thin target	$E_3 = KE_1$	(3.1)
Kinematic factor	$K = \left\{ \dfrac{(M_2^2 - M_1^2 \sin^2 \theta)^{1/2} + M_1 \cos \theta}{M_2 + M_1} \right\}^2$	(3.2)
Area under peak, thin film	$\Delta A = \dfrac{d\sigma}{d\Omega} \Delta\Omega\, N_M\, Q$	(3.3)
Average differential scattering cross-section	$\dfrac{d\sigma}{d\Omega} = \left(\dfrac{Z_1 Z_2 e^2}{16\pi\, \varepsilon_1 E} \right)^2 \dfrac{4}{\sin^4 \theta} \dfrac{\{[1 - ((M_1/M_2)\sin \theta)^2]^{1/2} + \cos \theta\}^2}{[1 - ((M_1/M_2)\sin \theta)^2]^{1/2}}$	(3.4)
Backscattered energy, thick target	$E_3'' = K(E_1 - \Delta E_1) - \Delta E_3 = KE_1 - \Delta E_{1,3}$	(3.5)
Energy loss, thick target	$\Delta E_{1,3} = K\Delta E_1 + \Delta E_3 = \dfrac{Kt}{\sin \alpha} S_1 + \dfrac{t}{\sin \beta} S_3 = St$	(3.6)
Depth or atom density scale	$t = (KE_1 - E_3)/S$ or $N = (KE_1 - E_3)/\varepsilon$	(3.7)
Stopping cross-section factor	$\varepsilon = \dfrac{K}{\sin \alpha} \varepsilon_1 + \dfrac{1}{\sin \beta} \varepsilon_3$	(3.8)
Spectrum height	$H = \dfrac{d\sigma}{d\Omega} \Delta\Omega\, Q\, \Delta E_c/\varepsilon \sin \alpha$	(3.9)
Spectrum height ratios; elemental films	$\dfrac{H_A}{H_B} = \left(\dfrac{Z_A^2}{E_A^2} \varepsilon_B \right) \Big/ \left(\dfrac{Z_B^2}{E_B^2} \varepsilon_A \right)$	(3.10)
Impurity content in elemental substrates	$N_i = (\Delta A_i Z_s^2 \Delta E_c)/(H_s Z_i^2 \epsilon_s)$	(3.11)
Impurity/substrate concentration ratio	$\dfrac{N_i}{N_s} = (H_i Z_3^2 \epsilon_i)/(H_S Z_i^2 \epsilon_s)$	(3.12)
Bragg's rule	$\varepsilon_{A_m B_n} = m\, \epsilon_A + n\, \epsilon_B$	(3.13)
Spectrum height ratios for compound	$\dfrac{H_A}{H_B} = m\, Z_A^2\, [\varepsilon]_B^{A_m B_n} / n\, Z_B^2 [\varepsilon]_A^{A_m B_n}$	(3.14)

beam as it traverses the film (e.g. a few monolayers of Si). From Equations (3.1) and (3.2), $K = 0.57$ and hence $E_3 = 1.14$ MeV. Assuming Si consists of a single isotope of mass 28, the RBS energy spectrum would contain a single peak, broadened primarily by the energy resolution of the detection system, which is typically 15 keV full width half maximum for

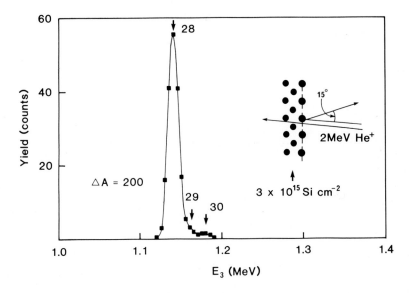

Fig. 3.2 A simulated RBS energy spectrum obtained with 2 MeV He$^+$ ions from a thin Si target consisting of 3×10^{15} Si atoms cm^{-2}.

a solid state surface barrier detector (see Section 3.3). However, Si has three isotopes, masses 28, 29, and 30, with abundances of 92.21%, 4.70% and 3.09%. Scattering of 2 MeV He$^+$ from masses 29 and 30 should give small peaks at $E_3 = 1.165$ and 1.185 MeV. In Fig. 3.2 the mass 29 and 30 isotopes of Si account for the high energy tail extending from the dominant mass 28 peak at $E_3 = 1.140$ MeV up to about 1.190 MeV. The small peaks are not well resolved as a result of finite detector energy resolution and the low abundance of masses 29 and 30.

The area, ΔA, under each peak in Fig. 3.2 is proportional to the number of target atoms (mass M_2) per unit area perpendicular to the incident beam direction, N_M, Equation (3.3). Since the detector solid angle, $\Delta \Omega$, and the incident ion fluence Q are given by the experimental conditions, N_M can be found directly using the differential scattering cross-section $(d\sigma/d\Omega)$ given by Equation (3.4). Thus, RBS can provide, in principle, an *absolute* measure of the number of atoms of a particular mass within a thin layer, without recourse to secondary standards. RBS cross-sections are tabulated for all elements in Chapter 14.3 for 2 MeV He$^+$ ion scattering and four typical geometries.

Assuming an average mass for Si of $M_2 = 28.086$, $(d\sigma/d\Omega)$, from Equation (3.4), is 1.10×10^{-25} cm^2 sr^{-1}. Given that the He$^+$ beam charge needed to collect the spectrum in Fig. 3.2 was 2.4μC cm^{-2} (i.e. $Q = 1.5 \times$

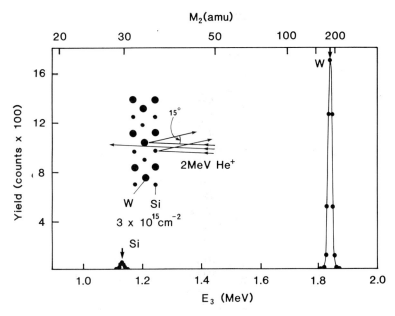

Fig. 3.3 A simulated RBS energy spectrum obtained with MeV He$^+$ ions from a thin target consisting of equal numbers of Si and W atoms (3×10^{15} cm^{-2} in each case).

10^{14} ions cm^{-2} incident on the target) and $\Delta\Omega = 4$ msr, then the number of Si atoms in the film (using $\Delta A = 200$ counts from Fig. 3.2) is $\sim 3 \times 10^{15}$ atoms cm^{-2}. This corresponds to an average Si film coverage of about three monolayers.

Fig. 3.3 shows an RBS spectrum from a thin target consisting of a Si/W mixture containing equal numbers of Si and W atoms (3×10^{15} cm^{-2}), equivalent to a few monolayers. The two peaks correspond to scattering from Si (with $E_3 = 1.140$ MeV) and W (with $E_3 = 1.851$ MeV). Equations (3.1) and (3.2) can be used to convert the backscattered energy scale into a mass scale, as shown in Fig. 3.3. Since the kinematic factor, K, varies approximately as M_2^{-2}, this mass scale is compressed at higher masses corresponding to higher backscattered energies. The mass resolution can be improved by selecting higher incident beam energies, higher mass projectiles or scattering angles close to 180°. Under most experimental conditions, the mass resolution, dM, is proportional to detector energy resolution, dE_d, and can therefore be enhanced by employing high energy resolution detection systems. Mass separation and mass resolution are discussed further in Section 3.3. Another noticeable feature of the spectrum in Fig. 3.3 is the much larger scattering yield (peak area) corresponding to scattering from W. Since the cross-section varies as Z_2^2,

Equation (3.4), the ratio of peak areas $[\Delta A_W/\Delta A_{Si}] = [Z_W/Z_{Si}]^2 = 27.9$. Thus, RBS is more sensitive to heavy elements.

3.2.2 Concentration Profiles

Thus far we have considered only thin films in which the energy loss in traversing the film is negligible in comparison with the system energy resolution. In thicker samples, energy loss of the incident ions can be utilised to provide depth information, as illustrated by the RBS spectrum in Fig. 3.4, which is taken from a 2000 Å layer of W on a thick Si substrate. The scattering yield from W appears as a broad straight-sided profile whose width (in energy) is related to the film thickness. The beam loses energy as it traverses the film and scattering from W at the W–Si interface is detected at an energy of 1.58 MeV compared with the surface scattering energy of 1.84 MeV. Similarly, scattering from Si at the interface is detected at an energy of 0.90 MeV compared with 1.14 MeV for surface scattering from Si. The displacement of the 'interface' signal from the 'surface' energy is almost identical for the two elements, a matter of some convenience for the analyst seeking to interpret features

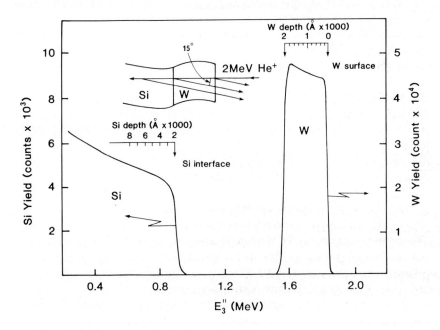

Fig. 3.4 A simulated RBS energy spectrum from a 2000 Å thick W film on Si obtained with 2 MeV He[+] ions.

of an 'unknown' spectrum. Further energy loss as the beam penetrates into the Si substrate gives rise to a continuous backscattered energy spectrum extending to zero energy.

The increase in height of the Si spectrum as the backscattered energy decreases (like that of the W signal) arises mainly from the E^{-2} dependence of the scattering cross-section and the effects of multiple scattering. If the stopping power is increasing with decreasing beam energy then the spectrum height can decrease as the backscattered energy decreases, particularly for geometries with small values of β (Fig. 3.1). It may be noted that the high energy W edge has a slope dictated primarily by the resolution of the detection system. The low energy W edge and the leading Si edge show a slightly increased slope as a result of the energy spread of the beam in penetrating the films. Further discussion of such energy straggling is given in Section 3.3.5, and an estimate of its influence on system resolution can be obtained using expressions in Chapters 12.2. and 14.2.

a. Energy to Depth Conversion

The conversion from spectrum energy to depth is straightforward, using the incident and exit energy loss components, Equation (3.5). Accurate methods of calculation, involve integration of Equation (3.5) along both the incident and outgoing paths to obtain the energy loss as a function of penetration depth as outlined in Chapter 12. Since the rates of energy loss or stopping powers, dE/dx, are given as a function of energy rather than depth (Chapter 14.1), this integration is best done by computer using RBS simulation programs such as RUMP (Doolittle, 1985) to provide depth scales. Computer generated tabulations of typical elemental depth scales obtained with 2 MeV He$^+$ ions for selected geometries are given in Chapter 14.3. A simple analytical approach is to use the surface energy approximation illustrated in Highlight 3.1.

b. Spectrum Height Ratios in Elemental Films

The general expression for RBS peak area, Equation (3.3), can be used to obtain the height of a spectrum, Equation (3.9), where ΔE_c is the energy width per spectrum channel and ε is given by Equation (3.8). Note that the value of H is dependent on the scattering species and the stopping power via the quantities $(d\sigma/d\Omega)$ and ε. Comparing Equations (3.3) and (3.9), the number of target atoms cm^{-2} represented by the counts in one channel is given by the term $\Delta E_c/\varepsilon \sin \alpha$, and is thus proportional to ΔE_c. Therefore, although the W peak *area* (Fig. 3.4) given by Equation (3.3) is independent of ΔE_c, its height would double if ΔE_c was doubled.

HIGHLIGHT 3.1
DEPTH SCALES USING SURFACE ENERGY APPROXIMATION

If constant stopping powers, $S_1 = (dE_1/dx)$ and $S_3 = (dE_3/dx)$ are assumed along the incident and exit paths, Equation (3.6) results where S_1 and S_3 are calculated at the surface energies of E_1 and E_3', respectively, and the scattering depth t is then given by Equation (3.7). This equation can be solved using tabulations of the stopping cross-section, $\varepsilon_s = (dE/dx)/N_o$ in units of eV cm^2 per atom, where N_o is the number density of the target. Tables of stopping cross-sections, which do not assume a knowledge of the density of the solid, may be found in Andersen and Ziegler (1977), and for 0.5 to 2 MeV He$^+$ ions in elemental targets the accuracy is better than about 5%. A more detailed discussion of stopping cross-sections, with illustrative tabulations, is given in Chapter 14.1. Using tabulated ε_s values the depth scale is given by Equations (3.7) and (3.8) in atoms cm^{-2}. If the density of the target is known independently (which frequently it is not) then a depth scale in length units can be obtained using $t = N/N_o$. Since the stopping cross-section varies rather slowly with He$^+$ energy, the surface approximation method is accurate to better than about 5% if the penetration depth is confined to less than a few thousand Å.

We illustrate the depth scale calculation by the example in Fig. 3.4. For W, ε_1 can be found from the tabulated energy loss for 2 MeV He$^+$ in W (1.04 x 10^{-13} eV cm^2 per atom) from Andersen and Ziegler (1977) or Chapter 14.1. Similarly, ε_3 for He$^+$ at 1.84 MeV, after scattering from W at the surface, is 1.07×10^{-13} eV cm^2 per atom. Using the measured energy width of the W signal, $\Delta E_{1,3} = (KE_1 - E_3'') = 110$ keV, in Equation (3.7) and (3.8), we find that the film consists of 1.27×10^{18} W atoms cm^{-2}. Taking the number density of W as 6.33×10^{22} atoms cm^{-3} this gives the W film thickness as 2000 Å, consistent with that initially stated. Similarly the depth scale within the Si substrate below the W layer can be found using the stopping cross-sections for Si, evaluated at the appropriate He$^+$ energies before and after scattering *at the interface*. In this case the energy loss of the He$^+$ beam in reaching the interface can be conveniently approximated as one half of the W energy width (i.e. 55 keV). Thus, a depth scale for Si (density 2.33 g cm^{-3}) of ~ 22 Å per keV decrease in backscattered energy, $\Delta E_{1,3}$, can be obtained.

When calculating the *ratio* of heights of elemental films (A and B) in a spectrum, the experimental parameters in Equation (3.9) cancel, giving Equation (3.10), where ε_A and ε_B are the respective stopping cross-section factors, Equation (3.8), and E_A and E_B are the average incident beam energies. For Si and W in Fig. 3.4, $H_W/H_{Si} = 9.3$, consistent with the experimental spectrum height ratio.

c. Impurity Concentration Profiles

The surface energy approximation and the principles of the previous sections can be applied in a very straightforward manner to obtain concentration versus depth profiles for low concentrations (up to a few atomic percent) of impurities in the near-surface region (several thousand Å) of elemental substrates. The method is illustrated using the example of ion implanted Au profiles in Si (Fig. 3.5) where the RBS spectrum shows the continuous Si substrate signal and the ion implanted Au profile at higher energies.

i. *Linear depth scales* can be assigned to both Si and Au portions of the spectrum from Equations (3.7) and (3.8), using, in both cases, stopping cross-sections in elemental Si since the low Au concentration will have a negligible effect on the energy loss. The Si scale is 90 Å/channel (for 5×10^{22} atoms cm^{-3} of Si and 4 keV/channel) and the Au scale (using the K and ε_3 values appropriate to scattering of He^+ from Au in Equation (3.8)) is 70 Å/channel. Alternatively, these depth scale values can be estimated from the tabulations and scaling curves given in Chapter 14.3. The location of the substrate surface for the Au profile is given by Equation (3.1).

ii. The *total impurity content* of Au can be estimated from the spectrum (in atoms cm^{-2}) without knowing the experimental parameters $\Delta\Omega$ and Q in Equation (3.3). Rather, the ratio of Au area to Si spectrum height can give the required value directly from Equation (3.11), where the subscripts i and s refer to the impurity (Au) and substrate (Si). Using the relevant experimental parameters with a Au area of 9600 counts and a Si height of 850 counts, the Au content is $\sim 1.2 \times 10^{16}$ cm^{-2}, in reasonable agreement with the nominal implanted dose.

iii. A *linear concentration scale* for the Au profile can be obtained using Equation (3.12) which expresses the ratio of impurity to substrate concentrations. The stopping cross-section factors ε_1 and ε_S are found from Equation (3.8) using Si stopping cross-sections only but with the respective impurity and substrate kinematic values. For Fig. 3.5, and Au peak height of 450 counts gives $A_{Au}/N_{Si} \sim 2$ atomic percent. Using $N_{Si} = 5 \times 10^{22}$ atoms

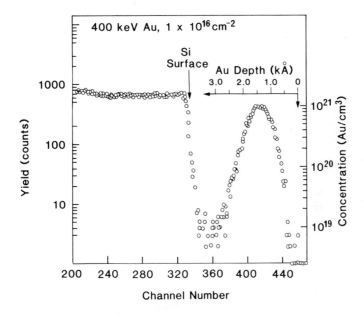

Fig. 3.5 RBS spectrum using 2 MeV He$^+$ of Au implanted Si showing the as-implanted Au concentration versus depth profile plotted on a log scale. Adapted from Poate *et al.* (1987).

cm^{-3}, the peak Au concentration is 10^{21} Au cm^{-3} and the linear Au concentration scale of 5×10^{18} Au atoms cm^{-3} per count is thus established.

This example illustrates the simplicity of quantitative, non-destructive depth profiling by RBS for cases in which a low concentration impurity is heavier than the substrate.

d. Composite Layers and Stoichiometry

We now consider a case where a single layer is composed of more than one major element. Fig. 3.6 shows the RBS spectrum from a 3-layer target consisting of 1000 Å of W over ~ 2500 Å of WSi$_2$ on an underlying Si substrate. This particular structure could have resulted from heating the W-over-Si system (Fig. 3.4) to cause the partial growth of WSi$_2$ from the original Si–W interface via a solid state reaction at 700°C. The spectrum clearly shows that the W and Si spectrum heights change markedly for WSi$_2$. The ratio of these new heights gives the relative atomic concentrations of these elements in the compound (i.e. the stoichiometry).

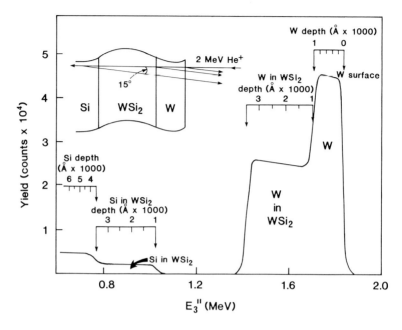

Fig. 3.6 A simulated RBS energy spectrum from a 3 layer target consisting of 1000 Å of W over ~2500 Å of WSi_2 on a Si substrate. Analysis with 2 MeV He^+ ions.

The lower height of the W spectrum in the silicide compared with that of elemental W arises, Equation (3.9), from a smaller stopping cross-section factor ε in the silicide. Thus, from Equations (3.8) to (3.10), this height change may be expressed as $(H_{WSi}/H_W) = (\varepsilon_W/\varepsilon_{WSi})$ (assuming the energies before scattering to be roughly the same) where ε_W and ε_{WSi} are the stopping cross-section factors for scattering off W in the elemental film and the silicide, respectively.

In order to compute ε_{WSi}, from Equation (3.8), it is necessary to use Bragg's rule, Equation (3.13), for additivity of stopping cross-sections to first evaluate ε_1 and ε_3 for the compound. For the WSi_2 case, $\varepsilon_{WSi} = 2\varepsilon_{Si} + 1\varepsilon_W$ can be used to give ε_1 or ε_3, enabling the energy to depth conversion for the silicide to be made using Equations (3.7) and (3.8). This is shown in Fig. 3.6, using $9.857\,\mathrm{g\,cm^{-3}}$ as the density of WSi_2 to give a depth scale. Note that the density of the compound cannot normally be found simply from the densities of the elemental constituents. The depth scales for the elemental W and Si layers in Fig. 3.6 are essentially the same as those in Fig. 3.4. Further details of the method for evaluating depth scales for compounds and multielement targets are given by Chu *et al.* (1978).

The general formula for the height ratio of a compound $A_m B_n$ is obtained by combining Equations (3.10) and (3.11) and is given in

Equation (3.12). In this case, $[\varepsilon]_A^{A_m B_n}$ and $[\varepsilon]_B^{A_m B_n}$ are the stopping cross-section factors for the compound $A_m B_n$ according to Equations (3.8) and (3.11) with K calculated for scattering by A and B atoms in the compound, respectively. Equation (3.12) can be used to calculate the stoichiometry (m/n) of a compound of mixture, using an iterative procedure. First assume that the ratio of $[\varepsilon]_W^{W_m Si_n}$ to $[\varepsilon]_{Si}^{W_m Si_n}$ is *unity* to give a first estimate of m/n as $(H_W Z^2_{Si})/(H_{Si} Z_W) = 0.47$ where the heights H_W and H_{Si} are measured from the spectrum (Fig. 3.6). This value of m/n can be used to obtain better estimates of $[\varepsilon]_W^{WSi}$ and $[\varepsilon]_{Si}^{WSi}$ from Equations (3.8) and (3.11). Equation (3.12) can then be employed to obtain a refined estimate of m/n. The first estimate of the m/n ratio, assuming equality of the stopping cross-section factors, is accurate to better than 10% (noting $m/n = 0.5$ for WSi_2) and a first iteration is often all that is necessary to provide sufficient accuracy for determination of the stoichiometry from profile heights.

e. Thick Multielement Targets

The procedure outlined above can also be adopted for determining the composition of multielement targets where the RBS signals from the various masses overlap. Fig. 3.7 shows the RBS spectrum from a bulk homogeneous machinable ceramic. The spectrum indicates a large number of steps of varying heights and application of Equation (3.1) identifies the presence of B, O, F, Mg, Al, Si and K. The relative proportions of these elements can be deduced from the ratio of step heights (an elaboration of Equation (3.14)) or by spectrum simulation to match the observed spectrum. Such analysis gives the composition as $(B:O:F:Mg:Al:Si:K) = (35:584:37:73:54:181:36)$. In fact, the material consists mainly of mixed oxides of B, Mg, Al, Si, and K. The step from the light mass B is very difficult to detect in the spectrum and the close proximity of Al and Si edges makes quantification of these elements difficult.

3.3 EXPERIMENTAL CONSIDERATIONS

3.3.1 Experimental Planning

a. Simulation of Spectra

Although RBS is highly informative for analysis of existing specimens (e.g. surface of a worn bearing), theoretical simulation of the expected RBS spectrum will often suggest ways to prepare a sample that will yield the clearest interpretation from RBS, while still studying the desired

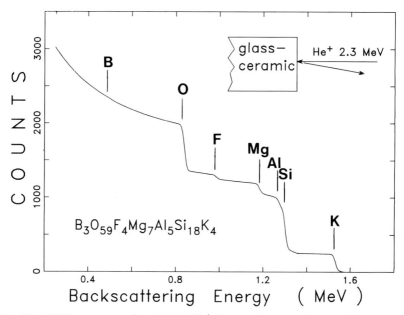

Fig. 3.7 A RBS spectrum using 2 MeV He^+ from a glass ceramic containing various elemental constituents.

material properties. For example, this may require tailoring thicknesses of test films or the use of a preferred substrate. Simulation can also be used to select optimum experimental conditions, such as the probing ion energy and the sample tilt. Programs prepared for generating simulated RBS spectra have been presented by Ziegler *et al.* (1976), Doolittle (1985) and Butler (1986). Each requires a data base of atomic masses and stopping power values, or algorithms with which to generate them. The usual approach is to subdivide the sample into very thin layers of specified composition and allow the program to follow the energy loss and scattering of the probing ions through successive layers. Synthesis of the spectrum utilises the equations listed in Table 3.1, together with expressions for energy straggling and detector resolution (see Chapters 12 and 14). Many of the RBS spectra in this chapter were, in fact, synthesised with such programs.

b. Reference Samples

An alternative to absolute calibration of RBS geometry and collection conditions is the use of reference samples prepared from a pure single element, whose RBS 'height' can be used to calculate the atomic concentrations of all elements in a test spectrum, after correction for stopping power differences between the standard and test films. Care must obviously be taken in preparing thin film reference samples to exclude all possible contaminants (e.g. oxides, H) which might inadvertently lower the yield. The sample should also be prepared so as to minimise channeling and texture effects. Ion implanted samples, having a known atoms cm^{-2} of a heavy impurity can also be used as standards. The use of standard samples is discussed in more detail by Amsel and Davies (1983).

3.3.2 Normal Experimental Conditions

In a majority of routine RBS applications to surface and thin layer problems, it is an advantage to use standard experimental conditions such as those given in Table 3.2. These simplified parameters are convenient defaults in display and analysis programs. Samples are generally mounted, several per load, on the table of a goniometer allowing three degrees of freedom, two in sample position and one in angle, so as to enable sample changing and manipulation without breaking vacuum. A liquid nitrogen-cooled shield placed around the target is a simple means of minimising C build up during analysis. Examples of target chamber arrangements, particle detection, pulse electronics and data storage are given in Chapter 2.

3.3.3 Mass Resolution

If the standard conditions will not produce separated (resolved) peaks or segments for each element in the sample, then some options for improving mass resolution, dM, may be explored. Equations (3.15) and

TABLE 3.2
Standard RBS analysis conditions

He^+ beam energy: 1.5 to 2.3 MeV
Beam spot size: 1 to 4 mm^2
Beam current: 10 to 50 nA
Ion dose to accumulate one spectrum: 10 to 40 μC
Incidence angle: 5 to 10° to the sample normal
Incident beam divergence: better than 3° (full angle)
Target vacuum: better than 10^{-6} Torr
Detector angle: scattering at 165 to 170° to the incident beam.
Detector solid angle: 3 to 5 msr.
Surface barrier detector area: 25 to 300 mm^2
Detector resolution: 15 keV
Spectrum channel width: 4 keV
Analyser data storage per spectrum: 512 channels
Typical accumulation time: 5 to 20 minutes.

(3.16) in Table 3.3 define the mass resolution (obtained from differentiation of Equation (3.1)) and the resolution can be improved in a number of ways:

    i. use higher beam energy to effectively expand the energy spectrum;

    ii. use heavier ions to provide better kinematic separation;

    iii. select a scattering geometry with θ as close to 180° as possible; and

    iv. reduce the system energy resolution, dE_s, by improving detector energy resolution, dE_d, and ensuring that the detector solid angle is small so that kinematic broadening of energy resolution is negligible.

The effectiveness of these steps may be gauged from Fig. 3.8 where dM is plotted as a function of E_1 for the ions H^+, He^+, C^+ and Ne^+. A Si surface barrier detector has been used, with an energy resolution of ~10 keV for MeV He^+ ions. The energy resolution of these detectors is poor for heavy ions. For this reason, the curves in Fig. 3.8 do not reflect the improvement in mass resolution for C^+ or Ne^+ ions that would be expected from kinematics alone. Only the use of higher energy resolution ($\Delta E/E \leqslant$ 0.1%) electrostatic, magnetic or time-of-flight analysers produces *unit* mass resolution at $M_2 = 100$ with 2 MeV He^+, or at $M_2 = 200$ with 2

TABLE 3.3
RBS mass, sensitivity and depth resolution relationships

Relationship	Equation	
Mass resolution (general)	$dM_2 = \dfrac{dE_s}{E_1}\left(\dfrac{dK}{dM_2}\right)^{-1}$	(3.15)
Mass resolution for $\theta = 180°$ and $M_2 \gg M_1$	$dM_2 = \dfrac{dE_s}{E_1}\left(\dfrac{M_2^2}{4M_1}\right)$	(3.16)
Sensitivity limit for impurities heavier than the substrate	$N_i/N_s \gtrsim (Z_s^2/Z_i^2) \times 10^{-3}$	(3.17)
Depth resolution (general)	$dt = \left(\dfrac{dE_3''}{dt}\right)^{-1}\Delta E_s$	(3.18)
Depth Resolution, surface energy approximation	$dt = \dfrac{dE_s}{(KS_1/\sin\alpha + S_3/\sin\beta)} = dE_s/S$	(3.19)
Total System Energy Resolution	$(dE_s)^2 = (\Delta E_d)^2 + (\Delta E_g)^2 + (\Delta E_{3m})^2$ $+ (\Delta E_{1s})^2 + (\Delta E_{3s})^2 + (\Delta E_{1m})^2$	(3.20)

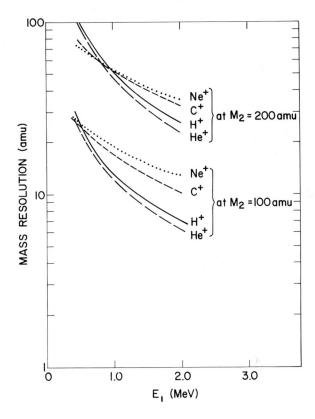

Fig. 3.8 Mass resolution available using H^+, He^+, C^+, Ne^+ ions and a detector with an energy resolution of 10 KeV (after Chu *et al.* 1978).

MeV C^+. The extremely small geometrical efficiency of such detectors and their inability to simultaneously analyse all backscattered energies are unacceptable in most practical applications unless extraordinary mass resolution is essential. Hence, in general usage, He^+ ions detected with good surface barrier detectors, offer the best practical performance, especially at high (say, 2 MeV) energy.

3.3.4 Sensitivity

The sensitivity of RBS for detecting trace impurities in bulk samples depends very much on the sample composition and the experimental conditions. For impurities heavier than the substrate, in principle, it would be possible in a system with no background counts, to measure infinitesimal amounts of impurities simply by increasing the integrated charge without limit. However, a finite background count level extends above substrate edges as a result of multiple scattering in the sample, pulse pile up and scattered incident beam reaching the detector from defining slits and chamber environs. Chu *et al.* (1978) have suggested a lower working limit for heavy impurity detection in lighter substrates as 10^{-3} of the substrate height. From Equation (3.12) this can be expressed approximately by Equation (3.17).

Examples of sensitivity limits are:

- 4×10^{18} atoms cm^{-3} (i.e. 3.1×10^{-3} atomic % or 500 μg cm^{-3}) of As in C
- 7×10^{18} atoms cm^{-3} (i.e. 5.4×10^{-4} atomic % or 241 μg cm^{-3}) of Pb in C
- 6×10^{19} atoms cm^{-3} (i.e. 9×10^{-2} atomic % or 7500 μg cm^{-3}) of As in Zn.

For surface impurities, the sensitivity can be improved by selecting grazing incidence and/or exit geometries, giving improvements by up to $1/\sin \alpha$ and/or $1/\sin \beta$ depending on the scattering geometry. This improved sensitivity results from the increased path length through the surface region containing the impurity. A useful monolayer limit is given by Equation (3.17) with 10^{-3} replaced by 2×10^{-2} monolayers. For example, the minimum detection limit for Pb on C would be $\sim 10^{-4}$ monolayers but for As on Zn it would only be $\sim 2 \times 10^{-2}$ monolayers.

Impurities which are lighter than the substrate are usually detected with very poor sensitivity by RBS and several atomic percent or several monolayers are needed for their detection. However, there are several alternative methods of enhancing the detection of light impurities and these are discussed in Sections 3.4 and 3.5.

3.3.5 Depth Resolution

The RBS depth resolution is given by Equations (3.18) to (3.20) in Table 3.3, the first two being derived from differentiation of Equation (3.5), where dE_s is the total system energy resolution, Equation (3.20). The standard conditions of Table 3.2 will typically provide depth resolutions of $\sim$300 Å.

Improvements to depth resolution may be obtained in a number of ways (by inspection of Equation (3.19)):

    i. maximising ε_1 and ε_3 by choice of either an energy close to the stopping power peak (Chapter 14.1) or heavier probe ions;

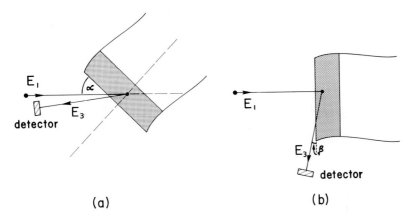

Fig. 3.9 Extending the effective RBS depth resolution by: a. Sample tilting; and b. oblique exit geometry.

ii. improving the energy resolution; and
iii. selecting a geometry (grazing beam incidence and/or exit) to optimise depth resolution normal to the surface.

Depth resolution is discussed in more detail in Chapter 12. We discuss here the simplest case of option (iii).

Geometrical scaling of depth resolution can easily be achieved by tilting the sample as shown in Fig. 3.9a, or by moving the detector to receive ions scattered at a grazing angle to the sample surface (Fig. 3.9b). In the first case the apparent thickness of the sample is approximately increased by a factor ($1/\sin \alpha$); in the second case the path of the scattered ion through the sample is extended by the factor ($1/\sin\beta$). The value of these approaches is illustrated in Fig. 3.10, where the RBS spectra from a uniform 150 Å Au film on Si are given for a series of sample tilt angles. As the apparent film thickness increases, the Au signal changes from being a low, resolution-broadened peak, to reach the full peak height for Au and finally develops a measurable flat-topped peak width. Ultimately, for very large apparent thicknesses of Au, the energy resolution broadening effects of energy straggling (ΔE_s), multiple scattering and lateral spread (ΔE_m), and detector acceptance angle (ΔE_g) of both the incident and exist ion trajectories begin to dominate the detector energy resolution, ΔE_d, and diminish the overall depth resolution. Such contributions can be approximately added in quadrature as shown in Equation (3.20) where the subscripts 1 and 3 refer to incident and exit trajectories, respectively. A series of resolution curves as a function of sample tilt angle is displayed

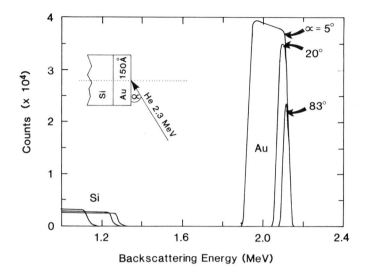

Fig. 3.10 Effect of sample tilt upon apparent depth resolution for RBS of 2.3 MeV He ions from 150 Å of Au on a Si substrate. The scattering angle is kept constant at $\theta = 165°$.

schematically in Fig. 3.11 (Williams and Möller, 1978), indicating that there is usually an optimum tilt angle for the analysis of smooth, continuous thin films. For analysis of the near-surface region, where detector energy resolution dominates the total system resolution, depth resolutions less than 20 Å are achievable. However, as the probing depth increases, the optimum resolution is degraded. Lateral spread due to multiple scattering and/or the geometrical term arising from finite detector acceptance angle dominate dE_s and limit depth resolution. The absolute depth resolution will also depend on the stopping power of the sample material, the probing ion species and beam energy. A more complete treatment of the factors limiting depth resolution and how to calculate them is given in Chapter 12.

For optically flat samples, grazing geometries to within 3° of the sample surface can be employed for near-surface analysis. If grazing incidence is chosen, the beam spot size must sometimes be reduced in order not to extend too far across the sample. Grazing geometry also demands detector collimation in order to limit resolution loss due to finite solid angle, $d\Omega$, which can cause kinematic broadening and path length variations. In addition, the higher depth resolution for grazing angle geometries enhances sensitivity to C build-up which can occur where the beam strikes the sample surface in a poor vacuum system.

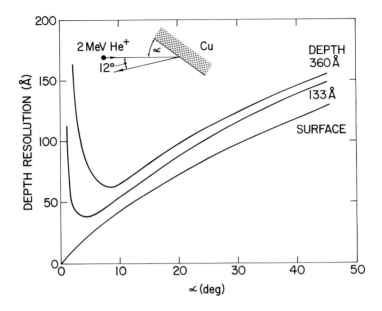

Fig. 3.11 Depth resolution calculated using the method outlined in Chapter 12 for Cu films at depths of 0, 133 and 360 Å, for backscattering of He$^+$ at 2.0 MeV (Williams and Möller, 1978).

In many practical cases, a test surface or interface can be microscopically rough and a layer may even contain pinholes or islands. In such cases, RBS produces a spectrum summed over all the features within the beam spot. Grazing geometries, in particular, need special care in interpretation. In the case of a surface covered with spherical 'islands', for example, the original spectral peak will not change with RBS geometry in the simple analytical way indicated for the Au film of Fig. 3.10. Such surface topographic and non-uniformity effects have a detrimental effect on depth resolution and are discussed in Section 3.4.2.

3.3.6 Experimental Difficulties

a. Insulating Samples

Dielectric materials can present the problem of charge build up during irradiation. This may damage the sample, degrade the beam energy resolution and involve beam current integration of breakdown pulses, which is seldom successful. Some materials such as alumina also fluoresce and light pulses spoil the resoluion of the surface barrier detector. These problems can sometimes be relieved by methods described in Chapter 2.

b. Crystallinity

Channeling in single crystal materials, see Chapter 6, can reduce the RBS spectrum height by as much as a factor of 30. However, channeling can occur to some degree in all polycrystalline materials, often without being recognised, and the diminished RBS yields give misleading results. A routine precaution against inadvertent channeling is to tilt the sample so that the beam enters 5 to 10° off the normal to the surface. Since most single crystal substrates are likely to have been cut within about 2° of a major crystal axis, this technique avoids potential channeling directions near 0°, 30°, 45° etc. A more reliable approach, which removes the chance of the beam entering along a planar channel, is to continuously rotate the tilted sample about its normal axis during data collection. Alternative methods for avoiding channeling are described in Chapter 6. It should be emphasised, however, that in crystalline material there is no single geometry that will give, in every case, the same RBS yield as that from an amorphous sample.

c. Radiation Damage

RBS is correctly considered to be a generally non-destructive analysis technique. The power density (0.1 W mm^{-2}) and dose (6×10^{13} He^{+} cm^{-2}) normally used for 2 MeV He^{+} analysis are small. The temperature rise produced by such a beam at the surface of a Si wafer can be estimated to be only a few degrees. The ion dose is typically three orders of magnitude below that needed to amorphise a single crystal Si layer; nevertheless, some radiation damage will occur in such a sample at, or near, room temperature. One means of reducing the risks of both damage and heat is to distribute the beam over a much larger surface area, by defocussing the ion beam and rotating or rastering the sample. Physical evidence of substrate damage can sometimes be seen after RBS analysis. Some materials such as garnets readily develop dark color centres, which disappear after annealing. In the case of glass or fused quartz substrates, a compacted region is produced after long exposures, appearing like a dimple in the surface < 1000 Å deep. The physical cracking caused by compaction permanently degrades the substrate surface.

Polymers and other organic samples require extraordinary care, being easily degraded by the ion beam. Polyamide, Teflon, Mylar and epoxy, for example, will all show discoloration and perhaps disintegration after irradiation with $\sim 10^{14}$ H^+ cm^{-2} or less. With all such materials RBS analysis requires large diameter beams and minimal exposures.

3.4 APPLICATIONS AND LIMITATIONS OF RBS

The major strengths of RBS are absolute measurement of composition and trace impurity concentration, through a precise knowledge of the Rutherford scattering cross-section, and non-destructive depth profiling using accurately known stopping powers for He^+ ions in solids. Principal applications include: bulk composition analysis and major element profiling; minor element identification and concentration versus depth profiling; thin film analysis for studying solid state reactions, interdiffusion, stoichiometry, film thickness, uniformity and impurity content. This section illustrates the practical application of both the basic principles of Section 3.2 and the techniques of Section 3.3. It also highlights limitations in mass resolution, sensitivity, depth profiling and for the analysis of laterally non-uniform samples.

3.4.1 General Analysis Considerations

Although RBS is not usually employed for the analysis of samples of totally unknown composition or containing a multitude of unknown trace impurities, *unexpected* detail revealed by careful RBS analysis often holds the key to successful analysis in such cases. It is not possible to prepare a universal set of rules which will allow one to optimise RBS analysis for all possible unknown samples or to interpret unexpected detail in RBS spectra. However, the considerations in Highlight 3.2 may be of some help.

HIGHLIGHT 3.2
ANALYSIS OF UNKNOWN SAMPLES

STEP 1: Preliminary inspection of sample.

    a. Is the surface rough (on a 1000 Å scale) or laterally non-uniform? Section 3.4.2 discusses the special precautions which must be taken for analysis using large area beams; alternate microbeam or complementary techniques may be advisable.

    b. Ensure that the surface is free of unwanted contamination. Section 3.4.1c illustrates how surface contaminants can be analysed.

    c. Is the material sensitive to severe radiation damage (e.g. a polymer) or is it an insulator? See Section 3.3.5 for precautions in such cases.

STEP 2: Inspection of an initial RBS spectrum (using the standard conditions of Table 3.2).

a. Does the spectrum indicate a bulk target or a layered structure? The initial clue to recognising layers containing more than one element will be similar (or complementary) profile *widths* and *shapes* for the elements concerned. Examples in Section 3.4.1 illustrate such features. A bulk spectrum will contain steps (at energies representative of particular masses) and plateaux continuing to zero energy as given by Fig. 3.7. Surface or interface impurities appear as sharp *peaks* (Section 3.4.1c).

b. Elemental profiles which have sloping or broadened high and low energy edges may result from resolution broadening, surface or interface roughness, non-uniform concentration or severe lateral non-uniformities. Section 3.4.2 indicates how such possibilities might be distinguished. Ultimately, such cases may require alternative (optical or electron microscopy) techniques to aid the interpretation of RBS.

c. Overlap peaks can occur in spectra from layered films containing similar masses (e.g. Al overlaying a Si layer) and these can often be misinterpreted as surface or interface impurities (Section 3.4.1b).

STEP 3: Supplementary RBS spectra and quantification.

a. Uncertainty about the sequence or composition of layers, peak identification, impurity depth distribution and the suspicion of overlap peaks can often be resolved by tilting the sample to obtain a repeat spectrum (Fig. 3.10). In particular, the dilemma between impurities located at the surface or in depth can be resolved by target tilt; analysis at two tilt geometries gives impurity identification and depth distribution (Section 3.4.1c).

b. The probable elements and layer compositions can be found from simple application of the equations of Table 3.1 starting from the surface. Simulation of spectra to match the experimental data is a preferred approach. Fig. 3.7 (Section 3.3.2) is an example of quantification of an unknown bulk target.

c. Elemental spectrum heights depleted from that expected for the pure element or compound can result from a light element which is undetected by RBS (Section 3.4.3).

a. Choice of Sample Configuration

In an ideal RBS spectrum, each of the elements and each of the layers (in a multilayer thin film structure) will be clearly distinguishable. In principle, if the constituents of a sample are known, overlapping components of the RBS spectrum can be separated by an iterative procedure such as that given by Brice (1973). However, if the aim is to study a specific thin film process it is often possible to tailor the experiment by choosing a layer sequence and thickness to produce isolated constituent peaks in the RBS spectrum. This greatly simplifies interpretation. Simulation of spectra before carrying out the experiment can greatly assist in choosing optimum sample configurations.

An example of a well-behaved spectrum is shown in Fig. 3.12a. For measurement of Al–Ag thermal reaction properties, successive films (first Ag then Al) were deposited on an oxidized Si substrate. By choosing a layer thickness < 2000 Å, overlap between the Al and Ag signals is avoided; the signal from Si in the SiO_2 is displaced from the surface position by approximately the sum of the metal layer widths. The oxygen of SiO_2 produces a minor plateau added to the substrate Si signal which extends all the way to zero energy. Thermal treatment causes the Al and Ag to interdiffuse and, in each case, the diffusion will be seen in an otherwise empty spectral region. By contrast, thicker Al and Ag layers would shift the Ag peak to the left until eventually it overlapped the surface-Al peak (Fig. 3.12b). Interdiffusion of Al and Ag would obviously be more difficult to study in this case.

b. Choice of Substrate in Thin Film Analysis

If the Al-Ag films shown in Fig. 3.12a had been thoughtlessly deposited on lead-glass substrates instead of SiO_2 the confusing spectrum shown in Fig. 3.13a would have resulted. Clearly, in the absence of other physical considerations the choice of a passive substrate of low (Z_1, M_1) is desirable, so as not to obscure the thin film signals with an irrelevant substrate profile. Incidentally, for the hapless analyst obliged to handle the lead-glass substrate of Fig. 3.13a, a plausible approach would be to flip over the sample and obtain a 'substrate' spectrum from the back side as shown in Fig. 3.13b. If this spectrum is then translated to the left, by the net energy width of the overlying Al and Ag metal films, and subtracted from the original spectrum (Fig. 3.13a), then the required 'clean' metal peaks may be obtained.

The use of vitreous C substrates is shown in Fig. 3.14 for the

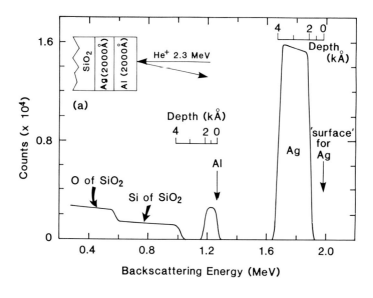

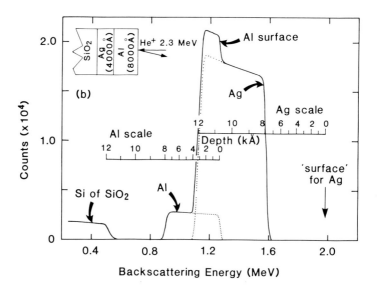

Fig. 3.12 RBS (2.3 MeV He$^+$) spectra from Al–Ag films on SiO$_2$. a. Element profiles are separated sufficiently to display possible interaction effects. b. Spectrum with overlapping profiles results from a bad choice of Ag and Al layer thicknesses.

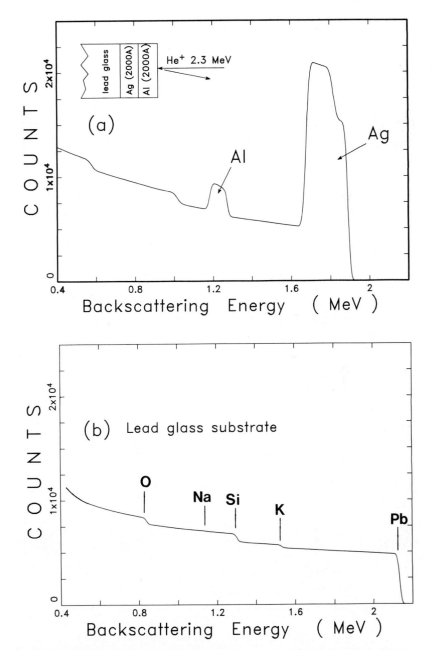

Fig. 3.13a. RBS spectrum for Ag-Al films deposited on a lead-glass substrate. b. Spectrum of a lead-glass substrate without a deposited film.

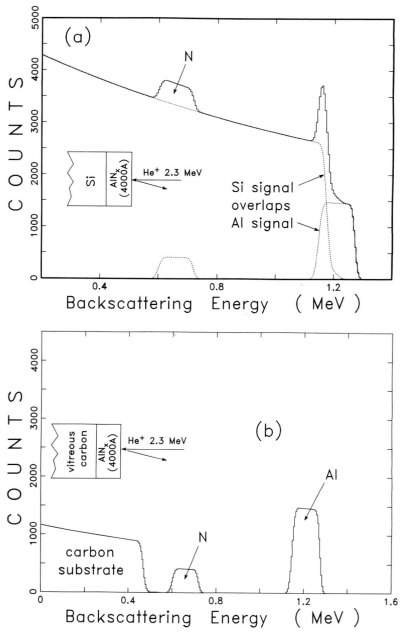

Fig. 3.14 Simulated RBS spectra from AlN (4000 Å) deposited on substrates of: a. Si; and, b. vitreous C. The dashed and dotted curves in a. are the partial Si and AlN spectra.

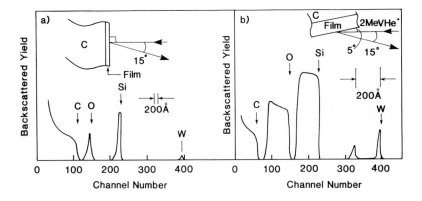

Fig. 3.15 RBS spectra from ~200 Å SiO_2 e-beam deposited on vitreous C using 2 MeV He^+: a. with a geometry of $\alpha = 90°$ and $\theta = 165°$; and b. optimised low angle geometry with $\alpha = 5°$ and $\theta = 165°$. From Williams (1976).

determination of small variations in the Al:N ratio in a series of sputter-deposited films ~ 4000 Å thick. If the films were placed on Si substrates (Fig. 3.14a), spectrum height ratios (or deconvolution) must be used in a region disturbed by resolution broadening and profile overlap of Si and Al signals at the AlN/Si interface. By using C substrates, areas of fully isolated Al and N peaks are observed and the total atoms cm^{-2} content for each element and film stoichiometry can be derived, free of substrate signals. Peak area ratios offer the simplest and most accurate quantification of thin film composition. (A practical hint: C surfaces tend to absorb moisture and such substrates should be heated in vacuum directly before coating, to avoid spurious substrate-oxgyen signals.)

c. Choice of Geometry in Impurity Identification

In many analysis situations, RBS geometry can be optimised for thin film determination and for identifying and locating impurities. Fig. 3.15 illustrates such applications for ~ 200 Å of SiO_2 on a vitreous C substrate, where e-beam deposition from a W boat was used. For a standard RBS geometry ($\theta = 165°$ and normal incidence of 2 MeV He^+) the RBS spectrum in Fig. 3.15a shows well resolved peaks corresponding to Si and O in the deposited film, together with a probable W impurity. However, the depth resolution of about 300 Å is insufficient to determine film parameters (thickness, uniformity and stoichiometry) and the depth distribution of the impurity. Fig. 3.15b illustrates the spectrum obtained for an optimised geometry with $\theta = 165°$ but $\alpha = 5°$. This geometry preserves the energy separation of C, Si, O and impurity signals but gives excellent depth resolution and improved sensitivity for impurity identification (Section 3.3.4). The sharp low energy cdges of the Si and O profiles indicate a continuous film (Section 3.4.2) and Equation (3.14) gives the film composition as $SiO_{1.95}$. Two impurity peaks are apparent, the higher energy peak coincides with the energy of that in Fig. 3.15a and thus identifies a W surface impurity. The lower energy peak in Fig. 3.15b corresponds to W at the film-substrate interface: it was not resolved in Fig. 3.15a because of the poor depth resolution.

3.4.2   Lateral Inhomogeneities

Many materials have finely featured patterns which ideally require a corresponding finely collimated and registered ion beam. Ion microbeams of a few microns diameter have been used for RBS analyses (see Chapter 10); however, the instrumentation required for such work is more complex than for 'standard' RBS analyses with beam spots of > 0.5 mm diameter. RBS can provide helpful insights even in situations where the beam area covers regions with gross lateral non-uniformity.

a. Mixed Phases

Consider a uniform layer of a homogeneous mixture of composition $Au_{20}Si_{80}$. It will be seen by RBS as a single layer whose composition could correctly be deduced from profile height ratios or from Si and Au peak areas as outlined in Section 3.2.2. Such a spectrum is shown in Fig. 3.16 (solid line). In contrast, if the same number of atoms of Au and Si were to be rearranged to form a layer of constant thickness but consisting of mixed phases of Au columns within a Si matrix, as shown in inset (b) of Fig. 3.16, then a very different RBS spectrum would result. (This spectrum is simulated assuming that each column is wide enough to be considered a separate entity within which an entire ion trajectory will be contained). A casual estimate of composition from height ratios would be totally wrong. However, the spectrum offers several clues for the analyst. Firstly, the profile widths for Si and Au from the column sample are not identical; secondly, the substrate profile is moved to a lower energy corresponding to the area of substrate covered by the Au columns (whose stopping power is greater than that of the matrix Si); and thirdly, the composition deduced from peak areas (20:80) would disagree with that obtained from profile height ratios. Thus, the possibility of a mixed structure could be inferred from experiment and the sample should be promptly taken to the optical microscope or SEM to check its homogeneity.

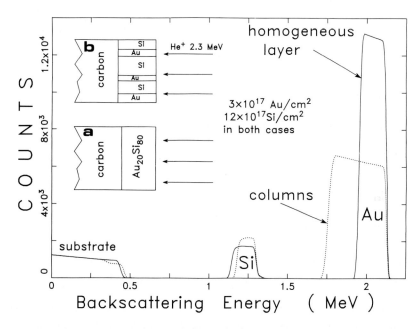

Fig. 3.16 RBS from an inhomogeneous layer. Simulated spectra are compared for an homogeneous layer of composition $Au_{20}Si_{80}$ (solid line) on a C substrate, and for a rearrangement of the same Au and Si atoms in a columnar-structured layer of constant thickness (dotted line).

In Fig. 3.17 we illustrate a more practical example of mixed phases and columnar structures. Samples made by heating Ir films on Si for 2h at 950°C produced different RBS spectra (Fig. 3.17a) depending on the beam position on any *one* sample. Lateral inhomogeneity was confirmed by microscopic examination of the surface (Fig. 3.17b). The key to interpretation was provided by X-ray diffraction spectra showing the co-existence of just two phases, $IrSi_{1.75}$ and $IrSi_3$. The RBS spectra are seen to be consistent with a model in which crystallites of the phase $IrSi_3$ grow extremely rapidly after nucleation, which occurs heterogeneously across the sample during heating. After 2h at 950°C, not all the silicide has been converted and, depending on the area 'averaged' by the RBS beam, a variable proportion of fully converted $IrSi_3$ (the 'steps' to the left of the profiles) contributes to the RBS spectrum. (After longer heating, conversion to $IrSi_3$ is completed.) This example clearly illustrates the need to complement RBS analysis with other techniques in difficult situations.

Samples with compositional non-uniformities or mixed compound phases are not normally well handled by RBS. However, if the constituent elements and their probable compound phases are known useful quantitative information can be obtained, particularly when RBS is

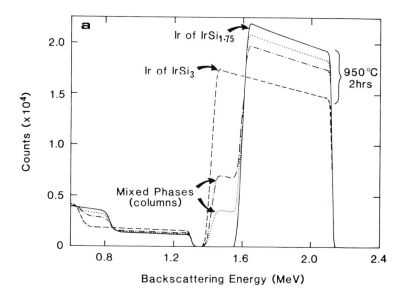

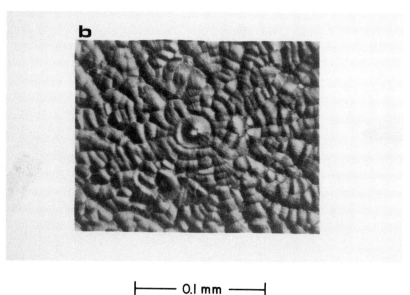

|—— 0.I mm ——|

Fig. 3.17a. RBS spectra (2 MeV He$^+$) from different regions of an Ir on Si sample reacted for 2 h at 950°C. b. An optical micrograph illustrating the lateral non-uniformities in silicide reaction. After Petersson *et al.* (1979).

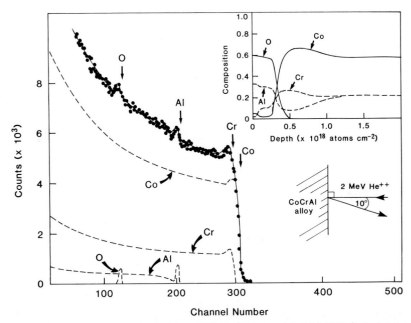

Fig. 3.18 RBS spectrum from an oxidized (CoCrAl) alloy using 2 MeV He$^+$. A simulated spectrum (solid curve) is shown for comparison with the experimental spectrum (full circles). Component (simulated) spectra of the individual elements are shown dashed and these are used to obtain the concentration versus depth profiles in the inset. After Johnston *et al.* (1984).

combined with spectrum simulation. We illustrate a depth profiling application in Fig. 3.18 for oxidation of a (Co Cr Al) alloy (Johnston *et al.*, 1984). The experimental spectrum (closed circles) is from a mixed oxide and the solid curve is a spectrum fit obtained using the program of Butler (1986). Since the starting film composition was known to be Co$_3$CrAl, this was used for the simulation and, by trial and error, O was added to the near-surface until the best fit to the experimental spectrum was obtained bearing in mind the equilibrium oxide phases Al$_2$O$_3$ and CrO$_2$. The individual signals (or partial spectra) from Cr, Co, Al and O, which were used to obtain the fit, are also shown in Fig. 3.18. These individual spectra, which were the outcome of the simulation, were used to provide the concentration versus depth profiles shown in the inset. The surface oxide is consistent with the expected mixed oxide phases of Al$_2$O$_3$ and CrO$_2$ with an additional small surface peak of Co. Scanning AES analysis confirmed that oxides completely covered the surface, but consisted of separated regions of a single oxide phase.

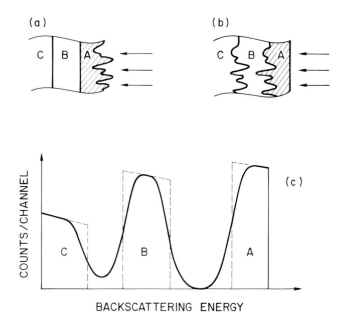

Fig. 3.19 The rough-coated sample shown in a. will appear to the arriving RBS beam as illustrated in b. The resultant spectrum c. always has a sharp profile edge corresponding to the outer surface, with all other profile features smeared out.

b. Rough Surfaces

RBS depth scales (see Section 3.2.2) are always calculated with reference to the sample surface. Hence it could be said that even when the sample surface is rough (Fig. 3.19a), the RBS technique (especially for near-normal incidence and detection) will behave as though the *surface* were flat and everything below it were rough (Fig. 3.19b). The RBS spectrum from such a sample is illustrated in Fig. 3.19c and shows a steep well-defined profile at the *surface*, and sloped profiles, apparently from ill-defined interfaces, for *all* the interfaces. This qualitative observation should normally alert the analyst to examine the surface of the sample microscopically. Provided that the layer is still continuous, normal RBS analysis may continue. Repetition of RBS spectra at a series of sample tilt angles can, in principle, yield further information about the shape of irregularities in such a rough surface (Edge and Bill, 1980; Bird *et al.*, 1983). However, microscopy will usually be preferred. Spectra showing sloping profile edges can also arise from sample features other than roughness. How then can the spectrum of Fig. 3.19c be distinguished from a spectrum showing, for example, interdiffusion of layers A and B?

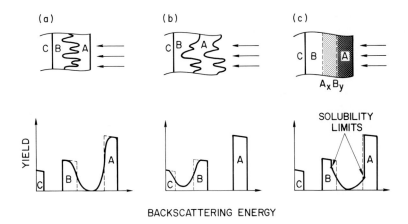

BACKSCATTERING ENERGY

Fig. 3.20 Response of RBS spectra to: a. a diffused or rough interface, b. a rough underlayer, or, c. an intermediate layer of physically limited composition, e.g. a solid solution.

After simple interdiffusion, no 'blurring' of the substrate C/film B interface should occur, whereas the left side of the B profile is indeed sloped when the surface layer or interfaces are rough.

c. Rough Interfaces

A rough interface in a 'multi-layer' sample can generally be recognised by the complementary slopes on the sides of the corresponding RBS layer profiles, as shown in Fig. 3.20a. This will not apply when the top layer (A), for example, has been deposited uniformly on an initially roughened surface (B) (Fig. 3.20b). As shown, the RBS spectrum from such a sample will be characterised by sharp profiles for the surface and A/B interface, and a sloped profile at the substrate/B interface. RBS profiles rather like those shown in Fig. 3.20a would also be expected following interdiffusion of layers A and B to form a solution or alloy. Analytic recognition of diffusion profiles from such spectra would be difficult. A clue would be provided in cases where the profile slope began at a concentration (yield) corresponding to a known solubility of A in B or B in A (Fig. 3.20c).

d. Discontinuous Layers

When a layer contains pinholes, exposing patches of substrate over $x\%$ of its area as in Fig. 3.21, the RBS spectrum will represent a simple sum of $[(100 - x)\%$ of the RBS signal for a continuous film of A) plus ($x\%$ of the

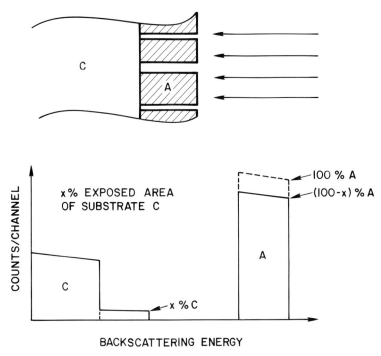

Fig. 3.21 Response of RBS to pinholes constituting *x*% of area of layer A, through which the ion beam "sees" the substrate surface partly exposed.

RBS signal from a fully exposed substrate)]. Similar reasoning will apply to islands of beads of A on an exposed substrate surface. As in the case of columnar mixtures shown in Fig. 3.16, it will be found that simulation for the mistaken model of an intimately mixed A-B layer will not exactly match the observed spectrum. This is due to the difference in stopping power between the pure A film and the A-B mixture. In the special case of a beaded surface, such as that shown in Fig. 3.22, which is the result of heating a very thin Cu film deposited on silica, RBS at normal incidence will produce an asymmetric profile of the beaded element (Cu). This arises since the thickness of Cu seen by an arriving ion may have any value from zero to the diameter of the bead, depending on where the ion enters. Without prior knowledge of the state of the layer, the analyst may read such a spectrum as a sign of a thin layer diffused into the substrate. However, this interpretation can be tested by tilting the sample, e.g. by 60°, and repeating the spectrum. A diffusion profile would be widened by this procedure in accordance with geometric scaling, whereas the bead signal would change very little (depending on the shape of the beads).

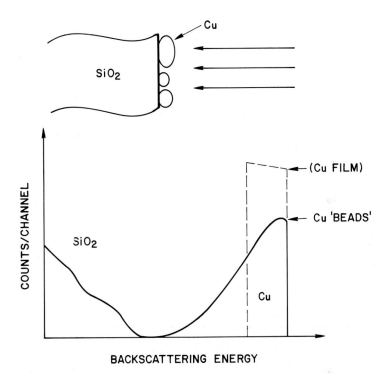

Fig. 3.22 Response of RBS to a beaded, discontinuous layer of Cu partly covering a smooth SiO$_2$ substrate.

e. Voids and Cavities

A schematic diagram of a cavity or blister in a thin film system is shown in Fig. 3.23a. Outside the area of the cavity, the RBS spectrum behaves normally. Somewhat surprisingly, the spectrum from within the cavity area will be identical since arriving ions see the full thickness of film, are then subject to no scattering or energy loss in traversing the cavity, after which scattering and energy loss processes resume in the substrate. A similar argument applies for isolated voids of any size occurring within a thick sample. Further, if a sample or film should be uniformly porous, or have abnormally low density, RBS will still deliver the same profile. RBS is thus 'blind' to cavities, voids and density variations within a sample; RBS alone has nothing to say on the existence or otherwise of such imperfections in a test specimen. Nevertheless, RBS data are frequently supplied with depth scales quoted in Å. Such linear dimensions are obtained only on the basis of an assumed average density for the layer.

(a)

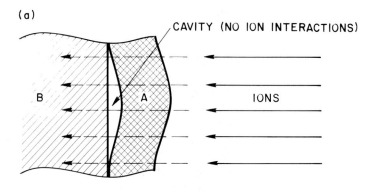

(b)

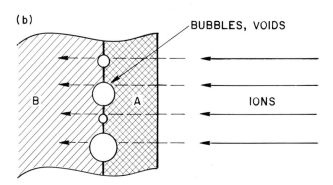

Fig. 3.23 a.A blister shown covered uniformly by film A will have no effect on an RBS profile. b. Cavities or voids like these will lead to an RBS spectrum implying a rough A–B interface.

Since RBS measures numbers of atoms, it cannot be directly employed to measure layer thicknesses in units of length, unless the average film density is known. A different situation prevails (Fig. 3.23b) when the voids or cavities are not uniformly covered (i.e. the film thickness is not uniform). The sample then appears to have a rough film/substrate interface, producing RBS spectra like those of Fig. 3.20.

f. Film Texture

Fibre texture is displayed by many vapor-deposited polycrystalline metal films. Crystallites may become aligned in a generally preferred direction with respect to the substrate plane. For example, Au films prepared on Si, may show strong texture, with alignment of grains spread over 15° or more about the substrate normal (Andersen *et al.*, 1978). The RBS yield for such a sample may be reduced by several percent even for slight tilt angles. Experimental testing for such an effect is the best resort if a set of routine RBS spectra from metal films seem to be showing inconsistency of RBS yield. An example of such an effect is given in Chapter 6.3.

3.4.3 Scattering Analysis of Light Elements

RBS becomes increasingly difficult when one seeks to detect lighter and lighter elements because of the low scattering cross-sections. This is compounded in most practical cases by the superimposed profile due to a heavier substrate material which obscures the already small signal from the low-mass material. The case is not hopeless, however, and several steps can be taken so that scattering analysis can handle a situation where O, N, C or B are involved. If these fail then other analytical techniques must be considered including ERA (Section 3.5).

a. Optimisation of Analysis Conditions

    i. *Improvement of count statistics.* A few atomic percent of O in a film supported by a Si substrate can generally be identified and quantified simply by accumulating data for much longer than one would normally require for heavier mass determination.

    ii. *Employment of substrate channeling.* The background from single-crystal substrates can often be greatly reduced by aligning the sample so that the ion beam enters in an axial channeling direction. Provided that the light element to be detected is not substitutionally incorporated within the crystal lattice, its signal will then be more easily observed. Chapter 6 further illustrates this procedure.

    iii. *Use of low-mass substrates.* In some cases an experiment can be deliberately configured using a light substrate chosen so as not to interfere with signals from O, N, C, etc. in a test film. High purity heated or sputter cleaned vitreous C substrates can often be used. Be is a possible alternative substrate material. Ideally, self-supporting thin films present no substrate problem. However, film preparation can be difficult and possible beam heating needs to be considered.

    iv. *Analysis of "missing mass".* The presence of a light element contained in a film of known heavier elements (e.g. O contaminating the bulk of a Cu film) will, according to Equation (3.14),

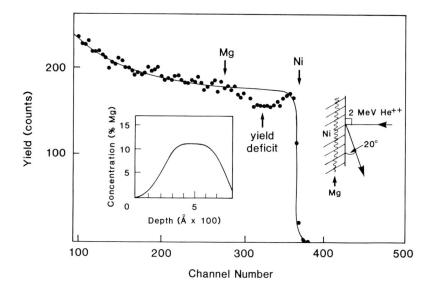

Fig. 3.24 RBS spectrum using 2 MeV He$^+$ of 1×10^{17} cm^{-2} implanted into Ni at 100 keV. The solid curve is a Ni spectrum without a Mg implant. The inset shows the Mg concentration profile extracted from the yield deficit. From Short *et al.* (1987).

cause a reduction in height of the RBS profile of the heavier elements from that expected for a pure (O-free) film. If the identity of the light component is independently known, then this *height reduction* (or yield deficit) can be used to derive the composition of the film without further efforts to quantify the RBS signal from the light component itself (Borgesen *et al.*, 1978). An example of such an analysis is given in Fig. 3.24 for the case of a high concentration of Mg (1×10^{17} cm^{-2}) implanted into Ni at 100 keV (Short *et al.*, 1987). The RBS spectrum shows a yield deficit in the near surface region of the Ni signal. In contrast, the direct scattering signal from the lighter Mg (channel 286 in Fig. 3.24) is masked by Ni scattering even at this Mg concentration (> 10 atomic %). The Ni yield deficit provides an 'inverse' profile of the Mg which can be found using Equations (3.13) and (3.14) or, more readily, using RBS simulation programs, by adjusting the Mg concentration to obtain a good fit to the data. The Mg concentration profile obtained by simulation is shown in the inset.

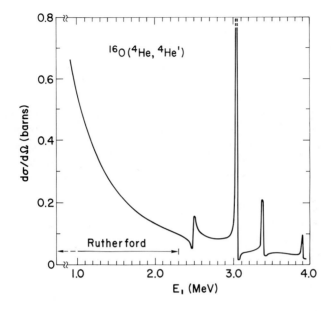

Fig. 3.25 The cross-section for He^+ ion scattering from O at $\theta = 168°$ as a function of energy. Adapted from Chu *et al.* (1978).

b. *Non-Rutherford Scattering*

He ions at 2 MeV, of all possible probing species, have the distinction that there is no resonant scattering from masses other than Be. Although there can be non-Rutherford contributions to the scattering cross-section, such disturbances are insignificant for scattering from masses heavier than Ne. There does remain a serious question as to whether scattering of He^+ off light isotopes such as ^{14}N, ^{15}N and ^{19}F is really Rutherford (Amsel and Davies, 1983). However, detailed measurements of scattering cross-sections for < 2 MeV He^+ scattering from O, Al and heavier targets have shown a close agreement with Rutherford cross-sections (MacDonald *et al.*, 1983). This means that the RBS cross-section, with its smooth energy dependence, can be used directly to derive absolute results from the RBS spectra. At 2.45 MeV, the first of a series of scattering resonances can disturb the spectra. For example, the ^{16}O (α,α) cross-section is shown in Fig. 3.25. To avoid scattering resonances in quantitative analysis, 2.3 MeV has often been chosen as an upper limit for He^+ RBS work. The situation for other probe ion species is not as 'clean', and the use of H^+, Li^+ or B^+ may be attended by nuclear resonances superimposed on the spectra.

A quite different means of selective light element analysis is to make use of narrow non-Rutherford resonances, such as the $^{16}O\,(\alpha,\alpha)$ resonance at 3.01 MeV shown in Fig. 3.25. Quantitative measurements are best achieved using standards since the non-Rutherford cross-sections are not particularly well established and do not have a simple dependence on scattering angle. The procedure for analysis involves comparing the scattering yield from a standard sample of known light element concentration with that from the test sample. Absolute measurements are simple if the light element is confined to a very thin layer such that energy loss of the incident ions is negligible in traversing this layer (or the energy loss is smaller than the width of the resonance). In this case, the incident energy is adjusted to maximise the count rate from resonant scattering events and the absolute light element content, c_T atoms cm^{-2}, in the test sample is obtained from:

$$c_T = Y_r c_s / Y_s \qquad (3.21)$$

where Y_r and Y_s are the light element resonant scattering yields from the test and standard samples and c_s is the known content in the standard sample.

When the light element is distributed more deeply such that energy loss is not negligible then stopping power corrections must be used and Equation (3.21) becomes:

$$c_T' = (Y_T c_s' \varepsilon_T)/(Y_s \varepsilon_s) \qquad (3.22)$$

where c_T' and c_s' are the respective atomic concentrations (atom %) of the light element and ε_T and ε_s are the stopping power factors, Equation (3.8), for test film and standard sample. Light element profiling can also be achieved using non-Rutherford scattering resonances by employing resonance scanning techniques as outlined in Chapters 4 and 12. Useful non-Rutherford resonances for absolute measurement and profiling of O and C are given in Chapter 14.4.

3.5 ELASTIC RECOIL ANALYSIS

3.5.1 Basic Principles

When $M_1 \geqslant M_2$, the incident ions are not scattered in a backward direction and RBS cannot detect such light target constituents. However,

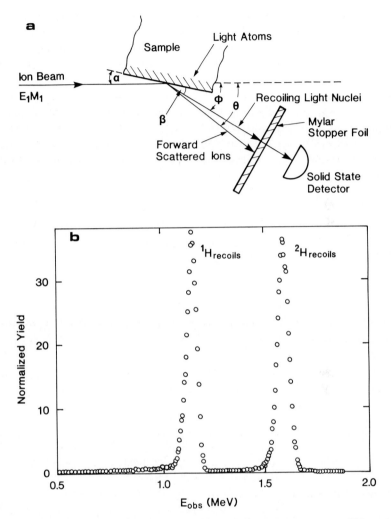

Fig. 3.26 a.The forward recoil geometry. b. The forward recoil spectrum of ^{1}H and ^{2}H (deuterium) from 3.0 MeV ^{4}He$^+$ ions incident on a thin (~200 Å) deuterated polystyrene film on Si. The detector is placed so that the recoil angle $\phi = 30°$ with a 10 μm Mylar film mounted in front of the detector. After Feldman and Mayer (1986).

forward scattering geometries can be used to advantage by detection of the recoiling light atoms which emerge after collisions by the heavier incident ions. The defining kinematic equations have been given in Chapter 1 (Highlight 1.1). The energy of the recoiling atom is:

$$E_4 = K'E_1 = 4M_1M_2E_1 \cos^2 \phi/(M_1 + M_2)^2 \qquad (3.23)$$

where ϕ is the recoil angle with respect to the incident ion direction. Maximum energy transfer occurs when $M_1 = M_2$ and $\phi = 0$. The recoil energy falls to zero as $\phi \to 90°$ (Fig. 1.7b). Chapter 14.3 tabulates recoil and scattering kinematic factors for forward geometries with θ, ϕ between 10° and 30°.

ERA relies on the ability to discriminate between forward scattered incident ions and recoiling light atoms. The typical experimental arrangement is shown in Fig. 3.26a. A Mylar foil is placed in front of the detector to block out the scattered incident ions but allow the lighter atoms which suffer considerably less energy loss to pass through to the detector. The observed energy is given approximately by:

$$E_{obs} = E_4 - f S_f \qquad (3.24)$$

where S_f is the average stopping power in the foil and f is the foil thickness. Typically, glancing geometries of θ, $\phi = 10 - 30°$ are employed to maximise E_4 and f is selected to just stop the scattered incident ions.

A typical recoil spectrum is shown in Fig. 3.26b using 3 MeV ^{4}He ions and θ, $\phi = 30°$ for the detection of ^{1}H and ^{2}H in thin hydrocarbon layers deposited on the surface of a Si sample. For ^{2}H, E_4 is ~ 2 MeV and for ^{1}H about 1.45 MeV (from Equation 3.22). The energy loss of these species in 10 μm Mylar is 300–400 keV (using stopping cross-sections given in Chapter 14.1) giving E_{obs} of ~ 1.15 MeV for ^{1}H recoils and 1.6 MeV for ^{2}H recoils, as shown in Fig. 3.26b. Note that 10 μm Mylar completely stops 2.6 MeV He$^+$ ions thus giving excellent energy discrimination between scattered ions and the various lighter recoiling species. Furthermore, the observed recoiling peaks are sharp since energy straggling in the foil introduces less than 50 keV energy broadening.

3.5.2 Depth Profiling

Depth profiling applications are illustrated in Fig. 3.27. The depth scale can be calculated using the energy loss method outlined in Chapter 12.2. If the surface energy approximation is used, together with an average stopping power S_f for the foil, the depth, t, at which the light ion suffered the recoiling collision can be related to the detected energy, E_{obs}, by:

$$t = (E_{obs} + f S_f - K'E_1)/(S_1/\sin \alpha + S_4/\sin \beta) \qquad (3.25)$$

where S_1 refers to the incident ion trajectory and S_4 to that of the recoiling atom. Normally, $S_4 \ll S_1$ and the depth scale is determined predominantly by energy loss of the incident ion.

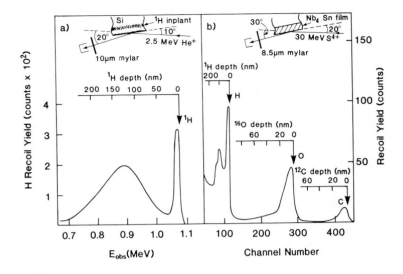

Fig. 3.27 Depth profiling using ERA. a. Profiling of a ^{1}H implant (3×10^{16} cm^{-2} at 15 keV) into Si using 2.5 MeV He$^+$ with $\phi = 20°$. After Williams (1987). b. Simultaneous profiling of ^{12}C, ^{16}O, and ^{1}H in a Nb$_4$Sn film using 30 MeV S^{4+} ions with $\phi = 30°$. Adapted from Nölscher *et al.* (1983).

In Fig. 3.27a, the ^{1}H recoil spectrum is obtained from a Si target implanted with 5 keV ^{1}H to a dose of 3×10^{16} cm^{-2}, using an incident 2.5 MeV He$^+$ beam at $\alpha = 10°$ with $\phi = 20°$. A thin photoresist layer was applied to the surface of the Si and partially removed by solvents to leave a trace surface coverage of a ^{1}H – containing film. The measured profile indicates the expected surface ^{1}H peak and a Gaussian-like implanted ^{1}H profile. Equations (3.23) and (3.24) give the detected surface energy for ^{1}H as 1.07 MeV, as observed in Fig. 3.27a, and Equation (3.25) was used to establish the depth scale.

Fig. 3.27b illustrates the simultaneous detection of several light impurities in a Nb$_4$Sn film using incident 30 MeV S^{4+} ions with $\alpha = 20°$ and $\phi = 30°$. An 8.5 μm Mylar foil stops the scattered S^{4+} ions but allows recoiling ^{1}H, ^{12}C and ^{16}O to pass through. However, the heavier ^{16}O recoils lose considerable energy in penetrating the foil and are detected at a lower energy than the lighter ^{12}C recoils. Again, Equations (3.23) to (3.25) provide surface energies and depth scales. Indeed, ERA using heavy ions can provide profiling of light isotopes up to ^{16}O.

The sensitivity of ERA depends on the experimental arrangement and the system-dependent background levels. Typically, 0.1 atom percent of ^{1}H is observable and from 0.1 to 1 atom percent of heavier atoms. Detection of surface ^{1}H above about 0.1 of a monolayer is also typical. Absolute measurements of light atom concentrations by ERA are best achieved by standards. For example, the ^{1}H implant in Fig. 3.27 (3 $\times$ 10^{16} cm^{-2}) gives a peak ^{1}H concentration of ~1 $\times$ 10^{21} cm^{-3} in Si (or 2 atom percent) and a surface ^{1}H coverage of ~4 $\times$ 10^{15} cm^{-2} (or about 4 monolayers).

REFERENCES

Amsel, G. and Davies, J.A. (1983). *Nucl. Instrum. Methods* **218**, 177.

Andersen, H.H., Tu, K.N. and Ziegler, J.F. (1978). *Nucl. Instrum. Methods* **149**, 247.

Andersen, H.H. and Ziegler, J.F. (1977). *In* 'Stopping Powers and Ranges in all Elements', Pergamon Press, New York, Vols II and III.

Bird, J.R., Duerden, P., Cohen, D.D., Smith, G.B. and Hillery, P. (1983). *Nucl. Instrum. Methods* **218**, 53.

Borgesen, P., Scherzer, B. and Behrisch, R. (1978). *Nucl. Instrum. Methods* **149**, 78.

Brice, D.K. (1973). *Thin Solid Films* **19**, 121.

Butler, J.W. (1986). *Nucl. Instrum. Methods* **B15**, 232.

Chu, W.K., Mayer, J.W. and Nicolet, M.A. (1978). *In* 'Backscattering Spectometry', Academic Press, New York.

Doolittle, L.R. (1985). *Nucl. Instrum. Methods* **B9**, 344.

Edge, B.L. and Bill, U. (1980). *Nucl. Instrum. Methods* **168**, 811.

Feldman, L.C. and Mayer, J.W. (1986). *In* 'Fundamentals of Surface and Thin Film Analysis', North Holland, New York, 61.

Johnston, G.R., Butler, J.W. and Cocking, J. (1984). unpublished.

MacDonald, J.R., Feldman, L.C., Silverman, P.J., Davies, J.A., Griffiths, K., Jackman, T.E., Norton, P.R. and Unertl, W.N. (1983). *Nucl. Instrum. Methods* **218**, 765.

Nölscher, C., Brenner, K., Knauf, R. and Schmidt, W. (1983). *Nucl. Instrum. Methods* **218**, 116.

Petersson, C.S., Baglin, J.E.E., Hammer, W., d'Heurle, F.M., Kuan, I.S., Ohdomari, I., deSousa-Pires, J. and Tove, P. (1979). *J. Appl. Phys.* **50**, 3357.

Poate, J.M., Jacobson, D.C., Williams, J.S., Elliman, R.G. and Boerma, D.O. (1987). *Nucl. Instrum. Methods* **B19/20**, 480.

Short, R., Bond, P., Nygren, E. and Williams, J.S. (1987). *Proc. 5th Australian Conf. on Nucl. Tech. of Analysis,* AINSE, Lucas Heights, 95.

Williams, J.S. (1976). *In* 'Ion Beam Surface Layer Analysis', (O. Meyer, G. Linker and F. Kappeler, eds), Plenum Press, New York, 223.

Williams, J.S. and Möller, W. (1978). *Nucl. Instrum. Methods* **157**, 213.

Ziegler, J.F., Lever, R.F. and Hirvonen, J.K. (1976). *In* 'Ion Beam Surface Layer Analysis', (O. Meyer, G. Linker and F. Kappeler, eds), Plenum Press, New York, 163.

4
Nuclear Reactions

J.R. BIRD

ANSTO Lucas Heights Research Laboratories
Menai, Australia

ION BEAMS FOR
MATERIALS ANALYSIS
ISBN 0 12 099740 1

The study of nuclear reactions, which began in the 1930s, quickly brought a realisation that very pure sample materials, and even separated isotopes, were needed to avoid the effects of interfering nuclear reactions. On many occasions, nuclear physicists were thus engaged in the analysis of their own target materials but it was not until 1957 that Rubin *et al.* (1957) published a paper on "Chemical Analysis of Surfaces by Nuclear Methods" which paved the way for the use of accelerated ion beams for Prompt Nuclear Analysis (PNA) of samples. From 1960 onwards, the use of nuclear reactions in prompt analysis began to grow rapidly. However, this is one of the most complex of the ion beam methods. Every isotope can undergo a variety of nuclear reactions, each having unique characteristics such as energy release, excited state energies, cross-sections and angular distributions. Applications to specific analytical problems require assessment of all these factors for the isotope of interest and for possible competing reactions in other isotopes present in a sample.

    It is convenient to consider reactions according to the type of product radiation since this determines the methods and equipment required. Four major categories are outlined in Highlight 4.1, viz. ion–ion,

ion–gamma, ion–neutron and activation analysis. The important features of Nuclear Reaction Analysis are:

- high selectivity in the determination of specific light nuclides;
- good sensitivity for many nuclides which are difficult to determine by other techniques;
- unique capabilities for non-destructive depth profiling of specific light nuclides, and
- accurate absolute determination of many nuclides.

Much is known about the nuclear physics of useful reactions but the knowledge of available reactions is still expanding — particularly in the field of heavy ion induced reactions. Applications of PNA have been recently reviewed by Peisach (1981), Deconninck (1978), Deconninck *et al.* (1981) and Borderie (1980). A bibliography of applications reported up to 1976 has been published (Bird *et al.*, 1978) and much information is contained in the proceedings of International Conferences on Ion Beam Analysis (see the table in the introduction to this book).

HIGHLIGHT 4.1
ENERGY RELATIONS IN NUCLEAR REACTIONS

Energy level diagrams are used to illustrate the energy relations in nuclear reactions. Typical examples involving proton or deuteron irradiation of a sample containing fluorine atoms are shown in Fig. 4.1. If a proton merges with the nucleus of a ^{19}F atom, a compound nucleus (^{20}Ne) is formed and an energy of 12.485 MeV is released. This is illustrated by placing the ^{19}F energy level diagram above the ^{20}Ne diagram by a distance proportional to the energy release (Q). The ^{20}Ne nucleus is formed in a highly excited state and may decay in a number of ways. Full details of possible reactions can be obtained from the references in Table 4.1. The following examples illustrate the use of energy level diagrams to display the energy relations involved in various types of nuclear reaction.

a. Ion-Gamma Reactions, e.g. ^{19}F(p,γ)^{20}Ne; Q = 12.845 MeV.
Suppose that the incident proton had an energy of 0.872 MeV. The ^{20}Ne nucleus is then formed with an excitation energy of 13.717 MeV ($= E_1 + Q$) (arrow (1) in Fig. 4.1). An excited nucleus may emit one or more gamma-rays as shown by wavy lines linking the initial state and low lying excited states within the ^{20}Ne level diagram. After the full excitation energy has been emitted as gamma-rays, a stable nucleus of ^{20}Ne remains,

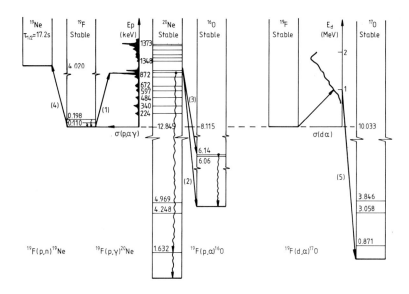

Fig. 4.1 Energy levels and cross-sections in nuclear reactions.

i.e. F has been transmuted to Ne. The gamma-ray energy spectrum has peaks at energies which are characteristic of the energy levels involved in this reaction and the peak intensities can be used to determine the concentration of fluorine in the sample. Gamma-ray emission may also accompany other reaction types. For example in (p, α) reactions, ^{16}O can be formed in an excited state which decays by gamma-ray emission to the ground state as shown by the wavy lines within the ^{16}O level diagram. Gamma-rays at 6 to 7 MeV characterise this reaction and can also be used for F determination. Note that the cross-section curve in Fig. 4.1. is for the (p, $\alpha\gamma$) reaction and not the (p, γ) reaction although the 872 keV resonance occurs in both reactions.

Coulomb Excitation, e.g. ^{19}F(p,p' γ) ^{19}F, is an important source of gamma-rays. For example, the first and second excited state of ^{19}F at 110 and 198 keV are readily excited by a passing proton and they decay by emission of gamma-rays of these energies. Once again these are a clear signature for the presence of fluorine in the sample.

b. Ion-Ion Reactions, e.g. ^{19}F(p,α) ^{16}O Q = 8.115 MeV.
If the ^{20}Ne compound nucleus decays by the emission of an alpha-particle, a nucleus of ^{16}O is formed (arrow (2) in Fig. 4.1). The energy of the emitted alpha-particle depends on the difference in energy of the

states in ^{20}Ne and ^{16}O (see section 4.1.1). In Fig. 4.1 arrow (2) shows the formation of the ground state of ^{16}O and arrow (3) shows the formation of an excited state. Each state contributes a separate energy group to the alpha-particle spectrum and any one of these can be used for the determination of ^{19}F in thin layers at the surface of the sample. Each group is designated by the sequence number of the excited state formed. Thus arrow (2) is the (p,α_0) reaction and arrow (3) is the (p,a$_2$) reaction.

Resonant Scattering, e.g. ^{19}F(p,p) ^{19}F Q = 0.
Proton emission by the compound nucleus can also occur and this leads back to the ^{19}F nucleus. This process is indistinguishable from Rutherford scattering except that the cross-section will now show features such as resonances which are typical of nuclear reactions. Resonant scattering is dominant for incident energies near and above the Coulomb barrier (Equation (1.9), Table 1.1) and can be significant at even lower energies. For example, there is a well-known resonance at 3.045 MeV in alpha-particle scattering from ^{16}O and this is commonly observed in backscattering experiments. Resonance scattering is discussed further in Chapter 3.

c. Ion-Neutron Reactions, e.g. ^{19}F(p,n)^{19}Ne Q = -4.020 MeV
The Q-value for this reaction is negative and high energy protons must be used before the reaction becomes energetically possible (see Section 4.1.2a). The position of the energy level diagram for ^{19}Ne is shown in Fig. 4.1 and arrow (4) shows the relative energy involved in the (p, n) reaction.

d. Particle Induced Activation Analysis (PAA), e.g. ^{19}F(p,n) ^{19}Ne $\beta^+_{\rightarrow}$^{19}F.
The nuclei formed in a and b above are stable but ^{19}Ne is radioactive with a half-life of 17.2 s. After sample irradiation has been completed, the beta radioactivity or any associated gamma-ray emission can be observed in the sample and can be used for sample analysis in the same manner as for the well known technique of Neutron Activation Analysis (NAA). Radioactive gamma-rays are sometimes observed during PNA measurements but usually with relatively low intensities.

4.1 PRINCIPLES

Nuclear reactions are governed by the general equations for energy kinematics and yields are given in Chapters 1 and 12. However, the application of these equations depends on information on energy levels, observed particle and gamma-ray groups and cross-sections which is given in nuclear physics publications (Table 4.1). A catalog of nuclear

TABLE 4.1
References to nuclear reaction data

Reaction Q-Values	
Gove, N.B. and Wapstra, A.H. (1972). Nucl. Data Tables **11**, 127.	

Energy Levels

A = 3	Nucl. Phys. A251 (1975)
A = 4	Nucl. Phys. A206 (1973)
A = 5–10	Nucl. Phys. A413 (1984)
A = 11–12	Nucl. Phys. A433 (1985)
A = 13–15	Nucl. Phys. A360 (1981)
A = 16–17	Nucl. Phys. A375 (1982)
A = 18–20	Nucl. Phys. A392 (1983)
A = 21–44	Nucl. Phys. A310 (1978)

Other Elements
Nuclear Data Sheets (Academic Press, San Diego)
Nuclear Level Schemes (1973), Ed., Nuclear Data Group, ORNL,
   Academic Press, New York.

Cross-sections
Bird, J.R. (1980). *Nucl. Instrum. Methods* **168**, 85.
Feldman, L.C. and Picraux, S.T. (1977). 'Ion Beam Handbook', eds, Mayer, J.W. and
   Rimini, E., Academic Press, New York, 112.
Jarjis, R.A. (1979). Nuclear Cross-section Data for Surface Analysis, Dept. Phys.,
   U. Manchester.

reactions and experimental parameters which are useful in analytical applications is given in Chapter 14.4.1.

4.1.1 Kinematics

The kinematics of nuclear reactions are dominated by the magnitude of the energy balance Q (Equation (1.21), Table 1.3). Most useful reactions have large positive values of Q (Table 4.2) and this results in relatively high energy product radiation. Those reactions which have the highest Q-values give products which are easy to observe as having the highest energies. Low Q-value reactions may well be masked by competing reactions. More extensive lists of Q-values can be found in Gove and Wapstra (1972). If the final nucleus is formed in an excited state of energy E', then a modified energy balance ($Q' = Q - E'$) must be used in kinematics calculations.

a. Thresholds

If Q is negative, this energy must be provided by the incident ion before a reaction is possible. In order to meet the additional requirements of

TABLE 4.2
Q-Values of nuclear reactions (keV)

Nuclide	p,α	d,p	d,α	³He,p	³He,α	α,p	p,n	d,n	t,n	³He,n	α,n
^{1}H											
^{2}H		4033	23848					3269			
^{3}He		18353	18354					−5153			
^{4}He		−3115						−4185			
^{6}Li	4021	5026	22374				−5070	3384			
^{7}Li	17347	−192	14230				−1644	15028			
^{9}Be	2125	4587	7151	10323	18913	−6886	−1850	4361	9559	7558	5702
^{10}B	1146	9231	17822	19695	12143	4063	−4433	6467	18930	1569	1060
^{11}B	8590	1145	8031	13185	9123	784	−2765	13733	12421	10182	157
^{12}C	−7551	2722	−1341	4779	1857	−4966		−281	4015	−1149	−8508
^{13}C	−4063	5952	5169	10666	15632	−7423	−3003	5326	9902	7124	2215
^{14}N	−2922	8609	13575	15243	10025	−1193	−5927	5067	14479	−969	−4735
^{15}N	4966	265	7688	8552	9745	−3980	−3542	9903	7788	5010	−6418
^{16}O	−5218	1918	3111	2033	4909	−8115		−1624	1269	−3196	588
^{17}O	1193	5822	9802	8321	16436	−5656	−3542	3384	7557	4301	−698
^{18}O	3980	1732	4245	6876	12532	−5601	−2438	5768	6112	13119	−1950
^{19}F	8115	4377	10033	11888	10148	1675	−4020	10620	11124	7557	−7216
^{20}Ne	−4129	4536	2797	5784	3713	−2377		206	5020	202	2553
^{21}Ne	−1738	8141	6466	11440	13817	−2179	−4331	4516	10677	6601	−480
^{22}Ne	−1675	2972	2702	8034	10213	−3533	−3625	6568	7270	12766	−2967
^{23}Na	2377	4735	6913	11305	8161	1820	−4839	9467	10541	6243	−7193
^{24}Mg	−6884	5107	1963	5920	4048	−1601		46	5156	76	2653
^{25}Mg	−3144	8869	7049	11646	13247	−1207	−5062	4082	10882	6054	33
^{26}Mg	−1820	4218	2915	8278	9485	−2865	−4787	6046	7514	12138	−2638
^{27}Al	1600	5501	6708	12341	7521	2372	−5592	9361	11577	6615	−8140
^{28}Si	−7715	6249	1429	6355	3401	−1916		523	5592	−566	−1526
^{29}Si	−4820	8385	6013	10189	12105	−2453	−5726	3375	9425	3965	−3492
^{30}Si	−2372	4364	3129	7515	9969	−2959	−5010	5073	6752	8443	−5648
^{31}P	1916	5712	8165	9789	3370	625	−6224	6639	9025	3423	−8612
^{32}S	−4200	6419	4901	6067	5490	−1866		53	5303	−759	

conservation of momentum, the threshold ion energy required to make a nuclear reaction possible is given by Equation (1.27). Examples of useful threshold reactions include Coulomb excitation and (p, n) reactions.

4.1.2 Reaction Yields

The yield of nuclear reaction products is dependent on the reaction cross-section (Chapters 1.1.1d and 12.1). Each reaction cross-section has its own characteristic dependence on energy and angle of observation (see Highlight 4.2). Nuclear physics models can be used to calculate many cross-sections but for analytical work, measured values are used. Tabulations are available in the references of Table 4.1 and cross-sections for the most useful reactions are presented in Chapter 14.4.

HIGHLIGHT 4.2
NUCLEAR REACTION CROSS-SECTIONS

---

a. Resonance Reactions

Resonances in nuclear reaction cross-sections arise from the effects of excited states in the compound nucleus. This is illustrated in Fig. 4.1 for proton irradiation of fluorine. A cross-section curve for the (p, $\alpha\gamma$) reaction, as a function of incident proton energy (plotted vertically), is shown between the energy level diagrams for ^{19}F and ^{20}Ne. A number of peaks (resonances) are observed in the cross-section and each occurs at an energy which corresponds to an excited state of the ^{20}Ne nucleus. The cross-section at a specific energy may involve the sum of a number of resonance terms, a smooth component and, if necessary, interference terms.

An isolated resonance can be exploited for the analysis of a very thin layer of a sample and for depth profiling by the resonance scanning method (Chapter 12.2). A typical example is the 0.872 MeV resonance in (^{19}F + p) reactions which produces a high yield of alpha-particles and gamma-rays from a thin layer in the sample for which the incident ion energy is close to the peak of the resonance. When the incident energy is high, many resonances will contribute to the yield from a thick sample and an averaged composition of the sample can then be derived. However, each resonance has a different peak cross-section and hence sensitivity. The measured total yield is the sum of unequal contributions from a set of isolated layers at which the incident ions reach the energy of each resonance in turn.

Resonances offer a number of special advantages in materials analysis, for example:

- simpler spectra may be observed than for a thick sample;
- for a thin layer, the probability for the desired reaction to occur can be maximised by choosing the incident energy to coincide with a cross-section peak; most of the other potential competing reactions will then have low probabilities unless they chance to have a resonance at the same energy — or at any lower energy if a thick sample is used;
- it is possible to minimise contributions from surface contamination by using an incident energy above the resonance energy which would then only give high yield from a layer within the sample (see Chapter 12.2); and
- in depth profiling, a very narrow resonance will give excellent depth resolution (e.g. 10 nm or better).

b. Direct Reactions

A different kind of cross-section curve is observed for reactions such as (d,p) (d,α) and (d,n) reactions (see curve to the left of the ^{17}O level diagram in Fig. 4.1). This shows that the cross-section for the ^{19}F(d,α) ^{17}O reaction, which has a Q-value of 10.033 MeV, changes smoothly as the incident deuteron energy increases. Such cases are useful in thick sample analysis and for depth profiling by the energy spectrum method (Chapter 12.2).

c. Coulomb Excitation

Another type of reaction with a smooth dependence of cross-section on incident ion energy is Coulomb excitation. The cross-section has a separate threshold for each excited state and increases rapidly above that energy. For incident energies well above threshold, the cross-section is related to the Rutherford scattering cross-section ($d\sigma_R$) by:

$$d\sigma = P \, d\sigma_R \qquad (4.1)$$

where P is the probability for excitation of a low-lying level in the target nucleus. The differential cross-section term introduces a dependence on $(Z_1 Z_2 / 4 E_1)^2$. Coulomb excitation cross-sections are larger than other reaction cross-sections for incident energies below the Coulomb barrier but at higher energies this is no longer true. The most common analytical application of Coulomb excitation involves the observation of de-excitation gamma-rays.

The differential cross-section ($d\sigma/d\theta$) is usually given as mb/sr for a specific angle (θ) relative to the incident ion direction. Cross-sections at other angles are related by the angular distribution function:

$$d\sigma/d\theta = f(E_1, \theta)\, \sigma(E_1) \qquad (4.2)$$

where $\sigma(E_1)$ is the total cross-section for the reaction. Reactions involving compound nucleus formation usually have angular distributions which are symmetric about 90° but which change from resonance to resonance. Direct reactions are strongly forward peaked.

If the detector solid angle is large compared with the rate of change of differential cross-section with angle, the observed yield involves an integral over angle. In the irradiation of thick samples, all energies from the incident energy down to zero are involved leading to averaging of angular distributions which usually change with incident energy. There is often a lack of published information for the specific energies and geometries chosen for analytical applications and a calibration using a standard sample is necessary.

Reaction yields are related to sample composition as described in Chapter 12.1. In the case of thin samples for which energy loss by the incident ion is not sufficient to change the cross-section or stopping power significantly, the relation is very simple (Equations (12.7) and (12.8)). This has similar advantages to the use of a narrow resonance (see Highlight 4.2) but there are also significant disadvantages. The yield of product radiation is usually very low and sample preparation requires great care, an accurate determination of sample thickness and, possibly, the use of a pure backing material which does not give competing reactions. Because of these disadvantages, thin sample measurements have only been used when the original material is a thin layer on a thin or inert backing (e.g. air filters, surface layers, corrosion products) or when the effort required to prepare thin samples and measure their thickness is worthwhile.

For thick uniform samples, the need to use integral yield equations (Chapter 12.1) is a significant complication, since the major element composition must be known in order to calculate the stopping power as a function of energy. This presents no difficulty for the determination of minor or trace elements in a known matrix but must be taken into account for more general analytical problems. Approximations which can usually be used to obtain results with sufficient accuracy are discussed in Chapter 12.1. In nuclear reaction analysis, it is customary to use standard samples to establish the cross-section or reaction yield for specific experimental conditions and then to use stopping power corrections to measurements with unknown samples (e.g. Equation 12.18). If the samples are similar in composition, the corrections are small and nuclear reaction methods are then suitable for absolute measurements requiring only the determination of yield and incident beam dose. No attenuation correction is required.

4.1.3 Depth Profiling

The variety of options available for non-destructive depth profiling is one of the most important features of prompt nuclear analysis. The principles used are described in Chapter 12.2. Their use in conjunction with nuclear reactions exploits the following features:

i. the high peak cross-section of a resonance gives high sensitivity;
ii. energy loss by the incident ion can be used to determine depth profiles by resonance scanning or yield curve unfolding but the maximum depth is limited by resonance spacing in the former case;
iii. energy loss by both incident and product ions can be used with ion–ion reactions for energy spectrum profiling; in this case the maximum depth is limited by the spacing of different product energy groups;
iv. thresholds in (p,n) reactions also provide a method for depth profiling with good resolution; and
v. glancing angle techniques and the effects of topography are important (Chapter 12.2).

4.2 ION–ION REACTIONS

Ion-induced nuclear reactions are inhibited by the Coulomb barrier which, according to Equation (1.9), increases with the atomic number of both the incident ion and target nuclide. Ion emission from an excited nucleus is further inhibited (relative to gamma-ray or neutron emission) because of the Coulomb barrier. At energies up to a few MeV, ion–ion reactions are consequently most probable for light ion irradiation of light nuclides. Reactions used for analysis and depth profiling are listed in Chapter 14.4.1. Features such as non-Rutherford (resonant) scattering and product nucleus recoil add to the choice of ion–ion methods of analysis.

4.2.1 Equipment

The equipment required for ion–ion analysis is similar to that for backscattering and is shown schematically in Fig. 4.2. If two detectors are used, both RBS and ion–ion reactions can be studied with the same equipment.

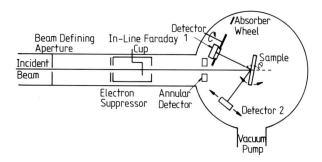

Fig. 4.2 Schematic layout of ion–ion analysis equipment.

a. Choice of Detector

A solid state detector is customarily used — or multiple detectors if higher count rates are required. However, a high count rate from scattered ions is observed at low energies unless a foil is placed in front of the detector. The thickness of the foil should be equal to the range of the scattered ions so that these are absorbed while the higher energy reaction products pass through. Mylar, which has a composition $H_{10}C_8O_4$, and thin metal foils, are commonly used for this purpose (see Chapter 14.4.1). It is important to measure the thicknesses and not just use the rated thickness.

The major disadvantage of the absorber foil technique, is that energy loss straggling occurs in the foil resulting in poor energy resolution in the measured spectrum. For example, a detector resolution which is typically 10 to 15 keV for a Si detector, may be degraded to 50 to 100 keV. This is not important if there is sufficient energy difference between ion groups from different reactions or different target nuclides but is a major limitation if depth profiles are to be measured. The problem of poor energy resolution is exacerbated by any inhomogeneities in thickness of the absorber foil. Variations of at least a few percent in absorber thickness must be expected and included in quadrature with other contributions to energy resolution.

If an absorber foil is mounted on a movable support, it can be removed when required so that backscattering measurements can be made with the same detector. This is a convenient way to calibrate the detection efficiency and solid angle using the absolute value of the backscattering cross-section. Angle adjustment of the sample and detectors from outside the chamber is needed for versatile operation.

High detection efficiency can be achieved by placing one or more

TABLE 4.3
Performance of ion detection methods

Detector	Reject Scattering	Sensitivity	Energy Resolution (keV)	Depth Resolution
Gas ionisation	no	good	1%	very poor
Ion identification	yes	good	1%	very poor
Surface barrier	no	good	10	medium
Low detector bias	partial	good	10	medium
Position sensitive	yes	medium	10	medium
Absorber foil	yes	good	>25	poor
Graded foil	yes	excellent	200	poor
Convergent beam	no	excellent	20	medium
Magnetic analyser	yes	poor	<10	excellent
Electrostatic analyser	yes	poor	<10	excellent

large surface barrier detectors close to the sample or by the use of an annular detector (see Fig. 4.2). A solid angle of 1 sr is possible but 10 to 100 msr and even lower values may be necessary to achieve the best energy resolution. Some typical count rates from commonly used reactions are included in Chapter 14.4.1.

b. Alternative Detectors

A comparison of various alternative detection techniques is given in Table 4.3. Solid state detectors suffer deterioration in resolution due to radiation damage especially when they are used to detect heavy ions. This can be avoided by the use of gas ionisation detectors (England, 1974) but they are larger and more complex than Si detectors. Either type can be used in detector 'telescopes' which use the rate of energy loss to determine the atomic number of each detected ion.

Magnetic analysers (see Chapter 2) can be used both to separate reaction products from scattered ions and to achieve better energy resolution than can be obtained with a solid state detector. However, solid angle and detection efficiency are low and the good energy resolution is only useful for reactions occurring near the sample surface. At depths beyond 10 to 20 nm, the depth resolution worsens because of the effects of energy loss straggling and a magnetic analyser no longer has any advantage. It is also possible to use an electrostatic analyser for the analysis of product ion energies but similar problems arise. The separation of scattered and product ions is discussed further in Section 4.2.2d. Time-of-fight techniques have been used in Rutherford Back-scattering and should also be useful for ion–ion reactions. Examples of

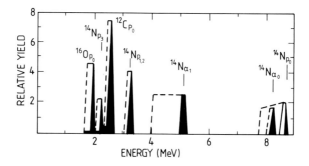

Fig. 4.3 Schematic energy spectrum from 972 keV deuteron irradiation of thin (full curve) and thicker (dashed curve) layers of C, N and O. Peaks labelled with the target nucleus and product particle (see text).

the use of different detection techniques to solve specific problems are discussed in the next section.

A typical spectrum observed in ion-ion reaction studies is shown schematically in Fig. 4.3 for the case of a very thin layer containing C, N and O which is irradiated with 972 keV deuterons. This energy is recommended for simultaneous analysis of these elements using (d,p) and (d,α) reactions (Davies *et al.*, 1983). Higher energies are preferable for greater sensitivity in N determination if this is the main interest. Each energy group in Fig. 4.3 is labelled with the nuclide and product particle involved. If a sample is used for which the product ion energy loss is greater than the detector resolution, each peak in Fig. 4.3 will be extended to lower energies as illustrated by the dashed curves. If the sample is too thick, peaks from reactions in different isotopes may overlap and analysis then becomes difficult.

4.2.2 Choice of Experimental Conditions

Analysis using ion-ion reactions involves the choice of reaction, suitable values of incident energy, reaction angle, detector angle and resolution, sample angle and thickness and various special geometries.

a. Reaction Angle

The change of product particle energy with reaction angle (θ) depends on the Q-value and atomic masses involved and is different for different reactions and for different excited state groups. This is illustrated in Fig. 4.4 for the reactions shown in Fig. 4.3. The dashed curves indicate the

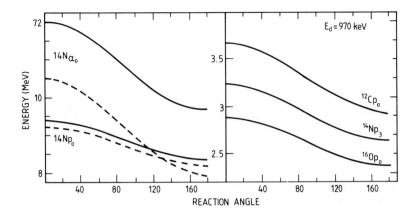

Fig. 4.4 The dependence of product particle energy on reaction angle for the reactions shown in Fig. 4.3. The dashed curves show the effects of an absorber used to remove scattered ions.

effect of an absorber foil on the energy of the high energy groups. There are three immediate consequences:

    i. the detector angle can often be selected so as to achieve optimum separation of an energy group from a required isotope from groups arising from competing reactions;

    ii. the change in energy over the solid angle subtended by the detector may be sufficient to degrade the observed energy resolution; and

    iii. variations in the ion path length occur within the sample and absorber introducing an additional energy spread.

An annular Si detector can be used (Fig. 4.5a) to give a larger solid angle and hence sensitivity (without increasing the energy spread) by detecting all ions emitted in a cone having the required reaction angle. However, it can only be used for reaction angles close to 180°. Further increases in solid angle can be achieved if the thickness of the absorber foil is graded according to radius to compensate for the changes in product ion energy (Fig. 4.5b). The use of such a system with the $^{18}O(p,\alpha)^{15}N$ reaction is shown in Fig. 4.5c. In this case, the alpha-particle peak width for a solid angle of 0.42 sr was improved from 250 to 200 keV (Lightowlers *et al.*, 1973). This is still inadequate for good depth profiling but was found useful for obtaining concentrations of surface or sub-surface O and B in semiconductors.

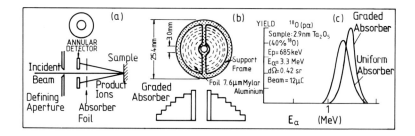

Fig. 4.5a. Annular detector arrangement; b. special purpose graded absorber; c. ^{18}O measurements with a graded absorber.

b. Sample Angle

A very important aspect of experimental configuration is the use of glancing angle techniques to enhance depth resolution. In reaction studies, the depth resolution is dependent on the rate of energy loss by both incident and product ions and is dominated by whichever has the higher Z. Thus, in the ^{16}O(^{6}Li,p) reaction the incident ion has the highest stopping power and the best depth resolution is achieved by placing the sample so as to have a glancing angle of incidence (Fig. 4.6a). On the other hand, in the ^{18}O(p,α) reaction, the stopping power of the product ion is greatest and a glancing emergence angle is preferable (Fig. 4.6b). This is discussed further in Section 4.2.5. The opposite choice should be made to minimise depth effects in sample analysis.

c. Special Geometries

If the beam is convergent when it strikes the sample, it is possible to arrange the sample and detector so that different ion paths involve the same reaction angle (Fig. 4.6c). Falk *et al.* (1976) showed that an improvement in energy resolution by a factor of three could be achieved using this method for the determination of C with the ^{12}C(p,α)^{9}B reaction. A position sensitive detector was used to allow corrections to be applied for changes in energy with reaction angle across the diameter of the detector. In this case, the sample must be thin enough to transmit the product ions with little energy loss.

d. Ion Separation

One method for removing scattered ions is to use a small magnetic analyser (Fig. 4.7a). A magnetic field of a few hundred mT applied over a

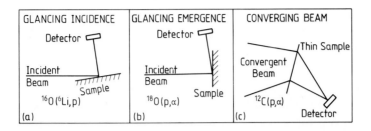

Fig. 4.6a. Glancing incidence geometry for heavy ion irradiation; b. glancing emergence for heavy product ions; c. special geometry to improve energy resolution with a large sample.

distance of 5 to 10 cm will cause enough spatial separation between scattered protons and alpha-particles from (p,α) reactions in ^{11}B, ^{18}O and ^{19}F. An aperture in front of the magnet limits the acceptance solid angle and a carefully placed aperture after the magnet will transmit only reaction products and a few multiple-scattered incident ions. Giles *et al.* (1977) used a position sensitive detector to help distinguish product ions from scattered ions. This geometry involves a reduction in acceptance solid angle by the order of a factor of two which can often be offset by an increased beam current.

A small electrostatic analyser can also be used. For example, plates 174 mm long, placed as shown in Fig. 4.7b with a potential difference of 10 kV, will separate scattered ions from reaction products from light isotopes in the sample (Moller *et al.*, 1977). Some results obtained with this technique are discussed in Section 4.2.5e.

e. Ion Identification

Ions of different mass and energy can be identified by rate of energy loss, or range (Goulding and Harvey, 1975). For example, alpha-particles can be detected in the presence of scattered protons by reducing the bias on a Si detector until the depletion depth is 100 μm or less. Scattered protons pass right through such a thin sensitive layer and give pulses corresponding to the fraction of their energy which is lost in the layer. Alpha-particles of similar energy have a range less than the layer thickness and hence give full energy pulses.

Ligeon *et al.* (1973) used this method with the ^{31}P(p,α)^{28}Si reaction to profile P in Si. The Q-value of this reaction is only 1.917 MeV and 3 MeV alpha-particles are produced for an incident energy of 1.892 MeV. The range of these alphas is much too low to be able to use an absorber foil to remove scattered protons. Pulse-height spectra from a 10 μm thick detector gives an isolated alpha peak (Fig. 4.8a) which is suitable for P

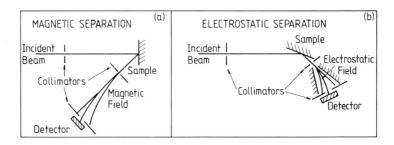

Fig. 4.7 Schematic arrangement for: a. magnetic, and b. electrostatic separation of scattered and reaction product ions.

determination in Si at levels down to the order of 10^{14} atoms cm^{-2}. With a 300 μm detector, pulses from scattered protons interfere with the alpha-particle peak.

A more sophisticated particle identification system consists of two detectors which can be either ionisation chambers or Si detectors (Fig. 4.8b). The first detector is made thin enough to give a signal for each ion which is proportional to energy loss (ΔE) and to completely absorb short range particles while transmitting long range particles to the second detector. For example, Thomas *et al.* (1975) used a 50 μm Si detector followed by a "thick" Si detector to detect C at the surface of a boron layer. The first detector is just thick enough to absorb products of (d,p) and (d,α) reactions in B while transmitting the protons from the ^{12}C(d,p) reaction. The full energy of the latter is obtained by summing the signals from both detectors and this allows normal resolution to be achieved (in this case 40 keV) rather than have the resolution degraded by at least a factor of two which would occur if a passive absorber foil was used.

f. Dual Ion Coincidence Detection

In suitable cases, it is possible to detect recoil atoms simultaneously with the detection of the light product ions. For reactions in light nuclides, there is no clear distinction between the two particles emerging from an ion–ion reaction. As an example, let us consider the ^{6}Li(p,^{3}He)^{4}He reaction at $E_1 = 1.9$ MeV. The Q-value is 4.021 MeV and if the ^{3}He ion is observed at 110° ($E_3 = 2.709$ MeV), then the ^{4}He ion must recoil at an angle of 48.1° ($E_4 = 3.212$ MeV) in order to satisfy the kinematic relation (Equation (1.21), Table 1.3). Two detectors can be placed as shown in Fig. 4.9a and typical spectra (from the forward detector) are shown in

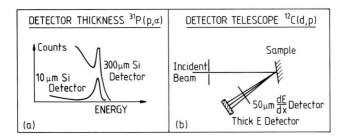

Fig. 4.8a. Schematic energy spectra from thin and thick solid state detectors for $^{31}P(p,\alpha)$ studies; b. Schematic layout of thin and thick detector telescope for product ion identification.

Fig. 4.9b. The top curve is the spectrum with no coincidence requirement and the bottom curve shows the dramatic simplification achieved by demanding that a particle is also observed in coincidence in the second detector (Pretorius and Peisach, 1978). The main limitation to such measurements is that the sample must be thin enough to transmit both product particles. Nuclides for which suitable reactions are available are listed in Table 4.4 (Coetzee *et al.*, 1975).

The coincidence circuit must use a time delay and resolution which take into account the difference in velocities of the two particles being sought. Energy windows can be used to further restrict the type of event which can give a coincidence. The sum of the pulses in both detectors will give a single peak (Fig. 4.9c) which is quite narrow, even if the acceptance

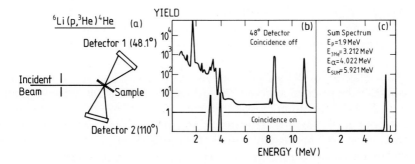

Fig. 4.9 Coincidence detection of charged particles from the $^{6}Li(p, {}^{3}He)^{4}He$ reaction in a thin sample. a. Detector geometry; b. pulse height spectra with coincidence off (top) or on (bottom); c. sum spectrum.

TABLE 4.4
Light element reactions for coincidence measurements

Reaction	p,d	d,p	d,α	³He,p	³He,d	³He,α	α,p	α,d
Nuclide								
²D		X						
³He		X						
⁶Li		X	X		X			
⁷Li	X			X		X		
⁹Be	X	X	X	X	X			
¹⁰B	X	X		X	X			X
¹¹B		X	X	X	X	X	X	

solid angle of the detectors is sufficiently large for kinematic broadening to be significant. This follows from the fact that:

$$E_1 + Q = E_3 + E_4 - \Delta E \qquad (4.3)$$

where ΔE is a constant sum of energy losses by the two particles in the sample material, irrespective of the depth at which the reaction occurs. The coincidence technique cannot be used if one of the products has a short half-life and decays before reaching the detector. For example, ^{8}Be, which is produced in a number of light element reactions, has a half-life of approximately 3×10^{-16}s and breaks up into two alpha-particles. These ions are responsible for the continuum count-rate at low energies shown in Fig. 4.9b but they do not satisfy the coincidence requirement and do not contribute to the final spectra.

4.2.3 Thin Sample Analysis

a. Narrow Resonance Reactions

As an example of the use of a narrow resonance to analyse thin samples, consider the resonance in the ^{19}F(p,α) reaction at 1348 keV (see Chapter 14.4.2). The maximum cross-section at 150° is 3.2 mb sr^{-1} and this gives 2.4 counts/μC using a detector solid angle of 0.12 sr and a sample containing 10^{15} atoms cm^{-2} of F (Dieumegard *et al.*, 1980). Of other elemental constituents, only Li can give rise to alpha-particles of the same or higher energy and F determination is thus almost interference free. The F content of dental enamel, anodic oxide layers and other materials can be measured in this way with a sensitivity better than 10^{13} atoms cm^{-2}.

b. Layer Thickness Determination

It was shown in Fig. 4.3 that a broad peak with an almost flat top is observed from the $^{16}O(d,\alpha)^{14}N$ reaction in a thick oxide layer. For an incident deuteron energy of 900 keV, 2.63 MeV alpha-particles are produced at 145° and their energy (E_3''), after emerging from the sample is related to the depth (t nm) at which the reaction occurred by Equation (12.35). Turos *et al.* (1973a) showed that the relation is approximately linear to depths of 1 μm and, for reactions in SiO_2:

$$t = 2.91 \, (2.63 - E_3'') \qquad (4.4)$$

Results for SiO_2 layers on Si were found to agree with measurements by ellipsometry to within 2% using this approach. Glancing angle geometry can be used to improve the precision of thickness determinations but it is limited by the relatively poor energy resolution available in most ion–ion reaction studies.

If the energy loss in the layer is much less than the width of the resolution function, the peak will not be flat topped and depth information is resolution limited. In this case an estimate of thickness (t .10^{-24} atoms cm^{-2}) can be obtained from the peak area (A_3):

$$t = A_3/d\sigma \, d\Omega \qquad (4.5)$$

If a narrow resonance is used, the differential cross-section can be replaced by the resonance area (A_r) which is given in nuclear physics tables as proportional to $(2J + 1) \, \Gamma_a \, \Gamma_b/\Gamma$ where J is the angular momentum quantum number, Γ_a and Γ_b are partial widths describing scattering and reactions respectively and Γ is the width of the resonance. Relative values of resonance areas are listed in Chapter 14.4.2. Observed cross-sections are not zero between resonances so that small corrections are needed when using resonance areas from nuclear physics tables (which are integrated from $-\infty$ to $+\infty$) with experimental peak areas taken over a finite energy range. The simplest approach is to obtain a value of the product of effective resonance area and detector solid angle by calibration with a standard sample.

4.2.4 Bulk Analysis

Bulk composition can be derived from ion–ion measurements on the assumption that the sample is homogeneous. This can be illustrated by the determination of nitrogen in steels using the combined counts from the (d,p_0) and (d,α_0) reactions (see Fig. 4.3). The cross-sections for these reactions increase rapidly with energy but this is also true for other competing reactions. Using 1.9 MeV deuterons, N can be determined at mg g^{-1} levels by counting protons and alphas above 8 MeV from standard and unknown samples (Olivier *et al.*, 1975). Greater sensitivity can be achieved by exploiting the higher count rate from these reactions at forward angles (e.g. 45°). Even with deuterons of only 1.2 MeV, N can then be determined below 100 μg g^{-1} provided that boron is not present at much higher levels (Olivier *et al.*, 1976).

4.2.5 Depth Profiling

The principles used in depth profiling with ion–ion reactions are described in Chapter 12.2. Reactions used, typical experimental parameters and performance figures are summarised in Chapter 14.4. Examples are given in this Section to illustrate the most important features of ion–ion profiling. The resonance scanning method which is illustrated in Section 4.3.2 is of only limited use with ion–ion reactions because narrow resonances are unusual in their cross-sections. Examples include the $^{18}O(p,\alpha)^{15}N$ reaction with a narrow resonance at 629 keV and the $^{3}He(d,p)^{4}He$ reaction with a broad asymmetric resonance at 430 keV. In the latter case, the resonance shape must be taken into account when deriving the depth profile (Pronko and Pronko, 1974) — something that is not usually necessary with resonance scanning.

The energy spectrum method is the main profiling technique used with ion-ion reactions and is described in Highlight 4.3. It is a relatively quick method since the necessary data are obtained during one irradiation with a fixed incident energy, in a very similar way to the use of Rutherford backscattering (Chapter 3). However, the product ion is usually different to the incident ion and higher in energy which has an important influence on the choice of experimental conditions and the performance achieved. The optimisation of depth resolution following the principles described in Chapter 12.2, is illustrated in Highlight 4.4. Most isotopes from ^{1}H to ^{19}F can be profiled with this method, the 'surface' resolution being typically in the range 10 to 100 nm. This is not as good as often achieved by resonance scanning with ion–gamma reactions but is nevertheless useful, especially when several isotopes can be profiled simultaneously (for example ^{12}C and ^{16}O).

4.2.6 Non-Elastic Recoil

It was pointed out in Section 4.2.2f that the recoiling nucleus can sometimes be detected after a nuclear reaction in a very thin sample. This can also be done if the beam is incident at a glancing angle to the surface

HIGHLIGHT 4.3
PROFILING WITH REACTION ENERGY SPECTRA

The simplest method for deriving depth distributions from the observed peaks in reaction energy spectra is to use reference samples and a

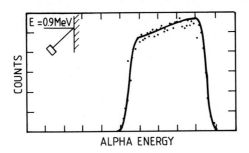

Fig. 4.10 Schematic illustration of a $^{16}O(d,\alpha)$ spectrum (points) and a calculated spectrum (full curve).

channel-by-channel ratio (N_u/N_s) of counts from unknown (u) and standard (s), Equation (12.43).

This is a good approximation provided that the stopping power curves for standard and unknown are approximately parallel (Gossett, 1980). The depth scale can be calculated with Equation (12.35).

a. Spectrum Simulation
Spectrum simulation is straightforward if the cross-section and concentration functions have a simple form. For example, there is only a small linear increase in cross-section of the $^{16}O(d,\alpha)^{14}N$ reaction between 800 and 900 keV. The calculated yield (Y_3) for alpha particle energy E_3 is given by (see Chapter 12.2):

$$Y_3 dE_3 = N_1 ed\Omega\, c(x)\, \sigma(E_1)\, dx/m_2 S(E_1) \qquad (4.6)$$

It is necessary to calculate the depth change (dx) for each increment in product particle energy (dE_3) using Equation (12.33) or (12.34). To include the effects of energy spread, Equation (12.41) should be used. The incident ion energy (E_1') at the beginning or centre of the depth interval is also needed (Equation 12.31) to evaluate the cross-section and stopping power. The calculation can be repeated with different forms of $c(x)$ until a satisfactory fit is obtained to the observed spectrum.

Fig. 4.10 shows the shape of calculated and measured yields for an 800 nm thick SiO_2 layer on Si (Turos *et al.*, 1973b). The concentration profile was assumed to be uniform with depth and the experimental

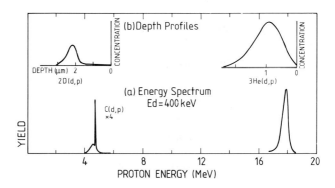

Fig. 4.11 Schematic illustration of: a. the proton energy spectrum from 0.4 MeV deuteron irradiation of Ni with implanted ^{2}D and ^{3}He; and b. ^{2}D and ^{3}He depth profiles obtained by deconvolution of the spectrum peaks.

resolution, as well as effects of straggling, were included to obtain the fit at low energies. It is desirable to include multiple scattering effects in spectrum simulation and suitable computer programs have been reported by Marcuso *et al.* (1983) and Simpson and Earwaker (1986).

Simulations can also be based on an expected shape for the concentration profile such as those involved in simple diffusion models. The first excited state group from the ^{16}O(d,p) reaction is useful for depth profiling because of its high yield and slowly varying cross-section between 800 and 920 keV. Calculated energy spectra for concentration profiles based on diffusion theory define the diffusion coefficient from a best fit to the measured spectrum (Amsel *et al.*, 1968).

b. Spectrum Deconvolution

As an alternative to simulation, the concentration profile $c(x)$ can be derived from an observed energy spectrum, $Y(E_3)$, by inverting Equation (4.6) and including a resolution function:

$$c(x) = k\, Y\,(E_3)\,(S_1\,(E_1)\,dE_3/dE_1)\,R(E_3)\,[d\sigma(E_1)]^{-1} \qquad (4.7)$$

Where $k = m_2/N_1\,e\,d\Omega$. The deconvolution of cross section, $\sigma(E_1)$ energy loss, $S(x)$ and resolution function, $R(E)$, can be carried out by using appropriate approximations or with numerical techniques (see Chapter 12.2). Examples are shown in Fig. 4.11. The lower curve shows schematically the ion energy spectrum from 0.4 MeV deuteron irradiation of Ni implanted with ^{2}D and ^{3}He. Peaks from (d,p) reactions in the implanted species and surface C are observed. The upper curves illustrate the depth profiles of ^{2}D and ^{3}He obtained by deconvolution after removal of the surface C peak (Möller *et al.*, 1977).

HIGHLIGHT 4.4
OPTIMISATION OF DEPTH RESOLUTION

The conditions for optimum depth resolution are:

- a reaction with high Z incident or product ions or both;
- a glancing angle to the surface for the high Z ion;
- a low beam diameter, divergence and energy spread;
- a high resolution detector with small solid angle and no absorber foil; and
- a low reaction angle if this is consistent with good separation of product energy groups.

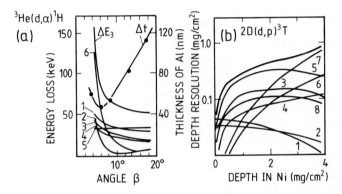

Fig. 4.12a. Energy and depth resolutions as a function of emergence angle (β) for the ^{3}He(d,α) reaction with $E_1 = 450$ KeV, $\theta = 90°$ and t = 230 nm; b. Depth resolution as a function of depth for the ^{2}D(d,p)^{3}T reaction with $E_1 = 2$ MeV, $\theta = 40°$, $\phi = 25°$, and $\psi = 65°$. 1 = Detector resolution; 2 = geometrical effects; 3 = multiple scattering of product ions; 4 = energy straggling of product ions; 5 = multiple scattering of incident ions; 6 = lateral spread of incident ions; 7 = lateral spread of incident ions; 8 = energy straggling of incident ions.

Having chosen values of these parameters appropriate to the reaction being used, it remains to assess how small a glancing angle can be used since the depth resolution improves as the sine of this angle until other factors intervene. This is shown in Fig. 4.12a for the ^{3}He(d,α)^{1}H reaction where the product energy resolution (ΔE_3) and the equivalent depth resolution (Δt) are plotted as a function of angle of emergence (β) of the product alpha particles (Bottiger, 1978). The components of the energy resolution are also shown, the most important being energy straggling and multiple scattering. However, at small angles, the contribution from geometrical effects and the lateral spread of the alpha particles rise rapidly and the depth resolution deteriorates below 8°. This is a typical result, but the optimum angle may vary from < 5° to > 10° depending on the reaction and experimental parameters involved. Of course, the maximum depth that can be analysed also decreases with sin β.

For fixed geometry, the depth resolution deteriorates rapidly as the depth increases. This is shown in Fig. 4.12b for the $^2D(d,p)^3T$ reaction (Möller *et al.*, 1977). Mulitple scattering of the incident ions is most important close to the surface, and their lateral spread is dominant at large depths. The depth resolution which is usually quoted (Chapter 14.4) is the 'surface' resolution with contributions from incident beam energy spread and geometry assumed to be small compared to the detector resolution. Energy straggling in an absorber foil is also ignored whereas this is likely to be the dominant factor if a foil is used.

---

of a thick sample. This is analogous to ERA (Chapter 3) but the energy gain available in many nuclear reactions gives the recoil nucleus greater energy. Conservation of momentum restricts recoil nuclei to travel within a cone in the forward direction and this can be exploited to obtain an enhanced yield close to the maximum angle defined by kinematic relations, Equation (1.30), Table 1.3. High mass incident ions increase the kinematic collimation and a variety of heavy ion reactions have been proposed for non-elastic recoil profiling of isotopes from H to F (Conlon and Parker, 1980).

4.2.7 Performance

Ion–ion reaction methods have been demonstrated for the determination of most isotopes from ^1H to ^{32}S.

The most used reactions are (p,α), (d,p) and (d,α) with ^3He reactions providing useful alternatives for the determination of isotopes such as ^2D, ^{12}C and ^{16}O. Alpha induced reactions have had limited use. Some ^6Li induced reactions have been tested and the $(^{11}B,\alpha)$ reaction has been used for hydrogen determination and profiling.

Incident ion energies from 0.5 to 2 MeV are most useful for minimising interference from reactions in heavy isotopes. In contrast, energies above 3 MeV make (d,p) reactions useful for the study of metals. High beam current densities can be used in this case so that microprobe scans are also practicable for beam diameters of the order of 10 μm and reactions which have been used in this way are listed in Table 10.5.

Reaction cross-sections of 10 to 100 mb sr^{-1} are observed for proton and deuteron induced reactions in light isotopes such as D, Li, Be and B. Sensitivities of the order of 10 μg g^{-1} or even less are then possible with measuring times of the order of tens of minutes. Three dimensional analysis (simultaneous depth profiling and microprobe scans) of ^7Li has been described by Heck (1988). For the important case of isotopes of C, N and O, similar sensitivities are possible. Cross-sections are smaller for higher Z isotopes and hence count rates are lower. Level densities also increase with Z and the choice of incident energy and reaction angle is crucial for minimising interference from unwanted groups in the observed spectrum. Fortunately, light elements such as Li, Be and B usually occur at low concentrations and, if this is the case, isotopes up to ^{32}S can be determined in heavier matrices at mg g^{-1} or per cent levels depending on the current that the sample will withstand.

Depth information is always available from the product ion spectrum if the depth exceeds that corresponding to the energy resolution. The energy interval to the next lower group in the spectrum sets the maximum depth that can be profiled. These parameters are unique for each reaction and are catalogued in Chapter 14.4. The maximum depth is usually limited to the order of 1 μm. Unfolding techniques can increase this range somewhat. The use of glancing incidence or emergence for heavy incident or product ions respectively can give typical depth resolutions at the surface of 10 to 100 nm.

At depths greater than approximately 100 nm, the effects of multiple scattering and energy loss straggling on both incident and product ions so degrade the resolution that there is no point in using low efficiency, high resolution detection systems. Most ion–ion reaction measurements are therefore made with a very simple experimental arrangement. Surface removal and wedge scanning have been used to extend ion–ion depth profiling to any depth with constant resolution for the profiling of ^{1}H, ^{2}D, ^{3}He, ^{11}B, ^{18}O, ^{23}Na, ^{22}Al and ^{31}P but the surface resolution is only of the order of 10 nm.

4.3 ION–GAMMA REACTIONS

Particle Induced Gamma ray Emission (PIGME or PIGE) is a versatile technique which complements other ion beam techniques for sample analysis and non-destructive depth profiling. It is the most common application of nuclear reaction analysis and has recently been reviewed by Borderie (1980), Deconninck *et al.* (1981) and Peisach (1981). Reactions used for analysis are included in Chapter 14.4.1.

4.3.1. Isotopic Analysis Methods

a. Experimental Arrangement

Gamma-ray detectors, with associated shielding, are usually too large to be fitted within a typical vacuum chamber for analysis work and this is

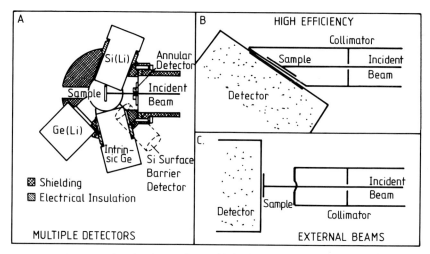

Fig. 4.13a. Compact chamber layout for multiple detectors; b. large solid angle arrangement for gamma-ray detection; and c. external beam system for PIGME analysis.

not necessary since gamma-ray attenuation in a suitable window material (e.g. 1 mm Al) is negligible for gamma-ray energies above 100 keV. However, detector efficiencies are relatively low and it is advantageous to use a small vacuum chamber so that the detector distance can be minimised. A compact arrangement for the use of up to five detectors can be seen in Fig. 4.13a (Giles and Peisach, 1976). A typical detector solid angle is 0.1 to 0.5 sr but this can be increased to at least 2 sr (sample to detector distance 1 cm or even less) with an arrangement such as that shown in Fig. 4.13b. In this case, corrections may be necessary for gamma-ray attenuation in the sample. A similar layout is also possible when using an external beam which is well suited to PIGME measurements (Fig. 4.13c). Gamma-rays produced in the window material (Raith *et al.*, 1980) can be minimised by choice of material, nickel being particularly suitable.

The choice of detector is discussed in Highlight 4.5. The detector angle is not critical provided that differential cross-sections or gamma-ray yields appropriate to the specific experimental geometry are used in concentration calculations. A uniform angular distribution of gamma-rays is usually observed with a thick sample but deviations can arise if the gamma-ray yield is dominated by one resonance. Such a case is the large resonance in the $^{24}Mg(p,p')$ reaction at 2.330 MeV. This resonance introduces a strong asymmetry into the angular distribution of the 1.368 MeV gamma-ray for ion energies of 2.4 to 2.5 MeV (Kenny *et al.*, 1980). The lower the incident ion energy the fewer resonances are involved in ion–gamma reactions and non-uniform angular distributions are more likely to be observed. The choice of sample angle affects the depth of ion penetration and hence the thickness of the layer which can be analysed. Low detector efficiency and hence low count rates can be offset to some extent by an increase in beam current provided that sample damage does not occur. For metals and other good thermal conductors it is possible to use a beam current of the order of 1 μA without special cooling arrangements. By mounting the sample on a cooled backing, at least 10 μA of beam can be used. Such high beam currents allow high sensitivities to be achieved. For poor thermal conductors, sample damage can be minimised by mounting a thin sample on a cooled backing. Even so, PIGME studies of biological materials are necessarily limited by the need to avoid sample damage — particularly when using a microbeam. Otherwise, there are few special requirements for sample preparation and satisfactory analyses can usually be made of untreated and unpolished specimens as well as pressed powders.

HIGHLIGHT 4.5
CHOICE OF DETECTOR

The probability of gamma-ray interactions depends on various powers of the atomic number of the detector material and the detector efficiency also depends on its size (see Chapter 2). Four types of detector have been used in analytical applications, viz.

- Bismuth Germanate (BGO) — highest efficiency; poorest resolution;
- Sodium Iodide (NaI) — high efficiency; poor resolution;
- Lithium Drifted Germanium (Ge(Li)) — low efficiency; good resolution; and
- Intrinsic Germanium (IG) — lowest efficiency; best resolution.

Comparisons of pulse height spectra from NaI and Ge(Li) detectors are presented in Figure 2.16 and the reasons for choosing one for analytical work are illustrated by the following examples.

For high sensitivity, a NaI detector (or BGO if available) can be used provided that the poor energy resolution does not lead to interference between peaks from gamma-rays emitted in different nuclear reactions. The best known example is the use of a large volume NaI detector (at least 10 to 15 cm diameter and length) to detect the 6.13, 6.92 and 7.12 MeV gamma-rays from the $^{19}F(p,\alpha\gamma)$ or $^1H(^{19}F,\alpha\gamma)$ reactions. These reactions have a high cross-section and the combined count rate from all

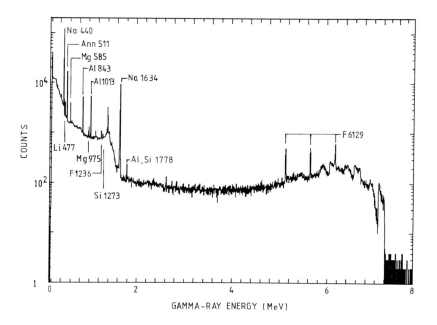

Fig. 4.14 Gamma-ray spectrum from 2.5 MeV proton irradiation of oil shale (USGS standard SGR1) using a 120 cm³ Ge(Li) at 135° and 155 mm from the sample.

three gamma-rays allows a fluorine detection limit of 10^{-7} g cm^{-2} to be achieved (Malmqvist *et al.*, 1982). Background reduction by using a plastic anticoincidence detector is useful for large detectors and has been applied for improving the performance with the ^{15}N(p,$\alpha\gamma$) reactions (Damjantschitsch *et al.*, 1983). It can also be advantageous to use a thin filter (2 to 5 mm Pb) to preferentially reduce the count rate from low energy gamma-rays so that the beam current can be increased to give a greater count rate for the high energy gamma-rays of interest.

A similar approach can be adopted with a Ge(Li) detector by summing all counts from, say, 5 MeV to 7 MeV. This has the advantage that the high resolution spectrum can be inspected to verify that no additional interfering gamma-rays are present. However, Ge(Li) detectors are much smaller than can be readily obtained using NaI or BGO and they are predominantly used for detecting low energy gamma-rays (up to the order of 1.5 MeV) which are most used for PIGME analysis. A typical spectrum is shown in Fig. 4.14 for 2.5 MeV proton irradiation of oil shale (USGS standard SGR1). The peaks are labelled according to the responsible target nuclide and the peaks and continuum at high energies are almost entirely from the ^{19}F(p,$\alpha\gamma$) reaction. Accurate energy determination and good separation of peaks for gamma-rays from different nuclear reactions are important — especially if incident energies above 3 MeV are used to increase gamma-ray yields.

Intrinsic Ge(IG) detectors have better resolution than Ge(Li) but they are only available with small volumes and so can only be used to advantage for gamma-ray energies up to a few hundred keV. Heavy elements usually have low-lying states which are readily excited by Coulomb excitation and it is sometimes advantageous to use an IG detector for their determination. For example, PIGME yields from Fe are relatively low and gamma-rays from minor or trace elements in steels can readily be observed using proton or alpha-particle induced reactions. The improved resolution and lower continuum observed with a small IG rather than a Ge(Li) detector allows greater sensitivity and precision for the determination of Cr, Mn, Mo, and W in steels (Peisach and Gihwala, 1981; Gihwala and Peisach, 1982).

---

b. Spectrum Analysis

Many analytical problems involve simple gamma-ray spectra with a few well-separated peaks. The effects of multiple scattering cause the continuum to be higher at the low energy side of a peak than just above the peak, and this step function in continuum is difficult to fit with automated peak search procedures. Since the step is caused by the same gamma-rays as contribute to the peak, the difference in counts summed over the channels containing the peak and in a suitable background region just above the peak is a good method of background subtraction.

Sophisticated peak search and fitting routines are widely used for computer analysis of complex pulse height spectra (see Chapter 2). Overlapping peaks require a fit of Gaussian or asymmetric peak functions and a suitable continuum function. Peak positions can be determined to an accuracy of the order of 0.1 keV and peak areas obtained which are reproducible to better than 1%. However, different peak fitting methods make different assumptions about the shape of the continuum under a peak and hence can give significantly different values of peak areas. It is most important therefore to use the same algorithm for determining peak areas during detector calibration as in measurements on unknown samples. The energy scale, including any non-linearity, can be determined using radioactive sources such as those listed in Table 2.9 or reaction gamma-rays (see Chapter 14.4.3). Sources with calibrated emission rates are available for the determination of detector efficiency as a function of gamma-ray energy (see Table 2.9).

TABLE 4.5a
Commonly observed background radioactivity

E_γ (keV)	Nuclide(s)	E_γ (keV)	Nuclide(s)
129	^{228}Ac	1120	^{214}Bi
186	^{226}Ra	1155	^{214}Bi
239	^{212}Pb, ^{214}Pb	1238	^{214}Bi
227	^{228}Ac	1378	^{214}Bi
285	^{214}Bi	1408	^{214}Bi
296	^{214}Pb	1461	^{40}K
328	^{228}Ac	1509	^{214}Bi
339	^{228}Ac	1588	^{228}Ac
352	^{214}Pb	1592	^{208}Bi,^{208}Tl (de)
463	^{228}Ac	1621	^{212}Bi
478	^{7}Be	1630	^{228}Ac
511	Annihilation	1661	^{214}Bi
583	^{208}Tl	1693	^{214}Bi (se)
609	^{214}Bi	1732	^{214}Bi
666	^{214}Bi	1764	^{214}Bi
727	^{212}Bi	1850	^{214}Bi
769	^{214}Bi	2103	^{208}Bi,^{208}Tl (se)
795	^{228}Ac	2210	^{214}Bi
861	^{208}Tl	2448	^{214}Bi
911	^{228}Ac	2614	^{208}Bi,^{208}Tl
935	^{214}Bi		
969	^{228}Ac		

de = double escape peak (E_γ − 1.022)
se = single escape peak (E_γ − 0.511)

Possible sources of interfering background gamma-rays include natural radioactivity in surrounding materials, reactions in collimators and other beam-line equipment which may be irradiated by the beam, reactions caused by neutrons produced during irradiation of the sample and collimator materials, and reactions produced by scattered ions. The first two contributions can be reduced by carefully positioned lead shielding and the last one by using low yield materials in chamber construction.

If neutrons are produced during sample irradiation they not only constitute a radiation hazard but they also create a high background in NaI detectors because of Na activation and neutron capture in iodine. Neutrons also cause increased background in Ge(Li) detectors and radiation damage which causes a loss of resolution. The onset of damage occurs for doses of the order of 10^8 to 10^{10} neutrons cm^{-2} for small and large detectors respectively and can be assessed by observation of the

TABLE 4.5b
Neutron induced background gamma-rays

E_γ (keV)	Reaction	Origin	E_γ (keV)	Reaction	Origin
53	^{73m}Ge	Detector	962	^{63}Cu(n,n′)	System
67	^{73m}Ge	Detector	985	^{27}Al(n,p)	System
91	^{27}Al(n,α)	System	989	^{27}Al(n,n′)	System
110	^{19}F(n,n′)	System	992	^{64}Zn(n,n′)	System
140	^{75m}Ge	Detector	1014	^{27}Al(n,n′)	System
186	^{65}Cu(n,γ)	System	1039(b)	^{70}Ge(n,n′)	Detector
197	^{19}F(n,n′)	System	1039	^{66}Zn(n,n′)	System
198	^{71m}Ge	Detector	1115	^{65}Cu(n,n′)	System
472	^{27}Al(n,α)	System	1201(de)	^{1}H(n,γ)	Shield
563(b)	^{76}Ge(n,n′)	Detector	1238	^{56}Fe(n,n′)	System
570	^{207}Pb(n,n′)	Shield	1268(se)	^{27}Al(n,γ)	System
596(b)	^{74}Ge(n,n′)	Detector	1369	^{27}Al(n,α)	System
608(b)	^{74}Ge(n,n′)	Detector	1464(b)	^{72}Ge(n,n′)	Detector
690(b)	^{72}Ge(n,n′)	Detector	1698	^{27}Al(n,p)	System
757(de)	^{27}Al(n,γ)	System	1712(se)	^{1}H(n,γ)	Shield
834(b)	^{72}Ge(n,n′)	Detector	1732	^{27}Al(n,α)	System
839	^{54}Fe(n,p)	System	1779	^{27}Al(n,γ)	System
844	^{27}Al(n,n′)	System	1809	^{27}Al(n,d)	System
867(b)	^{74}Ge(n,n′)	Detector	1811	^{56}Fe(n,n′)	System
885	^{73}Ge(n,α)	Detector	2223	^{1}H(n,γ)	Shield
894(b)	^{72}Ge(n,n′)	Detector	2243	^{27}Al(n,α)	System
			2754	^{27}Al(n,α)	System

b — broad high energy tail from recoil absorption
de — double escape peak ($E_\gamma - 1.022$)
se — single escape peak ($E_\gamma - 0.511$)

cumulative intensity of gamma-rays at 693 keV which arise from inelastic scattering of neutrons (Wilenzick, 1972). Because of these problems, most PIGME analysis is done with ions and energies chosen to involve negligible neutron production. Lists of background gamma-rays and neutron induced gamma-rays are given in Table 4.5.

c. *Determination of Concentration*

Peak areas from thick samples with known amounts of a particular nuclide can be used to derive estimates of concentration in an unknown sample using Equation (12.18) or the other methods described in Chapter 12.1. The stopping power correction (S_S/S) requires the evaluation of an integral of the cross-section as a function of incident energy (Equations (12.11) and (12.12)) (or a sum when the cross-section involves a set of

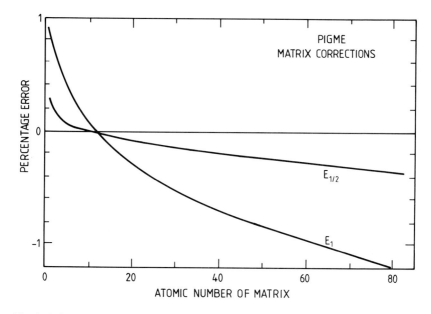

Fig. 4.15 Comparison of yield estimates using the surface energy (E_1) approximation and the $E_{1/2}$ rule as a percentage difference from the result of full integration.

well-separated resonances). Since the correction factor is close to unity unless the standard and unknown samples have major elements with very different Z, sufficient accuracy can be achieved with simple approximations. The stopping power ratio calculated at the incident ion energy (surface approximation) can lead to errors greater than 1% whereas that calculated at an energy for which the gamma-ray yield is one-half that at the incident energy is usually accurate to much better than 1% (Fig. 4.15). A reliable knowledge of the thick target yield curve is required to determine the half-yield energy and an iterative approach must be used to allow for the effect of the sought elements on stopping power. This method is adequate for most work but the 'effective mean energy' method (Ishii *et al.*, 1978a,b) involves negligible error whatever the matrix composition, although it requires detailed information on the relevant yield curves and reliable stopping power data. It should be noted that the approximations introduce systematic errors rather than random errors.

A possible important source of systematic errors is sample inhomogeneity. Most of the yield comes from the first part of the ion range where the ion energy is highest. The volume analysed is typically of the order of $10^3 \, d^2 \, \text{cm}^3$ (d = beam diameter in cm) for 2 to 3 MeV protons and less for heavier ions. Inhomogeneities with dimensions of the order of (100 d) μm will have a significant effect on observed concentrations. Lateral inhomogeneities can be easily checked by changing the beam position and observing the effect on scatter of results (Giles and Peisach, 1979). The presence of depth inhomogeneities can be checked by changing the ion type or energy or by using several gamma-rays from the same target nuclide which have different yield curves.

d. Isotope Ratios

Isotopes of one element have very different nuclear properties which can be exploited for the measurement of isotope ratios. Examples of such work are the measurement of $^{10}B/^{11}B$ (Olivier and Peisach, 1985), $^{13}C/^{12}C$ (Ricci, 1971) and $^{15}N/^{14}N$ (Xenoulis and Douka, 1979). The low abundance isotopes can be detected at or above natural levels by the choice of high yield nuclear reactions but, even so, the statistical accuracy in ion beam analysis is seldom better than 1% and no attempt has been made to observe the small changes of abundance which are of major importance in environmental and other studies.

4.3.2 Depth Profiling

It is useful to distinguish two types of gamma-ray transition involved in nuclear reactions. Primary gamma-rays are those which are emitted as a transition from a highly excited state in which a compound nucleus is formed to a low lying state. An example is the gamma-ray originating near the top of the ^{20}Ne level diagram in Fig. 4.1. The energy of primary gamma-rays changes as the incident ion energy is changed and this can be exploited for depth profiling by the energy spectrum method (Chapter 12.2). Secondary gamma-rays are emitted by low lying excited states such as the 1.632 MeV level in ^{20}Ne or the 6.14 MeV level in ^{16}O (see Fig. 4.1). The energy of secondary gamma-rays does not change with the incident ion energy and they cannot be used for energy spectrum profiling.

Measurements of the energy of primary gamma-rays have been used in profiling ^{12}C and ^{16}O via (p,γ) reactions. A broad resonance in $^{12}C(p,\gamma)$ occurs near 0.45 MeV and ^{16}O has a smoothly varying cross-section above 1 MeV. The observed gamma-ray peak shape can be converted to a depth profile using the methods described in Highlight 4.3. However, yields from these reactions are relatively low and little use has been made of this technique. Narrow resonances are a common feature of many gamma-ray producing reactions and either primary or secondary gamma-rays can be used in resonance scanning or yield curve unfolding (Chapter 12.2). Profiling by resonance scanning is described in Highlight 4.6 and some unusual effects which must be kept in mind are discussed in Highlight 4.7.

HIGHLIGHT 4.6
PROFILING BY RESONANCE SCANNING

A list of resonances used for depth profiling is given in Chapter 14.4.2. The equipment required for resonance scanning is the same as described in Section 4.3.1a plus a suitable system for frequent changes in the incident ion energy. In order to take advantage of narrow resonances to achieve good depth resolution, it is necessary to use an incident beam with a small energy spread—preferably of the order of 100 eV but usually no greater than 1 to 3 keV. Stabilisation and automated beam energy scanning are described in Chapter 2. The use of a large NaI detector is advantageous if the gamma-ray resolution is adequate, in order to speed up data taking and reduce the total time to complete a profile (e.g. Lenz *et al.*, 1987).

The procedure for profiling by energy scanning is to commence irradiation of the sample with a beam energy which is just below that of a resonance with a large peak cross-section. The yield of gamma-rays, which are known to be strongly excited at the resonance, is counted for a fixed dose (e.g. 1 μC or more). The beam energy is then raised step by step and the yield measured at each interval to obtain a yield curve as a function of incident energy. Two examples are shown in Fig. 4.16 where

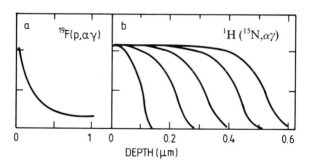

Fig. 4.16a. Depth profile for F diffusion into dental enamel measured using the 872 KeV resonance in $^{19}F(p,\alpha\gamma)$; b. H profiles in hydrated glass measured using the $^{1}H(^{6}Li,\alpha\gamma)$ reaction at 6.400 MeV.

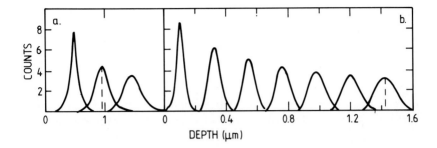

Fig. 4.17 The change in shape of the depth resolution function as a function of depth caused by energy straggling of the incident ions.

the incident energy scale has been converted to a depth scale with the resonance energy corresponding to zero depth. Fig. 4.16a shows the use of the 872 keV resonance in ^{19}F (p,$\alpha\gamma$) to determine the depth distribution of F in dental enamel (Bodart and Ghoos, 1980). The depth scale is obtained from the change in beam energy with depth, Equation (12.30), which requires a knowledge of the major element composition of the sample in order to calculate the stopping power. An approximate F depth distribution is obtained from

$$c(x) = KY_3(E_1)\,\bar{S}_1 \qquad (4.8)$$

where $Y_3(E_1)$ is the measured yield curve and K can be obtained by calibration with standard samples such as thin layers of pure elements, compound or implanted ions of known energy and dose. The second example is of H profiles, measured with the 6.400 MeV resonance in the ^{1}H(^{15}N,$\alpha\gamma$) ^{12}C reaction (Lanford, 1978). Fig. 4.16b shows a number of profiles which define the thickness of the hydration layer in obsidian for increasing hydration times.

In order to obtain accurate depth information it is necessary to take into account the change in stopping power with ion energy and the effects of energy straggling on depth resolution and profile shape. The effects of straggling are shown by the change in shape of profiles calculated for the 340 keV resonance in F at various depths (Fig. 4.17, Maurel *et al.*, 1982). For depths up to the order of 10 to 20 nm, Vavilov distributions must be used but at greater depths the straggling distribution can be assumed to be Gaussian (Chapter 14.2).

HIGHLIGHT 4.7
UNUSUAL EFFECTS IN DEPTH PROFILING

High resolution (i.e. near-surface) profiling can be subject to a number of processes which introduce unexpected structure into the depth profile. A 'surface' peak is often observed because of the presence of atoms adsorbed at the sample surface, especially for such common elements as H, C and O. An additional contribution to a peak at zero depth comes from the Lewis effect (Fig. 4.18). This arises because the incident ion loses energy in finite amounts (depending on the angle of small-angle scattering events) so that, even for a monoenergetic beam, less than 100% of the ions have particular value of energy once some energy loss has taken place. The Lewis effect is only observed if the resonance width and experimental resolution are of the order of 100 eV or less. The zero-depth peak can be reproduced by calculations based on stochastic theories of energy loss (Maurel *et al.*, 1982) but it may be ignored in the measurement of depth profiles below the surface.

Channeling (see Chapters 1 and 6) can also modify the rate of energy loss if the sample is crystalline. It is common, especially in semiconductor studies, to prepare a sample with a crystal axis colinear with the surface normal and to mount samples near perpendicular to the incident beam for nuclear reaction analysis. Channeling is likely to affect observed depth profiles in these circumstances and can be exploited in atom location studies (see Chapter 6). A rule-of-thumb, which is useful but not guaranteed, is to incline such samples at 7° to the direction of the incident beam in order to obtain results which approximate the case of a random direction of incidence. For a better understanding of the effects of crystallinity it is necessary to carry out a series of measurements with different angles of incidence.

If an insulating sample becomes positively charged during irradiation, it will cause deceleration of the incident ions with a reduction in their effective energy. This will lead to an apparent shift in the resonance energy and an incorrect depth scale. The most serious effect of this kind is produced by sparking which will change the effective energy randomly. Since surface potentials of thousands of volts can easily occur, any attempt to carry out depth profiling with good resolution can be completely destroyed by sparking. Various methods for preventing surface charging are discussed in Chapters 2 and 3.

Finally, any departure from a smooth flat surface can effect depth profiling. If ions are incident normal to the sample surface, a profile will

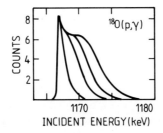

Fig. 4.18 Depth profiles of thin Ta oxide ratios measured using the 1167 keV resonance in the $^{18}O(p,\gamma)$ reaction. The peak at low energies is evidence for the Lewis effect.

be obtained which describes successive layers parallel to the surface even if this is not flat. For other angles of incidence, surface roughness or topography will cause a worsening of depth resolution (Chapter 12.2).

a. Multiple Resonances

If the incident ion energy is scanned over a range containing more than one resonance, the normal method for calculating a depth profile is subject to error. The contribution from successively higher energy resonances can be estimated from a knowledge of the depth distribution obtained at lower energies and suitable corrections can be made to extend the depth profile to greater depths, Equation (12.44). Various numerical methods have been developed for this purpose but, in practice, profiling is generally not continued far beyond the energy difference between adjacent resonances. However, profiles of F in teeth have been measured on 10 μm layers by applying corrections to the yield curve for proton energies from 0.8 to 1.5 MeV (Kregar *et al.*, 1979). Greater thicknesses were studied by the removal of successive 10 μm layers and repeated profile determination.

b. Yield Curve Unfolding

If there are many resonances in the reaction cross-section, the observed PIGME yield as a function of incident ion energy is a series of steps. If the cross-section is a smooth function of energy, then the yield curve is also smooth. The deconvolution of energy spectra and yield curves is discussed in Chapter 12.2 but a large computer is required and, to date, there has been little use of this approach for depth profiling — spectrum simulation being the preferred approach.

4.3.3 Performance

a. Detection Limits

Some typical values of detection limits for PIGME analysis are plotted in Fig. 4.19 showing the favourable performance for many isotopes with $A < 30$, $Z < 15$ and useful performance for higher Z. Such detection limits are illustrative only since, in practice, they depend on detector geometry and efficiency and other experimental factors including the possible effects of interfering reactions. Some of the factors that distinguish the PIGME method are:

- simultaneous measurements can be made of gamma-rays from a number of nuclides;
- cross-sections are highest for light isotopes ($A < 30$) so that these can be determined with good sensitivity (1 μg g^{-1} or less)
- the yield from Coulomb excitation of many heavy elements are sufficiently high that these can be determined if light elements are not present in high concentrations;
- low yield materials such as C, Co, Ni and Pb can be used in the target chamber to minimise gamma-rays produced by scattered ions;
- many reactions have narrow resonances which are suitable for depth profiling with a resolution of 1 to 10 nm; and
- The observed yield is not strongly affected by variation in surface angle or surface roughness which are important for low energy X-rays.

b. Proton–Gamma Analysis

Proton beams have been widely used for the determination of one or more isotopes of elements from Li to Cl (although not all simultaneously) and many heavier elements. For proton energies below 1 MeV, Coulomb excitation cross-sections are low and only light nuclides with low energy resonances give reasonable gamma-ray yields. In this region, the proton energy can often be chosen to optimise measurements on a particular nuclide. For example 500 keV protons give a good yield from ^{12}C and ^{16}O and this is suitable for their determination in biological materials (Demortier, 1974). Somewhat higher energies (e.g. 1.8 MeV) are needed for N determination.

For protons with energies from 1 to 3 MeV, the highest yields are from Li, B, F, Na and Al and these elements can be determined simultaneously in many cases. For example, a precision of better than 1%

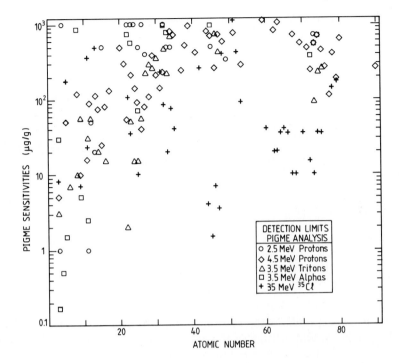

Fig. 4.19 Estimated detection limits in PIGME analysis using 2.5 MeV protons (Deconninck *et al.*, 1981), 4.5 MeV protons (Gihwala *et al.*, 1982), 3.5 MeV tritons (Borderie and Barrandon, 1978), 3.5 MeV alphas (Borderie and Barrandon, 1978) and 55 MeV ^{35}Cl ions (Borderie *et al.*, 1979).

can be obtained in F and Na determination with a measuring time of a few minutes. Fluorine is of particular interest, with very high sensitivity (0.1 μg g^{-1}) being available for an element which is difficult to determine by non ion beam techniques. The simultaneous use of PIXE data adds another 20 or more elements in many applications but, even so, the PIGME data remain valuable for the light elements and because of their precision.

At energies above 3 MeV, the gamma-ray yield from medium and heavy elements begin to compete with the light elements. For example, B, F, Na, Mg, Al, Si and Cu have been determined in potsherds at 4 MeV (Peisach *et al.*, 1982).

c. Deuteron–Gamma Reactions

Gamma-rays are produced by deuteron irradiation of most nuclides although light nuclides have the highest cross-sections. Thick sample

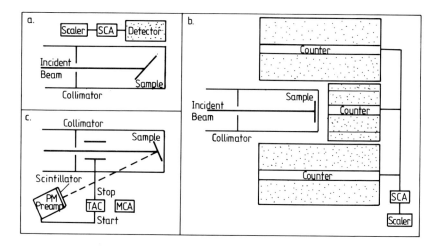

Fig. 4.20 Ion-neutron analysis equipment. a. simple configuration for neutron counting; b. large solid angle neutron counting system; c. time-of-flight measurement of neutron energies.

yields increase rapidly with deuteron energy, slowing down somewhat for light nuclides and energies above 3 MeV. In principle, therefore, deuteron induced gamma-ray emission could provide a versatile multi-element capability if it was not for the associated neutron flux from (d,n) reactions.

d. Triton and ^{3}He Reactions

The highest gamma-ray yields from triton irradiation are from Li, C, O and F. Estimated detection limits for these and other elements are included in Fig. 4.19. Only O has been analysed with a triton beam — taking advantage of the high yield of 937, 1042 and 1982 keV gamma-rays to achieve a detection limit of 0.13 μg cm^{-2} of oxygen for 1.9 MeV triton irradiation of steel (Peisach, 1972). Gamma rays from ^{3}He irradiation of Li, Be, B, C and O have been assessed (Deconninck and Demortier, 1973) but their usefulness for analysis of a range of sample materials has not been investigated.

e. Alpha-Induced Gamma Emission

The gamma-ray yields listed in Chapter 14.4 show that the sensitivity of alpha-induced gamma-ray emission is particularly good for Li, Be, B, N, F and Na. A number of medium and heavy elements also give reasonable sensitivity, including Mn, Ta, Br, As, Re, Rh, V, Ir, W, Ag, Ge and Ti in that order. On the other hand, essentially no gamma-rays are observed form C, S, Ca, Co, Ni, Ga, Sr, Y, Nb, In, Sn, Ba, La, Ce, Pr, Nd, Pb and Bi and these are favourable matrix materials in which other elements can readily be detected. For example, successful measurements have been made of a number of minor elements in steels (Gihwala and Peisach, 1980). An additional advantage of alphas is that few high energy gamma-rays are produced which increase the Compton continuum in the lower energy regions of the gamma-ray spectra. An important feature of the PIGME method is the depth to which the incident ions penetrate before the gamma-ray yield becomes insignificant. In gold this depth is 20 μm for 3 MeV protons and only 3 μm for 3 MeV alpha-particles. This has been exploited by Basutcu (1980) to investigate surface enrichment in coins and other objects containing Au, Ag and Cu. Copper is preferentially removed from the surface region of Au and Ag alloys and Ag is also preferentially removed from Au alloys. The yield of silver gamma-rays from two irradiations with different ions allows the determination of the degree of surface enrichment or depletion.

f. Heavy Ion–Gamma Analysis

The most notable application of heavy ion induced gamma-ray emission has been in the depth profiling of hydrogen using beams of ^{7}Li, ^{11}B, ^{15}N or ^{19}F ions (see Fig. 4.16b). Analysis of heavier isotopes can be carried out with heavy ions by exploiting Coulomb excitation (Borderie *et al.*, 1979).

4.4 ION–NEUTRON REACTIONS

Neutrons are produced in many nuclear reactions but are usually ignored rather than exploited. Like gamma-rays, neutrons emerge from sample and sample chamber without energy loss and usually with only very minor loss in intensity. Sample analysis can be carried out by two methods:

    i. by counting the number of neutrons produced—a method which is useful if it is known that only one nuclide is contributing to the neutron yield; and

    ii. by measuring neutron energy spectra which can reveal the existence of different energy groups from one or more reactions. Energy spectra also define depth profiles since the neutron energy is dependent on the energy of the incident ion at the depth at which the reaction occurs, through the usual kinematic equations (see Chapter 1). A list of ion–neutron reactions used for sample analysis and depth profiling and relevant data are included in Chapter 14.4.

4.4.1 Methods

a. Neutron Yield

The yield of neutrons from an ion-induced nuclear reaction can be measured by placing a hydrogenous scintillator(plastic, liquid or a crystal scintillator such as stilbene) near the sample chamber (Fig. 4.20a). A suitable range of pulse heights is selected for counting, depending on the Q-value of the reaction of interest and the kinematics. A pulse shape discriminator may be used to help reject background pulses caused by gamma-ray interactions in the scintillator. For a thin sample, the concentration of the isotope responsible for neutron production is given by Equations (12.7) and (12.8). For a thick sample Equations (12.11) and (12.12) must be used.

For example, the high cross-section of the $^7Li(p,n)$ reaction (20 to 80 mb/sr above 2.14 MeV) can be used to determine the lithium content of a sample at levels down to 1 μg g^{-1} provided that other nuclides with thresholds below 2 MeV are not present in large quantities. The Li content of diamond has been shown to be less than 2 μg g^{-1} (Sellschop et al., 1978) and the depth distribution of Li in glass has been determined by the wedge scanning method (see Chapter 1.7) with 2 MeV protons (Pomorski et al., 1976).

The efficiency of neutron detectors depends on the size and type of scintillator used and is generally in the range from 5 to 20%. However, if placed outside the sample chamber, the solid angle may be quite small. Higher counting rates can be achieved by surrounding the chamber with a large detector assembly such as a set of BF_3 or ^3He counters in a large paraffin moderator (Fig. 4.20b). The detection efficiency, including solid angle, for such a system can be at least 10% and this gives good sensitivity for the determination of nuclides with high reaction cross-sections. In this case Equation (12.1), involving the total cross-section rather than the differential cross-section for a specific detector angle, is used.

The only nuclides with positive Q-values for (α,n) reactions are ^9Be, ^{13}C, ^{10}B, ^{17}O, ^{11}B and ^{26}Mg in descending order. The product neutron energies follow the same sequence and this can be exploited for essentially interference-free determination of Be. For an incident alpha-particle energy of 2.6 MeV, chosen to use a local maximum of 30 mb/sr in the cross-section of $^9Be(\alpha,n)$ reaction at 0°, neutrons from other nuclides can be eliminated with a discriminator setting of 4.75 MeV. Concentrations of Be down to 200 μg g^{-1} can be readily determined by neutron counting and diffusion profiles can be determined by microbeam scanning (McMillan *et al.,* 1978). At this beam energy, the ^{9}Be(d,p) reaction is more sensitive but subject to interference from numerous light nuclides. The selectivity of the (α,n) reaction is of considerable value in avoiding problems from interference by C and other elements. A lower discriminator setting would permit ^{13}C determination and profiling in the same way providing that Be was known to be absent or demonstrated to be absent with higher discriminator runs. A somewhat higher alpha-particle energy improves the yield in ^{13}C measurements.

An important feature of neutron counting is the observation of accurately defined thresholds. Reactions with negative Q-values can only take place when the incident ion energy is greater than the threshold value given by Equation (1.27), Table 1.3. If the product nuclide is formed in an excited state the Q-value is correspondingly more negative and the threshold energy is even higher. For beam energies just above threshold, conservation of momentum requires that neutrons can only be emitted within a narrow cone about the incident beam direction, where the cone angle is given by Equation (1.29), Table 1.3. As the beam energy is increased, the cone angle increases until, at energy E_{max} (given by Equation (1.28), Table 1.3) all angles of emission are possible. The 0° neutron yield increases dramatically above the threshold energy and this can be exploited to identify the nuclide responsible for neutron emission, the concentration of that nuclide and, in suitable cases, its depth distribution (from the neutron energy distribution).

The kinematic collimation effect is stronger for heavy ion induced reactions which can be exploited for light element determination. A list of threshold energies and neutron energies is given in Table 4.6 for all proton and alpha induced reactions with $Z < 33$, $Q < 3$ MeV. Only some of these reactions have been used in analysis but the information can be used in considering possible competing reactions and radiation hazards.

b. Neutron Energy

A neutron energy spectrum can be obtained by unfolding the pulse height response function of a hydrogenous scintillator from the observed pulse height spectrum. However, the most common approach is to measure the time-of-flight of each neutron. A neutron detector is placed at a suitable distance and angle relative to the direction of the ion beam (Fig. 4.20c). Neutron and gamma-ray shielding is placed so as to minimise the detector background. A pulsed ion beam is used which has the smallest possible pulse width (e.g. 1 ns or less) and a repetition rate selected to suit

TABLE 4.6
Threshold energies for ion–neutron reactions ($Q < 3$ MeV, $Z < 31$)

(p,n) Reactions				(α,n) Reactions			
Nuclide	Abundance	Threshold (MeV)	Neutron Energy (keV)	Nuclide	Abundance	Threshold (MeV)	Neutron Energy (keV)
^{3}H	–	1.019	64	^{7}Li	92.4	0	
^{7}Li	92.4	1.880	30	^{9}Be	100	0	
^{9}Be	100	2.057	21	^{10}B	19.6	0	
^{11}B	80.4	3.017	21	^{11}B	80.4	0	
^{13}C	1.1	3.236	17	^{13}C	1.1	0	
^{18}O	0.2	2.574	7	^{17}O	0.04	0	
^{37}Cl	24.2	1.640	1.2	^{18}O	0.2	0.851	7
^{41}K	6.9	1.233	0.7	^{19}F	100	2.360	18
^{45}Sc	100	2.909	1.4	^{25}Mg	10.1	0	
^{49}Ti	5.5	1.413	0.6	^{27}Al	100	3.027	13
^{51}V	99.8	1.564	0.6	^{29}Si	4.7	1.736	6
^{53}Cr	9.5	1.405	0.5	^{33}S	0.8	2.244	7
^{55}Mn	100	1.032	0.3	^{43}Ca	0.2	0	
^{57}Fe	2.2	1.648	0.5	^{45}Sc	100	2.440	4
^{59}Co	100	1.888	0.5	^{47}Ti	7.3	0.347	0.5
^{64}Ni	1.1	2.495	0.6	^{49}Ti	5.5	0	
^{65}Cu	30.8	2.168	0.5	^{53}Cr	9.5	0.351	0.4
^{67}Zn	4.1	1.810	0.4	^{57}Fe	2.2	1.451	1.6
^{70}Zn	0.6	1.458	0.3	^{67}Zn	4.1	3.142	2.5

(d,n) reactions produce neutrons for all deuteron energies

the flight path and the lowest neutron energy that is to be measured. Overlap neutrons, which have a flight time greater than the time between beam pulses, contribute to the background and require careful attention. Detector pulses in a suitable range of pulse heights start a time to amplitude converter which is stopped by a signal from the next beam pulse. The shortest times then correspond to the greatest flight times and hence to the lowest neutron energies. A typical time-of-flight spectrum from an Fe sample irradiated with 3 MeV deuterons is shown in Fig. 4.21a. The neutron detector was placed at 6° to the incident beam direction. Peaks due to (d,n) reactions in ^{12}C, ^{14}N and ^{16}O were observed (Lorenzen, 1976). The ratio of background corrected areas of specific peaks in spectra from standard and unknown materials can be used to estimate the concentration of the relevant nuclide in the unknown sample provided that the stopping powers of the two samples are the same. Otherwise the methods of Chapter 12.2 must be used to calculate the concentration of each nuclide.

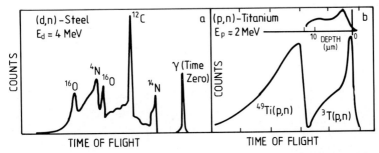

Fig. 4.21a. Neutron time-of-flight spectrum from 4 MeV deuteron irradiation of steel; b. neutron time-of-flight spectrum from 2 MeV proton irradiation of a tritiated Ti layer on Cu; above the high energy peak a depth profile of T derived from the measured spectrum is shown.

The relative positions and magnitudes of the peaks vary with the angle of observation which can therefore be chosen to suit particular problems. The possible contribution from heavier elements must also be kept in mind. For example, Cr, Fe and Ni as major elements may contribute a significant number of neutrons. Small dips can appear in time-of-flight spectra because of resonance absorption of neutrons in nitrogen and oxygen in the flight path between sample and detector. If necessary, this can be avoided by using a vacuum or He filled container rather than an air gap.

c. Depth Profiling

A depth distribution can be obtained from the shape of the time-of-flight spectrum of a specific energy group. The simplest method is to calculate a channel by channel ratio of counts from standard and unknown samples. If these have equivalent stopping powers, the result will be a depth profile but with a non-linear scale because of the square root relation between time of flight and neutron energy. The yield scale is also distorted becaused one channel corresponds to different depth intervals at different neutron energies.

The calculation of the true depth distribution of the nuclide of interest in a different matrix (m) to that of the standard (s) requires a different approach (Lefevre *et al.*, 1976; Lorenzen, 1976; Overley and Lefevre, 1976). The kinematic Equation (1.21), Table 1.3, and the time-of-flight Equation (1.7), Table 1.1, are used to calculate neutron energies, and hence time-of-flight intervals (Δt_m, Δt_s) corrresponding to equal depth intervals. The results are different for spectra from the standard and

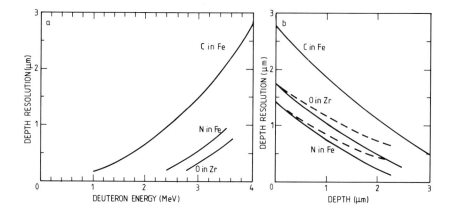

Fig. 4.22 Depth resolution in neutron time-of-flight profiling of C, N and O. a. Dependence on incident deuteron energy; b. dependence on interaction depth (full curves); the influence of energy straggling is included in the dashed curves.

unknown samples unless these have the same stopping power (ε). The ratio of yields in the calculated time intervals gives the required concentration at depth x:

$$f_m(x) = f_s(x) \, [Y_m(\Delta t_m)/Y_s(\Delta t_s)] \, [\varepsilon_s/\varepsilon_m] \qquad (4.9)$$

If the nuclide being determined is present in sufficient quantity to modify the stopping cross-sections, Equations (1.7), Table 1.1, and (4.10) must be integrated to obtain self-consistent values of the concentration profile and rate of energy loss.

A typical spectrum and corresponding depth profile of ^{3}H in a tritiated Ti layer on Cu is shown in Fig. 4.21b (Lefevre et al., 1976). In this case, neutrons are observed from (p,n) reactions in both ^{3}H and ^{48}Ti. If the total thickness of the ^{3}H containing layer was much greater, the two regions would overlap and analysis would become more complex.

The depth resolution ($\Delta x \ \mu m$), from time-of-flight measurements, depends on contributions from incident beam energy spread (ΔE_1), straggling (ΔE_{1s}) and the time resolution (Δt). Although these are not all necessarily characterised by Gaussian distributions, it is convenient to combine them in quadrature, Equations (12.37) and (12.38).

An unusual feature of the time-of-flight method is that the depth resolution depends on $E^{1.5}$(Chapter 12.2) and hence it improves with depth because of the reduced neutron energy—in spite of the effects of beam energy straggling. This is illustrated in Fig. 4.22a for (d,n) profiling of C, N and O (Lorenzen, 1976). It also improves at low incident beam energies or when using low Q-value groups, because lower energy neutrons are then produced. This is illustrated in Fig. 4.22b for the same reactions. Because the depth resolution is also dependent on the stopping cross-section of the sample, optimisation must be considered for each analytical problem using the methods described in detail by Lorenzen (1975). Measurements can be made of a depth profile at C, N and O levels of 1% in a period of 5 to 30 m using deuteron currents of 0.1 to 1 μA, depending on the concentrations. Detection limits are of the order of 100 μg g^{-1} and are best when using a heavy element matrix.

4.4.2 Performance

a. (p,n) Reactions

Nuclides with relatively low threshold energies and high cross-sections are ^{3}H, ^{7}Li and ^{9}Be (see Table 4.6). These and ^{11}B can be determined at levels of of 10 μg g^{-1} or less and have also been studied in a number of depth profiling applications. Typical depth resolutions are 0.3 to 3 μm depending on the length of flight path and other parameters. The maximum depth is 10 to 100 μm although in the ^{7}Li reaction corrections must be made for an excited state group with a threshold at 2.38 MeV.

At proton energies above 3 MeV, most nuclides will undergo (p,n) reactions and cross-sections increase rapidly. The minor isotopes ^{13}C, ^{15}N, ^{17}O and ^{18}O have lower (p,n) thresholds than the major isotopes of these elements. Advantage can be taken of this fact to detect the presence of minor isotopes in thin layers (Peisach, 1968). The Ca isotopes, ^{43}Ca and ^{48}Ca, have also been detected in this way.

b. (d,n) Reactions

Stripping reactions have mostly positive Q-values and relatively high and smooth cross-sections — particularly in light nuclides. Reaction yields generally increase with deuteron energy and broad maxima may be observed at various energies. The cross-sections usually have an angular distribution which is peaked at an angle near the incident beam direction. Nuclides which have been studied with (d,n) reactions include ^{2}D, ^{12}C, ^{14}N and ^{16}O. Neutrons are produced in various energy groups, which are broadened on the low energy side if a thick sample is used and so profiling is possible with a depth resolution of the order of 0.5 μm. The limiting thickness that can be profiled is set by the spacing between adjacent energy groups and is at least 10 μm (Lorenzen, 1976).

Measurements can be made of a depth profile at levels of 1% in a period of 5 to 30 min using beam currents of the order of 0.1 to 1 μA, depending on the concentrations. Detection limits are of the order of 100 μg g^{-1} and are best when using a heavy element matrix. All measurements are accompanied by neutrons from the $^2D(d,n)$ reaction arising from ^2D self-implanted into the sample during irradiation. At high deuteron energies, neutron backgrounds become quite high, being produced by heavy major isotopes in the sample as well as minor isotopes.

c. (t,n) Reactions

One such reaction has been used in the analysis of hydrogen, viz. $^1H(t,n)$. This has a high cross-section and can be used with similar depth resolution to that of (p,n) reactions. The main drawback is that neutron production can occur with most common materials used in the beam line and with hydrogen which is a common contaminant. Tests using a high Z target such as Au, have shown that background neutron levels can be equivalent to 3% H in the sample. Hazards arise in the use of triton beams from neutron production and from implantation of the ^3T into beam line components. Other methods of hydrogen profiling are thus generally preferred.

d. (α,n) Reactions

Overley *et al.* (1979) have pointed out that the $^{11}B(\alpha,n)$ reaction has a favourable cross-section and should give considerably better depth resolution in time-of-flight profiling than can be achieved with the $^{11}B(p,n)$ reaction.

e. Heavy Ion Reactions

One use of a heavy ion induced neutron reaction has been reported for He profiling with the $^4He(^{10}B,n)$ reaction. A liquid scintillator was used (Bottiger *et al.*, 1976) to determine the profile of implanted He with a resolution of 60 nm to a depth of 1 μm and a sensitivity of 1% of He.

4.5 ION ACTIVATION ANALYSIS

Activation analysis using an accelerated deuteron beam was demonstrated in the early stages of the development of nuclear techniques (Seaborg and Livingood, 1938). Many types of ions have since been used for activation analysis and the choice of type and energy of ion as well as the variety of reactions available make this a complex field of applications. However, because of the widespread availability of neutron activation analysis, ion activation is normally only considered if it offers special advantages for particular problems.

Hoste and Vandecasteele (1987) and Engelmann (1981) have reviewed ion activation analysis methods and applications and systematic studies for light ions include: protons — Barrandon *et al.* (1976), Debrun *et al.* (1976), Borderie *et al.* (1977); ^{3}He, ^{4}He — Borderie (1982), Engelmann (1981); and, heavy ions—Schweikert (1978, 1981).

In addition, there are many books and reviews which present the theory and practice of neutron activation analysis and much of this material is applicable to ion activation. The following sections therefore only give a brief description of ion activation.

4.5.1 Methods

a. Analysis

The acitivity (A) resulting from a specific nuclear reaction, which is present at the end of an irradiation, is obtained from Equation (12.11) for the reaction yield, with an additional term to account for the decay that takes place during irradiation:

$$A_4 = N_1(c_2N_0/M_1)\,\lambda_4(1 - \exp[-\lambda_4\tau])\int_o^E [\sigma(E_1)/S_m(E_1)]\,dE \quad (4.10)$$

where λ_4 is the decay constant of the product nuclide; and τ is the duration of the irradiation.

The saturation activity reached when τ is much greater than the half life of the product is:

$$A_4(SAT) = N_1(c_2N_0/M_1)\,\lambda_4\int_o^E [\sigma(E_1)/S_m(E_1)]\,dE \quad (4.11)$$

If adequate cross-section data is not available, especially for irradiation at relatively high energies (10 to 50 MeV), absolute determinations are not possible and the activity of known and unknown samples must be compared using the approximations discussed in Chapter 12.1.

b. Measurements

Although measurements are possible on rough surfaces, samples should preferably be smooth and flat. They may also need to be mounted on a cooled backing to prevent overheating during irradiation. Even then, the temperature of the sample surface may rise by many hundreds of degrees with beam currents of the order of 10 μA. This may be acceptable for some materials but it is desirable that they be baked before use to avoid excessive outgassing during irradiation. A conducting surface layer or foil may be needed when irradiating insulating materials to prevent sparking and consequent sample damage. The beam dose can be determined by current integration or by placing a thin standard foil in front of the sample and measuring the foil activation independently.

An important consideration is the beam uniformity since measurements of decay products are made with a different geometry to that involved in irradiation. Any changes in beam position or the distribution of current within the beam spot change the counting efficiency. It is therefore desirable to use such techniques as defocussing or oscillation to spread the beam and then collimate it to an accurately known diameter to ensure uniform irradiation of an area of the order of 1 cm^2. If the beam energy is higher than that required for sample activation, the energy can be reduced by energy loss in a filter foil of carefully chosen thickness. Samples are removed from the irradiation chamber and activities counted at successive intervals chosen to suit the half lives for the decay of specific radioisotopes. A light etch may be necessary to remove any activated surface contamination, including beam deposited carbon, if the best sensitivity is to be achieved. Surface removal techniques can be used to obtain depth profiles, for example in the study of the oxidation of metals (Perkins, 1977). It is also possible to remove a considerable thickness and only measure activities at depths below which a competing reaction is not energetically allowed (if the beam energy cannot be lowered).

Radioisotopes with short half-lives can be observed during irradiation and the background can be minimised by pulsing the beam or deflecting it so that measurements can be made while the beam is off the sample. It is also possible to use chemical separation to isolate specific radioisotopes before measuring their activities by gamma ray spectrometry.

Charged particle bombardment often leads to the creation of positron emitters which can be detected by beta counting or the measurement of annihilation radiation (511 keV). The identification of a specific product radionuclide is then only possible by the measurement of decay curves and the extraction of specific half life components. Internal conversion and electron capture are also common so that delayed X-ray emission can be used to extend the scope of ion activation methods (McGinley and Schweikert, 1976).

4.5.2   Performance

The advantages of ion activation include:

- sensitivities down to 1 ng g^{-1} for a different suite of isotopes to those for which neutron activation is most sensitive;
- the choice of ion type and energy provide a versatility for optimising sensitivity and selectivity to suit particular problems;
- radioactivity is only produced in a surface layer, the thickness of which depends on the ion energy and the variation of reaction cross-section with energy — the total activity produced is therefore low;
- incident particle fluxes can be very high (10 μA = 6.24 $\times$ 10^{13} ions s^{-1}) which offsets relatively low cross-sections; and
- accelerators are available in many laboratories which do not have facilities for neutron activation analysis.

There may also be disadvantages such as:

- sample heating when a high beam current is used may lead to changes or damage to the sample;
- cross-sections are generally lower than those for neutron activation analysis;
- counting times of 10 to 120 h are necessary to achieve high sensitivity and this reduces the number of samples that can be analysed; and
- ionisation energy loss prevents the irradiation of large volumes and so the activity produced is limited by the ion range.

Higher beam energies are generally used for activation analysis than for prompt nuclear analysis except for a few cases of light isotope activation. For example, proton energies from 5 to 20 MeV are needed to obtain a high enough cross-section for good sensitivity in PAA. However, the higher the energy the more likely it is that interfering reactions will create difficulties and the best choice of ion type and energy depends on the problem in hand. A catalog of some of the reactions available at energies up to 10 MeV is given in Chapter 14.4.5 and typical values of estimated detection limits are plotted in Fig. 4.23 for 3600 μC irradiations and up to 60 h counting times. PAA has excellent sensitivity for such light elements as B, C, N and O and is unique amongst activation techniques for the use of heavy ion beams for the determination of H and He with sensitivities below 1 μg g^{-1}. Elements such as Ti, Sr, Mo and Pb are also of special interest because they are difficult to determine by NAA but can be determined at similar levels by PAA. The fact that at least 70 elements

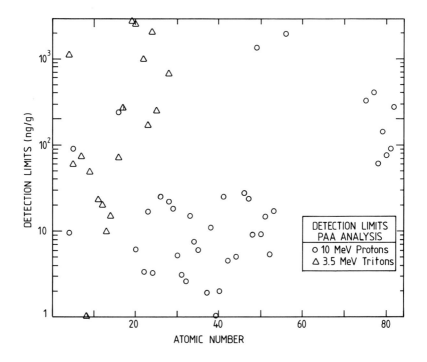

Fig. 4.23 Estimated detection limits in PAA using 10 MeV protons (Barrandon *et al.*, 1976; Debrun *et al.*, 1976) and 3.5 MeV tritons (Borderie *et al.*, 1977).

can be determined at these levels makes PAA a powerful technique when a sufficiently large accelerator is available.

4.5.3 Thin Layer Activation

Ion induced radioactivity occurs at depths less than the range of the incident ions so that only a thin layer is activated rather than the whole sample. Furthermore, a small beam area can be used to activate only that part of the surface of a large object which is of interest. These are major advantages in the study of surface removal processes since, although the activity in the region of interest may be reasonably high, the total activity can be so low that normal handling of the specimen is possible.

For example the $^{56}Fe(p,n)^{56}Co$ reaction has a threshold at 5.35 MeV and a broad cross-section maximum in the region of 10 MeV. Irradiation of the surface of an iron or steel object will therefore produce ^{56}Co activity in a layer of approximately 0.25 mm thickness at depths from 0

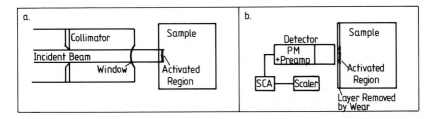

Fig. 4.24a. Thin Layer Activation by 15 MeV proton beam; b. measurement of activity after surface wear.

to 1 mm or more, depending on the proton energy used (Fig. 4.24a). The size and shape of the activated area can be controlled by changes in beam focussing and/or scanning and this area can be positioned as required on the object by suitable sample mounting systems.

A gamma-ray detector placed close to operating machinery will monitor the drop in radioactivity as wear takes place (Fig. 4.24b). A sensitivity of the order of 0.2 μm can be achieved in this way. Alternatively, material removed from the surface can be trapped in a filter (for example in oil or coolant stream) and a very high sensitivity can then be obtained, e.g. 10^{-7} cm^3.

Thin layer activation has been shown to be a versatile technique for the study of wear, corrosion and other surface degradation problems in engines, machine tools and many kinds of industrial components (Conlon, 1982; Jeanneau, 1983). Elements which have been listed as suitable for thin layer activation are shown in Table 4.7 although further development will undoubtedly increase this list.

TABLE 4.7
Some nuclear reactions used for thin layer activation

Matrix	Reaction	Product	E_1 (MeV)	Useful Depth (μm)	Material
^{48}Ti	(p,n)	^{48}V		25	Ti
^{52}Cr	(p,n)	^{52}Mn			
^{56}Fe	(p,n)	^{56}Co	13	300	steel
^{56}Fe	(d,n)	^{57}Co	9.2	150	cast iron
^{58}Fe	(^{6}Li,α)	^{58}Co	24	25	steel
^{92}Zr	(p,n)	^{92}Nb			

4.6 CHOICE OF REACTION

The catalog of nuclear reactions which have been used for sample analysis and depth profiling (Chapter 14.4.1) is far from a complete list of all possibilities but it shows that there is a considerable choice available for the study of any one isotope ($Z < 17$). The selection of a preferred reaction depends on the problem, the material and the equipment available but experience points to obvious advantages in the use of specific reactions in many cases.

- ^1H — the ($^{15}N,\alpha\gamma$) reaction gives the best depth resolution and yield for profiling at a moderate beam energy (> 6.385 MeV); the ($^7Li,\gamma$) reaction requires a lower energy and has a greater maximum depth while the ($^{19}F,\alpha\gamma$) reaction at either 6.418 or 16.586 MeV is often used for H determination or profiling.
- ^2D — the ($^3He,\alpha$) reaction gives the best depth resolution; the ($^3He,p$) and (d,p) reactions are useful for D determination while the ($^{15}N,n\gamma$) reaction has been shown to have better sensitivity.
- ^3T — the (d,α) reaction gives excellent sensitivity and moderate depth resolution.
- ^4He — two reactions have been shown to have reasonable depth resolution but poor sensitivity.
- ^6Li — the (d,α) reaction gives reasonable performance.
- ^7Li — both the (p,p'γ) and ($\alpha,\alpha'\gamma$) reactions give good sensitivity for Li determination and the (p,γ) reaction is useful for depth profiling.
- ^9Be — the ($\alpha,n\gamma$) reaction has excellent sensitivity (even being used with alpha sources for portable Be monitors); the (p,α) reaction is useful for depth profiling.
- ^{10}B — the (p,$\alpha\gamma$) reaction gives good sensitivity.
- ^{11}B — the (p,α) and (p,γ) reactions are useful for profiling and B determination.
- ^{12}C — the (d,p_o) reaction is most used but the ($^3He,p$) reaction has better depth resolution; both can be used in simultaneous C, N and O determination in thin layers; the (d,pγ) reaction can be used for thick samples or (p,p'γ) reactions at energies above 6 MeV.
- ^{13}C — the (p,γ), (d,p) and ($^3He,p$) reactions can be used for determining and profiling and C isotope ratios.
- ^{14}N — the (d,p_o) and (d,α_o) reactions can be used together for nitrogen determination but the (d,α) reaction is the best to use for simultaneous C, N and O determination in thin layers and for N profiling; other (d,p) energy groups in these reactions are also useful as is the (d,pγ) reaction for thick samples.

- ^{15}N — the (p,α) or (p,$\alpha\gamma$) reactions are useful for ^{15}N determination.
- ^{16}O — the (d,p) reaction is most used for profiling and simultaneous C,N and O determination in thin layers but the (^{3}He,α) reaction is also useful, the (d,pγ) reaction can be used for thick samples or (p,p'γ) reactions above 6 MeV; the (t,n) reaction can be used for activation analysis and autoradiography.
- ^{17}O — is little studied because of its low abundance.
- ^{18}O — the (p,α) reaction at the narrow 629 keV resonance or broad 846 keV resonance is widely used in stable tracer studies and profiling; the (d,p) or (d,α) reactions can also be used; activation analysis with the (p,n) reaction gives good sensitivity and spatial resolution by autoradiography.
- ^{19}F — the (p,p'γ) gives the best sensitivity for F determination and the (p,$\alpha\gamma$) reaction can be used for profiling or analysis.
- ^{20}Ne — can be profiled with the (p,γ) reaction and determined from the resulting ^{21}Na activity.
- ^{23}Na — excellent sensitivity is achieved with the (p,p'γ) or (α,$\alpha'\gamma$) reactions and the (p,α) or (p,$\alpha\gamma$) reactions have narrow resonances for depth profiling.
- ^{24}Mg — has relatively low reaction yields and hence poor sensitivity.
- ^{25}Mg — the (p,p'γ) reaction gives reasonable sensitivity for Mg determination.
- ^{26}Mg — can be determined with the (p,γ) reaction but there is serious interference from Al.
- ^{27}Al — can be determined with good sensitivity by the (p,p'γ) reaction and profiled with a narrow resonance in the (p,γ) reaction.
- ^{28}Si — can be determined with reasonable sensitivity using the (d,p) or (p,p'γ) reactions.
- ^{30}Si — has a narrow resonance in the (p,γ) reaction suitable for profiling.
- ^{31}P — the (p,α) reaction gives the best sensitivity, the (p,p'γ) reaction is useful above 3 MeV; both (p,α) and (p,γ) can be used for profiling.
- ^{32}S — the (d,p) and (p,p'γ) reactions have been used for S determination and profiling.

Heavier elements have mostly been analysed using PIGME techniques or by activation analysis with relatively high energy ions. The sensitivities for these techniques are illustrated in Figs. 4.19 and 4.23.

REFERENCES

Amsel, G., Beranger, G., De Gelas, B. and Lacombe, P. (1968) *J. Appl. Phys.* **39**, 2246

Barrandon, J.N., Benaben, P. and Debrun, J.L. (1976). *Analytica Chimica Acta* **83**, 157.

Basutcu, M. (1980). Thesis, Univ. Paris FRNC-TH-1015.

Bird, J.R., Campbell, B.L. and Cawley, R.J. (1978). 'Prompt Nuclear Analysis Bibliography', AAEC/E443.

Bodart, F. and Ghoos, L. (1980). *Int. Rad. Prot. Assoc.* **3**, 65.

Borderie, B. (1980). *Nucl. Instrum. Methods* **175**, 465.

Borderie, B. (1982). 'Analyse par Activation a l'aide de Tritons, Particules Alpha et Ions Lourds', Thesis, U. Paris-Sud Orsay.

Borderie, B. and Barrandon, J.N. (1978). *Nucl. Instrum. Methods* **156**, 483

Borderie, B., Barrandon, J. N. and Debrun, J.L. (1977). *J. Radioanal. Chem.* **37**, 297.

Borderie, B., Barrandon, J.N., Delaunay, B. and Basutcu, M. (1979). *Nucl. Instrum. Methods* **163**, 441

Bottiger, J. (1978). *J. Nucl. Materials* **78**, 161.

Bottiger, J., Picraux, S.T. and Rud, N. (1976). 'Ion Beam Surface Layer Analysis', Eds. Meyer, O., Linker, G. and Kappeler, F., Plenum Press, New York, 811.

Coetzee, P.P., Pretorius, R. and Pelsach, M. (1975). *Nucl. Instrum. Methods* **131**, 299.

Conlon, T.W. (1982). *Contemp. Phys.* **23**, 353.

Conlon, T.W. and Parker, D.J. (1980). *Nucl. Instrum. Methods* **177**, 199.

Damjantschitsch, H., Weisser, M., Heusser, G., Kalbitzer, S. and Mannsperger, H. (1983) *Nucl. Instrum. Methods* **218**, 129.

Davies, J.A., Jackman, T.E., Plattner, H. and Bubb, I. (1983) *Nucl. Instrum. Methods* **218**, 141.

Debrun, J.N., Barrandon, J.N. and Benaben, P. (1976). *Anal. Chem.* **48**, 167

Deconninck. G.(1978). 'Introduction to Radioanalytical Physics', Elsevier, Amsterdam.

Deconninck, G. and Demortier, G. (1973). 'Nuclear Techniques in the Basic Metal Industries', IAEA, Vienna, 573.

Deconninck, G., Demortier, G. and Bodart, F. (1981). *Atomic Energy Review* **19**, 151.

Demortier, G. (1974). *Radiochem. Radioanal. Lett.* **20**, 197.

Dieumegard, D., Maurel, B. and Amsel, G. (1980). *Nucl. Instrum. Methods* **168**, 93.

Engelmann, C. (1981). *At. En. Rev.* **19**, 107.

England, J.B.A. (1974). 'Techniques in Nuclear Structure Physics', Macmillan, London.

Falk, W.R., Abou-Zeid, O. and Roesch, L.P.(1976). *Nucl. Instrum. Methods* **137**, 261.

Gihwala, D. and Peisach, M. (1980). *J. Radioanal. Chem.* **55**, 163.

Gihwala, D. and Peisach, M. (1982). *J. Radioanal. Chem.* **70**, 287.

Giles, I.S. and Peisach, M. (1976). *J. Radioanal. Chem.* **32**, 105.

Giles, I.S. and Peisach, M. (1979). *Cement and Concrete Res.* **9**, 591.

Giles, I.S., Olivier, C. and Peisach, M. (1977). *J. Radioanal. Chem.* **37**, 141.

Gossett, C.R. (1980). *Nucl. Instrum. Methods* **168**, 151

Goulding, F.S. and Harvey, B.G. (1975). *Ann. Rev. Nucl. Sci.* **25**, 167.

Gove, N.B. and Wapatra, A.H. (1972). *Nucl. Data Tables* **11**, 128.

Heck, D. (1988). *Nucl. Instrum. Methods* **B30**, 486.

Hoste, J. and Vandecasteele, C. (1987). *J. Radio and. Nucl. Chem.* **110**, 427.

Ishii, K., Valladon, M. and Debrun, J.L. (1978a). *Nucl. Instrum. Methods* **153**, 213.

Ishii, K., Valladon, M., Sastri, C.S. and Debrun, J.L. (1978b). *Nucl. Instrum. Methods* **153**, 503.

Jeanneau, B (1983). *IEEE Trans. Nucl. Sci.* NS-30, 1614.

Kenny, M.J., Bird, J.R. and Clayton, E. (1980). *Nucl. Instrum. Methods* **168**, 115.

Kregar, M., Mueller, J., Rupnik, P., Ramsak, V. and Spiler, F. (1979). 'Nucl. Act. Tech. in Life Sciences', IAEA-SM-227, **83**, 407

Lanford, W.A. (1978). *Nucl. Instrum. Methods* **149**, 1.

Lefevre, H.W., Davis, J.C. and Anderson, J.D. (1976). Proc. 4th Conf. on Sci. Ind. Appl. of Small Accelerators, IEEE, NY, 225.

Lenz, T., Baumann, H. and Rauch, F. (1987). *Nucl. Instrum. Methods* **B28**, 280.

Ligeon, E., Bruel, M., Bontemps, A., Chambert, G. and Monnier, J. (1973). *J. Radioanal. Chem.* **16**, 537.

Lightowlers, E.C., North, J.C., Jordan, A.S., Derick, L. and Merz, J.L. (1973). *J. Apl. Phys.* **44**, 4758.

Lorenzen, J. (1975). AE-502, Studsvik, Sweden.

Lorenzen, J. (1976). *Nucl. Instrum. Methods* **136**, 289.

Malmqvist, K.G., Johansson, G.I. and Akselsson, K.R.(1982). *J. Radioanal. Chem.* **74**, 125.

Marcuso, T.L.M., Rothman, S.J., Nowicki, L.J. and Baldo, P. (1983). *Nucl. Instrum. Methods* **211**, 227.

Maurel, B., Amsel, G., and Nadai, J.P. (1982). *Nucl. Instrum. Methods* **197**, 1

McGinley, J.R. and Schweikert E.A. (1976). *Anal. Chem.* **48**, 429

McMillan, J.W., Hirst, P.M., Pummery F.C.W., Huddleston, J. and Pierce, T.B. (1978). *Nucl. Instrum. Methods* **149**, 83.

Möller, W., Hufschmidt, M. and Kamke, D.(1977). *Nucl. Instrum. Methods* **140**, 157.

Olivier, C. and Peisach, M.(1985). *S. Afr. J. Chem.* **38**,169.

Olivier, C., McMillan, J.W. and Pierce, T.B. (1975). *Nucl. Instrum. Methods* **124**, 289.

Olivier, C. Peisach, M. and Pierce. T.B.(1976). *J. Radioanal. Chem.* **32**, 71.

Overley, J.C. and Lefevre, H.W. (1976). 'Radiation Effects on Solid Surfaces', Ed. Kaminaky, M., *Adv. Chem. Ser.* **158**, 282.

Overley, J.C., Ebright, R.P. and Lefevre, H.W. (1979). IEEE *Trans. Nucl. Sci.* NS-26, 1624.

Peisach, M. (1968). 'Practical Aspects of Activation Analysis with Charged Particles', Ed.Ebert, H.G., EUR-3896, 65.

Peisach, M. (1972). *J. Radioanal. Chem.* **12**, 251.

Peisach, M.(1981). *J. Radioanal. Chem.* **61**, 24.

Peisach, M. and Gihwala, D. (1981). *J. Radioanal. Chem.* **61**, 37

Peisach, M., Jacobson, L., Boulle, G.J., Gihwala, D. and Underhill, L.G. (1982), *J. Radioanal. Chem.* **69**, 349.

Perkins, R.A. (1977). *J. Nucl. Mat.* **68**, 148.

Pomorski, L., Karcz, W. and Jarzmik, B. (1976). *Nukleonika* **21**, 1089.

Pretorius, R. and Peisach, M. (1978). *Nucl. Instrum. Methods* **149**, 69

Pronko, P.P. and Pronko J.G. (1974). *Phys. Rev.* **89**, 2870.

Raith, B., Wilde, H.R., Roth, M.S., Stratmann, A. and Gonsior, B.(1980). *Nucl. Instrum. Methods* **168**, 251

Ricci, E. (1971). *Anal. Chem.* **43**, 1866.

Rubin, S., Passell, T.O. and Bailey, L.E.(1957). *Anal. Chem.* **29**,736.

Schweikert, E.A.(1978).'From Idea to Application: Some Selected Nuclear Techniques in Research and Development', IAEA, STI(PUB) **476**, 1.

Schweikert, E.A.(1981). *J. Radioanal. Chem.* **64**, 195.

Seaborg, G.T. and Livingood, J.J. (1938). *J. Am. Chem. Soc.* **60**, 1784.

Sellschop, J.P.F., Annegarn, H.J., Keddy, R.J., Madiba, C.C.P. and Renan, M.J. (1978). *Nucl. Instrum. Methods* **149**, 321.

Simpson, J.C.B. and Earwaker, L.G. (1986). *Nucl. Instrum. Methods* **B15**, 502.

Thomas, J.P., Engerran, J. and Tousset, J. (1975). *J. Radioanal. Chem.* **25**, 163.

Turos, A., Wielunski, L. and Olenski, J. (1973a). *Phys. Stat. Sol.* **16(a)**, 211.

Turos, A., Wielunski, L., Barcz, A. and Olenski, J. (1973b). *J. Radioanal. Chem.* **16**, 627.

Wilenzick, R.M. (1972). *Nucl. Instrum. Methods* **103**, 429.

Xenoulis, A.C. and Douka, C.E. (1979). *J. Radioanal. Chem.* **54**, 205.

5
Ion Induced X-ray Emission

D.D. COHEN

Australian Institute of Nuclear Science and Engineering, Menai, Australia

E. CLAYTON

ANSTO, Menai, Australia

ION BEAMS FOR
MATERIALS ANALYSIS
ISBN 0 12 0997401 1

5.1 INTRODUCTION

This chapter is concerned with particle induced X-ray emission (PIXE), although the term PIXE is also commonly used to refer to just proton induced X-ray emission. The definitive review by Johansson and Johansson (1976) of PIXE was probably the beginning of an international PIXE explosion. Since then many papers have appeared on methods and applications, e.g. the proceedings of the last three international PIXE conferences (Johansson, 1977, 1981; Martin, 1984).

Ion induced X-ray emission is a novel and powerful tool for multi-element non-destructive trace element analysis of small samples, taking typically only a few minutes irradiation for each sample. It employs ion energies in the range 0.5 to 10 MeV/amu and solid state Si(Li) X-ray detectors. Most elements above Na can be analysed in the X-ray energy range 1 to 100 keV. With crystal spectrometers or windowless semiconductor detectors the range can be extended down to Be. The energy of the emitted X-ray is characteristic of the bombarded atom and the number of characteristic X-rays produced is proportional to the elemental concentration. Twenty-five to 30 elements can be analysed simultaneously with detection limits below 1 $\mu g\,g^{-1}$ in some instances. Fig. 5.1 shows a typical X-ray spectrum for an oil shale. Many elements are detected and the spectrum is relatively complicated.

In this chapter the concepts are explained and the details needed to construct a PIXE system given. Useful data can also be found in Chapter 14.5.

5.2 CONCEPTS AND THEORY

Ion induced X-ray emission is a multi-stage process; firstly the ion creates vacancies in the electron shells of the target atom and, secondly, these vacancies are filled by outer shell electrons and the excess energy taken

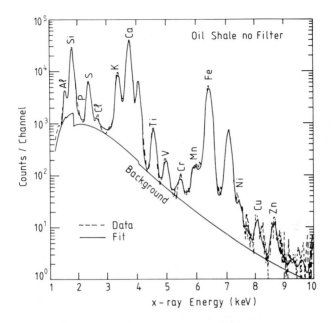

Fig. 5.1 A typical X-ray spectrum from an oil shale. The dashed lines are the data and the solid lines are a fit using a program discussed later.

away by either photons or Auger electrons. Fig. 5.2 shows the most commonly occurring transitions for an initial K or L shell vacancy in one of any of their subshells. Both the conventional (Siegbahn) and spectroscopic notations are given. The difference between these two notations arises in their naming of the target atom electron energy levels. The lines are generally grouped into three main subgroups, α, β and γ, according to their X-ray energies. The α lines are lower in energy and more intense than the β lines which are, in turn, higher in energy and more intense than the γ lines. There are some 13 K lines, 37 L lines and 39 M lines commonly observed although for lower atomic number elements only a detector of the highest energy resolution will observe all these lines.

5.2.1 X-ray Transitions

The conventional (Siegbahn) notation uses the symbols K, L_1, L_2, L_3, M_1, and so on, to label the electron subshells going outwards from the nucleus (decreasing in binding energy). The K shell has no subshells, the L shell has three subshells, labelled L_i (i = 1 to 3), the M shell has five

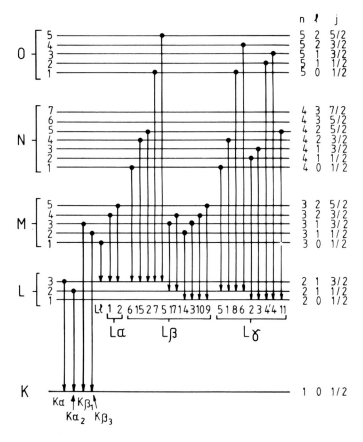

Fig. 5.2 Energy level diagram (not to scale), conventional (Siegbahn) notation on the left and spectroscopic notation on the right. Transitions giving rise to *K* and *L* series lines only are indicated.

subshells, labelled M_i (i = 1 to 5) and the *N* shell has seven subshells labelled N_i (i = 1 to 7). These letters are used to label the characteristic X-ray lines according to the electron transitions that produce them.

The spectroscopic notation uses the three quantum numbers n, 1 and j to label the electron subshells. The principal quantum number n = 1, 2, 3 . . . is analogous to the major shells *K, L, M,* . . . The symbols s, p, d, f, . . . are used to represent the subshells with orbital angular momentum 1 = 0, 1, 2, 3 . . . respectively, while the quantum number j = 1 + s, where s = 0 or 1/2 is the usual electron spin number. Each subshell is then labelled nlj, hence the L_3 subshell, in the spectroscopic notation, is written $2p_{3/2}$ and the $L\alpha_1$ X-ray is produced by the transition $2p_{3/2} - 3d_{5/2}$.

Allowable electron transitions have $\Delta n > 0$, $\Delta l = \pm 1$ and $\Delta j = 0$, ± 1; transitions not having these values do occur with very low probability and are called forbidden transitions.

Semiconductor X-ray detectors have an energy resolution greater than or of the order of 140 eV and because of entrance window limitations are generally very inefficient for X-ray energies below ≈ 1 keV. This limits the number of both detectable and resolvable X-ray lines to 2 to 3 K lines, 9 to 13 L lines and less than 6 M lines. A modern semiconductor detector, having significant efficiency for X-rays with energies between 1 and 60 keV, could observe K X-rays from elements Na to W, L X-rays from elements Zn upwards and M X-rays from elements Dy upwards. A Bragg crystal spectrometer X-ray detector with energy resolution ≈ 5 eV can be used to observe many more X-ray lines but only one can be tuned at any one time (Goldstein *et al.*, 1981).

5.2.2 Thin and Thick Target Yields

A thin target is one in which corrections for projectile energy loss and X-ray absorption are negligible, while a thick target is one in which the incident ion is completely stopped and corrections are needed for X-ray absorption. For thin targets it is easily verified that the X-ray production cross section, $\sigma_p^x(E_1)$, and the number of X-ray counts, $I_p(E_1)$, in a peak p for an incident ion energy E_1 are related by the simple expression (Cohen, 1984a):

$$I_p(E_1) = 4.968.10^{-17} \, N_2 x \, \sigma_p^x \, Q \, \Omega \, \varepsilon / \sin(\theta_i) \qquad (5.1)$$

where N_2 is the number of target atoms cm^{-3}, x is the target thickness (μg cm^{-2}), Q is the total charge hitting the target (μC), Ω is the detector solid angle (steradians), θ_i is the angle between the surface normal and the beam direction and ε is the total detection efficiency.

If (μ/ρ) is the X-ray mass attenuation coefficient, in cm^2 g^{-1}, for the peak p of X-ray energy E_x (keV) and ΔE is the energy lost by the ion of initial energy E_1 (MeV) in traversing the target of density ρ (g cm^{-3}) and thickness $t = \rho x$ (g cm^{-2}) and the X-ray emerges at an angle θ_0 to the target surface normal then, for thin targets, the projectile energy loss and X-ray absorption corrections are made to first order in the following manner. In Equation (5.1), E_1 is replaced by the mean ion energy $[E_{1-}(\Delta E/2)]$ and assuming all the X-rays are produced at the centre of the target x is replaced by $x \exp[-0.5 \, (\mu/\rho)t \sec(\theta_o)]$. Hence Equation (5.1) with projectile energy loss and emergent X-ray absorption corrections becomes:

$$I_p\,(E_1 - \Delta E/2) = 4.968.10^{-17}N_2\,\sigma_p^x\,Q\,\Omega\varepsilon x$$
$$\exp\,[-0.5\,(\mu/\rho)\,t\sec\,(\theta_o)]/\sin\,(\theta_i) \tag{5.2}$$

For targets less than $500\,\mu g\,cm^{-2}$ thick these corrections for X-ray energies greater than about 5 keV are less than 10%. Comparison with thick target calculations have shown Equation (5.2) to be surprisingly accurate even for targets as thick as a few mg cm^{-2}. A good rule of thumb is that Equation (5.2) is sufficiently accurate provided the target thickness x is less than one quarter of the ion range R in the target. Nielson et al. (1976) and Maenhaut et al. (1981) have discussed other techniques for thin target corrections of this type.

When a thick target is used, corrections for the slowing down of the incident ion and the absorption of the emergent X-ray, in the target, become excessive and a more rigorous treatment is required. Merzbacher and Lewis (1959) gave the following expression for the X-ray line intensity, $I_p\,(E_1)$, as a function of the X-ray production cross section (in barns) for a peak p:

$$I_p(E_1) \propto \Omega\varepsilon \int_{E_o}^{0} \sigma_p^x(E_1)\,\frac{dE}{S(E)}\,\exp(-\mu\,x) \tag{5.3a}$$

$$\mu\,x = (\mu/\rho) \int_{E_o}^{E} \frac{dE}{S(E)}\,\frac{\cos\,(\theta_i)}{\cos\,(\theta_0)} \tag{5.3b}$$

where $S(E_1)$ is the ion stopping power at energy E_1, in MeV cm^2 g^{-1} (Andersen and Ziegler, 1977), and θ_i and θ_0 are the ion and X-ray directions relative to the target surface normal. The integrals in Equation (5.3) are usually calculated numerically using stopping power tables (Andersen and Ziegler, 1977) and mass absorption coefficient tables (Mayer and Rimini, 1977).

The Merzbacher-Lewis relation, Equation (5.3), was derived for protons and, in order that it be valid for heavier ions, several assumptions must be fulfilled. Taulbjerg and Sigmond (1972) and Taulbjerg et al. (1973) have discussed the contributions to X-ray production by target recoil and ion straggling effects. Brandt and Laubert (1975) reported analytical relationships to estimate such contributions to the X-ray yield. It is always advisable to calculate the contributions of these effects, when using ions heavier than alphas before applying Equation (5.3). Because of uncertainties in the ionisation cross-section (5%), stopping powers (3%) and mass attenuation coefficients (3%), the accuracy of Equation (5.3) is in the range 10 to 15% (Clayton, 1983; Campbell, 1983). However each

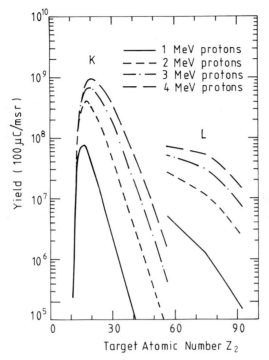

Fig. 5.3 Pure thick target X-ray yields versus target atomic number for protons of energy 1, 2, 3 and 4 MeV. Curves are for $K\alpha$ and $L\alpha$ X-ray yields.

type of sample should be examined separately as target current errors, and hence Q measurements depend strongly on target material.

Fig. 5.3 shows the thick target X-ray yield per $100\,\mu C$. msr versus the target atomic number for various proton energies, as calculated using Equation (5.3). Yields are for pure elements, for $K\alpha$ and $L\alpha$ peaks, assuming the detector efficiency of Fig. 5.9. The very high data acquisition rate possible for $10 < Z_2 < 92$ is one of the major advantages of the PIXE technique. When the $K\alpha$ yields become small ($Z_2 > 50$), $L\alpha$ yields take over and reasonable sensitivity is obtained over a large fraction of the periodic table. The sharp drop in yield near $Z_2 = 12$ is produced by the rapidly falling detector efficiency in this region. In fact the X-ray production cross-section may still be rising (depending on the ion energy) and better sensitivities can be obtained down to $Z_2 = 4$ if windowless Si(Li) or other more efficient detectors in the X-ray energy region less than 2 keV are used. The slower fall for $K\alpha$ yields above $Z_2 = 30$ and $L\alpha$ yields above $Z_2 = 80$ are produced by the falling ionisation cross-sections for a given ion energy.

a.

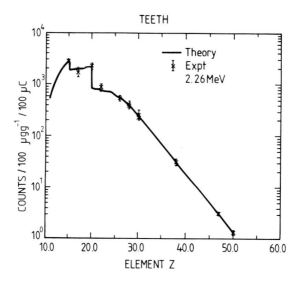

b.

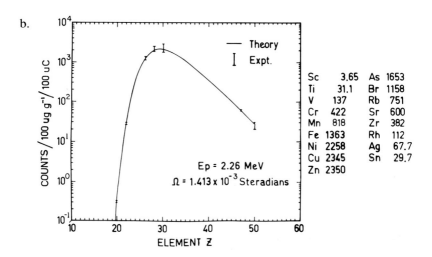

Fig. 5.4 The $K\alpha$ yield for a thick apatite matrix (a) without filter and (b) with 1.5 mm Perspex filter. Both curves are for 100 μC of 2.26 MeV protons and for 100 μg g^{-1} of trace element.

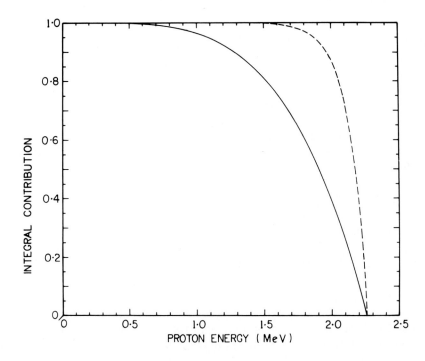

Fig. 5.5 The contribution to the total X-ray yield for Ca as a function of residual proton energy for C (solid line) and Si (dashed line) matrices. The incident energy is 2.26. MeV.

The X-ray yield increases by more than an order of magnitude when the proton energy is increased from 1 to 4 MeV. The maximum ionisation cross section occurs around an ion energy of $(134U^2n^4/Z_2^2)$ MeV/amu where U (keV) is the electron binding energy of the target electrons nth shell. This is called the velocity matching peak and occurs when ion and target shell electrons have similar velocities. For K shell ionisation of P ($U = 2.103$ keV, $Z_2 = 15$) and Ca (U = 3.691 keV, $Z_2 = 20$) this corresponds to 2.4 MeV/amu and 4.6 MeV/amu respectively. This is one of the reasons why the popular choice of ion energies for PIXE lies between 2 and 4 MeV/amu.

Generally, however one is not dealing with a pure matrix as shown in Fig. 5.3, but with a range of trace elements in a given matrix which for example may be silica or C. In these situations Equation (5.3) produces yield curves which are somewhat different to those shown in Fig. 5.3 and are characteristic of the matrix chosen. Such an example is shown in Fig. 5.4 (Cohen *et al.*, 1981), where the yield curve for $K\alpha$ X-rays for trace elements Na to Sn is shown for human teeth. Calcium hydroxy apatite, $Ca_{10}(PO_4)_6(OH)_2$ has been used to model the matrix for human teeth. The two figures correspond to (a) no filter and (b) 1.5 mm Perspex filter placed between the target and the detector. The two sharp discontinuities in the no filter case correspond to the P and Ca K absorption edges. The points shown are the experimentally measured ones and the ratio of experiment to theory was found to be (1.03 ± 0.07) over the trace element range $11 \leq Z_2 \leq 50$.

Fig. 5.5 shows the percentage of X-rays produced by a proton during slowing down to the energy shown as abcissa as a function of proton energy loss in a sample. Even for thick targets over 90 % of the yield comes before the incident proton has lost 30 % of it's energy. The contrast between the yields for C and Si show the importance of matrix composition to the yield. Yields for heavier ions are expected to scale approximately as Z_1^2, provided they are calculated at the same ion velocity (i.e. equal E_1/M_1). Hence heavier ions have larger X-ray yields. But there are two major disadvantages. Firstly, they tend to produce many more nuclear reactions (for ion energies above the Coulomb barrier) and hence many more gamma rays, increasing the photon background and reducing the PIXE sensitivity. Secondly, their range in the target material is approximately inversely proportional to Z_1^2 and hence they sample much less volume of the target material than protons, making them more sensitive to surface roughness and grain size effects.

5.2.3 X-ray Emission Rates and Line Intensities

As already pointed out a vacancy in a particular subshell can be filled by various transitions from different higher subshells. To calculate the probability of a particular transition i occuring we must know the ratio of the partial radiative width, $\Gamma_i(R)$, for that transition to the total radiative width, $\Gamma_s(R)$ of that subshell s. The ratio $S_i = \Gamma_i(R)/\Gamma_s(R)$ is called the relative X-ray emission rate for the transition i in the subshell s. Widths are defined in electron volts (eV) and transition rates in units of (eV/$\hbar$), where $1 \quad eV/\hbar = 1.5193 \times 10^{15} s^{-1} = 3.6749 \times 10^{-2}$ a.u.$^{-1}$. Scofield (1972) has calculated theoretical K and L shell radiative transition rates for elements from $Z_2 = 5$ to 102 using the relativistic Hartree-Slater potential for these atoms. Salem *et al.* (1974) compared these calculations with all experimental K and L subshell data up to 1974 from experiments with radioactive sources or samples ionised by photon or electron bombardment. Data obtained by positive ion bombardment were excluded since such bombardment creates multiple vacancies especially for the heavier ions.

The emission rates of lines within a subshell are important in identifying elements in quantitative X-ray analysis. The K family consists of two recognisable lines, $K\alpha$ (1) and $K\beta$ (0.1). The values in parentheses give approximate relative intensities since they vary with element and subshell ionisation cross-sections. The L series consists of $L\alpha$ (1), $L\beta_1(0.7)$, $L\beta_2(0.2)$, $L\beta_3(0.08)$, $L\beta_4(0.05)$, $L\gamma_1(0.08)$, $L\gamma_3(0.03)$, $Ll(0.04)$ *and* $L\eta(0.01)$. The M series consists of $M\alpha(1)$, $M\beta(0.6)$, $M\gamma(0.05)$, $M\zeta(0.06)$ and $M_2N_4(0.01)$. A full tabulation of L shell line intensities for the 16 commonly occurring lines, for most elements, has been given by Cohen and Harrigan (1986) for both proton and He ion bombardment. Ratios for lines originating from vacancies in different subshells are related by the subshell ionisation cross-sections, the fluorescence yields and the Coster-Kronig transition rates in a manner discussed in the next section.

5.2.4 Fluorescence Yields and Coster-Kronig Transitions

The comparison of the total X-ray production cross-section, σ^x, and the theoretically calculated total ionisation cross-sections, σ^I involves the radiative probability, that the initial vacancy will decay producing an X-ray. Thus, the fluorescence yield ω of a shell is equal to the number of photons emitted when vacancies in the shell are filled, divided by the number of primary vacancies N_s in the shell. For the K shell there is only one subshell and the fluorescence yield is written as ω_K while for the L and M shells there are three and five subshells and the fluorescence yields are written as ω_i where $i = L_1, L_2, L_3$ or M_1, M_2, M_3, M_4, M_5 for the L and M shells respectively. The K shell ionisation cross-section, σ_K^I, is related to the total K shell X-ray production cross-section by the K shell fluorescence yield ω_K by:

$$\sigma_K^x = \omega_K \sigma_K^I \qquad (5.4a)$$

where ω_K lies between 0 and 1. Some confusion may arise between the terms ionisation cross-section and X-ray production cross-section so it should be noted that Equation (5.4) defines the relationship between these two for the K shell.

The fluorescence yield generally depends on the target atomic number Z_2 and its initial charge state, but is independent of the projectile atomic number Z_1 and incident energy E_1. For light ions incident on heavy targets, i.e. $(Z_1/Z_2) \ll 1$, the fluorescence yield has generally been considered to be independent of the charge state of the target atom and neutral atom fluorescence yields are used. The X-ray production cross section for a particular peak p, σ_p^x, is related to the K shell X-ray production cross-section by multiplying it by the corresponding emission rate S_p. Hence for the $K\alpha$ peak we have:

$$\sigma_\alpha^x = S_\alpha\,\sigma_K^x \qquad (5.4b)$$

where S_α is the width of the $K\alpha$ transition relative to the total K shell width.

In studies of L shell ionisation, the relations between the observed X-ray intensities and theory become more complex because of the existence of additional decay mechanisms and of more than one subshell. The average s shell fluorescence yield $\bar{\omega}_s$ depends on how the shell was ionised, since different ionisation methods give rise to different sets of primary vacancies. Transitions between the subshells of an atomic shell having the same principal quantum number, n, make it possible for a primary vacancy created in one of the subshells to shift to a higher subshell before the vacancy is filled by an X-ray transition. These are called Coster-Kronig transitions and the probability of shifting a vacancy from a subshell i to a higher subshell j, both in the same shell s, is denoted by f_{ij}^s. When Coster-Kronig transitions are included the average fluorescence yield of the shell s becomes:

$$\bar{\omega}_s = \sum_1^3 N_i^s v_i^s \qquad (5.5)$$

where N_i^s is the relative number of primary vacancies in the subshell i of the shell s. The coefficients v_i^s are called the effective subshell fluorescence yields and represent the total number of characteristic s shell X-rays (from both radiative and non-radiative processes) that result per primary vacancy in the ith subshell of the shell s. For the L shell, the coefficients v_i and the subshell fluorescence yields ω_i are related as follows:

$$v_{L1} = \omega_{L1} + f_{12}\omega_{L2} + (f_{13} + f_{12}f_{23})\omega_{L3} \qquad (5.6a)$$

$$v_{L2} = \omega_{L2} + f_{23}\omega_{L3} \qquad (5.6b)$$

$$v_{L3} = \omega_{L3} \qquad (5.6c)$$

where the superscript s has been dropped for simplicity. Similar but more complicated expressions have been given by Bambynek *et al.* (1972) for the five M subshell effective fluorescence yields v_i ($i = M_1$ to M_5).

The limit on the mean L shell fluorescence yield $\bar{\omega}_L$ is set by the extreme primary vacancy distributions of $N_{L1}/N_{L2}/N_{L3}$. Namely $1/0/0$, $0/1/0$ and $0/0/1$. For these situations $\bar{\omega}_L$ becomes identical with v_i ($i = L_1, L_2$ and L_3); hence $\bar{\omega}_L$ must lie between the limits set by the effective subshell fluorescence yields v_{L1}, v_{L2} and v_{L3}. Krause (1979) has recently compiled experimental values of ω_K, ω_{Li} and f_{ij}. Only values pertaining to singly ionised atoms were adopted and state and chemical effects were not included.

The total L shell and L subshell ionisation cross sections are related to their X-ray production cross sections in the following way:

$$\sigma^x = \sigma_1^x + \sigma_2^x + \sigma_3^x \qquad (5.7a)$$

$$= v_{L1}\sigma_1^I + v_{L2}\sigma_{L2}^I + v_{L3}\sigma_{L3}^I \qquad (5.7b)$$

$$= \bar{\omega}_L\,\sigma_I^1 \qquad (5.7c)$$

where the effective subshell fluorescence yields v_i are given by Equation (5.6). Thus at least six quantities enter into a complete description of the decay of atomic states characterised by a single vacancy in the L shell. In a similar fashion to Equation (5.7) the L X-ray production cross-sections for a particular peak p are related to the L subshell cross-sections via the subshell emission rates. Hence for the commonly occurring L shell lines we have (Cohen and Harrigan, 1986):

$$\sigma_{Lp}^x = (\sigma_{L1}^I(f_{13} + f_{12}f_{23}) + \sigma_{L2}^I f_{23} + \sigma_{L3}^I)\omega_{Li}\frac{\Gamma_{Lp}}{\Gamma_{Li}} \qquad (5.8)$$

where the peak Lp originates from an initial vacancy in the L_i subshell. For $i = 1$ we put $\sigma_{L2}^I = \sigma_{L3}^I = f_{13} = 0, f_{12} = f_{23} = 1$; for $i = 2$ we put $\sigma_{L3}^I = f_{13} = 0, f_{23} = 1$, and for $i = 3$ all three L subshell ionisation cross sections and Coster-Kronig transitions are used. (Γ_{Lp}/Γ_{Li}) is the fractional radiative width of the Lp transition in the L_i subshell (Salem et $al.$ 1974; Scofield, 1974).

In the studies of M and higher shell ionisation processes the relations between X-ray production cross-sections and subshell ionisation cross-sections become considerably more complicated, because of the existence of five or more subshells. Furthermore, the low energies of M or higher X-rays limit experimental techniques to only the highest Z_2 and the large number and small energy separation of characteristic X-ray lines limits their detection to only the highest resolving power detectors.

TABLE 5.1
Coefficients a_i for least squares fits to the data of Krause (1979).

	a_0	a_1	a_2	a_3	Z(range)
ω_K	8.51051(-2)	2.63414(-2)	1.63531(-4)	$-1.85999(-6)$	15–100
ω_{L2}	1.70027(-1)	2.98746(-3)	8.70636(-5)	$-2.19916(-7)$	42–90
ω_{L3}	2.72447(-1)	$-9.47505(-4)$	1.37650(-4)	$-4.64780(-7)$	40–100
ω_L	1.77650(-1)	2.98937(-3)	8.91297(-5)	$-2.67184(-7)$	28–96
ν_{L1}	6.14390(-1)	$-1.67030(-2)$	3.75467(-4)	$-1.61631(-6)$	48–94
ν_{L2}	5.67613(-1)	$-1.56443(-2)$	3.78921(-4)	$-1.65536(-6)$	45–110
ω_M	$-3.25820(-1)$	9.01791(-3)	3.80983(-5)	$-4.62897(-7)$	60–100

In these high target atomic number regions most $M_1, M_2,$ and M_3 vacancies are transferred to the M_4 and M_5 subshells by Coster-Kronig transitions. For $Z_2 = 93$ and 96 over 95% of all M_1, M_2 and M_3 vacancies undergo Coster-Kronig shifts to higher M subshells before they are filled from higher shells. The measured M shell X-ray production cross-sections are therefore characteristic of the M_4 and M_5 subshells and quite insensitive to the initial M subshell vacancy distributions. The relations between the ionisation cross-sections and the X-ray production cross-sections for the M subshells is obtained in a manner similar to Equations (5.6) and (5.7) used for the L shell. The explicit subshell expressions have been given by Bambynek *et al.* (1972) and Johnson *et al.* (1979).

The ω_i ($i = K, L_2, L_3$ and M) and ν_i ($i = L_1$ to L_3) parameters, with the exception of ω_{L1}, are smooth increasing functions of Z_2 and are well fitted by empirical polynomial expressions of the form:

$$[\omega_s/(1 - \omega_s)]^{1/4} = \sum_{i=0}^{3} a_i Z_2^i \tag{5.9}$$

where $s = K, L$ or M and the coefficients a_i are the fitting coefficients, and are obtained from least squares fits to the K, L and M shell data of Krause (1979). They are given in Table 5.1, together with the range of Z_2 for which they reproduce the numbers of Krause (1979) to better than $\pm 3\%$. Polynomial fits of this type are most useful when calculating absolute theoretical X-ray yields.

For the L_1 subshell the variations of ω_{L1} with Z_2 are not so smooth and ω_{L2} values are much smaller than ω_{L2} and ω_{L3} values since the Coster-Kronig rate is the dominant component of the L_1 decay rate. The cutoffs and onsets of Coster-Kronig transitions produce discontinuities in the ω_{L1} versus Z_2 plot at $Z_2 = 40–41$, $Z_2 = 49–50$, $Z_2 = 74–75$ and $Z_2 = 90–91$. These L shell discontinuities are most clearly seen in Fig.

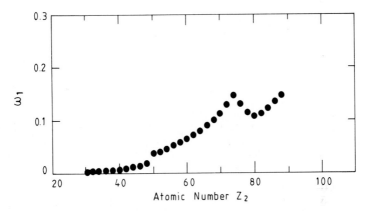

Fig. 5.6 A plot of the L_1 subshell fluorescence yield ωL_1 versus target atomic number Z_2.

5.6 where ω_{L1} has been plotted against the target atomic number Z_2 over the range $10 < Z_2 < 90$. Discontinuities of this type are brought about by energy considerations. For example, an L_2 vacancy can generally be transferred to the L_3 subshell through an L_2–L_3X Coster-Kronig transition, where X represents an outer shell. Such a transition is not possible, however, if the X electron binding energy exceeds the L_2–L_3X binding energy difference. Consequently, only N or higher shell electrons can be ejected by all atoms, while M_4 electron ejection is only possible for $Z_2 < 30$ and M_5 electrons can be ejected by atoms with $Z_2 < 30$ and $Z_2 < 91$.

5.2.5 Polynomial Fits to the Theoretical Ionisation Cross-sections

Several theories exist for the prediction of ionisation cross-sections σ^I. Brandt and Lapicki (1979, 1981) modified the plane wave Born approximation (PWBA) by incorporating polarisation and binding effects in the perturbed stationary states (PSS) approximation (Basbas *et al.,*1971) and by correcting for relativistic (R), energy loss (E), and Coulomb deflection (C) effects. This is generally referred to as the ECPSSR theory and has become one of the most successful theories for predicting K and L shell ionisation cross-sections for light ion impact on most targets.

Sufficient information has been given in the ionisation cross-section tabulations of Cohen and Harrigan (1985) to calculate ECPSSR ionis-

TABLE 5.2
Protons

ENERGY	K Shell ECPSSR		L Shell ECPSSR (barns)					M Shell CPWBA					
(MeV)	Z_2	K	Z_2	L_1	L_2	L_3	L_{TOTAL}	Z_2	M_1	M_2	M_3	M_4	M_5	M_{TOTAL}
0.5	15	1619	60	4.06	10.8	34.9	49.7	66	1811	3500	102E2	212E2	347E2	715E2
	20	96.5	70	1.12	1.19	5.13	7.44	74	322	844	2971	6948	117E2	227E2
	30	1.12	80	0.243	0.128	0.812	1.18	82	477	188	801	2258	3916	7240
	40	331E-4	90	432E-4	129E-4	0.134	0.190	90	28	40	214	708	1297	2288
1	15	6109	50	19.1	76.4	221	317	66	8282	104E2	272E2	447E2	721E2	163E3
	20	573	70	2.91	12.9	46.2	62.0	74	2447	3398	102E2	163E2	269E2	593E2
	30	11.6	80	1.05	2.19	10.6	13.8	82	570	988	3580	5936	101E2	211E2
	40	0.505	90	0.377	0.361	2.59	3.33	90	111	260	1187	2237	3898	7694
2	15	125E2	50	217	313	838	1367	66	161E2	183E2	446E2	711E2	133E3	263E3
	20	1903	70	30.8	69.4	219	319	74	6897	7731	207E2	301E2	486E2	114E3
	30	72.9	80	4.15	16.0	63.1	83.2	82	2626	3033	9193	125E2	206E2	480E3
	40	4.61	90	0.918	3.66	19.5	24.1	90	857	1084	3908	5185	8851	199E2
3	15	153E2	50	533	586	1503	2622	66	182E2	211E2	499E2	801E2	127E3	296E3
	20	2973	70	110	149	444	703	74	8988	9900	254E2	372E2	598E2	141E3
	30	167	80	19.7	39.2	142	201	82	4119	4463	126E2	169E2	175E2	656E2
	40	13.5	90	3.18	10.3	48.3	61.8	90	1710	1888	6032	7620	128E2	301E2
5	15	163E2	60	1065	1046	2556	4667	66	177E2	211E2	–	–	–	–
	20	4136	70	317	319	883	1520	74	9846	112E2	276E2	415E2	660E2	156E3
	30	364	80	85.7	98.3	322	506	82	5256	5723	153E2	211E2	304E2	813E2
	40	41.1	90	20.2	30.1	123	173	90	2673	2869	8197	105E2	175E2	418E2
10	15	138E2	60	1536	1585	3696	6817	66	–	–	–	–	–	–
	20	4586	70	640	616	1567	2822	74	–	–	–	–	–	–
	30	673	80	255	237	686	1179	82	5003	5806	–	–	–	–
	40	118	90	94.7	89.4	308	492	90	3001	3350	9013	123E2	199E2	475E2

TABLE 5.2 (cont.)
Helium

ENERGY (MeV)	K Shell ECPSSR		L Shell ECPSSR (barns)					M Shell CPWBA						
	Z_2	K	Z_2	L_1	L_2	L_3	L_{TOTAL}	Z_2	M_1	M_2	M_3	M_4	M_5	M_{TOTAL}
1	15	622	15	5.39	2.91	10.5	18.8	66	1122	3145	102E2	353E2	587E2	108E3
	20	23.7	20	0.855	0.247	1.22	2.32	74	403	657	2507	100E2	173E2	309E2
	30	0.225	30	0.126	219E-3	0.160	0.308	82	115	146	621	2516	4640	8037
	40	654E-5	40	170E-4	189E-5	229E-4	419E-4	90	19	38	173	551	1109	1889
2	15	5478	15	18.3	39.8	127	186	66	7786	149E2	432E2	884E2	145E3	299E3
	20	277	20	5.67	5.09	21.1	31.9	74	1428	3687	129E2	295E2	494E2	969E2
	30	3.45	30	1.50	0.690	3.99	6.18	82	350	853	3582	9883	171E2	318E2
	40	0.136	40	0.362	961E-4	0.842	1.30	90	138	195	1002	3231	5878	104E2
3	15	142E2	15	27.5	133	399	560	66	208E2	294E2	799E2	139E3	255E3	494E3
	20	944	20	9.38	20.6	78.0	108	74	5059	8608	275E2	486E2	807E2	170E3
	30	14.5	30	3.46	3.35	17.2	24.0	82	966	2257	8765	173E2	295E2	588E3
	40	0.631	40	1.11	0.557	4.18	5.84	90	230	558	2696	6349	112E2	210E2
5	15	326E2	15	154	462	1305	1921	66	451E2	538E2	136E3	218E3	351E3	804E3
	20	3281	20	18.8	88.0	300	407	74	156E2	193E2	556E2	835E2	137E3	311E3
	30	73.2	30	5.76	17.6	77.8	101	82	4421	6301	213E2	318E2	535E2	117E3
	40	3.72	40	2.55	3.55	22.0	28.1	90	1005	1856	7786	124E2	215E2	446E3
10	15	519E2	15	1349	1711	4505	7565	66	702E2	814E2	194E3	309E3	489E3	114E4
	20	9971	20	235	407	1249	1891	74	330E2	367E2	943E2	137E3	223E3	524E3
	30	435	30	36.2	102	383	521	82	142E2	156E2	452E2	605E2	997E2	235E3
	40	30.3	40	6.30	26.0	127	159	90	5377	6187	207E2	264E2	447E2	103E3

ation cross-sections. The PWBA ionisation cross-sections for an inner shell vacancy are proportional to Z_1^2 for a given target material and ion velocity. This means that approximate heavy ion cross sections can be obtained by multiplying the proton result at the same value of (E_1/M_1) by Z_1^2. Hence we would expect the ionisation cross-sections for D and He ions to be related to the proton cross-sections in the following manner:

$$\sigma_D = \sigma_p (E_1/2) \tag{5.10}$$

$$\sigma_{He} = 4\sigma_p (E_1/4) \tag{5.11}$$

For low ion energies ($< 2\text{MeV/amu}$) this will not be very accurate because of the various corrections applied to the PWBA theory to produce the ECPSSR theory and these have been discussed in detail by Cohen and Harrigan (1985). Typical ionisation cross-sections (barns) for K, L, and M shells are given in Table 5.2.

The Z_1^2 scaling for the direct ionisation cross-sections is a useful concept if an effective charge similar to that defined by Ziegler (1977) for stopping powers is considered. This effective charge is close to Z_1 for light ions near protons and decreases to about 80% of Z_1 for heavier ions near S for K shell ionisation. It is obviously also a function of ion velocity. Calculations by us for the K shell ionisation of Zn by heavy ions with energies near 4MeV/amu give:

$$\sigma_K(\text{ion}) = Z_1^2 R \sigma_K(\text{proton}) \tag{5.12}$$

at the same value of E_1/M_1 where:

$$R = \exp\left[\sum_{i=0}^{4} a_i Z_1^i \right] \tag{5.13}$$

and the coefficients a_i are 1.85476×10^{-2}, -1.25090×10^{-2}, -3.57718×10^{-3}, 2.79842×10^{-4}, -7.46116×10^{-6} respectively. This equation is valid to a few percent for ions from He to S provided ion energies are around 4 MeV/amu and are not too different for ions up to Zn. As expected for lower velocity ions the effective charge is somewhat lower and hence values of R are smaller. For example, for ions from He to S with energies of only 1 MeV/amu the coefficients of Equation (5.12) are 0.213751, -0.188137, -2.20145×10^{-2}, 2.02805×10^{-3}, -5.59197×10^{-5} respectively for these ions on a Zn target. For O ions on Zn this

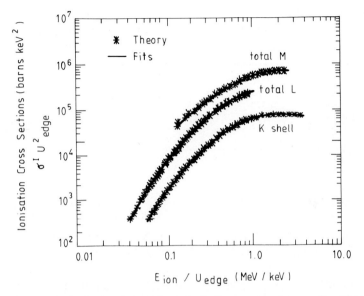

Fig. 5.7 The K, L (ECPSSR) and M (CPWBA) ionisation cross-sections as a function of the ion E/U for proton bombardment. U (keV) is the target atom absorption edge energy. The points represent the theory and the solid line the polynomial fits described in the text.

corresponds to a value of R of 0.153 at 16 MeV and 0.822 at 56MeV.

Since heavy ion cross-sections may be scaled up from proton cross-sections it is worthwhile to obtain accurate polynomial fits to the latter. Fig. 5.7 shows fits to the K and total L shell ECPSSR and the total M shell CPWBA ionisation cross-sections for proton bombardment of a wide range of targets. The ionisation cross-sections have been multiplied by the square of the K, L or M edge energy (in keV) and are expressed in (barns. keV^2) for the K, L and M shells. The mean L and M shell target edge energy is:

$$E_L = (L_1 + L_2 + 2L_3)/4 \tag{5.14}$$

for the total L shell and:

$$E_M = (M_1 + M_2 + 2M_3 + 2M_4 + 3M_5)/9 \tag{5.15}$$

for the total M shell. The proton energy has been divided by the K, L or M target edge energy and is expressed in MeV/keV. Table 5.3 gives the coefficients a_i to a fifth order polynomial fit to these ionisation cross-section of the form:

TABLE 5.3
The coefficients a_i, as given in Equation 5.16 for the fifth order polynomial fit to the theoretical ionisation cross sections for the K, L and M shells, for proton bombardment only.

Shell	a_0	a_1	a_2	a_3
K(ECPSSR)	10.99955	0.6078148	−0.4564251	3.312473(−2)
L(ECPSSR)	12.36572	0.5423245	−0.5570508	−5.193862(−2)
M(CPWBA)	6.301553	0.6173717	−0.4354831	9.984461(−3)

Shell	a_4	a_5	Z(range)
K(ECPSSR)	1.373860(−2)	1.481521(−3)	10–58
L(ECPSSR)	−7.355058(−4)	1.612867(−3)	50–93
M(CPWBA)	6.224392(−2)	1.773666(−2)	66–92

$$\ln (E^2 \sigma^I) = \Sigma a_i ln \ (E_1/E \ (\text{keV})) \qquad (5.16)$$

The fits of Fig. 5.7 were performed over the ion energy range $1 < E_1 < 4\text{MeV}$ and over the target range $10 < Z_2 < 58$ for the K shell, $50 < Z_2 < 93$ for the L shell and $66 < Z_2 < 92$ for the M shell. The polynomial fits and ECPSSR cross-section generally differ by less than $\pm 10\%$ and, over limited Z_2 and E_1 ranges, by less than $\pm 2\%$. Polynomial fits of this type are very useful for theoretical predictions in PIXE and, for proton energies $1 < E_1 < 4\text{MeV}$, avoids the lengthier calculations referred to by Cohen and Harrigan (1985).

Extensive comparisons with measured K shell ionisation cross-sections (Paul, 1983) show the success of the ECPSSR theory over a wide range of ion masses and energies. Paul also includes polynomial fits to the ratio of the experimental and ECPSSR theory which is most useful for computer simulation programmes requiring ionisation cross-sections. Deviations do exist, however, for the heavier ions at lower energies (< 200 keV/amu), where the ECPSSR theory overpredicts the experimental results for both the K and L subshells. The reader is referred to the work of Sarkadi and Mukoyama (1981) and Cohen (1984a,b) for further discussion.

5.2.6 Projectile Energy Loss and X-ray Attenuation

For an ion travelling in a thick target we need to consider the ion energy loss/unit distance travelled as well as the absorption of the induced X-rays in their passage out of the target. The mathematical significance of these two terms has been discussed in Section 5.2.2 for both thin and thick targets. Extensive compilations of the stopping power for various heavy ions in most materials have been compiled by Andersen and Ziegler (1977) and for protons by Janni (1982). The effect of target stopping power is not so much to reduce the X-ray yield but to reduce the total volume sampled by the incident ion beam. In near-surface applications this can be an advantage. The X-ray production cross-sections fall very rapidly with decreasing ion energy, and larger stopping powers lead to X-ray yield contributions from less than the first 10 μm in most cases.

The number of X-rays reaching the detector depends on the total amount of material between their source and the detector and on the X-ray energy; 10 mg cm^{-2} of C will reduce 5, 10 and 20 keV X-rays by 5%, 3% and 0.5% rrespectively, while 10 mg cm^{-2} of Cu reduces the same energy X-rays by 75%, 95% and 30% respectively. This X-ray absorption is exponential in nature and characterised by the mass attenuation coefficient (μ/ρ), measured in (cm^2/g). Several compilations of X-ray mass attenuation coefficients exist, the most recent being the tables of Leroux and Thinh (1977). They reviewed all published experimental and theoretical data between 1961 and 1974 for X-ray energies from 1 to 40 keV and for absorbers from He to Pu. The general equation (μ/ρ) = CE$_{abs}$ (12.3981/E$_x$)^n was used, where E_{abs} is the lower absorption edge energy (keV) of the two edges either side of the X-ray energy E_x (keV) and C and n are tabulated constants. These tables give (μ/ρ) for X-ray energies between the K, L_1, L_2, L_3, M_1, M_2, M_3, M_4, M_5, and N_1 absorption edges and for absorbers from He to Pu. The mass attenuation coefficients have large discontinuities at the K, L, M . . . absorption edges which can often be used to advantage in producing selective filtering.

5.2.7 Escape Peaks and Sum Peaks

There is a finite probability that an X-ray of energy E_x will interact with a Si atom in the detector and produce a K X-ray which will then completely escape from the sensitive volume. The escape probability increases with decreasing incident X-ray energy down to the K absorption edge of Si (1.838 keV). These events are lost from the full energy peak and reappear in the spectrum as a separate peak of energy ($E_x - 1.74$) keV. The ratio for the intensity of the escape peak to the parent peak (f_{esc}) can be fitted as a function of X-ray energy:

$$\ln (f_{esc}) = -3.0183 - 0.51995 \, E_x + 0.01344 \, E_x^2 \qquad (5.17)$$

where E_x is in keV. Fig.5.8 shows escape peaks for Zn K lines lying just to

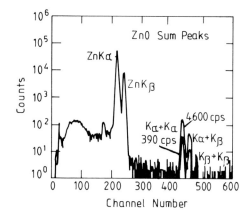

Fig. 5.8 Typical spectrum obtained by proton bombardment of ZnO. The Zn *K* escape peaks are to the left of the *Kα* peak and the sum peaks to the right. The broad bump on the low energy side of the escape peaks is produced by the secondary electron bremsstrahlung discussed in Section 5.4.2. Data for the smaller peaks were acquired at 390 Hz while the larger peaks were acquired at 4600 Hz. This shows the effect of higher count rates on spectrum sum peaks.

the left of the Zn *K* α and *K* β peaks. The peaks to the right are Zn *K* sum peaks. These sum peaks are produced by electronic pileup in the detector amplifier. Two pulses arriving in the amplifier at nearly the same time are seen as one pulse twice the height. This spectrum was acquired using 2 MeV protons on ZnO at a count rate between 390 and 4600Hz. Clearly the sum peaks are count rate dependent. The size of the sum peaks can be greatly reduced by using on demand beam pulsing. When an X-ray is detected the incident beam is deflected away from the target until processing of the detected pulse is complete. This technique also reduces the size of the dead time corrections.

5.2.8 Secondary Fluorescence

If the energy of characteristic radiation from element A exceeds the absorption edge energy for element B in a sample containing A and B, then characteristic secondary fluorescence of B by A occurs. This is quite distinct from the initial primary radiation induced by the ion interacting with A and B. The cross-section for ion induced X-ray production decreases with increasing absorption edge energy and the cross-section for X-ray induced X-ray production increases with increasing absorption edge energy, as long as the absorption edge is lower than that of the exciting X-ray. Therefore the enhancement is largest for elements with absorption edge energies just below that of an intense characteristic X-ray line, but it decreases rapidly with decreasing absorption edge energy.

Up to 40% of the total X-ray yield may be due to this process under very favourable conditions of K to K excitation. Reuter *et al.* (1975) give the following approximate expression for the ratio of secondary $K\alpha$ yield to primary $K\alpha$ yield:

$$\frac{Y_{sec,A}}{Y_{prim,A}} = R\, c_B \left[\frac{\mu_{K\alpha,B,A}}{\mu_{K\alpha,B,SPL}} \right] \left[\frac{a_A}{a_B} \right] \left[\frac{\sigma_{K\alpha,B}}{\sigma_{K\alpha,A}} \right] \omega_{K,B} \qquad (5.18)$$

where c_B is the weight concentration of element B, $\mu_{K\alpha,B,A}$ is the mass attenuation coefficient for $K\alpha$ X-rays of element B in matrix A, $\mu_{K\alpha,B,SPL}$ is the mass attenuation coefficient for $K\alpha$ X-rays of element B in the sample matrix, a_A and a_B are the atomic weights of elements A and B, $\sigma_{K\alpha,A}$, $\sigma_{K\alpha,B}$ are the $K\alpha$ X-ray production cross-sections for elements A and B and $\omega_{K,B}$ is the fluorescence yield for element B. R is related to the absorption edge jump ratios and is given as 0.44 for K to K and L to K fluorescence and 0.375 for K to L and L to L fluorescence. Reuter *et al.* (1975) consider Equation (5.18) to be valid only for secondary fluorescence effects up to 10%.

Consider the example of the apatite matrix described in Section 5.2.2 and shown in Fig.5.4; apatite is 39.9% Ca and 18.5% P by weight. We would expect the primary Ca$K\alpha$ ion induced X-rays to fluoresce all elements present below Ca in the matrix, producing slightly higher experimentally measured yields from these elements than theoretically expected. Substitution of the appropriate values into Equation (5.18) gives $Y_{sec}\,/\,Y_{prim} = 2.5\%$ for P and 11% for K for 2.3 MeV protons in an apatite matrix. This implies that the experimentally measured $K\alpha$ yield should be reduced by 11% before comparing it with the theoretical yield of Fig.5.4. Van Oystaeyen and Demortier (1983) have also used Equation (5.18) to estimate secondary fluorescence yields in Au/Cu/Ag alloys. They concluded that this approximation was quite satisfactory for most applications.

The photoelectric absorption of X-rays by the Si deadlayer in a Si(Li) detector results in the emission of Si K X-rays from this layer into the active volume of the detector and hence secondary fluorescence. This form of secondary fluorescence, which does not originate in the sample, appears in the spectrum as a small Si peak, called the Si internal fluorescence peak. For many quantitative analysis situations, this fluorescence peak corresponds to an apparent concentration of less than 0.2% by weight of Si in the specimen.

The continuum background radiation provides X-rays at all energies up to some cutoff value described in Section 5.3.2. For X-ray energies below this cutoff there will always be continuum induced secondary fluorescence. The problem can be minimised by using as light an atomic weight matrix as possible and low energy ions. Another aspect of secondary fluorescence that is generally not fully appreciated is that it may originate from a totally different volume of the matrix than that sampled by the impinging ion beam. This is brought about by the different ranges of X-rays and ions in solids. For heavier slow ions the secondary fluorescence volumes maybe much larger than the volume sampled by the ions. For inhomogeneous samples and for ion micro-beams this effect may need serious consideration (Heck, 1981).

5.3   INSTRUMENTATION FOR PIXE ANALYSIS

A basic experimental system which will meet most needs involves an ion beam, suitable target facilities and an X-ray detector with associated data processing capabilities (Chapter 2). Some features characteristic of X-ray analysis are discussed here.

5.3.1   Beam Characteristics

Protons having energies of between 1 and 4 MeV are a common choice for ion induced X-ray emission. The beam should be uniform and of known area. Uniformity can be achieved either by passing the beam through a thin diffuser foil or defocussing it and picking off the central portion. Beam sizes can vary from 10 mm diameter to micron spots. For thin targets currents of approximately 1 nA/mm^2 are used and large beam areas are needed to increase the count rate. For thick targets the beam current is limited by sample damage and count rate as fixed by the system dead time. Obviously there is little point in analysing a sample if the very act of measuring destroys the sample or removes volatile elements. (Sections 2.4.2, 10.3.3, 10.4.2). Typical target currents are in the 10 to 100 nA range, depending on the sample type and beam size.

a.   Target Chambers

Features that are important specifically for PIXE analysis may be found in Malmqvist *et al.* (1982) who have given a detailed description of their set up which warrants close study by anyone thinking of building a PIXE chamber. Cahill (1975) also has given a description of the facilities used by his group.

b. External Beams

It is possible to extract an ion beam through a thin window and perform analysis either in air or in a He environment. This removes problems associated with target charging and diminishes target heating but one problem is charge integration. Several groups collect charge directly on the target and window. A better method may be the use of a beam chopper (Hollis, 1972). A very good external beam design is that used by the Helsinki group (Anttila *et al.*, 1985). This combines a 7.5 μm Kapton window and a finely drilled graphite plate. By using N_2 cooling on both window and sample target currents up to 1.2 μA have been achieved. Two chambers at 10^{-3} torr and 10^{-1} torr and separated by small (0.10 mm) interconnecting holes and no window can be used but the external beam diameter is necessarily small.

5.3.2 X-ray Detectors

Table 5.4, from Goldstein *et al.*, (1981), gives a comparison of the major features of wavelength dispersive spectrometers and Si(Li) detectors.

The overriding feature in choosing which to use usually is the energy resolution required. For very high resolution work only the wavelength dispersive method will do. This uses a microanalyser crystal spectrometer with a gas proportional counter. Since the crystal spectrometers are normally kept under vacuum to eliminate absorption of the X-ray beam in air, it is usually necessary to support an ultra thin window of Formvar

TABLE 5.4
A comparison between wavelength and energy dispersive X-ray detection systems.

Operating Characteristic	WDS	EDS
Geometrical collection efficiency	Variable < 0.2%	<2%
Overall quantum efficiency	Variable < 30%	100% for 2 to 16 keV
	Detects $Z \geq 4$	Detects $Z \geq 11$
Resolution	Crystal dependent	Energy dependent
	5 eV	140 eV at 5.9 keV
Count rate (maximum)	50 kHz	5 kHz
Minimum useful probe size	0.2 μm	0.005 μm
Typical data collection time	10 min	5 min
Spectral artifacts	Rare	Major ones are:
		Escape peaks
		Pulse pileup
		Peak overlap
Energy for electron-hole pair creation	28 eV	3.8eV

or cellulose nitrate on a fine wire screen to withstand a pressure differential of 1 atmosphere.

By far the most common PIXE detector is the energy dispersive Si(Li) detector. It has a rather low efficiency for X-ray energies below 5 keV when compared with a standard gas flow proportional counter. Also it is very difficult to accurately measure window and Si deadlayer thickness (most Si(Li) detectors being sealed units to enable cooling to liquid nitrogen temperatures). The efficiency of a gas flow proportional counter is given by the transmission of the entrance window if the gas pressure is sufficiently high and if there is no dead gas layer. Window thicknesses are typically greater than 200 μg cm^{-2} giving detection efficiencies of about 90% at 1 keV. This is at least an order of magnitude better than the thinnest window (8 μm of Be) Si(Li) detector at 1 ke V. For this reason the gas flow proportional counter is favoured over the Si(Li) detector for soft X-ray (below 5 keV) detection.

5.3.3 Detector Resolution

The resolution of the wavelength dispersive crystal spectrometer gas proportional counter system is typically 5 eV. This is more than an order of magnitude better than the best Si(Li) detector resolution (140 eV at 5.9 KeV). For the Si(Li) or Ge(Li) type photon detectors there are two main contributions to the Full Width Half Maximum (*FWHM*) of the detected photon peaks:

$$(FWHM)^2 = \Delta_{el}^2 + 2.355^2(FeE_x) \qquad (5.19)$$

where Δ_{el}, the width associated with the electronics, is a function of the main amplifier shaping time constant, the Fano factor, F, is 0.11 ± 0.04 for Si and 0.13 ± 0.05 for Ge (Musket 1974), $e = 3.81$ eV for Si and 2.98 eV for Ge and is the average energy required to form an electron hole pair and E_x is the photon energy. The electronic component of the *FWHM* for most semiconductor detectors is between 50 and 250 eV for pulse shaping constants between 1 and 10 μs. Inspection of Equation (5.19) shows that semiconductor detector resolution increases with photon energy and can change by a factor of two or more over the useful photon energy range of most detectors. Si(Li) detectors have by far the best resolution of any semiconductor detector.

5.3.4 Detector Efficiency

Fig. 5.9 gives the efficiency curve for a typical Si(Li) detector as modelled by the techniques of Cohen (1980). This detector was 4 mm in diameter,

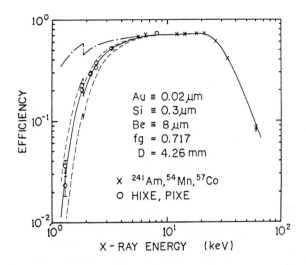

Fig. 5.9 The efficiency for a typical Si(Li) detector versus X-ray energy. The (x) are the experimental points from standard sources and the (o) are calculated points from PIXE and HIXE results using known theoretical ionisation cross-sections.

4.26 mm thick and contained an 8 μm thick beryllium window. The solid curve is a fit to the measured efficiency points and the dashed-dot curve the calculated efficiency using the manufacturer's specifications. It clearly demonstrates the need to measure each detector's efficiency before use especially in the lower energy region (< 5 keV). The low energy drop in efficiency is due to the window and any material between the source and the detector. The high energy drop is determined by the detector thickness, thicker detectors are more efficient at the higher X-ray energies.

5.3.5 Source Calibration of Detectors

Techniques required for the calibration of semiconductor X-ray detectors have been described by Campbell and McNelles (1975) and the half lives of useful sources have been listed in Mayer and Rimini (1977) and by Gray (1980). The calibrated source technique includes the effects of geometry and absorption automatically. One of the most useful sources is ^{241}Am since it has a long half life (432.9 ± 0.8) years and produces many photon lines over the energy range 3 to 60 keV. Source strengths around 400 kBq are commonly used. Fig. 5.10 shows a typical spectrum obtained from an ^{241}Am source using a Si(Li) detector. The 26 and 33 keV lines are gamma ray transitions while the M(3 keV) and L(11-22 keV) lines are X-ray transitions from Np. The Cr, Fe and Ni K transitions are

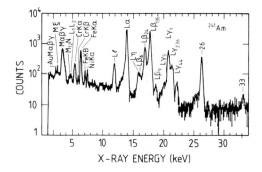

Fig. 5.10 Typical spectrum for an ^{241}Am source of strength 385 kBq distance 64 mm from a 4 mm diameter Si(Li) detector. The run time was 5000 s.

produced by secondary fluorescence of the source stainless steel backing. All the peaks sit on a fairly flat background produced by Compton scattered high energy gamma rays (> 100 keV) interacting with the Si(Li) detector. For very thin windowed detectors (< 8 μm) a further background maybe introduced into the spectrum by the 5 MeV alphas from this source.

The full energy peaks consist of Gaussian peaks and fairly flat low energy tails extending to zero energy. The tail arises from incomplete charge collection from certain regions of the detector. In order to obtain the true counts in the photo peak all events in this low energy tail should be added to those in the Gaussian peak. Obtaining areas from such modified Gaussian peaks will be discussed further in Section 5.4. Table 5.5 gives the line energies and the number of photons/decay for the most commonly observed lines in ^{241}Am sources. Particular care must be taken to make the radioactive source used as thin as possible to minimise self absorption losses, for low energy photons. Standard solution evaporation techniques and VYNS foil coverings enable sources of less than ≈15 μg/cm^2 thickness to be made. These have less than 1% absorption for X-ray energies greater than 2 keV. With considerable care these techniques are capable of calibrating a Si(Li) detector to ±3% over the photon energy range 5 to 60 keV.

5.3.6 Low Energy and Time Dependent Efficiency

The efficiency calibration of Si(Li) detectors in the low energy X-ray region (0 to 5 keV) must be treated separately, as standard radioactive sources for this low energy region do not exist. The most common method is by secondary fluorescence. This may include *K*, *L* or *M* shell

TABLE 5.5
The line energies and the number of photons/decay for the most commonly observed lines in ^{241}Am sources.

Transition	Energy (keV)	Photons/Decay		IAEA* (1985)
		Campbell and McNelles(1975)	Cohen(1980)	
NpM(total)	3.3		0.0543 ± 0.0032	0.0635
NpM($\alpha\beta\gamma$)	3.35		0.049 ± 0.003	
L_1L_3	4.82		0.0017 ± 0.0005	
L1	11.89	0.0086 ± 0.0008	0.0087 ± 0.0003	0.0085
$L\alpha$	13.94	0.132 ± 0.003	0.132 ± 0.003	0.13
$L\eta$	15.87		0.0038 ± 0.002	
$L\beta$	17.8	0.1925 ± 0.006	0.194 ± 0.004	0.193
$L\gamma$	20.8	0.0485 ± 0.002	0.0496 ± 0.002	0.0493
γ	26.345	0.024 ± 0.001	0.0236 ± 0.001	0.024
γ	33.119		0.0014 ± 0.0001	0.00103
γ	43.463		0.00057 ± 0.00018	0.00057
γ	59.537	0.359 ± 0.006	0.355 ± 0.009	0.357

* W. Bambynek (1985).

fluorescence for ion bombardment of targets from Na to U or secondary fluorescence of thick low atomic number targets (Na to Ca) by radioactive sources with primary lines whose energies are just above the target K edge energy.

Work of this type down to photon energies ≤ 1 keV has been performed by Lennard and Phillips (1979), Shima (1979), Shima *et al.* (1980), Cohen (1982) and Geretschlager (1982). Quantitative work in this region is difficult because of the errors associated with the fluorescence yields and with the large numbers of parameters necessary to describe the absolute efficiency. Typical one standard deviation errors for efficiencies near 1 keV are $\pm 30\%$. Geretschlager's paper showed that it was possible to use a standard surface barrier detector as an X-ray detector for energies $1 \leq E_x \leq 5$ keV if it was cooled to liquid N_2 temperature. It was also claimed that the physical parameters of the surface barrier could be measured with sufficient accuracy so that the efficiency could be determined with an error of less than 1%.

Work in this region has shown the time dependent nature of efficiency for 0 to 5 keV photons. Cohen (1982) measured a loss of efficiency of 2% per month at 1.5 keV for a Si(Li) detector over a two year period. This could be mainly attributed to build up of contaminants (ice, oil) on the detector front face which was cooled to liquid N_2 temperatures. These contaminants are not only a problem for windowless detectors but also for detectors which are cooled and have very thin Be windows (≤ 10 μm). It shows the need to maintain a very clean vacuum

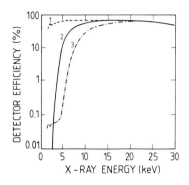

Fig. 5.11 A typical Si(Li) detector efficiency versus X-ray energy. The solid curve is for no filter, the dashed curve for 0.356 mm of Mylar, and the dash-dot curve is for a pinhole filter.

system on both sides of the detector window, in order to reduce contaminants on the detector front face produced by the cryopumping action of the cooled detector.

5.3.7 Filters

It is often necessary to remove unwanted X-rays by using a suitable filter. Several materials are suitable for filter construction. Low atomic number filters ($Z < 10$) have their K absorption edges well below 1 keV and simple high pass filters can be designed using 0.1 to 5mm of these materials between the target and the detector. Fig. 5.11 shows a typical Si(Li) efficiency versus X-ray energy for no filter, 0.356mm Mylar and a pinhole filter. For the Mylar filter the low energy cutoff is quite sharp. A reasonable transmission for Mylar can be obtained from the following:

$$T = \exp\left(-470.17x.E_x^{2.9897}\right) \tag{5.20}$$

where x is the thickness in mm and E_x is the X-ray energy in keV. This expression is quite accurate for $T > 0.01$.

When higher atomic number filter materials are used ($Z > 10$), discontinuities in the X-ray transmission appear at the K, L and M absorption edges. These manifest themselves as narrow X-ray regions of high transmission below the cutoff points of the high pass filters just discussed. Such a bandpass filter is often used to selectively reduce the count rate from a major element close to the element of interest. This is done by selecting a filter whose K edge energy lies just below that for the $K\alpha$ line of the element whose count rate must be reduced. For example a

TABLE 5.6
The effect of Cr filters on transmission.

| Element | Transmission(%) Cr Thickness (microns) | |
	6	12.5
Ca	33.4	10.2
Mn	73.4	52.5
Fe	13.9	1.5
Cu	33.0	10.2

thin Cr filter will reduce the count rate in an Fe $K\alpha$ peak. This is shown in Table 5.6 where the X-ray transmission (as a percentage) is given for the $K\alpha$ X-rays of Ca, Mn, Fe, and Cu. High pass or band pass may remove some useful X-rays as well. This then necessitates two analyses, one with a filter and one without. An alternative is to use a pinhole filter, which is a highpass filter with a small hole drilled through its centre to transmit a small fraction of the low energy X-rays. Careful design of the hole size and the filter thickness can make the count rate in different parts of the spectrum approximately the same. Fig. 5.12 shows a spectrum taken with such a filter (whose efficiency characteristics are given in Fig. 5.11) for an USGS oil shale bombarded by 2.5 MeV protons. The unfiltered spectrum is shown in Fig. 5.1; Fig 5.12a is the pinhole case and Fig. 5.12b shows the effects of 0.526 mm of Perspex. The pinhole consisted of 2 mm thick C with a 70 μm hole drilled through its centre. Clearly the higher elements (above Zn) which were not easily observable in the no filter case because of the high light element count rates, now become available in the one spectrum. The Perspex filter case has the effect of completely removing light elements X-rays from the spectrum.

5.4 SPECTRUM ANALYSIS

Quantitative PIXE analysis involves accurately measuring the intensity of the characteristic lines for elements in the sample and converting this information into concentration data. Analysis is complicated by the presence in the spectrum of sum and escape peaks, distortion in the basic Gaussian shape and interference between the characteristic peaks of different elements. In the worst possible case the spectrum can finish up looking like Fig. 5.13 which shows the spectrum taken from an NBS Glass 610 standard. This is a glass disc which has 60 elements each at approximately $500 \mu g\,g^{-1}$. Between 3 and 30 keV there are over 400 characteristic peaks contributing to the spectrum.

There are two methods of analysing pulse height spectra, namely

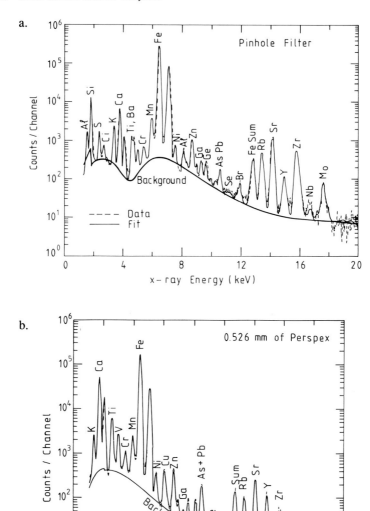

Fig. 5.12(a) Typical Si(Li) spectrum for the pinhole filter case, 2mm of carbon with a 70 μm hole through its centre. (b) Typical Si(Li) spectrum for 0.526 mm of Perspex. The solid lines are the background and Gaussian fits obtained as described in Section 5.4 and the dashed curves are the experimental data.

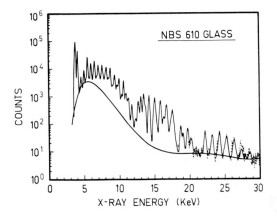

Fig. 5.13 The X-ray spectrum between 3 and 30 keV for a multielement glass standard. The smooth curve is the estimated background and the curve with peaks is the estimated fit.

spectrum stripping or spectrum synthesis. The former method is commonly used in analysing gamma spectra. For PIXE analysis, one program that uses this approach is RACE, which is described by Cahill (1975). First the background is estimated from the spectrum of a blank sample. This is subtracted and any residual background removed. Peaks are located by Gaussian correlation and multiplets are resolved by fitting to Gaussian functions. Elements are identified by comparison with tables of energies. Information from alpha and beta lines is utilised separately to determine element concentrations. Interferences are tentatively identified.

5.4.1 Spectrum Synthesis

The alternative is to calculate an X-ray spectrum and compare it with the measured spectrum. Thus we can write:

$$Y(E_i) = Back(E_i) + \sum_{j,k} A_j R_{jk} G_{jk}(E_i) \tag{5.21}$$

where *Back* is the background and G_{jk} is a function representing the peaks, A_j are a set of fitted peak heights and R_{jk} are a table of relative peak heights. A least squares fit to the experimental data provides estimates from which element concentrations can be calculated. One aims to minimise the reduced chi-squared value, defined by:

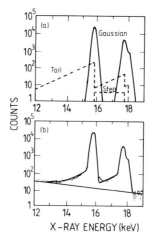

X-RAY ENERGY (keV)

Fig. 5.14 (a)   Components of the characteristic peak shape including the Gaussian peaks, with low energy tailing and a step function between the peaks. (b) A fit to the PIXE spectrum from a thick Sr sample.

$$\chi_R^2 = \frac{1}{(N-f)} \sum_{i=1}^{N} (y_{\exp} - y_i)^2 / y_{\exp} \qquad (5.22)$$

where $y_{\exp}$ are the measured data points, y_i are calculated from Equation 5.21, N is the number of channels and f is the number of parameters. A good algorithm is provided by Bevington (1969) as the subroutine CURFIT. Three computer programs HEX (Johansson, 1982), AXIL (Van Espen *et al.*, 1979) and BATTY (Clayton, 1983) which have recently been compared in a round robin on fitting PIXE spectra (Campbell *et al.*, 1986), illustrate the features necessary for accurate analysis. The computer programs all utilise a background that allows for X-ray absorption modifying some polynomial representing the bremsstrahlung contribution.

The second term in Equation 5.21 describes the X-ray peaks. It consists of a number of lines of known fixed relative intensities represented by the matrix R_{jk}. This is the intensity ratio of line k of element j relative to the reference line (normally the $K\alpha$ or $L\alpha$ line). This information is given by Salem *et al.* (1974). Before being used in the fit, R_{jk} has to be adjusted for detector efficiency and if necessary for self absorption effects. The characteristic peak shapes (G_{jk}) in Equation 5.21, to a first approximation are given by a Gaussian function. Deviations from a simple Gaussian can arise from a number of causes, including incomplete charge collection. These corrections are handled in different ways by each program. One representation of the characteristic peak spectrum for Sr is given in Fig. 5.14a and a spectrum fitted to experimental data is shown in Fig. 5.14b.

5.4.2 Background Radiation

An examination of PIXE spectra shows that the characteristic X-ray peaks are superimposed on a background which forms a limiting factor in sensitivity so that its origin should be understood.

a. Secondary Electron Bremsstrahlung

Folkmann *et al.* (1974) in studying the background, considered that it arose from three sources. Low energy background is largely due to the bremsstrahlung produced by the secondary electrons ejected from target atoms during irradiation. It was often six orders of magnitude larger than other sources of background. The maximum energy of a secondary electron is:

$$T = 4mM_1E_1/(m + M_1)^2 \tag{5.23}$$

where m is the electron mass, M_1 the ion mass and E_1 is the ion energy. A good approximation for $m \ll M_1$ is:

$$T(keV) = 2E_1(MeV)/M_1(amu) \tag{5.24}$$

The intensity of the bremsstrahlung background is angular dependent varying by approximately 50% over 180°, the minimum background being at backward angles to the incoming ion beam.

b. Projectile Bremsstrahlung

Ions decelerating in the target will also produce bremsstrahlung. The bremsstrahlung yield is a strong function of the ion mass and since m for electrons and M_1 for ions differ by several orders of magnitude, the ion bremsstrahlung is orders of magnitude less intense than the electron bremsstrahlung for X-ray energies below T_{max}.

c. Compton Scattering

Fluorine and Na have very high cross-sections for (ion, γ) reactions. Gamma rays produced by these reactions Compton scatter in the X-ray

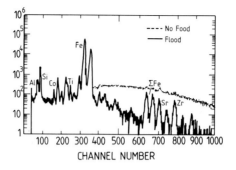

Fig. 5.15 The X-ray spectrum from an obsidian sample taken at 2.5 MeV with a pinhole filter. This figure shows the removal of the charging background of an insulating sample by using an electron flood.

detector producing large high energy X-ray backgrounds, usually much stronger than the ion bremsstrahlung background.

d. Insulating Targets

Insulating targets pose special difficulties because localised high voltages on the target surface accelerate free electrons, producing very high background up to tens of kev X-ray energy. This increased background not only reduces the detection sensitivity but may completely swamp smaller X-ray peaks in the spectrum. Current measurement is also quite difficult, especially at low currents. Various methods to overcome this problem are discussed in Chapter 2.3.3. Fig. 5.15 shows the effect of an electron flood on an insulating obsidian sample. The high energy continuum is reduced so that trace elements heavier than Fe become clearly visible.

5.4.3 Escape Peaks and Sum Peaks

Two further features that must be included in analysis codes are escape peaks and sum peaks. Escape peak intensities can be taken from Equation 5.17 and included in R_{jk} of Equation 5.21. These peaks are small but should be included to prevent possible error in element identification. Alternatively, the escape peak contribution can be stripped, channel by channel, from the spectrum prior to analysis. Sum peaks are greatly reduced by using on demand beam pulsing. Their presence must be balanced against the desire to increase the count rate to

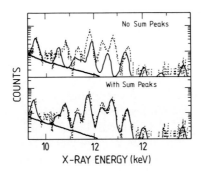

Fig. 5.16 The 10 to 14 keV region of the X-ray spectrum from a mineral sample containing high concentrations of Cr and Fe. In the top spectrum sum peaks are not included. The inclusion of sum peaks improves the fit in the bottom spectrum.

provide better counting statistics (Fig. 5.8). Johansson (1982) proposed an elegant way to model sum peaks in HEX and this has been modified for use in BATTY. For two X-rays of energy E_1 and E_2 the sum peak has a count rate at energy $E_1 + E_2$:

$$N = 2N_1N_2\tau \tag{5.25}$$

where τ is the minimum pulse pair resolution of the amplifier or logic circuit and N_1 and N_2 are the count rates of pulses E_1 and E_2 respectively. For two X-rays of the same energy E_1, the count rate at $2E_1$ is:

$$N = N_1^2\tau \tag{5.26}$$

Initial estimates for the peak heights (A_j in Equation 5.21) are used to form a SUM element before the least squares fitting is commenced. From the elements having the highest concentrations relative peak intensities and positions are calculated from Equations 5.25 and 5.26. These then form a SUM element which is added to the library of elements to be fitted. As R_{jk} has been corrected for the effects of detector efficiency and self absorption the sum peaks will also have the correct relative intensities. Fig. 5.16 shows the 10 to 14 keV region of the X-ray spectrum of a sample containing 6% Cr and 6% Fe. The top curve shows the fit without sum peaks and the bottom curve shows the much better fit obtained by including sum peaks (Clayton, 1983). Analysis incorporating all these features was done automatically on the BATTY system to produce the results in Fig. 5.12.

5.4.4 Yield Calibration

a. Thin Targets

Various methods are available to calibrate PIXE systems to provide the relation between peak areas and concentrations (Equations 5.1 and 5.2). These give a curve which relates the X-ray yield per unit concentration to the atomic number of the element being studied. Thin targets are normally calibrated by using measured yields from thin foils or from solutions evaporated onto a substrate. Johansson et al. (1981) use Micromatter foils (Heagney and Heagney, 1979) for thin target calibration. Their calibration is done in an iterative fashion using linear regression on nominal values for the system parameters and comparing predicted X-ray yields to those measured for the standards. They estimate that an individual element can be analysed with an accuracy of better than 5% for atomic numbers between 13 and 57 detected by K X-rays.

Maenhaut and Raemdonck (1984) carefully measured all the relevant system parameters and used ECPSSR cross-sections to estimate the expected yield from a set of accurate thin film standards. This absolute theoretical calculation gave accuracy to better than 4% for elements for Na to Sn. This result indicates that the physics data base for thin target PIXE analysis has good precision.

Thin foil calibrations can only be as good as the thin film standards. Mingay (1983) used a combination of RBS and PIXE analysis to study the accuracy of Micromatter foils. They concluded most standards were accurate to within the claimed 5%. The use of Mylar backing was preferred as Nucleopore backings tended to show diffusion of the element into the backing. Augustyniak et al. (1981) used solutions to calibrate their system for water analysis. A calibration curve giving the relative intensities of X-rays of different elements was estimated from three standard solutions containing 15 different elements. These were normalised to a known amount of Y added as an internal standard.

b. Thick Targets

If secondary effects can be ignored the thick target yield for an element of concentration c can be written (as in Equation 5.3):

$$Y = \frac{cN_0\Omega Q\varepsilon}{W4\pi e} \int_{E_p}^{0} \sigma^x(E)\frac{dE}{S(E)}T(E) \tag{5.27}$$

where $T(E)$ is the attenuation given by:

$$T(E) = \exp\left(-\mu\frac{\cos\theta_i}{\cos\theta_o}\left[\int_{E_p}^{E}\frac{dE'}{S(E')}\right]\right) \tag{5.28}$$

The thick target yield can be simply represented as

$$Y = cM \tag{5.29}$$

where cM is the calculated yield per unit concentration. In addition to the system parameters discussed for thin target stopping powers and mass attentuation coefficients also have to be estimated. Although the need to estimate a sample's major components seems to be a drawback the problem can be overcome. References such as ICRP (1975) and Bowen (1979) tabulate estimated compositions of a wide class of samples (Chapter 13). For instance a quick browse through Bowens's book reveals concentration data on aerosol composition, fresh and sea water, rocks, coals, soils, algae, fungi, plants, fish, blood and human tissues as well as other more exotic samples.

c. Internal Standards

Internal standards are commonly used. A known amount of a non-interfering element (commonly Y or Ru) is added to the sample and element concentrations are determined relative to the concentration of the standard, often by calculation of an expected yield. Equation (5.27) becomes, for an internal standard:

$$c_e = c_{st}Y_eM_{st}/Y_{st}M_e \tag{5.30}$$

where c_e and c_{st} are the concentration of the element and standard respectively, M_e and M_{st} the calculated yields from Equation 5.27 and Y_e and Y_{st} the measured X-ray yield. Using internal standards removes uncertainties in the current intergration and geometric factors such as solid angle.

d. External Standards

External standards can also be used to provide an accurate thick target calibration for PIXE analysis. Let the integral in Equation (5.27) be $I(E_1)$, so that:

TABLE 5.7
Elemental concentrations in NBS Bovine liver
(μg g^{-1}) by PIXE

Element	(a)	(b)	(c)	Reference
K	8603	9900	8500	9700
Ca	178	140	143	123
Mn	10.2	9.3	9.7	10.2
Fe	242	270	269	270
Cu	170	180	191	193
Zn	120	144	135	130
Se	<0.9	–	1.24	1.1
Br		9.7	9.6	8.7
Rb	17.5	18.8	19.1	18.3

(a) Pallon and Malmqvist (1981); (b) Biswas *et al.*
(1984); (c) Clayton *et al.* (1983).

$$I(E_1) = \int_{E_1}^{0} \sigma^x(E) \frac{dE}{S(E)} T(E) \tag{5.31}$$

then we can see that, if measured multielement yields are available for a known standard, then:

$$c_e = c_{st} Y_e I_{st}(E_1)/Y_{st} I_e(E_1) \tag{5.32}$$

This method has been extensively discussed by Khan and coworkers (Khaliquzzaman *et al.*, 1981, 1983; Biswas *et al.*, (1984). Its attraction is that it removes uncertainties in the detector efficiency, and to a large extent minimises uncertainty in both the cross section and attenuation coefficients.

An overview of different means of estimating PIXE yields is provided in Table 5.7, which shows measured concentrations for the NBS standard Bovine Liver given by several different methods. In general they show reasonable agreement, both with each other and with the certified values for different target types. Pallon and Malmqvist (1981) used acid digested sample, Biswas *et al.* (1984) used a thick pressed pellet and external beam and Clayton *et al.* (1981) used a thick pressed powder sample.

e. Absolute Measurements

Equation (5.27) has been used by the authors for a wide range of thick samples. Measured concentrations ranged from less than 1 μg g^{-1} to over

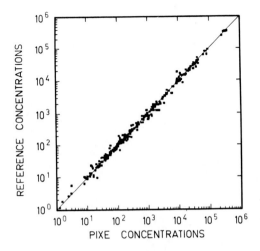

Fig. 5.17 Comparison between PIXE and reference values. Concentration values vary from 1 μg g^{-1} to over 20%. The solid line is the line of best fit.

20% for major components of geological samples. The success of this approach can be gauged from a comparison of our measured PIXE concentration values and data from standard reference materials (Fig. 5.17). Absolute estimates from Equation (5.27) and measurements relative to both internal and external standards are represented and the ratio PIXE1 Reference is 1.00 ± 0.20. Minimum detection limits (MDL) can also be quite low. Fig. 5.18 shows calculated MDLs for NBS Orchard Leaves and USGS SGR1 Oilshale under routine analysis conditions (2.5 MeV protons, 50 μC charge and 0.526 mm Perspex filter). Following Currie (1968) the 95% confidence level is chosen so that $N > 3.29\sqrt{B}$, where N is the number of counts in the X-ray peak and B is the background counts calculated over the range ± 3 standard deviations. By defining the background to be the total number of counts in that region of the spectrum minus the counts in the X-ray peak of interest one has a conservative estimate which gives the minimum detection limit in the presence of interfering elements.

5.5 APPLICATIONS

5.5.1 Other Techniques

Detection limits for PIXE, Fig. 5.18, are much lower than for electron induced X-rays (approximately 1 mg g^{-1}) owing to the much higher

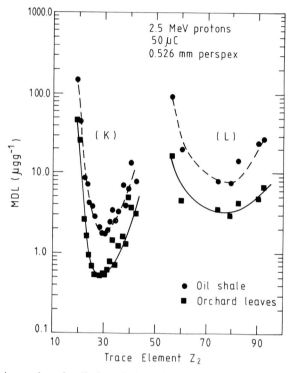

Fig. 5.18 Minimum detection limits calculated for routine PIXE analysis of a biological and geological sample.

background of the latter method. X-ray induced fluorescence (XRF) is the main alternative to PIXE for X-ray analysis. Either method can, in principle, meet the need for large scale routine analysis, and they should be regarded as complementary techniques. If large samples are available, XRF may be the preferred method, but for small samples or microprobe applications PIXE is more advantageous. In a study on thick target drill core samples Carlsson and Akselsson (1981) have compared both methods. They found that both had similar detection limits (Fig. 5.19). To cover the same range of elements as the PIXE analysis, three separate spectra were accumulated using three different targets in the XRF generator. Their calculations showed that PIXE was less sensitive to errors in the matrix composition than XRF.

The discussion in Section 5.2.2 showed that PIXE is a near surface technique, due to the small penetration of the incident protons and the absorption of the characteristic X-rays. Although the incident X-rays used in XRF have a greater range in the sample than protons, it also

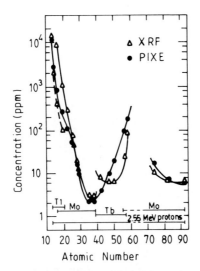

Fig. 5.19 Minimum detection limits for PIXE and XRF analysis of geological samples. The bars at the bottom represent the range of the trace elements covered by XRF target anodes of materials shown and by 2.55 MeV PIXE.

attenuates both the characteristic X-rays and the inducing radiation. Ahlberg (1977) has studied a wide range of samples from low Z (cellulose) to medium Z (stainless steel). He showed that the analysed layer was thicker for XRF than PIXE for low Z materials, but approached the same thickness as the sample Z increased. He further showed that the thickness analysed by PIXE was relatively constant, whereas with XRF there was a wide variation.

A few comparisons will indicate the differences between PIXE and the widely used neutron activation analysis (NAA) technique. System calibration is much simpler for PIXE as there is a systematic variation with element Z which can be easily modelled. With NAA, calibrations have to be made for each element that is being measured. The time of analysis is much shorter for PIXE as several analyses over a period of weeks may be needed for NAA. Minimum detection limits in general can be much lower in NAA than PIXE particularly for rare earth elements. On the other hand, elements such as Ca and Pb are much more easily measured by PIXE.

PIXE has been compared with standard chemical methods of analysis such as atomic absorption analysis by several authors, e.g. Clayton *et al.* (1983) and Bombelka *et al.* (1984). Fig. 5.20 shows a comparison of PIXE with flame atomic spectrometry on 40 hair samples for the element Ca (Clayton *et al.*, 1983). Both series of measurements

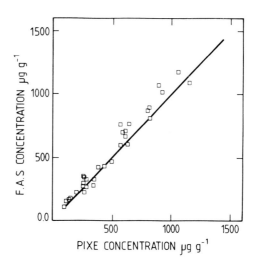

Fig. 5.20 A comparison of PIXE and Flame Atomic Spectrometry (FAS) results for Ca in 40 hair samples.

were routine analyses and the figure shows comparable accuracy for both methods. Overall, for six elements, only Zn showed a discrepancy between the two techniques. For Ca the overall agreement for the ratio PIXE/AAS was 0.92 ± 0.11 (Fig. 5.20).

5.5.2 Multitechnique Measurements

It is a simple matter in many cases to put additional detectors into the target chamber so that more data can be taken. Ge(Li) and surface barrier detectors can be placed so that simultaneous measurements can be made of PIGME and RBS spectra in conjunction with PIXE spectra.

Measurements such as ERA or NRA can be made simultaneously or performed separately to provide an extended range of elements, confirmation of a PIXE result or the major composition of a sample. For example Clayton *et al.* (1983) needed the major components of hair to calculate an X-ray yield curve from Equation 5.27. RBS gave information on C, N, O and S and He ERA gave data on H.

5.5.3 Depth Profiling with PIXE

The thick target yield for a depth distribution $n(x)$ of atoms in a near surface region is obtained from Equation (5.27) by replacing c by $n(x)$ and noting that the depth parameter x is also a function of the ion energy. The

X-ray yield is then a function of ion energy, incidence angle θ_i and exit angle θ_o. The depth profile $n(x)$ is in principle obtained by deconvolution of the integral in Equation (5.27), where the yield has been measured as a function of either ion energy or the incident and exit angles. Rossiger (1982, 1985) used these PIXE depth profiling techniques to study the near-surface Ar distributions in semiconductors and brass coatings on steel wire. Depths from 0.01 to 1.0 μm in Si were investigated with low energy protons (100 to 350 keV). The PIXE profiles were within 30% of those obtained by RBS for depths down to 1μm. Pabst (1974, 1975) used angle variation to study asymmetrical Gaussian implant distributions. In his first paper he showed that small variations in the calculated X-ray yield gave rise to strong oscillations in the computed profile, whereas small variations in the profile may not give a noticeable change in the calculated yield. To overcome these problems the yield for a number of angles was calculated for given profiles. Comparison of a measured curve with these theoretical curves gave an estimate for the actual concentration profiles, some of which could be distinguished by the characteristic shapes of their yield curves.

He ions have a shorter range than protons of the same energy and give better depth resolution in the same material than protons. Feldman *et al.* (1973) and Feldman and Silverman (1976) used He ions to study GaAs and Pt-Ga-As films. By examining the PIXE spectra at various angles before and after annealing of their samples, a simple depth profile was obtained. This gave concentration data for the film between the surface and a depth of approximately 0.3 μm. They concluded that qualitatively this technique could be a useful supplement to ion backscattering. Ahlberg (1975) used the variation of the $K\alpha/K\beta$ intensity ratio to determine whether an element was present on the surface or in the bulk of a sample. The technique was used to distinguish between 32 μg cm^{-2} of Fe on the surface and 6 mg g^{-1} of Fe in the bulk of an Al sample.

Geretschlager (1982) has determined the influence of errors of yield measurements on the determination of a depth profile using PIXE analysis. He concluded that these techniques are best at glancing incidence and exit angles (as expectecd). However under extreme glancing conditions ($< 10°$) errors of the order of 50% in the X-ray yield may occur because of surface roughness effects.

To sum up it is possible to obtain limited depth profile information using ion induced X-ray emission. The depth resolution is angle and ion energy dependent and is not as good as other techniques. However it does have very good mass resolution and is especially useful for profiling medium Z atoms in a low Z substrate. The optimal depth range for useful results appears to be between 0.1 and 10 μm. PIXE is generally not the best choice for most profiling jobs.

5.5.4 Surface Roughness Effects

Ahlberg and Akselsson (1976) estimated the effect of roughness on PIXE yield from teeth using a sawtooth model of the tooth surface. These calculations indicated that, for this matrix, height differences of about 4 mg cm^{-2} (13 μm) resulted in a decrease in yield of less than 5% for X-ray energies above 2 keV. Cookson and Campbell (1983) used the Harwell microprobe to follow regular surface roughness features of the order of 10 μm with a 3 MeV proton beam incident normally on the target and a 45° X-ray take off angle. The targets considered were C, Ca, Fe and Ag, each containing about ten trace elements. They concluded that triangular grooves and ridges gave similar but opposite effects. The dependence on the mass attenuation coefficient (μ/ρ) was complex. For (μ/ρ) ~500cm^{2}g^{-1} the X-ray yield increased strongly compared with the flat surface yield, intermediate values caused zero effect or a loss in yield. This loss reached a maximum and then fell to zero as (μ/ρ) fell to 10 cm^{2}g^{-1}. For high (μ/ρ) the gain in yield grew linearly with groove depth and saturated when the depth became comparable to a few attenuation lengths.

5.5.5 Heavy Ion and Chemical Effects

Regardless of how they are produced, X-ray spectra are sensitive to the chemical environment of the emitting atom and can yield information on atomic and electronic structure of host materials. High energy heavy ions are capable of producing multiple inner shell ionisation. The resulting spectrum of X-rays from particular target atoms is composed of a complex series of satellite lines. Environmental effects give rise to the redistribution of intensity from one group to another. These changes can be correlated with bulk chemical properties such as valence electron densities, effective target atom charges and covalencies. New chemical information from such samples as metal alloys and implanted materials can be obtained using a high resolution wavelength dispersive detector. Fig. 5.21 shows an example of high resolution PIXE spectra obtained with protons, He and O ions incident on solid Ti targets (Moore *et al.* 1972). The $K\alpha$ and $K\beta$ peaks split into several satellite peaks as Z_1 increases. In this figure n is the final number of $2p(L_{2,3})$ electrons remaining in the target atom.

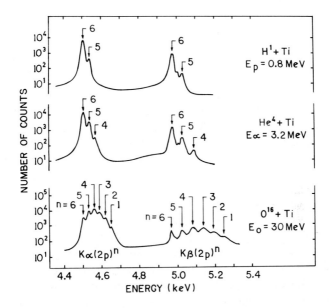

Fig. 5.21 The K X-ray spectra of ^{1}H, ^{4}He and ^{16}O on Ti; n is the number of $2p$ ($L_{2,3}$) electrons remaining after the ion collision.

The amount of multiple ionisation increases rapidly with ion energy to a maximum and then slowly decreases as the bombarding energy is increased further. Even for ions as light as He the amount of multiple ionisation (of the L shell and higher) can far exceed that of single ionisation. For the lighter target atoms, inner shell multiple ionisation of this type will be affected by the chemical bonds of these atoms. Fig. 5.22, taken from the work of Vane *et al.* (1984), shows satellite $SK\alpha$ spectra for two compounds under 36 MeV Ar ion impact. The intensity distributions of the $K\alpha$ satellites vary from target to target depending on the chemical environment of the emitting atom. These satellite peaks show significantly greater intensity and energy variations with chemical effect than do the normal $K\alpha$ lines obtained from proton bombardment.

The sensitivity of this technique to the chemistry is as yet too weak to allow specification of particular pure compounds, except for special cases where electron transfer from ligand atoms occurs resonantly. For light atoms, individual features of satellite spectra change more strongly and show higher correlation with chemical environment than does the average K or L shell ionisation probability. This field is still in its infancy,

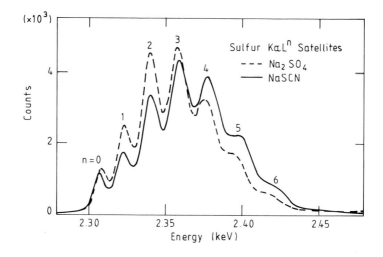

Fig. 5.22 Sulphur $K\alpha$ satellite spectra for Na_2SO_4 and NaSCN excited by 36 MeV Ar ions. The centre of the abscissae are at 2.37 keV and each frame represents 200 eV. Taken from the work of Vane *et al.* (1984).

the technical problems are difficult, since high resolution is required, and the needed theoretical basis for interpretation of the results is lacking. However, based on the results so far, it is reasonable to predict that the study of heavy ion induced satellite X-ray spectra will make a powerful contribution to the understanding of chemical bonding in complex chemical compounds.

For a given ion velocity the ionisation cross-sections, and hence the X-ray yields, are generally larger for heavier ions on a given target. This is clearly shown in Fig. 5.23 where the K shell ionisation cross-sections, for 4 MeV/amu heavy ions from protons to S ions, are plotted as a function of target atomic number Z_2. This effect is more pronounced for light targets ($Z_2 < 30$). The fact that the ionisation cross-sections may be, say, an order of magnitude larger for a particular heavy ion does not mean that the X-ray yield will also increase by a similar amount. The reason for this is that heavier ions, for the same velocity, have shorter ranges than lighter ions. For example, the increased ionisation cross section in going from proton induced X-ray production to He induced X-ray production is entirely offset by an equivalent decrease in He range. Hence for 1 to 4 MeV/amu protons and He ions, X-ray yields are surprisingly similar for most targets. The cross-section scaling laws of Section 5.2.5 and the range scaling laws of Chapter 14.1 can be combined to give a rough guide of expected X-ray yield changes when going from protons to heavier ions.

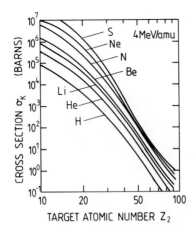

Fig. 5.23 K shell ionisation cross-sections (barns) versus target atomic number (Z_2) for protons, He, Li, Be, N, Ne and S ions of energy 4 MeV/amu.

5.6 SUMMARY

Methods and data for using K and L X-ray production by 1 to 4 MeV/amu protons are well established and thin or thick targets can be analysed with an absolute precision of 10% or better. Some work is going on using M shell X-rays or heavy ions but this work is still in its infancy. PIXE has advantages in its good sensitivity, multi-element capability and speed of analysis. It therefore is a preferred method for such applications as the analysis of 15 to 20 elements in thin samples such as air filters (~ 1 μg g^{-1}) and for automated major, minor and trace element analysis of large numbers of geological or archaeological samples. The low absolute detection limit (10^{-13} to 10^{-14}g) and good sensitivity for elements such as S, P, Cl, K and Ca make PIXE of great importance in biological and medical fields especially using the proton microprobe (Chapter 10).

ACKNOWLEDGMENTS

One of us (EC) would like to thank Dr P. Van Espen and Dr W. Maenhaut for discussions and help with parts of section 5.4 and Dr L. E. Carlsson for providing the information to produce Fig. 5.19.

REFERENCES

Ahlberg, M.S. (1975). *Nucl. Instrum. Methods* **131**, 381
Ahlberg, M.S. (1977). *Nucl. Instrum. Methods* **146**, 465

Ahlberg, M.S. and Akselsson, R. (1976). *Int. J. Appl. Rad. Isotop.* **22**, 279

Andersen, H.H. and Ziegler, J.F. (1977), 'Stopping and Ranges of Ions in Matter', Vols. I–VI. Pergamon Press, New York.

Anttila, A., Raisanen, J. and Lappalainen, R. (1985). *Nucl. Instrum. Methods in Physics Research* **B12**, 245

Augustyniak, S., Baranek-Lenczowska, J., Jarczyk, L., Slominska, D. and Strzalowski, A. (1981). *IEEE Trans. Nucl. Sci.* **28**, 1398.

Bambynek, W. (1985). *IAEA Technical Document,* **335**, 412.

Bambynek, W., Crasemann, B., Fink, R.W., Freud, H.U., Mark, H., Swift, C.D., Price, R.E. and Venugopala, R.P. (1972). *Rev. Mod. Phys.* **44**, 716.

Basbas G., Brandt W. and Laubert R. (1971). *Phys. Lett.* **34A**, 277.

Bevington, P.R. (1969). 'Data Reduction and Error Analysis for Physical Sciences', McGraw-Hill, New York.

Biswas, S.K., Khaliquzzaman, M., Islam, M.M. and Khan, A.H. (1984). *Nucl. Instrum. Methods in Physics Research* **B3**, 337.

Bombelka, E., Richter F.W., Ries H. and Watjen U. (1984). *Nucl. Instrum. Methods in Physics Research* **B3**, 296.

Bowen, H.J.M. (1979). 'Environmental Chemistry of the Elements', Academic Press, London.

Brandt, W. and Laubert, R. (1975). *Phys. Rev.* **A11**, 1233.

Brandt, W. and Lapicki, G. (1979). *Phys. Rev.* **A20**, 465.

Brandt, W. and Lapicki, G. (1981). *Phys. Rev.* **A23**, 1717.

Cahill, T.A. (1975). *In* 'New Uses of Ion Accelerators' (Ziegler, J.F. ed), Plenum Press, New York, pp 1–71.

Campbell, J.L. and McNelles, L.A. (1975). *Nucl. Instrum. Methods* **125**, 205.

Campbell, J.L. (1983). *Nucl. Instrum. Methods* **204**, 581.

Campbell, J.L., Maenhaut, W., Bombelka, E., Clayton, E., Malmqvist, K., Maxwell, J.A., Pallon, J. and Vandenhaute, J. (1986). *Nucl. Instrum. Methods in Physics Research* **B14**, 204.

Carlsson, L.E. and Akselsson K.R. (1981). *Advances in X-ray Analysis* **24**, 313.

Clayton, E. (1983). *Nucl. Instrum. Methods* **218**, 221.

Clayton, E. Cohen, D.D. and Duerden, P. (1981). *Nucl. Instrum. Methods* **180**, 541.

Clayton, E., Chapman, J.F. and Wooller, K.K. (1983). *IEEE Trans. Nucl. Sci.* **NS-30**, 1323.

Cohen, D.D. (1980). *Nucl. Instrum. Methods* **178**, 481.

Cohen, D.D. (1982). *Nucl. Instrum. Methods* **193**, 15.

Cohen, D.D. (1984a). *J.Phys. B:Atom. Mol. Phys.* **17**, 3913.

Cohen, D.D. (1984b). *Nucl. Instrum. Methods* **B3**, 47.

Cohen, D.D., Clayton, E. and Ainsworth, T. (1981). *Nucl. Instrum. Methods* **188**, 203.

Cohen, D.D. and Harrigan, M. (1985). *Atom. Data. and Nucl. Data Tables* **33**, 255.

Cohen, D.D. and Harrigan, M. (1986). *Atom. Data. and Nucl. Data Tables* **34**, 393.

Cookson, J.A. and Campbell, J.L. (1983). *Nucl. Instrum. Methods* **216**, 489.

Currie, L.A. (1968). *Anal. Chem.* **40**, 586.

Feldman L.C., Poate J.M., Ermanis F. and Schwartz B. (1973). *Thin Solid Films* **19**, 81.

Feldman L.C. and Silverman P.J. (1976). 'Ion Beam Surface Analysis', Plenum Press, New York, **2**, 735.

Folkman, F., Gaarde, C., Huus, T. and Kemp, K. (1974). *Nucl. Instrum. Methods* **116**, 487.

Geretschlager, M. (1982). *Nucl. Instrum. Methods* **192**, 117.

Goldstein, J.I., Newbury, D.E., Echlin, P., Joy, D.C., Fiori, C. and Lifshin, E. (1981). 'Scanning Electron Microscopy and X-ray Microanalysis', Plenum Press, New York.

Gray, T.J. (1980). *In* 'Methods of Experimental Physics' (L. Marton and C. Martin, eds.). Vol. **17**, Academic Press, NY, 193.

Heagney, J.M. and Heagney, J.S. (1979). *Nucl. Instrum. Methods* **167**, 137.

Heck, D. (1981). *Nucl. Instrum. Methods* **191**, 579.

Hollis, M.J. (1972). *Nucl. Instrum. Methods* **103**, 337.

ICRP (1975). 'Report of the task group on Reference Man', ICRP Publication 23, Pergamon, Oxford.

Janni, J.F. (1982). *Atom. Data. and Nucl. Data Tables* **27**, 147.

Johansson G.I. (1982). *X-Ray Spectrom.* **11**, 194.

Johansson, G.I., Pallon, J., Malmqvist, K.G. and Akselsson, K.R. (1981). *Nucl. Instrum. Methods* **181**, 81.

Johansson, S.A.E. (ed.). (1977). 'Proc. Int. Conf. on Particle Induced X-ray Emission and its Analytical Applications, Lund, Sweden, 1976'. *Nucl. Instrum. Methods* **142**.

Johansson, S.A.E. (ed.). (1981). 'Proc. Int. Conf. on Particle Induced X-ray Emission and Its Analytical Applications, Lund, Sweden, 1980'. *Nucl. Instrum. Methods* **181**.

Johansson, S.A.E. and Johansson, T.B. (1976). *Nucl. Instrum. Methods* **137**, 473.

Johnson, D.E., Basbas, G. and McDaniel, F.D. (1979). *Nucl. Data Tables* **24**, 1.

Khaliquzzaman, M., Zaman, M.B. and Khan, A.H. (1981). *Nucl. Instrum. Methods* **181**, 209.

Khaliquzzaman, M., Lam, S.T., Sheppard, D.M. and Stephens-Newsham, L.G. (1983). *Nucl. Instrum. Methods* **216**, 481.

Krause, M.O. (1979). *J. Phys. Chem. Ref. Data* **8**, 307.

Lennard, W. and Phillips, D. (1979). *Nucl. Instrum. Methods* **166**, 521.

Leroux, J. and Thinh, T.P. (1977). 'Revised Tables of X-ray Mass Attenuation Coefficients'. Corporation Scientifque Claisse Inc., Quebec, Canada.

Maenhaut, W. and Raemdonck, H. (1984). *Nucl. Instrum. Methods in Physics Research* **B1**, 123.

Maenhaut, W., Selen, A., Van Espen, P., Van Grieken, R. and Winchester, J.W. (1981). *Nucl. Instrum. Methods* **181**, 399.

Malmqvist, K.G., Johansson, G.I. and Akselsson, K.R. (1982). *J. Radioanal. Chem.* **74**, 125.

Martin, B. (ed). (1984). *Nucl. Instrum. Methods* **B3**.

Mayer, J.W. and Rimini, E. (eds) (1977). 'Ion Beam Handbook of Material Analysis', Academic Press, New York. p.450.

Merzbacher, E. and Lewis, H.W. (1959). *In* 'Encyclopedia Physics' (S.Flugge, ed.). Springer-Verlag, Berlin, New York, Vol. **134**, 166.

Mingay, D.W. (1983). *Radioanal. Chem.* **78**, 127.

Moore, C.F., Senglaub, M., Johnson, B. and Richard, P. (1972). *Phys. Lett.* **A40**, 107.

Musket, R.G. (1974). *Nucl. Instrum. Methods* **117**, 385.

Nielson K.K., Hill, M.W. and Mangelson, N.F. (1976) *In* 'Advances in X-ray Analysis', **19**, 511.

Pabst, W. (1974). *Nucl. Instrum. Methods* **120**, 543.

Pabst, W. (1975). *Nucl. Instrum. Methods* **124**, 143.

Pallon, J. and Malmqvist, K.G. (1981). *Nucl. Instrum. Methods* **181**, 71.

Paul, H. (1983). *Nucl. Instrum. Methods* **214**, 15.

Reuter, W., Lurio, A., Cordone, Г. and Ziegler, J.F. (1975). *J. Appl. Phys.* **46**, 3194.

Rossiger, V. (1982). *Nucl. Instrum. Methods* **196**, 483.

Rossiger, V. (1985). *X-ray Spectrom.* **14**, 62.

Salem, S.I., Panossian, S.L. and Krause, R.A. (1974). *Atom. Data and Nucl. Data Tables* **14**, 91.

Sarkadi, L. and Mukoyama, T. (1981). *J. Phys. B. Atom. Molec. Phys.* **14**, L225.

Scofield, J.H. (1972). UCRL-51231.

Scofield, J.H. (1974). *Atom. Data and Nucl. Data Tables* **14**, 121.

Shima, K. (1979). *Nucl. Instrum. Methods* **165**, 21.

Shima, K., Umetani, K. and Mikumo, T. (1980). *J. Appl. Phys.* **51**, 846.

Taulbjerg, K. and Sigmond, P. (1972). *Phys. Rev.* **A5**, 1285.

Taulbjerg, K., Fastrup, B. and Laegsgaard, E. (1973). *Phys. Rev.* **A8**, 1814.

Van Espen, P., Nullens, H. and Maenhaut, W. (1979). *In* 'Microbeam Analysis' (D.E. Newbury ed.). San Francisco Press. p.265.

Van Oystaeyen B. and Demortier G. (1983). *Nucl. Instrum. Methods* **215**, 299.

Vane, C.R., Hulett, L., Kahane, S., McDaniel, F.D., Milner, W.T., O'Kelley, G.D., Raman, S., Rossell, T.M., Slaughter, G.G., Varghesse, S.L. and Young, J.P. (1984). *Nucl. Instrum. Methods* **B3**, 88.

Ziegler, J.F. (1977). 'Stopping Powers and Ranges of Ions in Matter', Vol. **4**, Pergamon Press, New York.

6
Channeling

J.S. WILLIAMS

Microelectronics and Materials Technology Centre, RMIT
Melbourne, Australia

R.G. ELLIMAN

Microelectronics and Materials Technology Centre, RMIT
Melbourne, Australia

ION BEAMS FOR
MATERIALS ANALYSIS
ISBN 0 12 099740 1

261

6.1 INTRODUCTION

The phenomenon of ion channeling in crystalline solids was first predicted by Stark (1912) but was not discovered until the computer simulation studies of Robinson and Oen (1963). Experimental verification of enhanced penetration of ions in solids when incident along open (low index) crystallographic directions was established by several groups (Lutz and Sitzmann, 1963; Piercy *et al.*, 1963; Thompson and Nelson, 1963). Subsequent experimental and theoretical studies sought to develop an understanding of the channeling phenomenon and Lindhard's comprehensive theoretical treatment in 1965 established a firm foundation. By 1967, the first papers heralded the application of ion channeling for analysing crystal damage and atom location (Bogh, 1967). Since that time rapid progress has been made, stimulated by interest from the semiconductor industry and the suitability of channeling for probing ion implanted layers (Mayer *et al.*, 1970). More recent applications have covered the analysis of a wide range of materials and thin film structures. This chapter outlines the principles of channeling, illustrates how channeling can be applied to the near-surface analysis of materials, and gives selected examples. A more comprehensive treatment can be found in Feldman *et al.*, (1982).

Most applications of channeling have concentrated on the determination of near surface crystal disorder or atom location using MeV light ions. The particular choice of ion type and energy depends upon the interaction product being detected (e.g. backscattered ions, nuclear reaction products or X-ray emission), but most channeling analyses involve Rutherford backscattering (RBS) of MeV He^+ ions, a method called RBS/channeling (RBS-C).

6.2 BASIC PRINCIPLES

As outlined in Chapter 1, channeling in a crystalline solid occurs when an ion beam is well aligned with a low-index crystallographic direction. The fraction of channeled ions may be as high as 95–98%, giving rise to a corresponding reduction in the yield from small impact parameter interaction processes. The reduction in yield occurs because channeled ions mostly experience large impact parameter collisions with substrate atoms. Furthermore, this interaction yield is strongly influenced by crystal disorder and is sensitive to very small displacements of atoms from crystalline lattice positions. These processes underpin the most important applications of channeling. The treatment below follows Lindhard (1965), Gemmell (1974) and Feldman *et al.* (1982).

When an ion beam enters a crystal parallel to a channeling (low index) direction, the beam can be resolved into two distinct components, Fig. 6.1a, a random component χ_R whose path through the crystal is not influenced by the regular atomic arrangement; and a channeled component $(1 - \chi_R)$ which is steered along the open crystal direction by correlated collisions with the regular array of lattice atoms. In a perfect crystal, the interaction yield from small impact parameter processes results almost entirely from interactions of the random beam component with lattice atoms. Backscattered ions χ_S represent a third, but relatively small ($< 10^{-4}$), component of the ion beam. The backscattered component contains information about the identity and depth distribution of substrate atoms (Chapter 3) and exhibits a strong dependence on channeling, which makes it a useful tool for studying channeling phenomena.

Fig. 6.1a schematically depicts RBS spectra for two extreme cases of incident beam alignment. The spectrum labelled 'random' illustrates a situation in which the incident beam is not aligned with any specific crystallographic direction and there is no channeled component of the ingoing beam. In principle, this is identical to a conventional RBS spectrum of an amorphous target material as treated in Chapter 3. The spectrum labelled 'aligned' illustrates the other extreme in which the beam is incident along a low index crystallographic direction, so that $\sim 98\%$ of the ingoing beam is channeled. The normalised RBS yield then reflects the small, $\sim 2\%$, random component of the beam. As the beam penetrates into the solid, more and more of the channeled ions are lost from the channel and are said to be dechanneled. This is reflected as an increasing RBS yield with penetration depth.

The small peak observed at the surface in the aligned RBS spectrum of Fig. 6.1a arises from scattering of incident ions from the sample surface. The probability of small impact parameter collisions from the first layer of atoms is independent of the incident ion alignment. In an ideal situation the surface peak yield from the first monolayer would be

---

Note

In ion scattering texts the usual notation for incident and scattered ion energies employs the subscripts 0 and 1 respectively. Subscripts 1 and 3 are used here for consistency with the more general notation applicable for nuclear reactions. Subscripts 2 and 4 refer to the target atom and recoil atom respectively.

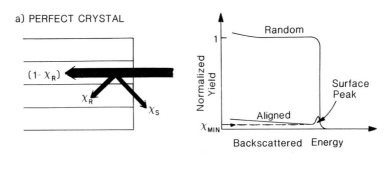

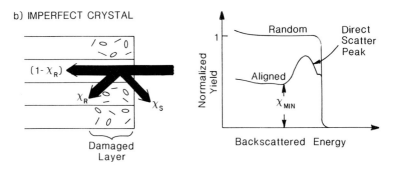

Fig. 6.1 Schematic illustrations of ion channeling, dechanneling (χ_R) and direct scattering (χ_s) in a) a perfect crystal and b) an imperfect crystal. Typical random and aligned spectra are also shown, indicating the minimum yield (χ_{min}).

identical to that of the random spectrum. However, energy and depth resolution limitations generally reduce this peak to the form shown in Fig. 6.1a. The surface peak area is conserved and is proportional to the number of surface atoms available for large-angle scattering (Chapter 9). The normalised yield immediately behind the surface peak corresponds approximately to the minimum random component of the beam as it enters the crystal and is termed the *minimum yield*, χ_{min}.

Because the channeling process is a consequence of the regular atomic arrangement in crystalline solids, it is sensitive to small departures from crystallinity. Fig. 6.1b illustrates the interaction of a channeled beam with crystal defects which give rise to two distinct effects:

i) they can increase the random component of the beam, χ_R, by increasing the rate of dechanneling; and

ii) they can increase the direct small impact parameter collision yield, χ_S, by introducing lattice atoms into the path of the channeled beam.

Fig. 6.1b shows a random RBS spectrum and an aligned spectrum which exhibits a higher yield than that from a perfect crystal, reflecting both an increased direct scattering yield (the broad peak) and an increased random fraction (increased χ_R).

Examination of a crystal model reveals that the open 'channels' in crystals have two distinct forms: axial channels, where the channel is defined by rows of atoms; and planar channels where the open direction is defined by parallel atomic planes. These structural differences are depicted in Fig. 6.2a, and typical backscattering spectra for axial and planar alignment of an incident beam are shown in Fig. 6.2b. Planar alignment is observed to result in a relatively high backscattered yield ($\sim 30\%$) and the spectrum exhibits distinct yield oscillations in the near-surface region. Axial alignment, on the other hand, has a low minimum yield ($\sim 3\%$) and exhibits only a strongly damped yield oscillation. In perfect crystals, low index axes typically have minimum yields around 1–5% whereas low index planes have minimum yields around 10–50%.

Channeling processes are equally possible for ions commencing their trajectories from within a crystal. Two important aspects of this process are *double alignment* and *blocking*. Double alignment refers to the situation in which ions are incident along a channeling direction and scattered ions are also detected along a channeling direction. This reduces the backscattered yield to less than 0.1% of the random level in perfect crystals and increases the sensitivity to lattice disorder and atom location. Blocking refers to the case in which an ion commences its trajectory from a crystal lattice site. This may arise, for example, from the spontaneous decay of an unstable lattice atom or from some form of ion beam interaction. Observed from outside the crystal, the ion yield from such processes is a minimum when viewed along certain channeling directions due to shadowing or 'blocking' by the crystal lattice. In this section the theory of channeling is developed in some detail for axial channeling and relevant equations are summarised for both axial and planar channeling in Section 6.5.

6.2.1 Channeling Concepts

To understand the channeling effect and its consequences it is first necessary to understand how an energetic ion interacts with a regular three dimensional array of atoms. Such problems can readily be tackled by computer simulation (e.g. using Monte Carlo techniques) but it is more informative, at least initially, to investigate such problems analytically.

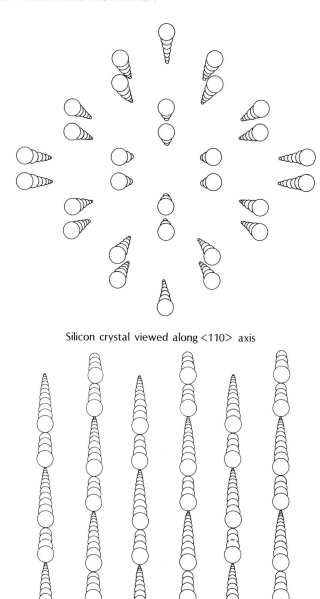

Silicon crystal viewed along <110> axis

(110) planar channel

Fig. 6.2a. Schematic illustrations of a ⟨110⟩ crystal axis and a set of (110) planes in a cubic crystal.

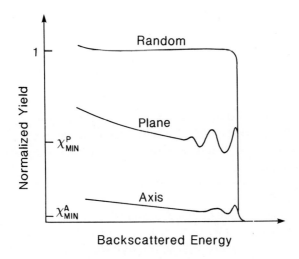

Fig. 6.2b. RBS-C spectra showing the result of planar and axial channeling.

a. Continuum Model

For a closed-packed structure (Fig. 6.2a), the channeling process can be modelled in terms of ion scattering from atomic strings (axial channeling) and planes (planar channeling). Such interactions can be considered to take the form of a sequence of ion–atom collisions as illustrated in Fig. 6.3. The continuum model of channeling, due largely to Lindhard (1965), further asserts that ion-string or ion-plane scattering can be approximated by scattering from a string or plane of uniform (continuum) potential. The discrete nature of the atoms is assumed to be unimportant since each steering collision is an average of many individual ion–atom collisions.

For axial channeling, where scattering can be modelled in terms of string potentials, the continuum potential is given by:

$$U(r) = \frac{1}{d}\int_{-\infty}^{\infty} V[(z^2 + r^2)^{1/2}]dz \qquad (6.1)$$

where V is the interatomic potential and d, r and z are defined in Fig. 6.3a. If the 'Lindhard standard potential' (see Chapter 1) is employed,

Equation (6.1) can be solved analytically for an ion scattering from a static row of atoms to yield:

$$U(r) = \left(\frac{2Z_1 Z_2 e^2}{d}\right)\ln\left[\left(\frac{Ca}{r}\right)^2 + 1\right] \qquad (6.2)$$

where Z_1 is the atomic number of the incident ion, Z_2 is the atomic number of the target material, e is the electronic charge ($e^2 = 14.4\,\text{eV.Å}$), C^2 is taken to be 3 and a is the Thomas–Fermi screening distance, Equation (1.4).

The total potential experienced by an ion in a channel is the sum of ion-string and ion-plane potentials. The continuum potential is, however, a strong function of distance and falls off rapidly as the ion-string or ion-plane distance increases. Consequently, the potentials for certain critical phenomena are well approximated by a single dominant ion-string or ion-plane potential. This simplifies the evaluation of such effects.

a)

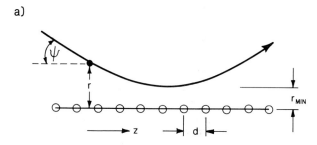

b)

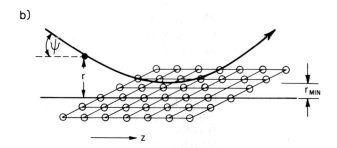

Fig. 6.3 Ion scattering from an axial string (a) and a plane (b) giving the appropriate parameters used in the continuum approach (see text).

The magnitude of the string potential is of the order of interatomic potentials. For example, a He^+ ion at a distance of 0.1 Å from a ⟨100⟩ silicon string ($d = 5.431$ Å) experiences a potential, given by Equation (6.2), of 97 eV. This reduces to 32 eV at a separation of 0.5 Å. Therefore, for distances of the order of the thermal vibrational amplitude of lattice atoms, 0.1 Å (Chapter 14.6), the continuum potential is of the order of 100 eV.

b. *Transverse Energy*

Since both the total ion energy and the ion velocity along a channel remain essentially constant during a steering collision, the energy associated with motion in the plane normal to the channel must also remain constant. Resolving the ion motion into these two components, motion along the channel and motion in the plane normal to the channel, reduces the analysis of ion motion in a channel to a problem equivalent to a particle in a two dimensional potential well.

The plane normal to the channel is generally referred to as the *transverse plane* and the ion energy associated with motion in this plane is referred to as the *transverse energy*. The transverse energy consists of a kinetic term, derived from the transverse component of the ion momentum, and a potential energy term arising from the interaction of the ion with atomic strings and planes. The transverse energy of an ion at a point r in a channel, travelling at an angle ψ to the channel direction is therefore given by:

$$E_T = E\psi^2 + U(r) \tag{6.3}$$

where E is the ion energy and $U(r)$ is the potential experienced by the ion at r. (It has further been assumed that the incident angles are sufficiently small that $\sin(\psi) \sim \psi$.) It is apparent from Equation (6.3) that a uniform distribution of ions incident upon a crystal will acquire a distribution of transverse energies as a consequence of their different entry positions into the channel. Those that approach strings and planes closely attain the greatest transverse energy and those entering near mid-channel attain the lowest transverse energy. The requirement for a stable channeling trajectory can be qualitatively understood in terms of the transverse energy. In particular, the transverse energy of an ion must not exceed the potential energy barrier of the confining potential which, as shown above, is of the order of 100 eV.

c. Critical Angle

When the transverse energy of an ion is comparable in magnitude to the potential that confines it to the channel, it approaches the atomic strings and planes closely. Under such conditions the discrete nature of the atomic strings and planes becomes evident and invidual ion–atom collisions begin to dominate the scattering process. The limiting condition for the applicability of the continuum approach can be expressed in terms of a critical approach distance, r_c, within which non-continuum scattering is probable. This is related to a *critical angle* for channeling given by:

$$E\psi_c^2 = U(r_c) \tag{6.4}$$

The thermal vibration of lattice atoms sets a lower limit to the value of r_c, and so a reasonable estimate of the critical angle can be obtained by the substitution of $r_c = \rho$ in Equation (6.4), where ρ is the root mean square (RMS) thermal vibrational amplitude normal to the axis. (Note that $\rho = \sqrt{2}u_1$, where u_1 is given in Section 6.5). For the 'standard' potential this gives:

$$\psi_c(\rho) = \frac{\psi_1}{\sqrt{2}} \left\{ \ln\left[\left(\frac{Ca}{\rho}\right)^2 + 1 \right] \right\}^{1/2} \tag{6.5}$$

where

$$\psi_1 = \left(\frac{2Z_1 Z_2 e^2}{E_d} \right)^{1/2} \tag{6.6}$$

From Equation (6.3), the initial transverse energy distribution of incident ions is a function of their angle of entry into the channel. Fig. 6.4a shows the effect on aligned RBS spectra of increasing the incident angle and hence the transverse energy of the analysing beam. The RBS yield increases rapidly with increasing misalignment, a feature which is most evident in a plot of the yield from a given depth as a function of angular deviation from the channel direction. Angular yield profiles corresponding to the depth interval shaded in Fig. 6.4a are shown in Fig. 6.4b for typical low index axial (solid curve) and planar (dashed curve) channels. Both profiles exhibit a minimum yield, corresponding to perfect alignment of the incident ion beam with the channel, and increase rapidly as the beam is misaligned. The minimum yield is generally higher and the angular half width smaller for planes than for

a)

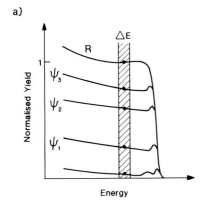

b)

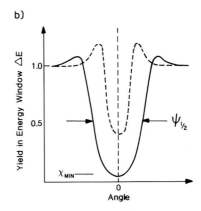

Fig. 6.4 a) RBS-C spectra taken at various degrees of misalignment from an axial channel (R is the random spectrum). b) The angular yield about an axial channel (solid curve) and a planar channel (dashed curve) indicating the channeling half angle ($\psi_{1/2}$).

TABLE 6.1
Comparison of calculated and measured values of $\psi_{1/2}$ for axial channeling. (After Barrett, 1971).

Target	Direction	Ion	Energy (MeV)	$\psi_{1/2}$ (degrees) Calculated	Measured
C (dia)	$\langle 011 \rangle$	H	1.0	0.53	0.54
		He	1.0	0.75	0.75
Al	$\langle 011 \rangle$	H	0.4	0.84	0.90
	$\langle 011 \rangle$		1.4	0.45	0.42
Si	$\langle 011 \rangle$	H	0.25	1.03	1.02
			0.5	0.73	0.68
			1.0	0.51	0.53
			2.0	0.36	0.36
			3.0	0.30	0.26
		He	0.5	1.03	1.10
			1.0	0.73	0.75
			2.0	0.51	0.55
Ge	$\langle 011 \rangle$	He	1.0	0.93	0.95
W	$\langle 111 \rangle$	H	3.0	0.83	0.85
		He	6.0	0.83	0.80
	$\langle 001 \rangle$	H	2.0	0.95	1.00
		He	2.0	1.34	1.39
		C	10.0	0.98	1.10
		O	10.0	1.13	1.20
		Cl	10.0	1.60	1.82
Au	$\langle 011 \rangle$	Cl	20.0	0.84	1.10
		I	60.0	0.80	1.10

TABLE 6.2
Comparison of calculated and measured values of $\psi_{1/2}$ for planar channeling. (After Barrett, 1971)

Target	Direction	Ion	Energy (MeV)	$\psi_{1/2}$ (degrees) Calculated	$\psi_{1/2}$ (degrees) Measured
Si	{100}	H	3.0	0.07	0.07
	{110}	H	3.0	0.10	0.09
Ge	{110}	He	1.0	0.31	0.30
	{100}	He	1.0	0.23	0.18
	{211}	He	1.0	0.18	0.20
W	{100}	H	2.0	0.25	0.22
	{110}	H	2.0	0.32	0.26
	{100}	He	2.0	0.35	0.27
	{110}	He	2.0	0.45	0.38
	{100}	C	10.0	0.25	0.20
	{110}	C	10.0	0.32	0.28
	{110}	O	10.0	0.36	0.33
	{110}	Cl	10.0	0.40	0.30
Au	{100}	Cl	20.0	0.27	0.31
	{110}	Cl	20.0	0.21	0.24
	{111}	Cl	20.0	0.30	0.32

axes. Angular half widths for commonly studied substrates are given in Tables 6.1 and 6.2.

The critical angle ψ_c is the maximum angle of incidence for which an ion will be specularly reflected from a string or plane potential. It is a theoretical parameter which is not directly accessible experimentally but which is closely related to the angular half width at half height, $\psi_{1/2}$, of angular scan profiles such as that in Fig. 6.4b. In general, the estimate given by Equation (6.5) will be within 20% of $\psi_{1/2}$ determined experimentally, and will predict the correct dependence on Z_1, Z_2, E and temperature. A more precise estimate of $\psi_{1/2}$ can be achieved from computer simulation studies, and an expression due to Barrett (1971) is given by Equation (6.24) in Section 6.5. The order of magnitude of ψ_c can readily be estimated. In particular, since continuum potentials are typically of the order of 100 eV, the critical angle for channeling of a 1 MeV He^+ ion is calculated from Equation (6.4) to be of the order 0.01 radian, or 0.60°.

d. Minimum Yield

The minimum yield, as defined in Figs. 6.2 and 6.4b, is an important channeling parameter for characterising crystal perfection. The magnitude of the axial minimum yield can readily be estimated for a perfect crystal by a simple geometric model of the crystal channel. From Fig. 6.5, the fractional area of a channel blocked by atoms is $Nd\pi p^2$, where N is the atomic density, d is the distance between atoms along the string, and p is the effective radius of the shadowing atoms. The effective atomic radius may be approximated by the Thomas-Fermi screening radius, or more realistically by the RMS thermal vibrational amplitude normal to the axis, p. Assuming the latter approximation, the minimum yield becomes:

$$\chi_{min} = Nd\pi p^2 \qquad (6.7)$$

which is independent of beam parameters, being determined solely by the properties of the crystal. An additional contribution can also be added to account for surface effects, such as atomic reconstruction or surface contamination. In practice, χ_{min}, and the surface peak yield, are determined not only by the top surface layer but by scattering contributions from subsurface layers. Experimental χ_{min} values are typically three times greater than predicted by Equation (6.7), and a more realistic estimate is given by Equation (6.26) in Section 6.5.

e. Energy Loss

The energy loss of high energy (MeV) ions is dominated by inelastic scattering from target electrons. For light ions incident on an amorphous target material this energy loss is well described by the Bethe-Bloch formalism described in Chapter 1. However, channeled ions execute a particular type of trajectory in which the ion is confined to a region well away from atomic nuclei. As a consequence of this trajectory, channeled ions encounter a lower than average electron density and consequently experience a reduced rate of energy loss. This has important implications

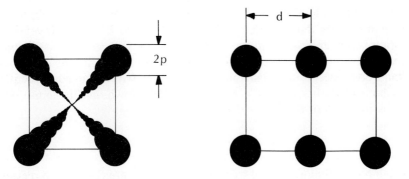

Fig. 6.5 The geometry of an axial channel used to derive χ_{min}.

for channeling analysis since it may limit the accuracy with which depth scales can be determined.

Lindhard (1965) extended the Bethe-Bloch formalism to include the effect of varying electron density. Such a treatment shows that well channeled ions may exhibit a rate of energy loss only half that of the equivalent randomly incident ions. The precise energy loss is a function of the electron density encountered by the ion and depends on its trajectory. A knowledge of the trajectories is therefore important.

f. Flux Distribution

The potential energy of a channeled ion is a function of its position in the transverse plane of the channel, as shown in Fig. 6.6a. It follows that when a uniform flux of ions enters a crystal close to a channel direction, the steering potential of the atomic rows or planes causes the ion trajectories to oscillate back and forth between the channel walls as the ions penetrate into the crystal. Large amplitude trajectories all have about the same wavelength in a planar channel, and the mid-channel crossover point and the points of closest approach to the channel walls coincide for such particles. Since large amplitude trajectories are also the ones most likely to result in close nuclear encounters, the yield from small impact parameter processes also displays a periodicity with depth (Fig. 6.2). Depth oscillations are quite pronounced for planar channeling because ions are limited to an essentially one dimensional oscillatory trajectory between planar walls. In the case of axial channeling the ion can spiral along the axis, executing a more complex two dimensional oscillatory trajectory (Fig. 1.17) where the oscillation wavelength is a function of impact parameter. Consequently yield oscillations in RBS spectra are not generally observed beyond the first maximum in the case of axial channeling (Fig. 6.2). As the penetration depth increases, depth oscillations are damped due to dispersion of the different wavelength components and to random fluctuations in the steering collisions. Eventually a statistical equilibrium is achieved in which ions of a given transverse energy, are distributed uniformly within an equipotential region called the *accessible area* $A(E_T)$, as shown in Fig. 6.6a. The accessible area of the ion, is defined by the inequality:

$$E_T(r,E,\psi) \geqslant U(r) \tag{6.8}$$

which means that all ions of transverse energy E_T are confined to regions where the potential energy $U(r)$ is less than E_T. From a knowledge of the transverse energy distribution of the incident ions it is possible to determine the equilibrium ion flux distribution within the channel. The equilibrium flux distribution for axial channeling of particles incident at an angle ψ is obtained by integration of all particles over their accessible areas. That is:

$$F(r,\psi) = \int_{E_T \geqslant U(r)} dA(r_{\text{IN}})/A(E_T) \tag{6.9}$$

where $U(r_{\text{IN}}) + E\psi^2 \geqslant U(r)$ defines the accessible area for particles incident at a distance of r_{IN} from the atom rows. The flux has been defined such that $F(r,\psi) = 1$ for a uniform flux. Equation (6.9) can be solved analytically by assuming cylindrical symmetry for the crystal channel and by taking the small r approximation to the standard potential, Equation 6.1. If the accessible area is taken to be $\pi r_0^2 - \pi r_{\text{IN}}^2$, the resulting flux distribution is given by:

$$F(r,\psi) = \exp\left(\frac{2\psi^2}{\psi_1^2}\right)\ln\left(\frac{r_0^2}{r_0^2 - r^2}\right), \psi \leqslant \psi_M \tag{6.10a}$$

$$F(r,\psi) = \exp\left(\frac{2\psi^2}{\psi_1^2}\right)\ln\left(\frac{1}{1 - \exp\left(\frac{-2\psi^2}{\psi_1^2}\right)}\right), \psi \geqslant \psi_M \tag{6.10b}$$

where ψ_1 is given by Equation (6.6) and $\psi_M = \psi_1\sqrt{\ln(r_0/r)}$. From Equation (6.10a) for $r \ll r_0$, $F(r, 0) \approx r^2/r_0^2$.

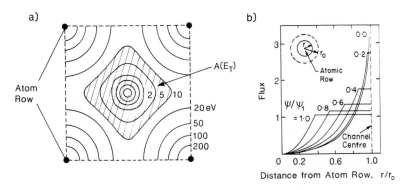

Fig. 6.6 a) Equipotential contours of the axial continuum potential for He^+ in the $\langle 100\rangle$ channel of Si. The accessible area contours $A(E_T)$ (shaded) are shown for particles with transverse energies $\leqslant 10\text{eV}$. b) Flux for axial channeling as a function of distance from atom rows for various values of incident angle (ψ), calculated from Equation (6.10). After Feldman *et al.* (1982)

The flux distribution is peaked at the point of minimum potential energy, since this area is accessible to all ions, and rapidly broadens as the incident angle is varied, as shown in Fig. 6.6b. This non-uniform flux distribution is referred to as flux peaking and it has important implications for atom location experiments as discussed in Section 6.2.3.

g. Dechanneling

The assumption that the transverse energy of channeled ions is conserved, is reasonable for shallow channeling depths but is not the case as ions penetrate deeper into the crystal. The average transverse energy increases with increasing penetration depth due to multiple scattering from target nuclei and electrons. Eventually some ions acquire sufficient transverse energy to surmount the potential barrier of the channel and escape into the crystal as part of the random beam component. Such ions are said to be *dechanneled*. To estimate the fraction of ions dechanneled at a given depth it is necessary to calculate the transverse energy distribution as a function of depth and establish a critical transverse energy for dechanneling.

The average change in transverse energy with increasing penetration depth can be estimated from Equation (6.3) by:

$$\frac{dE_T}{dz} = E\frac{d\psi^2}{dz} + \psi^2\left(\frac{dE}{dz}\right) \tag{6.11}$$

where statistical equilibrium is assumed and the averages are taken over the area, $A(E_T)$, accessible to ions of transverse energy E_T. The first term represents the increase in transverse energy due to multiple scattering and can be written as the sum of nuclear and electronic terms. The second term represents the decrease in transverse energy resulting from the decreasing ion energy. This latter contribution is relatively small and is often ignored in practice.

The increase in transverse energy due to multiple scattering processes can readily be calculated (Lindhard, 1965; Gemmell, 1974). For well channeled ions in a defect free crystal the increase is initially dominated by multiple scattering from electrons. This is a result of the fact that well channeled ions remain close to the centre of the channel and do not approach the surrounding atoms. As the average transverse energy increases, channeled ions approach the atomic nuclei more closely so that nuclear multiple scattering increases and eventually dominates.

Several different approaches have been employed to determine the dechanneled fraction of ions from the transverse energy distribution.

One such approach asserts that the initial transverse energy distribution broadens in a manner analogous to a random walk process and therefore can be described by a diffusion equation with a functional dependence between the 'diffusion coefficient' and $\langle dE_T/dz \rangle$. In a second approach, referred to as the steady-increase model, the transverse energy of each ion is assumed to increase at a rate given by $\langle dE_T/dz \rangle$. This approach is computationally simpler than the diffusion approach and gives similar results for random fractions less than ~ 0.2. Matsunami and Howe (1980) have compared a modified diffusion model of dechanneling with experimental data and found good agreement between theory and experiment for χ_{min} values up to ~ 0.6 if the electron density around the centre of the channel is used as a fitting parameter. The gradual increase in dechanneling with depth in perfect crystals is not often important in practical analysis situations. However, the rate of dechanneling is significantly enhanced by the presence of crystal defects and is a useful measure of crystal disorder.

6.2.2 Crystal Disorder

The relative magnitude of direct scattering and dechanneling processes depends on the nature and concentration of defects. Interstitial atoms, stacking faults and grain or twin boundaries, which introduce atoms directly into the path of the channeled ion, result directly in small impact parameter scattering events (Fig. 6.7a). In this case the direct scatter yield is proportional to the concentration of defects at a given depth, increasing and decreasing with the concentration of defects as shown in

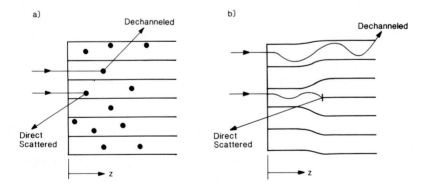

Fig. 6.7 Schematic illustrations of direct scattering and dechanneling at a) point defects and b) dislocations (extended defects).

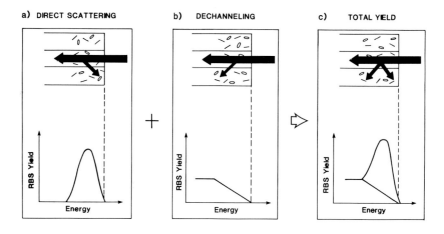

Fig. 6.8 Schematic illustration showing how a) direct scattering and b) dechanneling components of RBS-C spectra are additive to give the total yield (c).

Fig. 6.8a. On the other hand, defects such as dislocations give rise to lattice distortions which have the effect of increasing the transverse energy of the channeled ions (Fig. 6.7b) and hence increasing the dechanneling rate (Fig. 6.8b). The dechanneled fraction is a monotonically increasing function of penetration depth because ions, once dechanneled, remain so. Thus the dechanneled fraction at a given depth is related to the total number of defects encountered in reaching that depth. The experimentally measured yield (Fig. 6.8c) is the sum of the yield from channeled ions scattered directly from displaced atoms (Fig. 6.8a) and the yield from dechanneled ions scattered from all atoms in the target (Fig. 6.8b). These contributions can, in principle, be deconvoluted from the measured backscattered yield. They depend on the nature and distribution of the defects and on the ion flux distribution within the channel.

If f is the probability for direct scattering by the defect and σ is the dechanneling probability (or cross-section) the contribution of direct scattering to the measured yield, χ_S, is given by:

$$\chi_S(z) = [1 - \chi_R(z)]\, f\, n(z)/N \qquad (6.12)$$

where $n(z)$ is the defect concentration, $[1 - \chi_R(z)]$ is the channeled fraction of the beam at depth z and N is the total number of atoms per unit volume. Here, f, is given by:

$$f = (\int F(r)\, n(r)\, dr) \,/\, (\int n(r)\, dr) \qquad (6.13)$$

a) DECHANNELING AND DIRECT SCATTERING

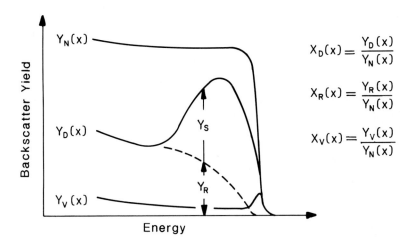

$$X_D(x) = \frac{Y_D(x)}{Y_N(x)}$$

$$X_R(x) = \frac{Y_R(x)}{Y_N(x)}$$

$$X_V(x) = \frac{Y_V(x)}{Y_N(x)}$$

b) DECHANNELING

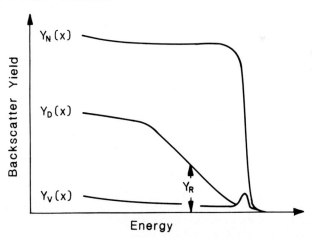

Fig. 6.9 RBS-C spectra illustrating a) the normal situation of dechanneling yield (Y_R) and direct scattering yield (Y_S) components to total yield; and b) the simple case of dechanneling only.

where $F(r)$ and $n(r)$ are the ion flux and scattering centre distributions in the transverse plane, respectively. Both the ion flux and the defect density distributions are suitably normalised so that if either the ion flux or the scattering centres are uniformly distributed in the transverse plane then f *is unity*. The corresponding dechanneled component is given by:

$$\chi_R(z) = \chi_V(z) + [1 - \chi_V(z)] \left\{ 1 - \exp\left(-\int_0^z \sigma n(z')\, dz' \right) \right\} \quad (6.14)$$

where $\chi_V(z)$ is the random fraction for a defect–free crystal. The total normalised backscattered yield for ions channeled in a crystal containing damage, χ_D, is the sum of the direct scattering contribution and the dechanneling contribution and is given by:

$$\chi_D(z) = \chi_R(z) + [1 - \chi_R(z)]\, f n(z)/N \quad (6.15)$$

The relationships between the parameters in Equations (6.12) to (6.15) and the experimentally measured RBS-C spectra are shown in Fig. 6.9. From such data and Equations (6.14) and (6.15), the damage distribution ($n(z)$) can readily be obtained once the defect scattering and dechanneling probabilities have been determined. For most defective crystals the precise determination of the scattering and dechanneling probabilities is non-trivial (Pathak 1982; Feldman *et al.*, 1982; Wielunska *et al.*, 1981a,b) and it is often necessary to employ suitable approximations and/or semiempirical methods (Walker *et al.* 1977; Tognetti, 1981). General methods for determining defect concentrations and distributions are summarised in Highlight 6.1. Some relatively simple treatments are given below.

a. Point Defects

The dechanneling cross section for randomly distributed point defects is readily determined since dechanneling results from simple Rutherford collisions. Moreover, for low defect concentrations dechanneling is dominated by single scattering events. The dechanneling cross section can then be determined by integrating the Rutherford cross section from the measured critical angle ψ_c to π. This gives:

$$\sigma(z) = \frac{\pi d^2 \psi_1^{\,4}}{4\psi^2_{1/2}} \quad (6.16)$$

where ψ_1 is given by Equation (6.6) and d is the atomic spacing along the crystal axis. Using this value of σ and the assumption that $f = 1$ (i.e.

randomly distributed defects), Equations (6.14) and (6.15) can be solved iteratively to give the defect concentration profile, as outlined in Highlight 6.1.

For large defect concentrations dechanneling is dominated by multiple scattering events. This is treated in detail by Sigmund and Winterbon (1974), who give numerical results for the angular distribution of ions as a function of defect density. The dechanneling cross-section can be determined iteratively, using Equations (6.14) and (6.15), by integrating the appropriate distributions to find the fraction of ions outside the measured critical angle ψ_c. The extraction of a damage distribution from RBS-C data is illustrated in Fig. 6.10 (Tognetti, 1981). Fig. 6.10a depicts a $\langle 100 \rangle$ aligned RBS spectrum from a GaAs sample implanted with 40 keV N^+ at 40K to a total fluence of 10^{14} cm^{-2}. The dechanneling contribution has been determined by applying the method outlined in Highlight 6.1. The damage distribution resulting from such analysis is shown in Fig. 6.10b. The measured profile is found to be in good agreement with calculated damage distributions.

b. Extended Defects

The lattice distortions which surround extended defects usually give rise to a high dechanneling rate in RBS-C spectra, which, in most cases, dominates over the direct scattering from the defects themselves. The simplest analytical treatment neglects the direct scattering component by assuming that the direct scattering probability from the defects, f, is zero. From Equation (6.15), the measured yield $\chi_D(z)$ is then equivalent to the random component of the beam $\chi_R(z)$. If the dechanneling cross section, σ, can be assumed constant over the depth of interest, Equation (6.14) can be inverted to yield an expression for $n(z)$ given by:

$$n(z) = \frac{1}{\sigma} \frac{d}{dz} \left\{ \ln \left(\frac{1 - \chi_R(z)}{1 - \chi_V(z)} \right) \right\} \tag{6.17}$$

If σ is known, $n(z)$ can be determined directly from the experimental data. However, a precise analysis should include a direct scattering component, Equation (6.15), with a non-zero value of f.

It is interesting that the product $n(z).\sigma$, Equation (6.17), is dependent only on experimentally measurable variables and can be used to identify defect type. Different defects exhibit different and characteristic dechanneling cross-sections, often with specific energy dependencies. It is

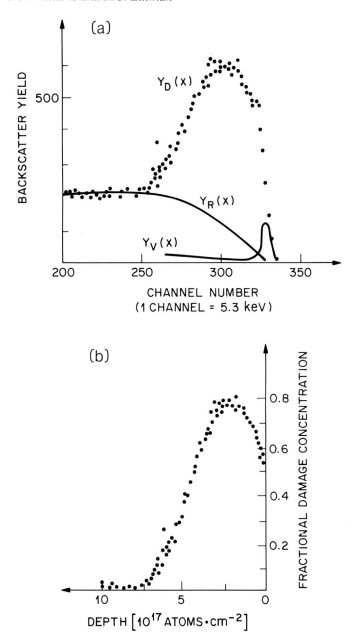

Fig. 6.10 a) RBS-C spectrum from the ⟨100⟩ axis of GaAs after irradiation at 40 K with 40 keV N^+ ions to a fluence of 1×10^{14} cm^{-2}. b) Extracted damage distribution from data in a). After Tognetti (1981).

TABLE 6.3
Energy dependence of dechanneling at defects (after Quéré, 1968)

Defect Type	Energy Dependence of σ.
Point Scattering Centres	E^{-1}
— interstitial atoms	
— amorphous clusters	
Discontinuities	E^0
— stacking faults	
— voids	
— gas bubbles	
Strain	$E^{1/2}$
— dislocation lines	
— dislocation loops	

therefore usually possible to deduce both the nature and depth distribution of defects from energy dependent dechanneling measurements. Analytical cross-sections have been estimated for many defects, Quéré (1968), and such analysis suggests that dechanneling cross-sections varying as $E^{-1/2}$ indicate dislocations, an energy independent cross-section indicates the presence of gas bubbles or voids and an E^{-1} dependence the presence of interstitials. The energy dependence of the dechanneling cross-section is tabulated for commonly encountered defects in Table 6.3.

More precise treatments of dechanneling (and direct scattering) at extended defects (Pathak 1982; Wielunska *et al.*, 1981a,b) provides better agreement with experimental energy-dependent dechanneling measurements. For example, for dislocations Wielunska *et al.*, (1981a,b) and Wielunski *et al.*, (1986) have shown, by including planar channeling effects in the analysis, that dechanneling occurs mainly at larger distances from the dislocation core than the Quéré estimate. Close to the dislocation, the dechanneling probability is very low, with many particles merely crossing the channel wall to remain channeled between adjacent planes (or axial strings). This process leads to a higher probability for direct scattering where the number of scattering centres is proportional to the particle energy. This energy dependence is superimposed on that given by the lattice distortion about dislocations ($E^{1/2}$ from Table 6.3). The superposition of direct scattering and dechanneling components is difficult to separate in RBS-C spectra even when detailed measurements at different ion beam energies are made. Knowledge of the defect type obtained by an alternative technique such as transmission electron

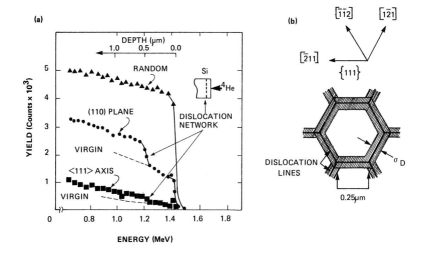

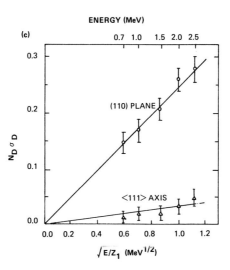

Fig. 6.11 a) RBS-C spectra for 2.5 MeV He incident on Si with edge dislocation network at fixed depth of ~ 4500 Å below the surface. b) Schematic of the dislocation structure giving rise to a) and c). σ_D is the extent of the zone of dechanneling. c) Experimentally measured product of defect density and dechanneling factor vs. square root of beam energy divided by beam atomic number for (110) planar and $\langle 111 \rangle$ axial dechanneling in Si with network of edge dislocations as in a) and b). After Picraux *et al.* (1982).

microscopy (TEM) can allow more precise fitting of experimental data. Further RBS-C measurements of well characterised defects are needed before channeling analysis alone can provide reliable information on defect type, concentration and distribution in cases where no prior knowledge of the defect types is available.

Fig. 6.11 illustrates defect identification by dechanneling (Picraux *et al.*, 1982). Fig. 6.11a depicts aligned RBS spectra for ⟨111⟩ axial and (110) planar alignment in a Si substrate containing an edge dislocation network at a fixed depth of ∼ 4500 Å below the surface. It is immediately apparent from this figure that planar channeling is more sensitive to the defects than axial channeling; the defects disrupt planar channels more than the axial channels. The nature of the dislocation network is shown in the schematic in Fig. 6.11b which is derived from TEM studies.

The product $n(z)$, σ can readily be determined from the experimental data in Fig. 6.11a (Highlight 6.1) and the energy dependence of this product provides insight into the nature of defects. The observed $E^{1/2}$ dependence (Fig. 6.11c) is consistent with the presence of dislocations.

c. Strained Lattices

Layered crystals are particularly amenable to channeling analysis and show a range of interesting channeling effects. Strained layer super-lattices (SLS) which are grown from alternating layers of lattice mismatched materials are of particular interest. These structures can be grown epitaxially in a highly strained configuration without the presence of extended defects, provided the layer thickness is not too great. The structure of a simple SLS is illustrated schematically in Fig. 6.12. It is apparent from this figure that the alternating compressive and tensile strain of the SLS can result in periodic distortions of axial and planar channels which are inclined to the surface normal. These structures can be probed by ion channeling techniques and such analyses offer further insight into channeling and dechanneling processes.

In Section 6.2.1 we showed that ions entering a channel, particularly planar channels, exhibit an oscillatory trajectory in which the wavelength of the oscillation is initially similar for a large fraction of the ions. Consequently, ions initially uniformly distributed across the channel become spatially localised at some depth beneath the crystal surface. For ions incident parallel to the channel direction this focal point occurs mid-channel, whereas if the incident beam is inclined at some angle to the channel the focal point is off centre. Once the ions have passed this focal

HIGHLIGHT 6.1
PROCEDURES FOR DETERMINATION OF DEFECT
CONCENTRATIONS AND DISTRIBUTIONS

DIRECT SCATTERING ANALYSIS

Where an obvious damage peak is observed in the RBS-C spectrum the direct scattering probability, f, is non-zero. As a first approximation f is usually taken to be unity. This simplifies equation (6.15), which can be inverted to give the following expression for the defect distribution:

$$n(z) = \left(\frac{\chi_D - \chi_R}{1 - \chi_R} \right) N \qquad (6.18)$$

A : Estimate Total Displaced Atoms
A simple estimate of the total number of displaced atoms can be obtained from Equation (6.18) by assuming a simple functional form for the dechanneled or random component of the beam $\chi_R(z)$. The simplest approximation is linear dechanneling obtained by drawing a straight line between the minimum virgin yield at the surface and the dechanneling level behind the damage peak. The number of defects can be estimated from Equation (6.18) for each channel. The total number of displaced atoms is the sum of $n(z)$ over all channels in the damage peak. This procedure provides a reasonable estimate of the total number of defects but not of the defect depth distribution.

B : Damage–Depth Distribution
To accurately determine the defect–depth distribution a more realistic estimate of the dechanneling contribution is required. This is generally achieved by applying Equations (6.14) and (6.18) iteratively, beginning at the surface where χ_R is assumed equal to χ_V and proceeding channel by channel until $n(z) = 0$ (Fig. 6.10). In cases where σ is not accurately known but can be assumed constant as a function of depth over the region of interest the value of σ can be used as a fitting parameter which is adjusted until the predicted dechanneling level converges to the experimentally measured value.

DECHANNELING ANALYSIS

Where RBS–C measurements give an obvious increase in yield but no obvious disorder peak, the direct scattering contribution is small and consequently $f \sim 0$. This enables the defect type and depth distribution to be determined from Equation (6.17).

A : Defect Type

By performing channeling measurements at a variety of incident ion energies, the nature of defects can be determined by evaluating $n(z).\sigma$ from Equation (6.17) at a fixed depth. The energy dependence of $n(z).\sigma$ is determined and compared with known dependencies, such as those in Table 6.3 (Fig. 6.11). If the incident ion energy does not vary significantly in traversing the defect layer, i.e. thin layers, the dechanneling cross-section, σ, can be assumed constant as a function of depth. This enables Equation (6.14) to be integrated with respect to z to yield a simplified expression for $n_T.\sigma$, where $n_T(z)$ is the total number of defects encountered by the ion beam in reaching depth z.

B : Depth Distribution

Where the dechanneling cross-section is known and/or can be assumed constant over the depth of interest, Equation (6.14) can be evaluated to give the defect–depth distribution. The precise defect density as a function of depth requires accurate determination of σ as well as taking account of any direct scattering contribution. In cases where σ is not known but can be assumed constant, the shape of the depth distribution can readily be determined.

---

point they separate, becoming focussed again at some greater depth. This behaviour continues until eventually the coherence of trajectories is lost. In a SLS the average wavelength of the oscillating ion trajectories can be chosen to match or mismatch the modulated strain field in the target, as

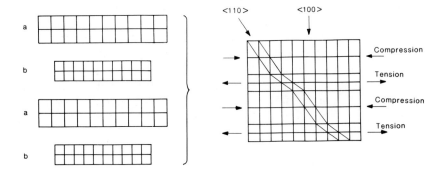

Fig. 6.12 Schematic showing how thin layers of different lattice spacing (a and b on the left of figure) come together in the solid with the same lattice spacing parallel to the surface to give a strained layer superlattice.

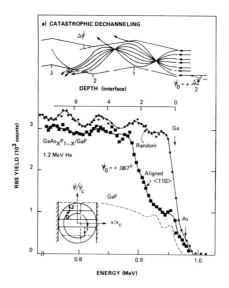

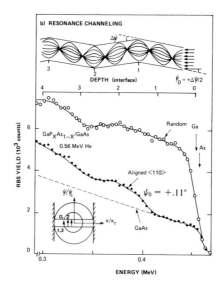

Fig. 6.13 RBS-C spectra illustrating (a) catastrophic dechanneling and b) resonance channeling in strained-layer superlattices. See text for explanation. After Picraux *et al.* (1986).

shown at the top of Figs. 6.13a and 6.13b. This leads to two extreme channeling situations called *resonance channeling* and *catastrophic dechanneling* (Picraux *et al.*, 1986). As shown in the lower portions of Figs. 6.13a and 6.13b, these processes lead to dramatic differences in planar channeling spectra. In particular, catastrophic dechanneling within the third layer of a $GaAs_xP_{1-x}/GaP$ superlattice (Fig. 6.13a) results in a sudden increase in aligned RBS yield to a near random level. Under resonance channeling conditions, on the other hand, dechanneling is minimised and a significant fraction of the beam remains channeled well into the SLS (Fig. 6.13b).

The relative distortion of inclined channels through successive layers of a SLS can be conveniently measured by ion channeling. This is achieved by comparing the angular yield profiles from the different layers. Fig. 6.14 a depicts such scans about the $\langle 110 \rangle$ axis for a $GaAs/In_xGa_{1-x}As$ SLS (Picraux *et al.*, 1983). The lower scan was obtained by setting the RBS energy window to correspond with scattering from Ga + As in the first (GaAs) layer: the middle scan corresponds to In in the second $In_xGa_{1-x}As$ layer; and the upper scan represents an average of the Ga plus As yields from the fifth and sixth layers. The angular distortion of successive layers can be obtained directly from the relative angular displacement of yield minima (Fig. 6.14). Strain and lattice parameter changes can then be determined.

Dechanneling processes in SLSs, which arise before statistical equilibrium has been achieved, can readily be understood with the aid of a simple phase rotation model (Chu *et al.,* 1984). This model is most applicable to the case of planar channeling (c.f. Fig. 6.13) where oscillating ion trajectories persist deep into the crystal before developing the equilibrium flux distribution discussed in Section 6.2.1. The model is constructed, in its simplest form, by assuming that the planar continuum potential is given by $U(r) = (1/2)dx^2$ (harmonic approximation) where x is the distance from the centre of the scattering planes. The motion of channeled ions in the transverse plane, described by Equation (6.3), is given by:

$$E_T = E\psi^2 + 1/2\alpha x^2 \qquad (6.19)$$

This equation defines a circle in $\psi - x$ phase space, as shown in Fig. 6.15 and in the lower left insets of Figs. 6.13a and 6.13b. Ion motion is described as a phase rotation in this plane, the model being equally applicable to ion motion in perfect crystals and SLSs.

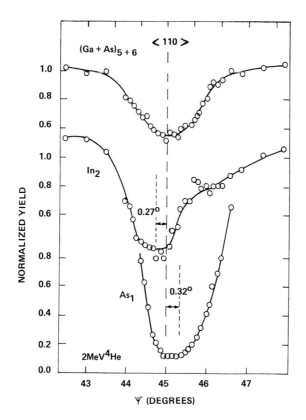

Fig. 6.14 Angular scans through [110] axis for 1st layer (As scattering signal), 2nd layer (In scattering signal), and 5th plus 6th layers (Ga and As scattering signals, respectively) of $In_xGa_{1-x}As/GaAs$ (38/38 nm) thick layered SLS. Angle ψ is measured with respect to the [100] axis which is normal to the surface. After Picraux *et al.* (1983).

The transverse energy of well collimated ions entering the crystal at a small angle ψ_0 to the channel can be represented on this diagram as a horizontal line (Fig. 6.15). Ions with trajectories which fall outside the unit circle (of radius $\psi/\psi_c = 1$) are not channeled. The rotation of this line in $\psi - x$ phase space represents the ion trajectories as they penetrate into a perfect crystal. The non-equilibrium flux distribution is represented schematically and in $\psi - x$ phase space at three depths in Fig. 6.15; horizontal lines represent near-uniform ion flux distributions across the channel (antinode points), whereas vertical lines represent localisation of ions at crossover points (nodes). The arrowed trajectory and rotation point in Fig. 6.15 represents the motion of ions entering at a particular position, x_0, with incident angle ψ_0.

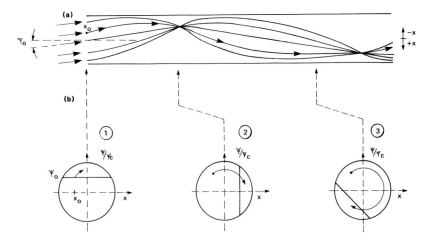

Fig. 6.15 A schematic illustration of the phase rotation model for describing the oscillatory motion of a flux of ions in a planar channel of a perfect crystal. See text for explanation.

In the case of SLSs the transverse energy of channeled ions changes abruptly as they enter successive layers. This has the effect of abruptly changing ψ/ψ_c for ions entering the new layer, which is represented on the $\psi - x$ phase diagram as a vertical displacement. Such displacements are shown in Figs. 6.13a and 6.13b for the extreme cases of catastrophic dechanneling and resonance channeling. Channeling conditions have been selected such that the channeled ions are uniformly distributed across the channel (antinodes) at the layer interfaces. This leads to a vertical displacement of horizontal lines in the $\psi - x$ phase diagram. If the displacements are progressively away from the centre of the circle as shown in Fig. 6.13a, catastrophic dechanneling will eventually occur when all ion trajectories are displaced outside the unit circle. Conversely, if the vertical displacements are towards the centre of the $\psi - x$ phase diagram, such that no ions are dechanneled (Fig. 6.13b), resonance channeling will occur.

6.2.3 Foreign Atom Location

The concept of employing ion channeling for atom location of foreign atoms within a crystal lattice was introduced in Chapter 1: substitutional impurity atoms are shadowed by the crystal lattice, whereas impurities occupying interstitial sites are exposed to the channeled ion flux. In this section, procedures are developed for precise atom location based on the channeling principles outlined in section 6.2.1. Highlight 6.2 summarises these procedures. Fig. 6.16a illustrates the atom location process for substitutional As atoms in a Si substrate. Typical RBS spectra are depicted in Fig. 6.16b for axial channeling (A) and random incidence (R). To a first approximation, the fraction of impurity atoms located along a particular channel can be determined from such spectra by comparing the impurity yield for aligned and random conditions. The fraction of shadowed atoms is, by analogy with Equation (6.18) in Highlight 6.1:

$$S = \frac{1 - \chi_I}{1 - \chi_R} \qquad (6.20)$$

where χ_I and χ_R are the ratios of the aligned and random yields from the impurity and host material respectively, determined at equivalent depths (W_{As} and W_{Si} in Fig. 6.16b).

The fraction of atoms shadowed along a given crystallographic direction is not necessarily the same as the fraction that occupy substitutional lattice sites. Shadowed atoms may occupy sites between atoms composing the strings or planes of the axial or planar channel. Such atoms will be shadowed when viewed along one channeling direction but will be visible along other directions. Precise location of impurity atoms must therefore involve a triangulation procedure (Fig. 1.21, Chapter 1) appropriate to the particular crystal structure under consideration. These channeling directions are readily determined by examining a crystal model.

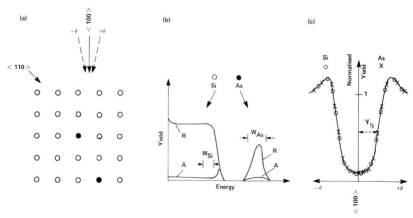

Fig. 6.16 Atom location of substitutional (ion implanted) As in Si. a) Schematic of the crystal and incident beam aligment. b) Typical random (R) and axial aligned (A) RBS-C spectra c) Angular yields for Si and As with respective window settings as in b), W_{As} and W_{As}.

The use of Equation (6.20) assumes that either the distribution of impurity atoms and/or the channeled ion flux distribution is uniform across the channel. In reality this may not be the case. The fraction of unshadowed impurity atoms in the channel should be determined from an equation analogous to Equation (6.15), with f representing the impurity scattering factor. In addition to comparing the aligned and random yields for various crystallographic directions, angular scans through low index axial and planar channels can provide more accurate atom location within the individual channels. This ability arises from the flux peaking effect discussed in Section 6.2.1. Impurity atoms residing in a channel will be exposed to different particle fluxes as the angle of the incident beam is varied. Fig. 6.16c compares a normalised angular yield profile from an ideal substitutional impurity (crosses) with that from the host crystal (circles). The angular yield profiles are observed to be identical in this case.

In general, the angular yield profiles exhibit features characteristic of the atom positions in a channel. Fig. 6.17 depicts 'signature' scans for commonly encountered lattice site locations of impurities. The centre of the figure shows foreign atom positions (numbered) with respect to the host atom rows (large circles) bounding an axial channel. Normalised angular scans are simplest for pure substitutional atoms (scan 1), where host and impurity profiles are identical and for impurity atoms randomly distributed (scan R) where the yield is unchanged during the scan. Slight displacement from substitutional lattice sites (scan 2) results in a narrowing of the angular scan from the impurity with respect to the host. Characteristic angular scans are also observed for other well defined impurity sites which may represent stable impurity sites or result from impurity–defect interactions. The specific shape of the scans is a result of convolution of the atom location within the channel and the ion flux distribution (Andersen, 1967; Swanson, 1982). Indeed, it is the flux peaking effect that enables such precise atom location. This is particularly evident for impurities occupying interstitial positions in the channel (scan 4). In this case the impurity yield for exact alignment exceeds the random yield (normalised to 1) due to flux peaking. Scan 3 shows a double peak resulting from atoms displaced midway between host atom rows and the channel centre and scans $1 + 2$ and $1 + 4$ depict particular cases of 50% occupation of each of two precise sites.

The above discussion has concentrated on determining the location of impurity atoms with respect to a substrate lattice. In some crystalline compounds, channeling analysis can also provide information about impurity atom location within specific sublattices. This is achieved by measuring angular yields for the impurity under channeling conditions

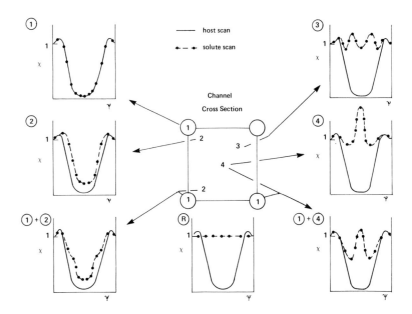

Fig. 6.17 Signature angular scans about an axial channel for both the host lattice (solid curve) and the solute impurity (solid circles) for various precise atom locations of the solute as indicated by the schematic cross section of a bcc $\langle 100 \rangle$ channel. Position 1 is a substitutional impurity, position 2 is an impurity slightly displaced from a substitutional site, position 3 represents an impurity midway between the strings and the channel centre and position 4 is an interstitial impurity at the channel centre. Adapted from Swanson (1982).

where one sublattice shadows the other. An example of this type of analysis is given in Section 6.5.

By combining angular scans and triangulation, channeling measurements can be employed to determine the crystallographic location of impurity atoms with a resolution approximating the thermal vibrational amplitude of the atom (~ 0.1 Å). Such measurements are often difficult and may be subject to interpretational difficulties as discussed in Section 6.3.4. In precise atom location experiments it is often necessary to study models of specific impurity locations in the host lattice and employ computer simulations to match experimental angular scans. Some typical impurity sites in fcc and bcc crystals are given in Chapter 14.

HIGHLIGHT 6.2
ATOM LOCATION PROCEDURES

---

1. If the impurity atoms and/or the ion flux distribution can be assumed to be uniformly distributed across the channel the shadowed fraction can be determined directly from Equation (6.20). Futhermore, when impurity atoms are highly soluble and are known to replace host atoms on regular lattice sites, comparison of the random yield with a channeling measurement along one low index direction is often sufficient to estimate the fraction of substitutional solute atoms. If, on the other hand, solute atoms occupy regular interstitial sites, a minimum of two directions are required for unambiguous lattice location. Successively less symmetric positions require additional channeling measurements and a combination of different lattice sites may require detailed analysis with model fitting.

2. In situations where impurity atoms are located at specific crystallographic sites and the ion flux distribution is known to be non–uniform, more detailed analysis is required. Angular yield profiles need to be obtained and compared with signature scans such as those depicted in Fig. 6.17. Analysis along different crystallographic directions enables a more accurate determination of atom positions.

3. For detailed atom location it is necessary to employ triangulation procedures in combination with computer simulation and matching of angular scans. Angular scans can be predicted either analytically by calculating the ion flux distribution at given sites as a function of beam tilt angle or by Monte Carlo simulation of the channeling process (Sections 6.3.3 and 6.4.3).

6.3 EXPERIMENTAL CONSIDERATIONS

6.3.1 Experimental Arrangements and Planning

To perform a channeling experiment it is necessary to align an incident ion beam with a specific crystallographic direction in the target to within an accuracy better than the critical angle, ψ_c. The orientation of the target crystal must, therefore, be adjustable with respect to the incident ion beam (using a goniometer with at least two rotational degrees of freedom as shown, for example, in Chapter 2, Fig. 2.9). The angular dispersion, or collimation of the beam, must be smaller than the critical angle and this is achieved using appropriately spaced beam-defining apertures. Highlight 6.3 summarises a set of basic operating conditions for channeling analysis with a He^+ beam. Backscattered particles exiting along a random direction are analysed by a solid state detector. For a simple system with fixed scattering geometry, a scattering angle close to 180° provides the most generally applicable arrangement. Channeling analysis requires a method to indicate the relative alignment of the incident beam with respect to the crystal axes and planes. This can be achieved by monitoring the backscattered ion yield, relative to the incident beam current, using a single channel analyser and rate meter. To extract quantitative information from channeling spectra, accurate charge integration is necessary to directly compare yields from channeled and random spectra.

Alternate experimental arrangements for channeling can be employed. Indeed, for specific applications, the yield from any ion–atom interaction process (e.g. nuclear reaction products or X-ray emission) can be monitored under channeling conditions. More sophisticated channeling arrangements can include both sample and detector manipulators which provide additional degrees of freedom. This may, for example, allow sample translation with respect to the incident beam and the ability to more readily access channeling directions inclined to the surface normal. The ability to independently vary the sample and detector geometries allows channeling/blocking (double alignment) configurations to be employed and experimental parameters such as mass resolution and depth resolution to be optimised, as discussed below.

The choice of scattering geometry has important consequences for channeling experiments. The most commonly employed geometries and their advantages and disadvantages are indicated in Fig. 6.18. For geometries in which the channeled ion trajectory is a significant fraction of the total ion path length (Fig. 6.18a and b) it can be difficult to assign accurate depth scales to RBS spectra. This arises because the inelastic energy loss or electronic stopping power of ions is different for channeled and random trajectories (Section 6.2). This effect can be quite significant as shown in Fig. 6.19. In this case, a damage layer was produced at a depth of ~1.6 μm below the surface of an otherwise crystalline Si sample by 1.5 MeV Ne^+ ion irradiation (Elliman *et al.*, 1986). Fig. 6.19a shows a TEM micrograph of a sample cross-section which clearly depicts the damage layer. An RBS spectrum from this sample, employing near 180° scattering geometry with the incident beam channeled along the normal ⟨001⟩ direction, is shown in Fig. 6.19b. The stopping power along the channeled ingoing path had to be reduced to 0.55 of the random value in order to match the depth scale of the RBS spectrum (Fig. 6.19b) with the actual depth obtained from TEM analysis. Thus, although the geometry in Fig. 6.18a is commonly employed for channeling applications and provides optimum mass resolution and access to analysis depths exceeding 1 μm, its applicability for depth profiling is limited by depth scale uncertainties and poor depth resolution (Chapter 3).

Double alignment geometries (Appleton and Feldman, 1972), in which both the incident and exit beams are aligned with low index crystallographic directions (Fig. 6.18b), can reduce the yield from substrate atoms to less than 0.1% of the random yield. This provides

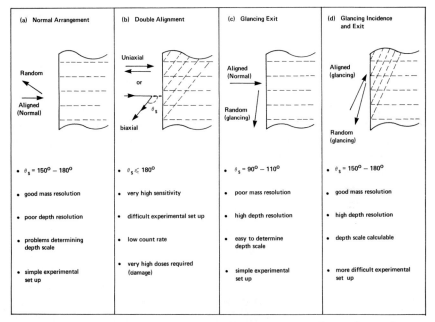

Fig. 6.18 Schematic showing 4 possible channeling geometries, listing attributes of each.

much enhanced sensitivity for the detection of crystal disorder and light elements, such as O and C (Section 6.3.6). However, such restrictive geometries usually require elaborate sample and/or detector manipulation and significantly increase the data collection time.

Geometrical enhancements of depth resolution, see Chapter 3, can also be employed with channeling configurations (Figs 6.18c and d). For these geometries, the channeled path length is small compared to the total ion trajectory, making depth scale determination relatively straightforward. However, the poor mass resolution for the case shown in Fig. 6.18c and the restricted analysis depth of both these glancing angle geometries may not suit all analysis situations. Results employing these different geometries are shown later in this chapter.

The choice of the most appropriate channeling configuration for an analysis task is an important part of experimental planning. Sample preparation and the structural nature of the sample are also important considerations since they often determine the feasibility or otherwise of channeling analysis. Almost all channeling experiments require single crystal samples (or at least samples in which the crystallites are larger than the beam area). Samples with appreciable mosaic spread may not be suitable for channeling; over the beam area the mosaic spread must be significantly smaller than the critical angle (say < 0.1°). This requirement is much more stringent than for many other analysis methods which probe crystal structure, e.g. TEM and X-ray diffraction (XRD). In addition, surface preparation is crucial for channeling analysis since damage in even the first few atom layers can significantly degrade and/or interfere with the channeling process. Usually, very careful mechanical polishing procedures must be employed, culminating in submicron grit size, followed by chemical polishing to remove the last traces of surface damage. Soft (metallic) single crystals must be handled with absolute care (e.g. avoiding stress during preparation and mounting) to minimise the introduction of stress-induced defects and lattice strain. Such surface

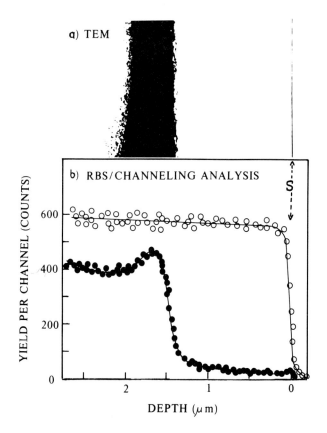

Fig. 6.19 (a) A TEM M-section micrograph and (b) 2 MeV He$^+$ RBS-C spectra showing a damaged layer 1.6 μm below the surface of (100) Si. Matching of the depth scales was used to find the incident ⟨100⟩ channeled stopping power of He$^+$ in Si. After Elliman *et al.* (1986).

preparation procedures are often considerably more demanding than are usually required for X-ray or TEM analysis.

Additional problems which may arise include:

i. beam-induced perturbations to the structural features under analysis; and

ii. non–uniformities of features in the sample which could lead to possible misinterpretation of the spectra.

Such considerations are discussed in some detail in Sections 6.3.3 and 6.3.4.

HIGHLIGHT 6.3
TYPICAL BEAM PARAMETERS FOR CHANNELING

Ion species: He
Ion energy: 1.0–3.0MeV
Beam current: 1–50 nA
Collimation: < 0.05° FWHM (e.g. 1 mm initial aperture separated by 2 m from a 1 mm defining aperture)
Beam size: 1 mm diameter

SAMPLE MANIPULATION

Degrees of freedom: At least two adjustable orthogonal rotational axes
Angular resolution: < 0.05°
Rotational limits: Minimum of ± 90° on each of two rotational axes

OTHER CONSIDERATIONS

Scattering angle: Near 180°
Special electronics: Single channel analyser, Rate meter
Charge integration: Reliable to better than 5%
Analysis time: 3–30 min

6.3.2 Sample Alignment

The alignment of an incident ion beam with a specific crystallographic direction to within an angle of ~0.1° can be achieved with the procedures summarised in Highlight 6.4, using goniometers with two orthogonal rotational axcs. It is a distinct advantage if all rotational axes intersect at the point of impact of the beam on target, to avoid altering the point of beam impact on the surface as rotational adjustments are made. In principle, alignment procedures for most applications rely on the fact that yield minima exist for alignment of the incident ion beam with both planar and axial directions in the crystal. The most common procedures for aligning an incident beam with a specific crystallographic direction involve relating yield minima, observed at well defined sample orientations, to the crystal symmetry. Fig. 6.20 illustrates a general alignment procedure for a Si crystal with near (100) surface orientation. In this case the crystal symmetry is most readily mapped by tilting the crystal by approximately $\theta = 5°$ off normal incidence and monitoring the interaction yield as the sample is rotated azimuthally (ϕ in Fig. 6.20a) through 360°. The RBS yield from a particular near-surface energy window can then be plotted as a function of azimuthal angle (Fig. 6.20b). The yield minima, usually reflecting the χ_{min} from low index planes, are typically 20–60% of the random yield. Minima of similar magnitude can generally be assumed to arise from like planes. Furthermore, the angular position and regularity of the measured yield minima can be related directly to the crystal geometry by depicting the minima on a polar plot, as shown in Fig. 6.20c. Comparison of the major dips in the resulting plot with a standard stereogram for the known crystal structure and orientation (Fig.

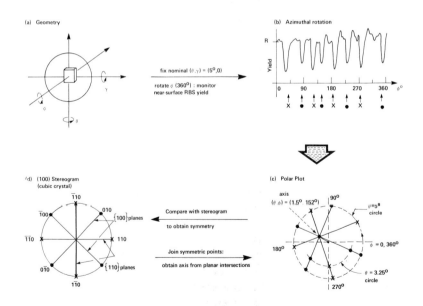

Fig. 6.20 Schematic illustration of a method of alignment of an ion beam with a near-normal axial channnel using azimuthal (ϕ) rotation and θ tilt. See text for explanation of method.

6.20d) can then be employed to identify the major planes and locate the angular position of the desired axis. This is achieved by joining appropriate dips in the polar plot to reflect the symmetry of corresponding major planes in the stereogram. Ambiguity in joining dips can be overcome by repeating the above procedure for one or more additional angular tilts (e.g. at $\theta = 3.25°$ in Fig. 6.20c). The θ-ϕ position of the intersection of the major planes then indicates the angular position of the desired axis. Finally, precise axial alignment is achieved by slight θ-ϕ adjustments to minimise the yield.

It is important to note that, for a two axis goniometer, with θ, ϕ or γ, ϕ rotations, the incident beam must be aligned colinear with the rotation axis to avoid the inaccessibility of some channeling directions near-normal to the sample surface. This problem can be overcome by mounting the sample on a wedge so that the sample normal is a few degrees off the ϕ axis or by choosing a goniometer with two tilt axes only (θ-γ in Fig. 6.20a). In the latter case, minima are depicted on a rectangular θ-γ plot (for small angular variations). The θ and γ rotations are individually varied by about $\pm$ 5° about nominal zero whilst keeping the other angle constant at $+5°$ and $-5°$. Symmetries and axis location are then determined as described above. Both θ-ϕ and θ-γ alignment procedures can be partly or completely automated by using computer control of the goniometer motions and appropriate software. When the structure and approximate orientation of the crystal are known it is usually more expedient to employ manual alignment procedures (Highlight 6.4). However, computer control is particularly useful for tedious or repetitive operations such as angular scan determination or difficult alignment situations. There is no recommended alignment procedure which covers all experimental situations; the analyst must choose the method most appropriate for a given situation. Stereograms are a necessary aid to alignment for all but the simplest cases. Stereograms for major directions in cubic lattices and angles between major directions are given in Chapter 14.6.

6.3.3 Beam Perturbations

In Chapter 3 it was indicated that RBS analysis using light ions can, even at low fluences, sometimes cause structural (and compositional) modifications to a solid target. Fortunately, metals and semiconductors are not particularly prone to compositional change under light ion bombardment; it is usually only insulators (e.g. ionic solids and plastics) which present some difficulties for RBS analysis. However, channeling analysis, which probes the atomic arrangements in the solid, is also additionally

---

HIGHLIGHT 6.4
SAMPLE ALIGNMENT PROCEDURES

A. Exactly-normal Axis
1. Employ independent alignment of the sample normal with respect to the ion beam (eg. optically). Optimise angular settings to minimise yield.
2. Tilt sample normal away from the incident beam direction employing either ϕ or γ (Fig. 6.20a). Vary θ through 360° and record yield minima. Compare with stereogram and locate desired axis from intersection of planes as described in text.
3. Locate a major plane in the vicinity of the axis by tilting one axis. Since major planes pass through the desired axis, alignment can be achieved by alternative adjustment of θ and ϕ (or θ and γ) rotations to continuously alter the sample orientation with respect to the beam whilst maintaining the low yield associated with planar alignment. A further large reduction in yield indicates axial alignment. This method is colloquially referred to as 'walking' along a plane towards the axis.

B. Near-normal Axis
1. Proceed as in A(2) or A(3).

C. Non-normal Axis.
1. If the sample contains a known near-normal axis proceed as described in A(2) above to map crystal symmetry and identify near-normal axis and major planes. Comparison with stereogram and known angles between major axes (Chapter 14.6) will provide the approximate angular position of the desired non-normal axis. Then optimise angular settings to minimise yield.
2. For a sample with no near-normal major axes or of unknown orientation it is strongly advised that alternative methods (eg. XRD) be employed to provide the approximate angular positions of major axes.

D. Planar Alignment
1. Proceed as for C. Select approximate angular position of desired plane using stereogram and minimise yield to optimise position.

E. Angular Scans
1. For axial scans, align as for A, B and C. It should be noted that specific scan directions, relative to predetermined planes and axes, are most readily obtained with a three axis goniometer. To scan in a specific direction it is desirable to align the scan

direction with a goniometer tilt direction. This can be achieved by an appropriate rotation, while maintaining alignment with the channel using θ-γ tilts. Alternatively, the desired scan direction can be obtained by simultaneous adjustment of two goniometer motions. This is the only method applicable for a two axis goniometer unless, by chance or by appropriate sample mounting on the goniometer, the desired scan direction is colinear with a goniometer rotation. Because the axis represents the intersection of a number of planes, it is usually difficult to choose an axial scan direction which gives both a symmetric angular yield profile and for which the yield approaches the random level beyond the dip. Symmetric scans can, however, be obtained by scanning through an axis along a planar direction. For routine angular scan measurements computer control is desirable.

2. For planar scans, align as for D. Proceed as for axial scans.

F. Random Alignment

1. With the aid of a stereogram it is usually possible to select an angular position approximately 5–10° from a channel direction which approximates a high index or 'random' direction. It should be noted that a truly random direction does not exist in a crystal lattice. For a random spectrum obtained along a fixed direction it is advisable to monitor the collected spectrum and count rate to avoid partial planar alignment; small adjustments in angle may be necessary to maximise the count rate to give an acceptable 'random' direction.

2. A reproducible 'random' spectrum can be obtained during continuous azimuthal (ϕ) rotation of the sample, usually by 360° or greater, at a fixed tilt (θ or γ) of 5–10° from the channel direction. Care must be taken to ensure that possible changes in scattering geometry do not have a significant effect on depth scales and kinematics.

3. Use an amorphous target or a fine grain polycrystalline sample of the same material as a control from which a random spectrum can be obtained.

---

sensitive to ion-induced structural changes (e.g. lattice damage). In Table 6.4, the typical damage behaviour for 1 to 2 MeV He$^+$ analysis of metals, semiconductors and insulators is summarised. Clearly, metals and semiconductors present few problems for channeling analysis but insulators can be damaged to a much greater extent. This arises from the fact that electronic energy loss can directly result in displacement damage in insulators through ionisation processes. For example, alkali halides dissociate when exposed to ionising radiation leading to gross damage and often severe sputtering (Townsend and Kelly, 1973). Nevertheless, if minimum ion fluences are employed for analysis and/or the beam is frequently moved to new positions on the crystal surface during analysis, even the most difficult insulator can be analysed without significant damage. This may not be the case for microbeams (Chapter 10) where very high beam densities (and fluences) are unavoidable.

In semiconductors and metals, damage produced by an ion beam arises predominantly from nuclear collisions (Chapter 1). The modified Kinchin-Pease formula, Equation (1.37), can be used to estimate the fraction of displaced target atoms for a particular ion fluence. Since MeV He^+ ions lose > 98% of their energy by electronic processes which do not directly produce damage, displacement damage is small. For example, for 2 MeV He^+ ions incident on Si, the fractional number of Si atoms displaced in the near-surface region is ~ 0.0015 for a beam fluence of $1 \mu C \ mm^{-2}$. For typical probe ion fluences this represents of the order of 10% of the minimum yield and should not be detected in a channeling spectrum. However, since damage is produced along the He^+ ion track at very much higher fluences, it should cause a rise in the dechanneling level of a backscattering energy spectrum. This situation is illustrated in Fig. 6.21, where integrated fluences of the order of $100 \mu C \ mm^{-2}$ are needed before an appreciable increase in the dechanneling level is observed. The damage produced during channeling analysis of Si is further reduced by the fact that only a small fraction of He-induced displacements survive as stable defects at room temperature (Picraux, 1981). A similar situation exists for analysis of metals where most irradiation-produced defects are mobile at room temperature and measured damage levels are thus much smaller than those predicted from the simple Kinchin-Pease displacement model. Analysis with heavier ions can result in higher damage levels, particularly close to the end of the ion range where nuclear energy loss dominates. For 1.5 MeV Ne^+, gross end of range damage was produced in Si at a temperature of 318 C for a fluence of $50 \mu C \ mm^{-2}$ ($3 \times 10^{16} \ Ne^+cm^{-2}$) Fig. 6.19. The same fluence at room temperature effectively amorphises the silicon to a depth corresponding to the Ne^+ range. Fortunately, ions heavier than He are not usually employed for channeling analysis. Indeed, as we have already indicated, He ion channeling can itself be employed to measure heavy ion damage produced during ion implantation.

Beam perturbations during atom location experiments can be more of a problem than damage to the host crystal lattice for two reasons:

TABLE 6.4
Beam perturbations for channeling analysis in typical classes of materials

Materials	Damage at $\phi = 10\ \mu C\ mm^{-2}$	Fluence for Detectable Damage	Fluence for Location Perturbations	References:
METALS				
Typical				
Al	usually negligible negligible	$> 500\ \mu C\ mm^{-2}$	usually negligible but some systems can be subject	Feldman *et al.* (1982) Picraux (1981)
W	negligible	$> 500\ \mu C\ mm^{-2}$	to special problems	Chu *et al.* (1978)
SEMICONDUCTORS				
Typical				
Si	barely detectable negligible	$> 500\ \mu C\ mm^{-2}$	$10\ \mu C\ mm^{-2}$	Feldman *et al.* (1982) Picraux (1981) and Wiggers and Saris (1979)
GaAs	negligible	$> 500\ \mu C\ mm^{-2}$	$10\ \mu C\ mm^{-2}$	Picraux (1981), and Wiggers and Saris (1979)
INSULATORS				
Typical				
SiO$_2$ (quartz)	often considerable moderate	$2\ \mu C\ mm^{-2}$		Matzke (1982) Williams (1973) Townsend and Kelly (1973)
NaCl	very high	$0.1\ \mu C\ mm^{-2}$		
C (diamond)	low	$10\ \mu C\ mm^{-2}$	$1\ \mu C\ mm^{-2}$	Hartley (1982)
Al$_2$O$_3$ (sapphire)	low	$10\ \mu C\ mm^{-2}$	$\geqslant 10\ \mu C\ mm^{-2}$	McHargue *et al.* (1984)

The fluence at which ion damage or perturbations to atom location are detected in RBS-C spectra of 1–2 MeV He$^+$ have either been obtained directly from the literature or estimated from the heavy-ion damaging behaviour of the material.

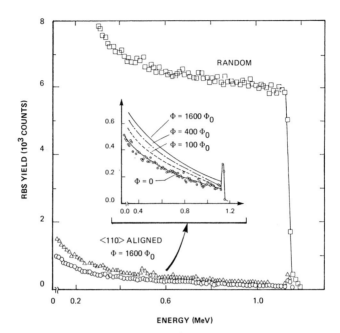

Fig. 6.21 Backscattering spectra for 2-MeV ^{4}He incident along the $\langle 100 \rangle$ aligned and random direction of Si. Inset shows the increase in aligned yield with increasing doses for 2-MeV He bombardment along a random direction due to radiation damage by the beam. Normal fluence for a backscattering measurement is Φ_0 to $10\Phi_0$ where $\Phi_0 = 1\ \mu C/mm^2$ ($\sim 6 \times 10^{14}$ He/cm^2). After Picraux (1981).

i. atom location measurements usually involve detailed angular scans with high total ion fluences; and

ii. irradiation-induced defects are often trapped at impurities, displacing them from their unperturbed lattice sites.

Beam perturbation effects in atom location measurements are also summarised for different materials in Table 6.4. For atom location in Si, He$^+$ fluences below $10\ \mu C\ mm^{-2}$ are usually acceptable. If higher fluences are needed during angular scans, then several fresh areas must be irradiated on the sample during data accumulation to minimise beam-induced damage-impurity interactions. Not all substitutional impurities are displaced from lattice sites under ion irradiation; the phenomenon is very much substrate-impurity specific. Metal substrates also experience similar defect-impurity interactions, particularly for the case of H and D impurities (Besenbacher *et al.*, 1982). Care must be exercised in any atom location experiment to detect beam perturbations. This can most easily be achieved by monitoring aligned impurity yields as a function of ion fluence.

6.3.4 Interpretational Difficulties

a. Measurements of Damage and Structure

In Section 6.2 we indicated that particular defects could result in either direct backscattering of the channeled beam or increases in dechanneling. Interpretation can be difficult when more than one type of defect is present, as illustrated in Fig. 6.22a. This particular example, from Csepregi *et al.* (1976), shows disorder distributions in ⟨111⟩ Si following ion implantation with 100 keV ^{28}Si$^+$ at two temperatures. The main feature of interest here is that the dechanneling level behind the disorder peaks is higher for the case of the > 100°C Si implant even though the area of these disorder peaks (B and C) is smaller than that of the corresponding implant at − 195°C (Curve D). If a single defect type (e.g. displaced atoms or amorphous regions) was present in both cases one would not expect this behaviour. In fact, the disorder from the − 195°C implant is almost certainly in the form of a continuous amorphous layer. In this case the dechanneling level could be estimated, using the method described in Section 6.2.2 to calculate the fraction of the incident beam scattered through an angle greater than the critical angle (with respect to the underlying crystal) in traversing the amorphous layer. In contrast, the disorder distribution for the > 100°C implant most probably results from both displaced atoms (amorphous regions B) and crystalline defects (A and C). In this case, the direct scattering contribution from displaced atoms is small but the dechanneling contribution is large as a result of strain surrounding crystal defects. Studies of the energy dependences of the scattering peaks and the dechanneling level in this latter case could reveal the probable nature of the crystalline defects (Section 6.2.2). However, the use of TEM is a more direct method to identify the defect type (Fig. 6.22b). This cross-sectional TEM micrograph (Seidel and Sheng, 1977) illustrates the defect layers which are more likely to correspond to the regions A-C in the spectra of Fig. 6.22a. In many situations, RBS-C cannot be employed to determine defect type and alternative techniques must be used. For example, amorphous layers and randomly-oriented polycrystalline layers give identical RBS-C spectra. The use of ion channeling combined with complementary analysis techniques is often a very powerful approach to materials analysis.

Under some circumstances, polycrystalline layers can lead to further misinterpretation of spectra. Fig. 6.23 illustrates the analysis of an Au film deposited on a single crystal Si substrate. The Au film is polycrystalline but, in this case, has some large grains preferentially aligned with the Si surface normal. Under ion channeling conditions with the beam

a.

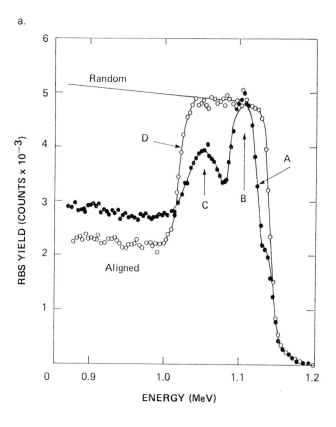

b.

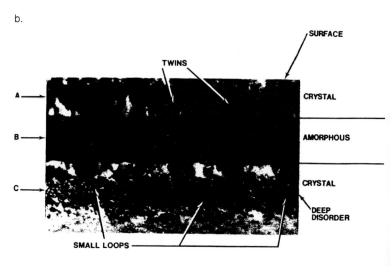

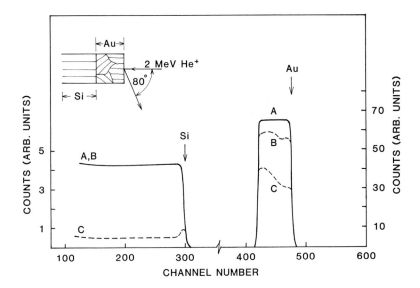

Fig. 6.23 RBS-C spectra illustrating the difficulty in obtaining a true random spectrum from thin films preferentially aligned with a single crystal substrate (Au on Si).

aligned with the underlying Si, the RBS spectrum shows partial alignment in the Au film (curve C). Even when a fixed random direction is located in the Si lattice, some alignment prevails in the Au (curve B). Only when the sample is rotated during analysis (Section 6.3.2) does the Au profile correspond to the true random yield (curve A). Such behaviour is common and care must be exercised when measuring thin film composition in order to eliminate the influence of film texture on the RBS yield.

b. Atom Location and Precipitates

Considerable care must be exercised when employing planar alignments and planar angular scans to determine the precise atom location of impurities in crystals, since yield oscillations can complicate the interpretation of the results. This can be important even for well defined

◀ **Fig. 6.22** a) Backscattering random and $\langle 111 \rangle$ aligned 2 MeV He$^+$ spectra for high (> 100°C; full symbols) and low (− 195°C; open symbols) temperature implants of 100 keV Si into Si to doses of 2×10^{16} cm^{-2} respectively. From Csepregi *et al.* (1976). b) Cross-sectional transmission electron micrograph view of high temperature implant into Si showing regions A, B, and C corressponding, respectively, to crystalline with twins, amorphous, and crystalline with dislocation loops and other clusters of defects (10^{16}Ar/cm^2, 200 keV, 0.5 mA). After Seidel and Sheng (1977).

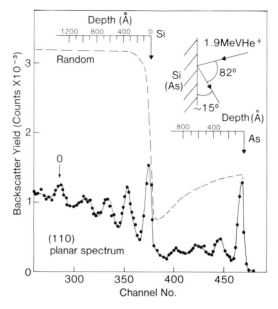

Fig. 6.24 RBS-C spectra illustrating (110) planar oscillations in both host and impurity for substitutional As in Si (100).

substitutional impurities (Fig. 6.24). In this example, planar oscillations are observed in the As impurity profile as well as the Si matrix for planar alignment along (110). In order to compute the depth distribution of substitutional As it would be necessary to obtain the As and Si χ_{min} values at exactly the same depth. This is not a simple task and would involve channel-by-channel fitting of the data. In cases where the impurity is entirely located at a well defined depth corresponding to a maximum or minimum in planar yield, the depth resolution must be sufficient to clearly resolve the yield oscillations, otherwise incorrect locations may be inferred. If the impurity is located at sufficient depths to ensure that the oscillations are damped, then such complexities do not arise. Wagh and Williams (1978) give further clarification of these features. However, if the analyst is aware of such potential problems and exercises care in selecting analysis geometries, planar alignments can be very valuable in atom location studies.

Fig. 6.25 (Appleton *et al.*, 1983) illustrates a further difficulty which can arise during atom location, in this case for Tl in Si. In some systems, impurities can precipitate within the host crystal structure in a coherent or semicoherent fashion. Fig. 6.25 shows angular scans (both Si and Tl yield) about the (110) and (111) directions. No attenuation in Tl yield is

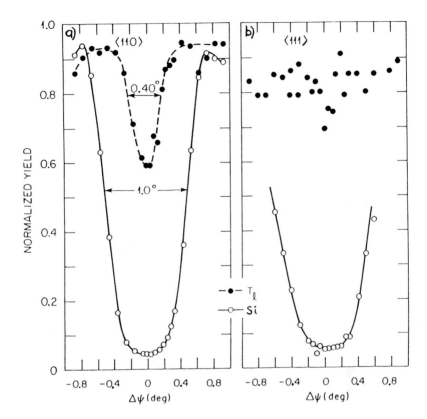

Fig. 6.25 ⟨110⟩ and ⟨111⟩ axial scans (through (100) and (110) planes respectively) for both Tl implanted impurity (full circles) and Si host (open circles). TEM indicated that the Tl had semicoherently precipitated in the Si.

indicated along the ⟨111⟩ direction but a significant dip is apparent along the ⟨110⟩ direction. Attempts to identify the Tl atom location from these and other scans gave confusing and inconsistent results. TEM solved the puzzle by indicating that the Tl had precipitated semicoherently with respect to the Si lattice. Thus, it is important in atom location experiments which are not clear cut to use complementary techniques. It should be stressed that atom location can never be inferred from a single axial or planar scan; many scans about different directions are required.

Fig. 6.26 illustrates the difficulties which can arise when an impurity is not located at a unique lattice position in the host lattice, or when it is slightly displaced from a lattice site. In this example of Bi⁺ implanted into Si (Picraux *et al.*, 1972), angular scans are shown in which the Bi dip widths are not identical to those of Si. In addition, the Bi χ_{min} is about

20% compared with ~ 5% for Si. Scans about other directions gave similar results, but with considerable variation in the Bi dip widths and minimum yields. Attempts to precisely identify the Bi location led to the following possibilities:

  i. Assuming the Bi atoms had identical location in the lattice, computer simulations gave a best fit to the experimentally measured Bi scans for Bi displaced 0.2 Å along the $\langle 111 \rangle$ direction. This simulated result is indicated by the dot-dashed curve in Fig. 6.26a. It does not adequately reflect the shape of the Bi dip, indicating that the Bi atoms may not all be located at precisely the same site.

  ii. The Bi was located in more than one well defined lattice site, each of which has an associated fraction of total Bi. It is impossible to arrive at a *unique* set of locations and associated fractions which fit the measurements; computer simulation can help to provide solutions. For example, the dashed curves in Fig. 6.26 provide reasonable fits to the experimental data and were calculated for 50% of Bi located exactly on Si substitutional sites and 50% of the Bi displaced 0.45 Å along the $\langle 110 \rangle$ directions (c.f. signature scans Fig. 6.17).

  iii. A situation in which a fraction of the Bi is located on precise sites (say exactly substitutional), with the remainder of Bi displaced by *varying* amounts, such as might be expected if the Bi were bound up in defect complexes with Si atoms. Again a multitude of possible solutions could be found to fit the data.

It is interesting to note that subsequent studies of the Bi-Si system (Wagh *et al.* 1981) have indicated that case (iii) is the most probable situation. In fact, the location of Bi in Si is particularly subject to beam perturbation during analysis. Part of the originally substitutional Bi fraction 'interacts' with defects introduced by the analysis beam (Wiggers and Saris, 1979). The Bi (in the example of Fig. 6.26) may have been substitutional initially, with the more complex angular scans arising from the formation of Bi-defect complexes during analysis. Such impurity-defect complexes can also be studied by channeling (Section 6.2) but additional (complementary) analysis techniques are usually needed to provide a complete description of the defect structure.

6.3.5 Compounds

Channeling in compounds can introduce added complexity to the extraction of structural information, atom location and data interpret-

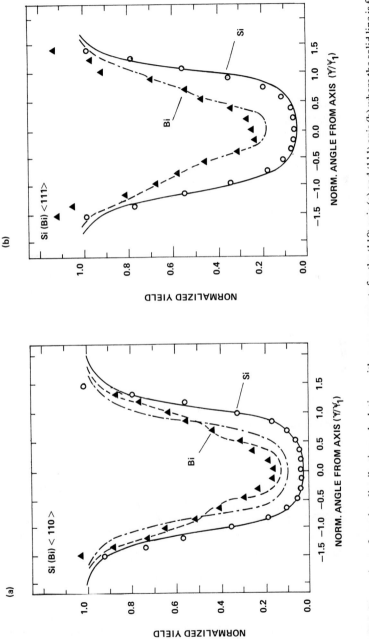

Fig. 6.26 Comparison of angular-distribution calculations with measurements for the ⟨110⟩ axis (a) and ⟨111⟩ axis (b) where the solid line is for Si, the dot-dashed line is for Bi assuming all of the Bi atoms displaced 0.2 Å along the ⟨110⟩ directions, and the dashed line is for 50% of the Bi displaced 0.45 Å along the ⟨110⟩ directions and 50% on substitutional lattice sites. *After Picraux et al.* (1972).

ation. However, the possibility in channeling analysis of discriminating between the various elemental constituents of compounds can be used to advantage. In this section we discuss some special aspects of channeling analysis of compounds.

The example of channeling analysis of In implanted sapphire (Al_2O_3) (Fig. 6.27, Mouritz *et al.*, 1987) illustrates typical spectra from crystalline compounds containing near-surface damage. The damage (i.e. displaced Al and O atoms in the crystal), caused by In ion implantation, appears as separated damage peaks corresponding to each sublattice, as indicated by the damage peaks A and B in Fig. 6.27, corresponding to Al and O respectively. The extraction of channeling data (e.g. χ_{min}) and damage distributions must be made for each sublattice. The extraction of data from the Al sublattice can be obtained in the same straightforward manner as for elemental crystals previously described in Section 6.2. However, the backscattering contribution from Al must first be subtracted from that of the O signal before χ_{min} and damage distributions can be obtained for this sublattice.

In addition to analysis of damage or structural modifications to compound sublattices, channeling can also provide more specific information on atom location and near-surface composition. The atom location possibilities are discussed in more detail in Section 6.4. Fig. 6.28

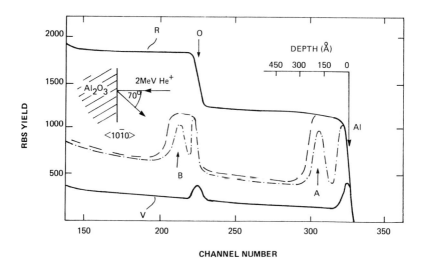

Fig. 6.27 RBS-C spectra for In implanted $(01\bar{0}1)$ sapphire (Al_2O_3) showing damage peaks in both Al (A) and O (B) sublattices. The dot-dashed $\langle 01\bar{0}1 \rangle$ spectrum is for a dose of 1×10^{15} In cm^{-2} and the dashed spectrum is for 1×10^{16} In cm^{-2}. An unimplanted spectrum (V) and a random spectrum (R) are shown for comparison. After Mouritz *et al.* (1987).

treats the surface composition of GaAs. Since Ga and As are closely spaced in mass, only a small backscattered energy separation exists between these elements (e.g. 38 keV for 2 MeV He$^+$). This separation is reflected only as a small step in the front edge of the random spectrum of Fig. 6.28 and small changes in surface composition would normally not be detected. However, when channeling conditions are employed, surface composition can be more readily studied as indicated by the

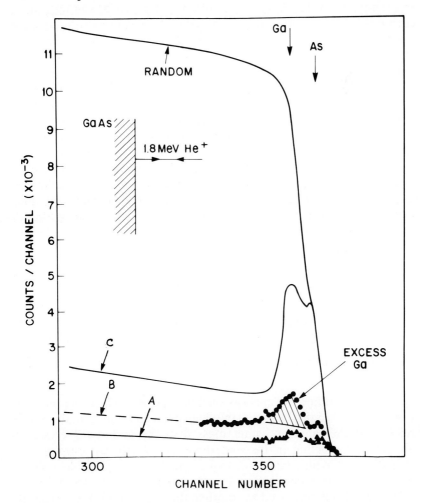

Fig 6.28 RBS-C spectra from (100) GaAs showing surface Ga and As peaks after various laser annealing treatments: (A) no laser, (B) low power showing excess Ga, (C) high power showing gross surface damage. Adapted from Barnes *et al.* (1978).

aligned spectra in Fig. 6.28. Spectrum A indicates the well defined Ga and As surface peaks which are obtained from stoichiometric GaAs. The peak areas are identical and indicate equal surface concentrations of Ga and As. Morgan and Bógh (1972) have employed changes in the Ga and As surface peak areas to measure surface stoichiometry following various GaAs cleaning procedures. Such applications are more relevant to the surface studies outlined in Chapter 9. Spectrum B was obtained following laser treatment of the GaAs surface (Barnes *et al.*, 1978) and indicates a much larger Ga peak compared with that of As. Using scanning electron microscopy techniques this enhanced Ga peak could be attributed to free surface Ga metal. Consequently, the 'surface' concentration of free Ga could be obtained from the peak area (after background subtraction) using Equation (3.12), Chapter 3. For spectrum B, the shaded area gives $\sim 1.5 \times 10^{16}$ Ga cm^{-2}. Finally, spectrum C was taken following higher power laser treatment. The Ga and As surface peaks overlap in this case and it is no longer possible to extract information on surface stoichiometry (except possibly by employing deconvolution techniques). The total surface concentration of Ga and As which are non-aligned with the GaAs substrate can, however, be found. In this example, higher energy analysis (e.g. with 5 MeV He^{+}) would have provided improved resolution of Ga and As peaks. In Fig. 6.28 the random spectra corresponding to each of the aligned spectra are essentially identical indicating the vast improvement in sensitivity under channeling conditions.

6.3.6   Light Elements and Thin Films

Channeling can be employed to enhance the sensitivity for

  i. light element detection; and
 ii. thin film analysis in cases where the underlying substrate is a single crystal.

Fig. 6.29a illustrates the analysis of thin (100 Å) SiO_2 layers on single crystal (100) Si. In this example, the elemental constituents of the film (Si and O) are the same or lighter mass than the underlying crystalline substrate. Channeling in the underlying crystal clearly simplifies the extraction of thin film data. Note that a grazing-exit angle geometry (with beam incident in a channeling direction approximately normal to the surface) has been employed to improve the depth resolution. From the channeled spectrum it is possible to extract the film thickness and average stoichiometry (Si/O atomic ratios) of the film (using the shaded areas and employing Equations (3.8) and (3.14), Chapter 3). There will

a.

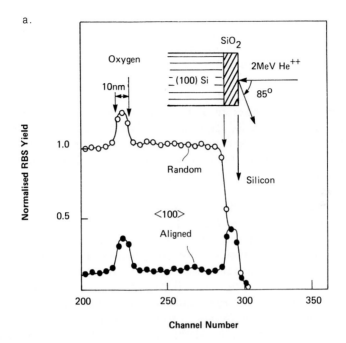

b.

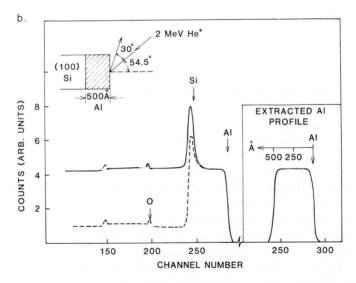

Fig. 6.29 Examples of thin film analysis using channeling a) SiO₂ on Si (100). After Sai Halasz *et al.* (1985). b) Al on Si (100) analysed along ⟨111⟩ showing Si/Al overlap peak and about 5% oxygen in the Al film with O peaks at both the Al surface and at the Al/Si interface.

also be a contribution to the Si peak from the first few layers of Si in the crystalline substrate. Extraction of thin film stoichiometry should normally account for this non-oxide contribution. However, in this case it is expected to be small compared with the Si scattering contribution from the oxide.

The analysis geometry ($\theta_s \sim 95°$) employed in Fig. 6.29a does not give particularly good mass resolution. Some applications, such as that illustrated in Fig. 6.29b, require both mass and depth resolution to be optimised by employing a geometry such as that in Fig. 6.18d. However, channeling conditions restrict the choice of incidence angle to a major channeling direction. In Fig. 6.29b the analysis of 500 Å of Al deposited on (100) Si is achieved by channeling along the $\langle 111 \rangle$ direction in Si at 54.5° to the normal and employing a grazing-exit angle of 5.5° to the surface plane (i.e. $\theta_s = 150°$). This geometry allows sufficient mass resolution to discriminate between the Si and Al yields. Since the underlying Si film is heavier, a Si–Al overlap peak appears close to the back edge of the Al signal. Under channeling conditions (with a low Si background) it is easier to extract both the Al profile (by substraction of a scaled Si channeled spectrum) and the profile of O contamination in the Al film. Twin oxygen peaks result from Al_2O_3 at the surface and SiO_2 at the Al–Si interface. Also, the Al film has $\sim 5\%$ oxygen content throughout its thickness, as calculated from the height ratio of Al to O.

6.4 FURTHER APPLICATIONS

6.4.1 General Applicability of Channeling

Table 6.5 summarises the extensive range of applications of channeling which have been employed in materials science along with selected references. Many examples of such applications have been given in previous sections and studies of epitaxial crystal growth, specific defects in crystals and bulk structural changes are illustrated in Section 6.4.2. Bulk structural (or phase) changes such as hydriding or martensitic transformations are very accurately and conveniently monitored by ion channeling. Furthermore, the ability of ion channeling to locate foreign impurities can be used to study a number of phenomena in crystalline solids. Further illustrations of these uses of ion channeling as well as ion atom location (concentrating on nuclear reactions and PIXE) are given in Section 6.4.3.

TABLE 6.5
Applications of MeV ion channeling

APPLICATION	LITERATURE
Radiation Damage	
Semiconductors	Mayer *et al.* (1970)
Metals	Poate (1978)
Insulators	McHargue *et al.* (1984)
Specific Defects	
General	Feldman *et al.* (1982)
Dislocations	Picraux *et al.* (1982)
Stacking faults	Mory and Quere (1972)
Twins	Foti *et al.* (1977)
Point defects	Eisen (1973)
Surfaces and Interfaces	
General	Feldman (1981)
Atomic arrangements	Bogh (1973)
Adsorbed atoms	Feldman *et al.* (1982)
Thin films	
General	Feldman *et al.* (1982)
Texturing	Andersen *et al.* (1978)
Background reduction	Chu *et al.* (1978)
Epitaxial Layers	
General	Feldman *et al.* (1982)
Crystal growth	Csepregi *et al.* (1977)
Strained superlattices	Chu *et al.* (1984)
Phase Changes	
Bulk transformations	Besenbacher *et al.* (1983)
Phase separation	Appleton *et al.* (1983)
Foreign Impurities	
General	Davies (1973)
Precise atom location	Swanson (1982)
Defect trapping of impurities	Swanson *et al.* (1980)
Solid solubility in semiconductors	Williams (1983)
Solid solubility in metals	Poate (1978)

6.4.2 Further Structural Studies

a. Epitaxial Growth

Ion channeling was employed to obtain the first measurements of the kinetics of solid phase epitaxial crystallisation of amorphous layers formed by ion implantation. This was achieved by forming a thin ($< 1\ \mu m$) amorphous layer on the surface of the bulk semiconductor by ion implantation and then measuring the extent of crystallisation by ion channeling after various stages of heat treatment (Csepregi *et al.*, 1975). Fig. 6.30 shows 2.0 MeV He$^+$ spectra illustrating the epitaxial growth of (100) Si from an amorphous layer implanted with 3×10^{15} cm^{-2} Sb$^+$ ions at 100 keV. The aligned spectrum corresponding to the as-implanted case shows the initial amorphous layer of about 1200 Å in thickness. The spectrum labelled 1 illustrates the reduction in amorphous layer thickness for annealing at 550°C for about 20 min. Spectrum 2 indicates the complete epitaxial crystallisation for annealing at 550°C for 60 min. This spectrum is essentially identical to that obtained from a perfect Si crystal. The corresponding Sb portions of the spectra, in which the aligned yield from Sb decreases with increasing amounts of epitaxial growth, indicate the incorporation of the Sb impurity onto Si lattice sites during near-perfect crystal growth. Such studies have allowed the influence of impurities on the crystal growth kinetics to be accurately measured (Williams, 1983).

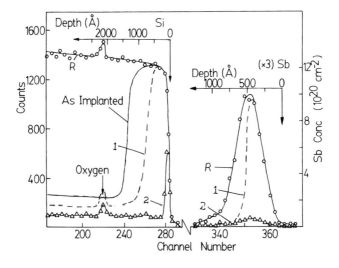

Fig. 6.30 RBS-C spectra from 3×10^{15} cm^{-2}, 100 keV Sb-implanted (100) Si. The as-implanted ⟨100⟩ aligned spectrum indicates ~ 1200 Å of amorphous Si. Spectra 1 and 2 represent 20 min and 60 min annealing at 550°C and illustrate solid phase epitaxial growth. Note the progressive decrease in the Sb yield with epitaxial growth. The random spectrum R is identical in all cases. Adapted from Pogany *et al.* (1983).

b. Defect and Precipitation Observations

Channeling can be used very successfully to study defect formation and precipitation phenomena in crystals, particularly when used with complementary analysis techniques. We illustrate such an investigation in Fig. 6.31, taken from the work of Pogany *et al.* (1983). The random (R) and aligned (A) spectra in Fig. 6.31a were taken following annealing at 850°C for 30 min of the Sb-implanted Si sample which gave rise to spectrum 2 in Fig. 6.30. It is clear, by comparing the respective spectra for virgin (V) and implanted Si in Fig. 6.31a with spectrum 2 in Fig. 6.30, that both the aligned Si and Sb yields have increased substantially after the further 850°C heat treatment. The increase in Sb yield indicates movement of Sb off Si lattice sites but the appearance of a peak (double-arrowed) in the aligned Si yield, at a depth corresponding to the peak of the Sb distribution, was unexpected. TEM, shown in Fig. 6.31b, indicated that the microstructure consisted of small interstitial Si dislocation loops and Sb precipitates lying at a precise depth (about the peak of the Sb distribution) in a plane parallel to the surface. It is this particular arrangement of a high density of small interstitial loops located at a precise depth which gives rise to the well defined direct scattering peak from Si in Fig. 6.31a. This result should be contrasted with that illustrated in Fig. 6.11, where a well defined direct scattering peak is not observed for channeling measurements of an edge dislocation network. Thus, the particular defect structure (and density) in addition to the chosen analysis conditions can determine whether direct scattering or dechanneling components dominate channeling spectra from dislocations. As discussed in Section 6.2.2, both components must be included in a complete analysis of data.

c. Measurements of Phase Changes

Fig. 6.32 illustrates the application of ion channeling for accurate measurements of phase changes in bulk single crystals. In this example, the specimen is placed on a cooled stage and channeling conditions set up in the crystal either above or below the phase transition temperature. Ion channeling spectra (or minimum yields) can then be monitored as the temperature is varied in the vicinity of the transition temperature. There is a dramatic change in the minimum channeling yield along $\langle 112 \rangle$ directions as a single crystal of austenitic stainless steel is cooled through the martensitic transition temperature of -98°C for a composition of 70% Fe, 17% Cr and 13% Ni (Besenbacher *et al.*, 1983). The advantage of channeling over other techniques, e.g. XRD, is that channeling is not

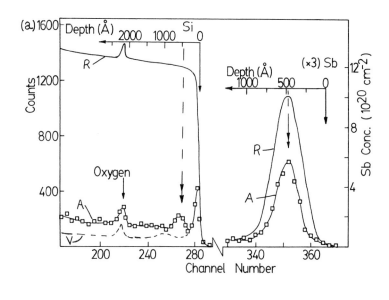

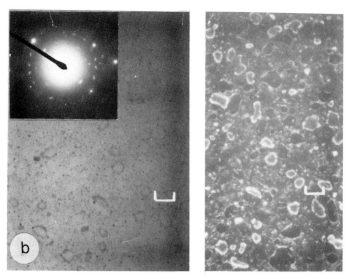

Fig. 6.31 a) RBS-C ⟨100⟩ aligned (A) and random (R) spectra from 3×10^{15} cm^{-2} Sb-implanted Si after 850°C annealing for 30 min. Note the non-substitutional Sb (compared with Fig. 30) and the appearance of a Si damage peak (double-arrowed). An unimplanted case (V) is shown for comparison. b) Bright field (left), dark field (right) and selected area diffraction pattern (upper left) TEM micrographs from the sample referred to in a). Note the dislocation loop array and the precipitates (spots) in the bright field image. Sb spots are observed in the diffraction pattern. After Pogany *et al.* (1983)

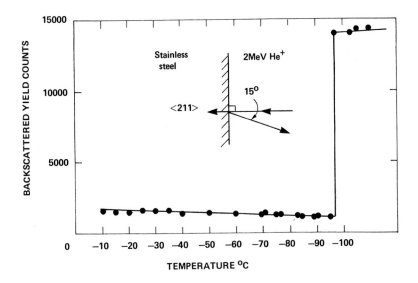

Fig. 6.32 Backscattered yield of 2 MeV He from an aligned ⟨112⟩ single crystal stainless steel sample as a function of decreasing target temperature. After Besenbacher *et al.* (1983).

particularly sensitive to the small lattice parameter changes and atomic rearrangements which take place as the transition temperature is approached. It is primarily affected by the gross structural changes occuring precisely at the transition temperature. Ion channeling is thus an ideal probe for accurate determination of phase transition temperatures and can be used as a rapid surveying technique to study the influence of composition and other parameters on transition temperature.

6.4.3 Further Foreign Atom Location Studies

Channeling is suitable for measuring the solubility (substitutional or interstitial) of impurities in a crystalline matrix. Such measurements have been made in both semiconductor (Williams and Short, 1982 a,b) and metallic systems (Sood, 1978). For example, the data of Figs. 6.30 and 6.31 are taken from a study of the metastability of supersaturated solid solutions of Sb in Si (Pogany *et al.*, 1983). In most cases the incident beam is channeled and the backscattered yield from the impurity is measured. However, other ion-atom interaction processes can be usefully employed for atom location as we illustrate below.

a. Channeling and PIXE

In cases where impurities are lighter than the host crystal atoms, it is possible to employ characteristic X-ray emission for atom location studies. Fig. 6.33a illustrates the principles for preferential location of impurities with respect to the Ga or As sublattices in GaAs (Andersen *et al.*, 1981). The measurement makes use of the fact that the ⟨100⟩ axis in a zinc-blende structure is made up of atomic strings containing only one of the two matrix atoms. When an ion beam is incident parallel to the (110) plane but at an angle $+\theta$ with respect to the ⟨110⟩ axis, it has a higher probability of initially striking an A (As) row than a B (Ga) row, as a result of shadowing. The opposite will occur if the beam is incident at $-\theta$ to the ⟨110⟩ axis. Thus, a complete angular scan should indicate an asymmetry in channeling yield from near-surface A and B atoms. Furthermore, when a substitutional impurity preferentially occupies an A or B site, its angular yield should also exhibit asymmetry.

Angular scans for 120 keV S⁺ implanted GaAs are illustrated in Fig. 6.33b, using a 330 keV H⁺ beam scanned across the ⟨110⟩ axis parallel to the {110} plane (Bhattacharya *et al.*, 1983). The intensity from S-$K\alpha$, Ga-$K\alpha$ and As-$K\alpha$ X-rays was monitored as a function of tilt angle

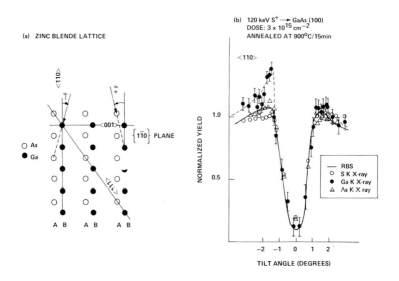

Fig. 6.33 a) Positions of Ga and As atoms in the {100} plane of the GaAs lattice. Two possible incident beam conditions, $+\theta$ and $-\theta$ with respect to the ⟨110⟩ axis, will shadow one or the other kind of atoms. b) Angular scans about ⟨110⟩ axis parallel to the {1̄10} plane. After Bhattacharya *et al.* (1983).

TABLE 6.6

Values by which the lattice constant must be multiplied to calculate the interatomic spacing d in axial directions and the interplanar spacings d_p for planar directions in (monatomic) cubic structures and in common diatomic compounds (atoms labelled A and B) having cubic structures[a].

Structure	Atoms per unit cell	Axis			Plane		
		⟨100⟩	⟨110⟩	⟨111⟩	(100)	(110)	(111)
fcc	4	1	1/√2	√3	1/2	1/2√2	1/√3
bcc	2	1	√2	√3/2	1/2	1/√2	1/2√3
fcc (diamond)	8	1	1/√2	√3/4,3√3/4	1/4	1/2√2	1/4√3,√3/4
Rocksalt (like NaCl)	4A+4B	ABAB... 1/2	pure 1/√2	ABAB... √3/2	mixed 1/2	mixed 1/2√2	pure ABAB... 1/2√3
Fluorite (like CaF₂)	4A+8B	pure 1(A),1/2(B)	pure 1/√2	BAB...BAB... √3/4,√3/2	pure ABAB.. 1/4	mixed 1/2√2	pure BAB..BAB.. 1/4√3,√3/4
Zinc blend (like ZnS)	4A+4B	pure 1	pure 1/√2	AB...AB... √3/4,3√3/4	pure ABAB.. 1/4	mixed 1/2√2	pure AB..AB.. 1/4√3,√3/4

[a] For the axial case, the term pure indicates that each row contains only one atomic species. For rows containing both species, the ordering in the row is given. For the planar case, the term pure indicates that each sheet of atoms contains only one atomic species and the way in which the sheets are ordered is shown. The term mixed refers to cases where each planar sheet of atoms contains both atomic species. After Gemmell (1974).

(Fig. 6.33b). The solid curve was obtained by monitoring the As + Ga near surface yield from RBS spectra, where there is insufficient mass resolution to discriminate between As and Ga and the scan is symmetric. In the case of X-ray emission, the As and Ga yields show only a very slight asymmetry in the shoulders of the angular scan at about ± 1.5°. Note that the As yield is slightly higher on the negative tilt side of the scan and similarly for Ga on the positive tilt side. This lack of marked asymmetry for As and Ga results from the fact that the X-rays arise from depths up to 1.5 μm. The preferential interaction of the incident beam with the different sublattice rows occurs over the first 1000 Å, before statistical equilibrium (Section 6.2.1) of the channeled beam is reached. Since the profile of the implanted S is confined to the first 2000 Å and 95% is substitutional in the GaAs lattice, one would expect a strong asymmetry in the X-ray angular yield profile from S if it is preferentially located on either As or Ga sites. Indeed, Fig. 6.33b clearly indicates (by the strong peak at $-1.5°$) a preference for S to occupy As sites in the GaAs lattice.

b. Channeling and Nuclear Reaction Analysis

Nuclear reactions provide a very selective mode of analysis for specific elements (isotopes) within a host material, particularly for identification and evaluation of light elements in heavier substrates (Chapter 4). When nuclear reaction analysis is combined with channeling in single crystals, it is possible to obtain additional information concerning the precise atom location of such impurities. Fig. 6.34 illustrates the atom location of N in single crystal stainless steel (Whitton *et al.*, 1982). Single crystals of austenitic stainless steel of composition (Fe:Cr:Ni = 70:17:13) were implanted with 40 keV ^{15}N to a dose of $5. \times 10^5$ cm^{-2}. The samples were then analysed using ion channeling either with 2 MeV He$^+$ (RBS), to monitor the lattice damage, or with 800 keV H$^+$ to determine the N location from the ^{15}N (p, α)^{12}C reaction. The backscattered particles or α-particles from the nuclear reaction were collected by an annular surface barrier detector mounted coaxial to the incident beam ($\theta_s \sim 175°$ to 180°). For the nuclear reaction, the elastically scattered protons were filtered out by a Mylar foil placed over the detector (Chapter 4).

The RBS (He$^+$) yield and the nuclear reaction (α-particle) yield are shown (normalised) in Fig. 6.34 for angular scans through the $\langle 100 \rangle$ and $\langle 110 \rangle$ axes of fcc stainless steel. The angular dips for the He$^+$ scattering (before implantation) and nuclear reaction yields (after implantation) have been compared on the same plot since the respective $\chi_{\min}$ and $\psi_{1/2}$ values for 2 MeV He$^+$ and 800 keV H$^+$ differ by less than 10% in the un-

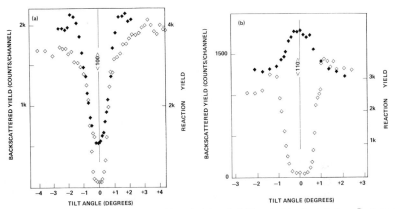

Fig. 6.34 Angular scans through the ⟨100⟩ axis (a) and ⟨110⟩ axis (b) of a stainless-steel single crystal before implantation (2 MeV He$^+$ backscattering data ◇) and after implantation with 40 keV ^{15}N$_2^+$ to a dose of 5 × 10^{15} cm^{-2} [^{15}N(p,α)^{12}C reaction]. After Whitton *et al.* (1982).

implanted crystal. In Fig. 6.34a, the N dip is pronounced along the ⟨100⟩ axis and suggests that the N is located along ⟨100⟩ strings. The high χ_{min} for nitrogen, compared with that of the unimplanted crystal results from implantation damage. In Fig. 6.34b, a similar scan along the ⟨110⟩ axis shows no dip in the N yield but a yield enhancement. This flux peaking effect indicates that the N predominantly resides within ⟨110⟩ channels of the stainless steel. It is interesting that the width of the flux peak is somewhat greater than would be expected for N atoms residing at the centre of ⟨110⟩ channels (c.f. Section 6.2.3 and Fig. 6.17) and this may suggest that some of the N is located off the channel centre. Comparison of the results in Fig. 6.34 with the expected positions of various interstitial impurities in fcc crystals (Fig. 6.17 and Chapter 14.6) suggests that N is preferentially located in octahedral interstitial sites in fcc stainless steel. Confirmation of this was obtained from further scans across ⟨111⟩ and ⟨112⟩ axes in which the former scan was similar to that in Fig. 6.34a and the latter similar to Fig. 6.34b.

6.5 CALCULATION OF CHANNELING PARAMETERS

This section follows a treatment in the review article by Gemmell (1974). It provides formulae, graphs and tables of relevant quantities to enable quick estimates of the channeling half angle and minimum yield for both axial and planar channeling conditions. The formulae are semi-empirical expressions derived by Barrett (1971) from a comparison of experimental results and Monte Carlo computer simulations. For further details about the derivation of these equations the reader is referred to the original references.

The expressions given below are valid for non-relativistic ion energies greater than E', where E' is given by:

$$E' = 2Z_1 Z_2 e^2 d / a^2 \tag{6.21}$$

and Z_1 and Z_2 are the atomic numbers of the incident ion and substrate atoms respectively, e is the electronic charge ($e^2 = 14.4$ eV/Å), d is the atomic spacing along the channeling direction, and a is the Thomas-Fermi screening length which is required for evaluation of both Equation 6.21 and subsequent expressions. Its value is given by:

$$a = 0.4685 Z_2^{-1/3} \tag{6.22a}$$

if the incident ion is fully ionised and by:

$$a = 0.4685 (Z_1^{1/2} + Z_2^{1/2})^{-2/3} \tag{6.22b}$$

if the ion is neutral or partially ionised.

The one-dimensional thermal vibrational amplitude, u_1 is also

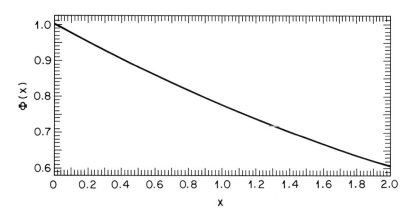

Fig. 6.35 The Debye function $\Phi(x) = \dfrac{1}{x} \displaystyle\int_0^x \frac{\delta d\delta}{e^\delta - 1}$.

required for the estimation of channeling parameters. Its magnitude can be estimated from the Debye approximation, which gives:

$$u_1 = 12.1([\Phi(x)/x + 1/4]/M_2\theta_D)^{1/2} \text{ (Angstrom)} \qquad (6.23)$$

where $\Phi(x)$ is the Debye function shown plotted in Fig. 6.35, M_2 is the atomic weight (amu) of the substrate atoms, θ_D is the Debye temperature (K) and $x = \theta_D/T$ where T is the substrate temperature (K). Values of a and u_1 are given in Table 14.25 for commonly studied crystals.

6.5.1 Axial Channeling

The axial half angle is given by:

$$\psi_{1/2} = 0.8F_A(1.2u_1/a)\,\psi_1 \quad \text{(degrees)} \qquad (6.24)$$

where $x=1.2u_1/a_1$ and $F_A(x)$ is plotted in Fig. 6.36 and defined in Gemmell (1974), and the angle ψ_1 is given by:

$$\psi_1 \sim 0.307(Z_1Z_2/d.E)^{1/2} \quad \text{(degrees)} \qquad (6.25)$$

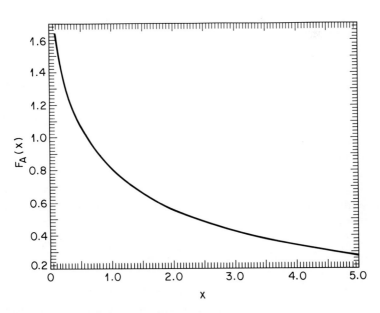

Fig. 6.36 The function $F_A(x)$ as defined by Gemmell (1974).

where d is the atomic spacing along the atomic row (in Å) and E is the ion energy (in MeV). Typical values of d are given in Table 14.25. (The factors 0.8 and 1.2 in Equation (6.24) are the semi–empirical fitting parameters determined by Barrett (1971)).

The minimum yield can be estimated from:

$$\chi_{min} = 18.8 N d u_1^2 (1 - \xi^{-2})^{1/2} \tag{6.26}$$

where N is the atomic density of the substrate (in atoms/cm^3) and

$$\xi = 126 u_1/(\psi_{1/2} d) \tag{6.27}$$

with $\psi_{1/2}$ in degrees from Equation (6.24).

6.5.2 Planar Channeling

The planar half angle is given by:

$$\psi_{1/2} = 0.72\, F_p(1.6 u_1/a, d_p/a)\psi_p \quad \text{(degrees)} \tag{6.28}$$

where the function $F_p(x,y)$ is plotted in Fig. 6.37, d_p is the interplanar spacing, and:

$$\psi_p = 0.545\, (n_p Z_1 Z_2\, a/E)^{1/2} \quad \text{(degrees)} \tag{6.29}$$

where n_p is the atomic density of the planes (atoms/Å^2) and a is the Thomas-Fermi screening length, given by Equation (6.22). The atomic density of the planes, n_p, is given by $N d_p$, and d_p can be calculated for common crystal structures using Tables 6.6 and 14.25.

The minimum yield for planar channeling is not accessible to the semi–empirical fitting procedure employed above. An approximate expression is given by

$$\chi_{min} = 2\sqrt{2}\, u_1/d_p \tag{6.30}$$

This expression tends to underestimate the magnitude of χ_{min} but illustrates its linear dependence on the thermal vibrational amplitude u and its independence of ion energy.

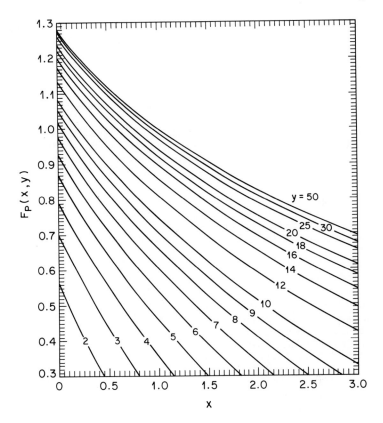

Fig. 6.37 The function $F_p(x,y)$ as defined by Gemmell (1974).

REFERENCES

Andersen, H.H., Tu, K.N. and Ziegler, J. (1978) *Nucl. Instrum. Methods* **149**, 247.
Andersen, J.U. (1967) *Mat. Fys. Medd. Dan. Vid. Selzk.* **36**, 1.
Andersen, J.U., Chechenin, N.G. and Zhang, Z.H. (1981) *Appl. Phys. Lett.* **39**, 758.
Appleton, B.R. and Feldman, L.C. (1972) *J. Phys. Chem. Solids* **33**, 507.
Appleton, B.R., Narayan, J., White, C.W., Williams, J.S. and Short, K.T. (1983) *Nucl. Instrum. Methods* **209/210**, 239.
Barnes, P.A., Leamy H.J., Poate, J.M., Ferris, S.D., Williams, J.S. and Celler, G.K. (1978) *Appl. Phys. Lett.* **33**, 965.
Barrett, J.H. (1971) *Phys. Rev.* **B3**, 1527
Bcscnbacher, F., Bottiger, J. and Myers, S.M. (1982) *J. Appl. Phys.* **53**, 3536
Besenbacher, F., Nielsen, B.B. and Whitton, J. (1983) *Nucl. Instrum. Methods* **218**, 551
Bhattacharya, R.S., Pronko, P.P. and Ling, S.C. (1983) *Appl. Phys. Lett.* **42**, 880
Bogh, E. (1967) *Phys. Rev. Lett.* **19**, 61.

Bogh, E. (1973) In 'Channeling' (D.V. Morgan, ed.) Wiley, New York, p. 435.
Chu, W.K., Mayer, J.W. and Nicolet, M.A. (1978) 'Backscattering Spectrometry', Academic Press, New York.
Chu, W.K., Ellison, J.A., Picraux, S.T., Biefeld, R.M. and Osbourn, G.C. (1984) Phys. Rev. Lett. 52, 125.
Csepregi, L., Mayer, J.W. and Sigmon, T.W. (1975) Phys. Lett. 54A, 157.
Csepregi, L., Kennedy, E.F., Lau, S.S., Mayer, J.W. and Sigmon, T.W. (1976) Appl. Phys. Lett. 29, 645.
Csepregi, L., Kennedy, E.F., Mayer, J.W. and Sigmon, T.W. (1977) J. Appl. Phys. 49, 3906.
Davies, J.A. (1973) In 'Channeling' (D.V. Morgan, ed.) John Wiley, New York, p. 392.
Eisen, F.H. (1973) In 'Channeling' (D.V. Morgan, ed.) John Wiley, New York, p. 415.
Elliman, R.G., Williams, J.S., Johnson, S.T. and Pogany, A.P. (1986) Nucl. Instrum. Methods B15, 439.
Feldman, L.C. (1981). Crit. Rev. in Sol. State and Mat. Sci. 10, 143.
Feldman, L.C., Mayer, J.W. and Picraux, S.T., (1982) 'Materials Analysis by Ion Channeling', Academic Press, New York.
Foti, G., Csepregi, L., Kennedy, E.F., Pronko, P.P. and Mayer, J.W. (1977) Phys. Lett. 64A, 265.
Gemmell, D.S. (1974) Rev. Mod. Phys. 46, 129.
Hartley, N.E.W. (1982) In 'Metastable Materials Formation by Ion Implantation', (S.T. Picraux and W.J. Choyke, eds.) North Holland, Amsterdam, p. 295.
Lindhard, J. (1965) Mat. Fys. Medd. Dan. vid. Selsk. 34, 1.
Lutz, H. and Sitzmann, R. (1963) Phys. Lett. 5, 113.
Matsunami, N. and Howe, L. (1980) Rad. Eff. 51, 111.
Matzke, H.J. (1982) Rad. Eff. 64, 3.
Mayer, J.W., Eriksson, L. and Davies, J.A. (1970) 'Ion Implantation in Semiconductors', Academic Press, New York.
McHargue, C.J., White, C.W. Appleton, B.R., Farlow, G.C. and Williams, J.M. (1984) Proc. Mat. Res. Soc. 27, 385.
Morgan, D.V. and Bogh, E. (1972) Surf. Sci. 32, 278.
Mory, J. and Quere, Y. (1972) Rad. Eff. 13, 57.
Mouritz, A., Sood. D.K., St. John, D., Swain, M.V. and Williams, J.S. (1987) Nucl. Instrum. Methods B19/20, 805.
Pathak, A.P. (1982) Rad. Eff. 61, 1982.
Picraux, S.T. (1981) unpublished.
Picraux, S.T., Brown, W.L. and Gibson, W.M. (1972). Phys. Rev. B6, 1382.
Picraux, S.T., Knapp, J.A. and Rimini, E. (1982).
Picraux, S.T., Dawson, L.R., Osbourn, G.C., Beifeld, R.M. and Chu, W.K. (1983) Appl. Phys. Lett. 43, 1020,
Picraux, S.T., Chu, W.K., Allen, W.R. and Ellison, J.A. (1986) Nucl. Instrum. Methods B16, 306.
Piercy, G.R., Brown, F., Davies, J.A. and McCargo, M. (1963) Phys. Rev. Lett. 10, 399.
Poate, J.M. (1978) J. Vac. Sci. Tech. 15, 1636.
Pogany, A.P., Preuss. T., Short, K.T., Wagenfeld, H.K. and Williams, J.S. (1983) Nucl. Instrum. Methods 209/210, 731.
Quere, Y. (1968) Phys. Stat. Sol. 30, 713.
Sai Halasz, G.A., Short, K.T. and Williams, J.S. (1985) IEEE Electron Device Lett. 6, 285.
Robinson, M.T. and Oen, O.S. (1963) Phys. Rev. 132, 2385.
Seidel, T.E. and Sheng, T.T. (1977) Appl. Phys. Lett. 31, 256
Sigmund, P. and Winterbon, B. (1974) Nucl. Instrum. Methods 119, 541.

Sood, D.K. (1978) *Phys. Lett.* **68A**, 465.

Stark, J. (1912) *Phys. Z.* **13**, 973.

Swanson, M.L., Howe, L.M. and Quenneville A.F. (1980) *Phys. Rev.* **B22**, 2213.

Swanson, M.L. (1982) *Rep. Prog. Phys.* **45**, 47.

Thompson, M.W. and Nelson, R.S. (1963) *Phil. Mag.* **8**, 1577.

Tognetti, N.P. (1981) *Rad. Effects Lett.* **58**, 151.

Townsend, P. and Kelly, J.K. (1973) 'Colour Centres and Imperfections in Insulators and Semiconductors', Sussex Univ. Press, Brighton.

Wagh, A.G. and Williams, J.S. (1978) *Phys. Lett.* **65A**, 235.

Wagh, A.G., Bhattacharya, P.K. and Kansara, M.J. (1981) *Nucl. Instrum. Methods* **191**, 96.

Walker, R.S., Thompson, D.A. and Poehlman, S.W. (1977) *Rad. Eff.* **34**, 157.

Whitton, J.L., Ferguson. M.M., Ewan, G.T., Mitchell, I.V. and Plattner, H.H. (1982) *Appl. Phys. Lett.* **41**, 150.

Wielunska, D., Wielunski, L. and Turos, A. (1981a) *Phys. Stat. Sol. (a)* **67**, 413.

Wielunska, D., Wielunski, L. and Turos, A. (1981b) *Phys. Stat. Sol. (a)* **68**, 45.

Wielunski, L.S., Hashimoto, S. and Gibson, W.M. (1986) *Nucl. Instrum Methods* **B13**, 61.

Wiggers, L.W. and Saris, F.W. (1979) *Rad. Eff.* **41**, 141.

Williams, J.S. (1973) Ph.D. Thesis, Univ. of NSW.

Williams, J.S. (1983) *In* 'Surface Modification and Alloying', (J.M. Poate, G. Foti and D.C. Jacobson, eds), Plenum, New York, p. 133.

Williams, J.S. and Short, K.T. (1982a) *Mat. Res. Soc. Symp. Proc.* **7**, 109.

Williams, J.S. and Short, K.T. (1982b) *J. Appl. Phys.* **53**, 8663.

7

Depth Profiling of Surface Layers during Ion Bombardment

R.J. MACDONALD
B.V. KING

Department of Physics, University of Newcastle,
Newcastle, Australia

ION BEAMS FOR
MATERIALS ANALYSIS
ISBN 0 12 099740 1

7.1 INTRODUCTION

Many of the surface properties of materials depend on both the elemental composition and the structure of the surface layer. It is often essential to have some knowledge of the spatial distribution of specific elements both in the surface plane and as a function of depth normal to the surface. This chapter will concentrate on ways of analysing spatial distributions of near-surface atoms using ion beam sputtering. The sputtering beam (usually < 10 keV) can itself generate the analysis signal through the use of SIMS, SIPS, SNMS or LEIS. Alternatively, surface analysis probes such as AES, XPS or laser ionisation techniques can be used in conjunction with sputtering. The depth resolution is determined by a combination of:

- bulk effects such as radiation damage and enhanced thermal diffusion;
- surface effects such as preferential sputtering;
- the formation of surface topography;
- the escape depth of the detected species; and
- instrumental effects.

Equipment requirements for sputter profiling are described in Chapter 2 and the production and use of ion microbeams are considered in Chapter 10. This chapter will provide an overview of the principles of sputter profiling and its limitations.

7.2 ION–SURFACE INTERACTIONS

Sputtering from surfaces has been observed since the early days of gas discharge experiments but there was a major upsurge in interest in the early 1950s when it was realised that ion beams could be used to simulate and speed up radiation damage processes of the type important in fission reactors. Later, the importance of the plasma-first wall interaction in the operation of fusion machines was realised. From these studies, a number of ion-bombardment based surface analysis tools have been developed. The theoretical modelling of ion–surface interactions involved in analysis methods such as SIMS, SIPS and LEIS has not proceeded as quickly. For this reason, most profiling applications have been based on comparison with standards or are concerned with relative rather than absolute changes in elemental concentration.

The interaction between a low energy ion and surface atoms is introduced in Chapter 1.1.3. Both scattering and secondary ion production give information on the elemental composition of the surface. Low

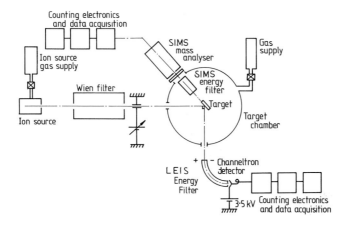

Fig. 7.1 Schematic diagram of a typical ion source and analysis system, complete with energy analyser for LEIS and SIMS.

energy ion scattering (LEIS) is discussed in Chapter 8. Both the ion which penetrates and the surface atom which is knocked-on as a result of a scattering event at the surface, initiate collision cascades in the solid which dissipate the incident ion energy within the lattice creating damage. Some cascades may intersect the surface with sufficient energy to cause ejection of atoms from the surface. This is sputtering and is the basis of sputter profiling. The sputtered atoms may be neutral, excited or ionised. If neutral, sputtered atoms may be subsequently ionised or excited by collisions with electrons or ions outside the solid as in the analysis technique called SNMS (Oechsner and Stumpe, 1977) or by photon absorption as in the laser induced fluorescent techniques (Pellin *et al.*, 1981; Husinsky, 1985). If sputtered atoms are ionised or excited (or both) in the passage through the surface they form the basis of SIMS. Sputtered excited atoms can emit photons due to radiative decay outside the surface and this is the basis of SIPS.

Most of the information available from these methods of analysis concerns the elemental composition of the surface. In SIMS, the mass spectra usually include a number of molecular ion species. The molecular ion yields are related to the proximity of atoms to each other and contain information of a statistical nature on the chemical state of the surface. Both positively and negatively charged ions are formed and the mass spectra may contain molecular species which include elements of the target and of contaminants. The analysis system must at least be able to distinguish between ions of the form M^+ and MH^+. In order to evaluate methods of analysis utilising these emissions it is necessary to have some idea of their energy and angular distribution and also of the mechanisms responsible for the ionisation, excitation and survival of each species. We consider first the general description of sputtering and then more explicit models of ionisation and excitation. Fig. 7.1 shows a schematic of the system used in our laboratory for ion-surface studies. Addtional information on low energy ion beam production and detection is given in Chapter 2.

7.3 THE SPUTTERING EVENT

In physical sputtering (Thompson, 1981), the basic event is an atomic collision cascade. Atoms of the solid are knocked off equilibrium sites and move through the lattice undergoing further collisions and causing further displacements. Eventually, atoms are ejected from surface sites. At very low energies of incidence, the ion will not impart enough energy to lattice atoms to initiate sputtering. However, under certain conditions, a chemically active ion species forms an unstable compound with the surface and surface layers are removed in a chemical sputtering process (Sanders *et al.*, 1984; Kolfschoten *et al.*, 1984). Such a process has not been used in sputter profiling but is used to increase the rate of sputter etching in the semiconductor industry. It will not be considered further here.

The main parameter of interest to sputter profiling is the sputtering yield — the mean number of atoms of the target ejected per incident ion. The yield determines the rate of removal of surface layers as a function of ion fluence. Another factor of interest is the depth of origin of the sputtered atom. If the atom comes from the outermost surface, the surface will be sectioned uniformly. If it comes from below the surface, the sectioning will not be ideal. In the latter case the SIMS or SIPS signal might not be representative of the surface concentration, but rather of a concentration averaged over the escape depth. Finally, the deposited energy distribution is important as this will determine the depth to which the cascade penetrates and consequently the depth to which atoms are put into motion. This is a measure of the depth to which the composition may have been altered by atomic mixing or radiation enhanced diffusion. Other important parameters are the energy and angular distributions of sputtered particles. These are most important in the measurement of the profile using SIMS or SIPS.

Whatever the mechanism of ionisation or excitation, the energy distribution $N'(E)$ of ionised or excited atoms of a particular species will be related to the energy distribution $N(E)$ of sputtered neutral atoms by an ionisation or excitation cross-section according to:

$$N'(E) \, dE = \sigma(E) \, N(E) \, dx \qquad (7.1)$$

The physical sputtering yield becomes measurable only for incident ion energies greater than about 20 eV. Above this threshold, the sputtering yield increases with ion energy to a value dependent on the ion–target atom combination and the angle of incidence of the ion. It is typically in the range from 1 to 10 atoms ejected per incident ion, for medium mass (Ne, Ar) ions incident with energies of the order of 10 keV. Above this energy the sputtering yield slowly decreases. Compilations of sputtering yield data are available (Matsunami *et al.*, 1984).

This form for the sputter yield arises from the energy dependence of the cross-section for elastic scattering between the incident ion and the target atom. The cross-section of most use in collision cascade theory is the differential cross-section for a particle of incident energy E_1 to produce a scattered target atom with transferred energy T:

$$T = \gamma \, E_1 \cos^2\theta = 4M_1 M_2 E_1 \cos^2\theta \, / \, (M_1 + M_2)^2 \qquad (7.2)$$

where M_1 is the mass of the incident ion, M_2 the mass of the scattered (target) atom and θ is the scattering angle. For the energy ranges of interest to sputter profiling, the interaction potential between the scattering particles is of the form:

$$V(r) \propto r^{-1/m} \qquad (7.3)$$

where r is the distance between colliding particles and m varies from $m = 1$ (high energies where the interaction is Coulombic) to $m = 0$ (very low energies where the collisions are something like hardsphere collisions).

The differential scattering cross-section $d\sigma(E_1,T) \, / \, dT$ is a function of E_1, T, M_1, M_2, Z_1 and Z_2 and may be approximated by:

$$d\sigma(E_1, T) \simeq c_m E_1^{-m} T^{-(1+m)} \, dT \text{ for } 0 < T < T_m \qquad (7.4)$$

where:

$$c_m = (\pi/2) \, \lambda_m \, a^2 \, (M_1/M_2)^m \, (2Z_1 Z_2 e^2/a)^{2m} \qquad (7.5)$$

$T_m = \gamma E_1$ is the maximum energy transfer; Z_1 and Z_2 are the atomic numbers of the incident and scattered particles; a is a screening radius associated with the interatomic potential, Equation (1.4), and λ_m is a dimensionless function of m which varies from 0.5 ($m = 1$) to 24 ($m = 0$).

The total cross-section for scattering is:

$$\sigma = \int d\sigma(E_1, T) \qquad (7.6)$$

which will usually be a divergent quantity. However, limits may be imposed by the fact that we require atoms to be displaced from lattice sites to contribute to the collision cascade and that the maximum energy transfer is T_m. The limits on the integral may be taken from some low value corresponding to the displacement energy ($E_d = 10$ to 25 eV) to T_m. At low energies the scattering cross-section is large but the average energy transfer is small. As the energy increases the cross-section decreases and the average energy transfer increases. The former means the mean free path to the first collision in the solid will increase, but the amount of energy deposited close enough to the surface to produce sputtered atom ejection will also increase. The sputtering yield therefore increases. As the cross-section decreases further (increasing energy), the first collisions leading to cascade formation occur deeper in the target and there is less chance of the cascade reaching the surface to produce sputtering. Thus at high energies the yield will fall.

The sputtering yield may also be expressed in the terms of the mean energy $\overline{\Delta E}$ transferred in elastic collisions over a path length x:

$$\overline{\Delta E} = N x \int T d\sigma(E_1, T) = N x \, \varepsilon_n(E) \qquad (7.7)$$

where N is the target atom density and $\varepsilon_n(E)$ is the nuclear stopping cross-section (see chapters 1 and 14). The sputtering yield is approximately proportional to ε_n and can be approximated using the linear cascade model of Sigmund (1969).

The region of importance for determination of the sputtering yield is the region in which Sigmund (1969) developed a linear cascade model. In this model, linear means that a struck atom of the solid is stationary before the collision occurs. The cascade is treated using the Boltzmann transport equation, with the cascades consisting of a series of binary collisions governed by the cross-sections and interatomic potentials outlined above. A variety of assumptions, including no inelastic loss and isotropic motion within the cascade leads to the backsputtering yield for normal incidence, $Y(E)$, being given by:

$$Y(E) = 0.042 \, \alpha(M_1/M_2) \, \varepsilon_n(E)/E_b \qquad (7.8)$$

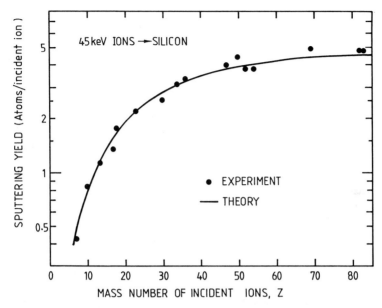

Fig. 7.2 Sputtering yield of Si due to bombardment with a variety of ions.

where E is the energy of sputtering ions, $\alpha(M_1/M_2)$ is a function of the mass ratio only and E_b is the surface binding energy. Agreement with experiment can be very good, as illustrated by the results of Andersen and Bay (1975a) for the sputtering of Si with various 45 keV ions (Fig. 7.2). The $\alpha(M_1/M_2)$ function can, for reasonably small values of M_2/M_1 (say < 5), be approximated by a universal function, as illustrated in Fig. 7.3 (Andersen and Bay, 1975b). For sputter profiling, this universal function is suitable for almost all experimental situations.

The assumption of isotropic motion in the cascade implies that the sputtered material should be distributed according to a cosine law, even in cases where the incident ion beam is not incident normal to the surface. In practice a preferred emission in the forward direction is observed for non-normal incidence. The cascade theory of Sigmund (1969) gives the energy spectrum of sputtered atoms in the form:

$$N(E) \, dE \propto E \, dE \, / \, (E + E_b)^{(3-2m)} \tag{7.9}$$

which has a maximum at $E_b/2(1-m)$. When the sputtered atoms are the result of predominantly low energy collisions ($m = 0$), the maximum value of $N(E)$ occurs at $E = E_b/2$ (Fig. 7.4). There is good agreement

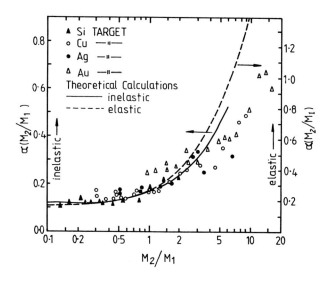

Fig. 7.3 Experimental values for the factor $\alpha(M_2/M_1)$ of Equation 7.8.

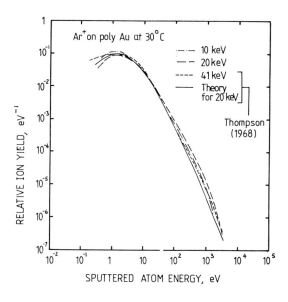

Fig. 7.4 Experimental measurement of the energy spectra of atoms sputtered under various conditions. The results are compared with the theory of Thomson (after Thomson, 1968).

between theory and experiment for the shape and peak position of sputtered energy distributions provided that $M_2 < 2M_1$.

The depth of origin, Δx, of the sputtered atom can also be extracted from the linear cascade theory. Very importantly, the higher the energy of the sputtered atom the more likely it is to come from deep inside the solid. Thus, when using an excited or ionised sputtered atom to monitor the composition, it is more indicative of surface composition when the low energy part of the sputtered particle emission is sampled. If the sputtered energy, E, is of the magnitude of E_b, Δx is of the order of 1 nm.

7.3.1 Momentum Focussing and Channeling in Single Crystal Lattices

The sputtering theory outlined above refers to amorphous targets but is also applicable to polycrystalline, elemental targets with very small grain size. In materials science applications of sputter profiling, most of the targets, particularly metals, will be of this type. In applications to semiconductors, the material to be analysed is a single crystal, although ion bombardment of most semiconductor targets at low temperature results in the surface layer becoming amorphous (MacDonald, 1970). A single crystal lattice is characterised by long range order and this property affects the way the deposited energy is distributed through the lattice. Two effects dominate the observations namely focusing of the collision cascade to produce preferential ejection along low index directions in some single crystal lattices (Fig. 7.5a) and channeling of the incident ion to produce the collision cascade more deeply inside the target than would be expected from a non-periodic lattice, with consequent reduction in the sputtering yield (Fig. 7.5b).

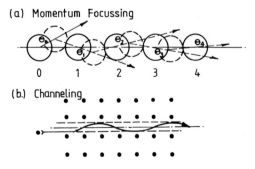

Fig. 7.5 Collision effects associated with an ordered atom structure. a. Momentum focussing b. Channeling.

The collision sequence shown schematically in Fig. 7.5a leads to momentum focussing resulting in an anisotropic distribution of sputtered ions (Dennis and MacDonald, 1972). If SIMS is used to monitor the elemental concentration of interest, exaggerated concentration values might result. Channeling effects on particle yield are also evident in the secondary ion and photon yields (Onderlinden, 1966 and 1968; Zwangobani and MacDonald, 1973). Channeling is likely to distort the conversion of the profile to a depth scale; use of polycrystalline sputtering yields distorts the scale badly if the incident beam is inadvertently channeled in a single crystal target.

7.3.2 Surface Topography Changes

Unfortunately, the surface does not retreat uniformly under the influence of sputtering and the ion bombarded surface shows a wealth of topographical features associated with sputter erosion (Carter *et al.*, 1985). The mechanisms for the initiation of the often dramatic array of surface cones, pits and contour changes are not fully understood. Some at least are due to impurity clusters which shelter the underlying target regions. If the impurity has a lower sputtering yield the net result is a protuberance on the surface; if the sputtering yield of the impurity is higher the net result is a pit. However, there is no simple criterion which would allow one to estimate the extent to which a surface might depart from the topographic ideal. When considering the depth resolution of sputter profiling, the effect of surface features developed as a result of ion bombardment can be crucial. The authors' experience suggests that wide scale features are observed more often when the bombardment involves active gas either in the beam or with an inert gas ion beam in a chemically active environment. Since these conditions often exist when using SIMS or SIPS to monitor changes in surface composition, the experimentalist must always consider the possibility of surface topographic features affecting the results. The SEM and a surface profilometer are valuable aids for this purpose.

7.4 SPUTTERING OF MULTICOMPONENT TARGETS — PREFERENTIAL SPUTTERING

The effects described in the previous two sections also occur in compound targets. In addition, there is a more complex nuclear energy loss situation, since the atoms participating in the cascade are of different mass. Even in an elastic collision, there may be significantly different distributions of available energy over the constituents of the cascade. Of most importance, is the possibility that the sputtering might be aniso-tropic with respect to the target components, i.e. that the differential or total distributions of sputtered material might not reflect the constitution of the target. This would be accompanied by a change in the composition of the surface layer from that of the bulk. The situation may be further complicated by the fact that in some compounds, the binding energy of a molecule to the surface may be considerably less than that of individual atoms e.g. in GaAs (Szymonski and Bhattacharya, 1979).

Andersen (1984) has reviewed the sputtering of multicomponent alloys and compounds concentrating on the question of the possible production of non-stoichiometric layers on a surface, as a result of preferential sputtering of one component. The considerations involved are summarised in Highlight 7.1. Preferential sputtering should be taken into account in interpreting sputter profiles whether the composition of the sputtered flux or the composition of the sputtered surface is measured. The sputtered flux can be collected and subsequently analysed by a method such as RBS or a component of the flux may be monitored during sputtering using SIMS, SIPS, etc. The ionisation or excitation mechanisms may perturb the results, e.g. the ionisation yield is related to the ionisation potential, so for components of differing ionisation potential the signal strengths may differ significantly for equal concentrations.

If the surface composition is measured, the information depth becomes important. Techniques which average over many layers are obviously not as useful for determining surface concentration as techniques sensitive to the composition in only one or two layers. Discrepancies between reported results can often be traced to differences in the information depth factor for different techniques. The most common experimental techniques are:

- LEIS — information depth one or two layers;
- SIMS — information depth one or two layers; and
- AES — information depth one or two nm.

Linear cascade theory (Sigmund, 1969) suggests that the depth of origin of sputtered atoms is equivalent to only the uppermost layers of the surface. Unless there are other factors contributing, such as radiation enhanced diffusion, composition changes should only occur in the outermost few layers, even though considerably more than this depth must be removed before equilibrium is established.

The main factor in Equation (7.20) is the ratio of the binding energies so that changes in isotope ratios at the bombarded surface, arising from the mass ratio term should be small. One recent study (Cox, 1982)

HIGHLIGHT 7.1
PREFERENTIAL SPUTTERING

When the target consists of a mixture, compound or alloy, effects such as momentum focussing, preferential ejection, channeling and the development of surface topography may influence the composition of the sputtered flux relative to the surface composition. Andersen (1984) defined preferential sputtering as a difference between the composition of the flux of sputtered particles (averaged over the emitted directions) and the composition of the outermost layers of the sample. This gives rise to an enhancement factor defined by Equation (7.10), Table 7.1. All observations indicate that preferential sputtering is a transient phenomenon varying between the initial surface concentration and an equilibrium concentration at large fluence. An equilibrium concentration different from that of the bulk solid can only be maintained by the sputtered flux having a composition equal to that of the bulk, Equation (7.13). This requires a lowering of the surface concentrations of the more easily sputtered components. The enhancement factors at time zero, Equation (7.12), and equilibrium, Equation (7.14), are only the same if they are independent of surface composition — in which case Equation (7.15) applies. At least one experiment (Goto *et al.*, 1978) indicates that the sample must be at least five times thicker than the depth required for the transient effect to reach equilibrium and the equations of Table 7.1 to apply. Thinner samples are gradually depleted of the preferentially sputtered components. Andersen (1984) and Betz (1980) provide a comprehensive list of preferential sputtering surface composition experiments.

---

describes experiments using medium energy ion scattering to study changes in the isotope concentration ratios in elemental samples. Changes of up to 10% of the bulk ratio were often detected, though the experimental results were not corrected for possible effects of different neutralisation rates associated with scattering from different isotopes (MacDonald and O'Connor, 1983). Nevertheless there are strong indications of isotope ratio changes considerably in excess of those predicted by Equation (7.20).

Okano *et al.* (1985) have studied changes in isotope ratios in several targets, due to sputtering with 12 keV O_2 beams. Changes of only a few per cent were observed, with the transition to the equilibrium ratio taking some tens of minutes corresponding to the removal of hundreds of atomic layers. They interpret their results not as preferential sputtering but as preferential recoil implantation of the lighter isotope. This

TABLE 7.1
Preferential sputtering relations

Enhancement Factor:	$f_{AB} = (Y_A/Y_B)/(c_A^s/c_B^s)$	(7.10)
For a homogeneous sample:		
At time zero:	$c_A^s/c_B^s = c_A^b/c_B^b$	(7.11)
	$Y_A^0/Y_B^0 = f_{AB}(c_A^b/c_B^b)$	(7.12)
At equilibrium:	$Y_A^\infty/Y_B^\infty = c_A^b/c_B^b$	(7.13)
so that:	$c_A^{s,\infty}/c_B^{s,\infty} = (c_A^b/c_B^b)/f'_{AB}$	(7.14)

and, if the enhancement factor is independent of surface concentration (i.e. $f'_{AB} = f_{AB}$):

$$c_A^{s,\infty}/c_B^{s,\infty} = (Y_B^0/Y_A^0)(c_A^b/c_B^b)^2 \qquad (7.15)$$

On the very dubious assumption that:

$$Y_A^0/Y_B^0 = [Y(A)/Y(B)](c_A^b/c_B^b) \qquad (7.16)$$

then,

$$c_A^{s,\infty}/c_B^{s,\infty} = [Y(B)/Y(A)][c_A^b/c_B^b] \qquad (7.17)$$

and since $c_A^b + c_B^b = 1$,

$$= [Y(B)/Y(A)][c_A^b/(1 - c_A^b)] \qquad (7.18)$$

So that, if $c_A^b \ll 1$, we have approximately:

$$c_A^{s,\infty}/c_B^{s,\infty} = [Y(B)/Y(A)]c_A^b \qquad (7.19)$$

Using linear cascade theory, Sigmund (1974) predicts that:

$$f_{AB} = (M_B/M_A)^{2m}(E_{bA}/E_{bB})^{(1-2m)} \qquad (7.20)$$

where m is in the range 0 to 0.2 and E_{BA}, E_{bB} are the bulk binding energies of the two components.

c_A and c_B are the fractional concentrations of components A and B which give sputter yields Y_A and Y_B; superscripts: s = surface, b = bulk, 0 = zero time, ∞ = equilibrium.

example highlights the uncertainty which can arise in developing definitive experiments to determine which processes contribute to the equilibrium surface concentration. The situation is complicated even further by the fact that many experiments use a sputtering beam whose energy is typical of sputter cleaning and etching ($\sim$ 1 keV). In such cases, there is doubt that the sputtering event is then truly in the linear cascade region. The results are often interpretable simply in terms of single ion–atom collision events, as in the work of Taglauer and Heiland (1978) on heavy metal carbides.

Andersen (1984) makes the very important point that, while experiments do measure differences between surface and bulk concentrations, these differences are not necessarily due to preferential sputtering. This is also shown by the work of Okano *et al.* (1985). Firstly, we would expect purely on thermodynamic grounds that equilibrium concentrations in the bulk might not be the same as those at the surface. Secondly, there are a host of solid state effects which can modify the surface concentration during irradiation. These include radiation enhanced diffusion, segregation of one component, implantation of the incident ion and recoil sputtering of one component in preference to another. For most cases of depth profiling, it is only necessary to be aware of these possibilities; only if quantitative concentration profiles are required is it necessary to attempt to correct for such effects. Recoil implantation is a problem when attempting to profile small concentrations of one element in another. There are also good examples in the literature of cases in which radiation enhanced diffusion has actually moved the concentration distribution about in the solid (e.g. Hart *et al.*, 1975). These effects are considered further in later sections but corrections are, in many cases, very difficult.

7.5 ANALYSIS OF ELEMENTAL CONCENTRATIONS

In order to obtain the concentration of a given element as a function of depth in the solid, it is necessary to combine sputter profiling with a probe which measures the surface concentration of a given element. Few such probes, however, provide absolute measurements of the elemental concentration or are sensitive to the few atomic layers we have indicated as the best depth of origin for the probe signal.

The most commonly used methods are:

- SIMS and SIPS which utilize the sputtering process to provide the analysis signal;
- Electron based techniques such as AES, UPS or XPS which, as well as providing elemental concentration profiles can also detect changes in near-surface chemical bonding; or
- LEIS, MEIS and RBS, which use the scattering process to provide the analysis signal; the depth of origin of the signal ranges from about one atomic layer in LEIS to microns in RBS.

RBS does have the advantages of quantitative analysis and freedom from recoil mixing or radiation enhanced diffusion (Myers, 1980) but it will not be considered further in this chapter.

7.5.1 SIMS

The monitoring signals provided by the sputtering ion beam, through SIMS and SIPS, are obviously attractive for use in compositional profiling. Secondary ion and photon yields must be related to the mechanisms for ion formation and excitation but these are not well enough understood to enable truly quantitative measurements to be made of elemental concentrations using SIMS and SIPS. Further, the target matrix and the extent of active gas absorption during ion bombardment affect the ion and photon yields markedly, so that one must also be very careful in evaluating relative signal strengths in terms of changes in the elemental concentration unless the experimental conditions are well controlled.

a. Angular and Energy Distributions of Secondary Ions

Very little information exists in the literature on the angular distribution of secondary ions from polycrystalline targets. One would expect that the angular distribution would be similar to that of the sputtered atoms, but modified by the angular dependence of the ionisation and neutralisation processes occurring along the outgoing path. The angular distribution of the ionisation follows an approximately cosine function with angle to the normal. The probability of post-ionisation neutralisation would be expected to increase with time spent in the vicinity of the surface and hence will have little effect on the angular distribution for angles close to the normal. For single crystal target materials, preferential ion ejection similar to that observed for neutrals is observed in sputtered ion distributions. Channeling effects are also observed. The theoretical variation of sputtered atom yield with angle of incidence about the low index direction, can be used with only small adjustments to also describe the variation of secondary ion yield (Onderlinden 1966, 1968; Zwango-bani and MacDonald, 1973).

A SIMS spectrum contains elemental ions and may include considerable yields of molecular ions which have a much narrower distribution of energy than the elemental ions. The energy distribution decreases in width with increasing number of atoms in the molecular ion and this can be used to distinguish between elemental and molecular peaks in the spectrum.

In the case of single crystals, the effect of focussed ejection along low index directions can be seen in sputtered ion energy spectra (Bukhanov *et al.*, 1970; Dennis and MacDonald, 1972). These results formed the basis for an early phenomenological attempt to model the dependence of the ionisation coefficient on ejected ion energy (MacDonald, 1974). Most studies of the energy spectra of secondary ions have been carried out on polycrystalline targets, using a variety of different methods. Each method requires some correction for ion optical effects. As a result, quantitative comparisons of energy spectra obtained in different laboratories are difficult. Wittmaack (1982a) has recently reviewed most of the common configurations of equipment used for SIMS work, particularly those using quadrupole mass spectrometers.

In general, the energy spectrum obtained from a material such as an alkali halide or an alkali metal sputtered from a metal surface has an inverse square dependence on energy above 10 eV i.e. similar to the energy spectrum of sputtered atoms (Bayly and MacDonald, 1977). For metals and other targets, neutralisation effects alter the energy distribution parameters to the form:

$$N(E)\, dE \propto E^{-n}; \qquad 1 \le n \le 2 \tag{7.21}$$

The form of the sputtered ion energy spectrum can change with time. In UHV, as the surface of the transition metals is cleaned, the yield in the low energy part of the spectrum decreases much more quickly than that in the high energy part, and the result tends towards a spectrum peaking at some hundred eV or so, with a very low yield at low energy (Bayly and MacDonald, 1977). Light elements such as Al and Mg do not show this behaviour, retaining instead the low energy peak tailing as E^{-n}. The reason for this difference in behaviour is not known.

The behaviour of the energy spectrum is important because most SIMS machines operate with a window to pass only ions of energy ($E \pm \Delta E$). If this window is of the order of 10 eV, changes in the observed ion yield may be due to changes in elemental concentration or changes in the form of the energy spectrum. This possibility must always be considered, when interpreting SIMS data.

b. Neutralisation Effects in SIMS

The yield of secondary ions in the energy range E to ($E + dE$) and the angular range θ to ($\theta + d\theta$) from the surface normal can be written:

$$Y(E,\, \theta)\, dE\, d\theta \propto N(E)\, \mathrm{f}(\theta)\, R(E_1,\, \theta)\, F(E_1,\, \theta)\, dE\, d\theta \tag{7.22}$$

where $N(E)\, dE$ is the energy spectrum of sputtered atoms, $f(\theta)\, d\theta$ is the angular distribution of sputtered atoms, $R(E,\theta)$ is the probability that the atom sputtered in the energy range E to ($E + dE$) and angular range θ to ($\theta + d\theta$) will be ionised and $F(E,\theta)$ is the probability that the ion ejected in the same energy and angular ranges survives its passage to the detectors in an ionised state. We have assumed that the ionisation process has a negligible effect on the energy of the sputtered atom. Models of neutralisation probabilities, as discussed in Chapter 8.2.2, give a qualitative picture of the type of process likely to have an important influence on the yield of secondary ions. Many experimental observations in both secondary ion emission and low energy ion scattering confirm the importance of neutralisation processes and allow the various models to be verified.

Alteration of the surface electronic structure is sometimes cited as the cause of sputtered ion yield variations. The alteration may be due to adsorption of active gases on the metal surface, changes to the electron exchange process or a combination of both. Adsorption can produce yield changes of up to two orders of magnitude even when the total adsorbed layer coverage is less than one monolayer. It is difficult to accept that this is due to the small changes in electronic structure associated with such adsorption. However, UPS shows that the adsorbate often perturbs the electronic structure by localising some of the electron distribution into molecular bonding states a few eV below the Fermi level. This must correspond to a localised electron density at the adsorbate site which may enhance the yield associated with electron–atom collisions or increase inelastic loss to electron excitation in the adsorbate layer. An alternative explanation involves excitation or ionisation of the atoms during quasimolecular formation and subsequent dissociation either in the collision or during the ejection process (Tsong, 1981).

c. Models of Secondary Ion Emission

There are a variety of models of the ionisation process occurring during sputtering. In the case of ejection from a clean surface, they range from descriptions of the bombarded region of the surface as a plasma in local thermodynamic equilibrium (Anderson and Hinthorne, 1973) to models involving inelastic excitation of particles in the final collisions leading to ejection (Newns *et al.*, 1985). Models attempting to account for the enhancement effects of active gas adsorption or bombardment usually involve 'bond-breaking' concepts (Tsong, 1981). Most uses of SIMS involve relative changes in ion yields, so that knowledge of the exact ionisation mechanism is not important, provided the mechanism does not change in the course of profiling. It is much more important to be aware of those factors which can contribute to a change in the individual ion yield and, in depth profiling, to ensure that changes in ion yield are related only to changes in elemental concentration and not to any other factors. Exact models of the ionisation event will only be required when absolute quantitative analysis is developed.

7.5.2 SIPS

The photons emitted and detected during sputtering in general come from the region in front of the target not from within the target. Photons are emitted from inside the target but their yield is small and they are not characteristic of the elements being sputtered. The photons are detected using an optical monochromator and photomultiplier in conjunction with a sputter profiling system (Fig. 2.2).

a. Photon Yields from Sputtered Atoms

The photon spectrum resembles very closely that observed from an arc in thermal equilibrium and conforms to a Boltzmann distribution function over excited states (Martin and MacDonald, 1977). Photon intensities are small and of similar magnitude to those of sputtered ions (Chapter 14.7). This has led to the suggestion that ionisation and excitation are due to the same process, though some experimental results are interpreted as refuting this suggestion (Williams *et al.*, 1980). The observed photon yields are subject to the same or very similar electron exchange processes as those believed to affect the secondary ion yield. The yields are enhanced by active gas adsorption at the surface, in a similar way to the ion yields, the results being used to distinguish between models of excitation and ionisation (Dzioba *et al.*, 1980). The angular distributions of atoms sputtered in excited states, leading subsequently to photon emission, have not been studied. However, some aspects of energy distributions of sputtered excited atoms have been the subject of experimental investigations showing that their energy distribution is not very different to the energy spectrum of all sputtered atoms.

7.5.3 Instrumental Effects When Using SIMS or SIPS for Depth Profiling

Accurate correlation of signal strength with uniform layer removal using sputter profiling is usually obtained by rastering the focussed beam across an area of the target and arranging for the detection system to accept only the signal from a central region of the rastered area. This 'electronic gating' is almost universally used in sputter profiling, particularly as it can be adapted to produce elemental image maps of the bombarded and gated region. Fig.7.6 illustrates the effect of electronic gating on a depth profile.

The determination of a depth profile requires the conversion of a signal variation with time into a signal variation with depth, a conversion which requires a knowledge of the sputter yield from the gated region of the rastered area. If the sputter yield is constant, the conversion can be obtained from the depth of the crater at the end of the profiling run. If the sputter yield changes with concentration, as well it might in layered structures, oxide–metal interfaces or ion-implanted layers, the depth

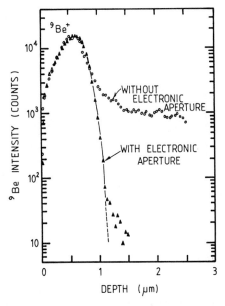

Fig. 7.6 An illustration of the effect of rastering and electronic gating on the depth profiling of a Be implant in Si. Without gating, the edge effects of the beam profile dominate the results (after Wittmaack, 1982b).

profile is considerably harder to quantify. There are several other factors affecting the accuracy of sputter profiling, particularly when SIMS or SIPS are used to monitor changes in elemental concentrations. Many of the possible effects have been discussed by Wittmaack (1981, 1982b, 1985a and 1985b).

a. Enhancement of SIMS and SIPS Signals

Enhancement of the SIMS or SIPS signal may be achieved by bombarding in an active gas environment or bombarding with active gas ions. In order to use the enhanced signal effectively, the enhancement must be constant and this requires that the concentration of adsorbate or implanted active gas must be constant over the profiled depth. Using an active gas beam it is necessary to build up an equilibrium concentration of implanted atoms in the near-surface region. Until equilibrium is reached the enhancement factor varies with adsorbate coverage and so does the signal strength. A similar, but not as serious variation, will be observed when an active gas environment is used. In this case, the recoil implant distribution must stabilise. If the elemental profile is much

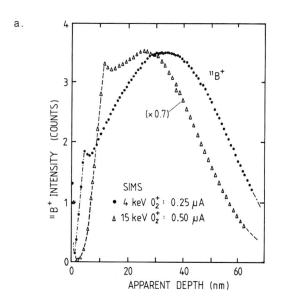

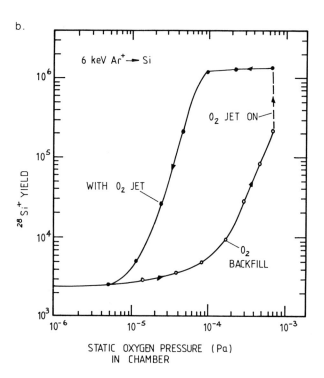

deeper than the implant profile, the initial variation might not be important.

Fig. 7.7a (Wittmaack, 1982b) shows an example of the effect of using O beams of different energies to profile identical samples. The differences in the depth profile of B implanted in Si are associated with changes in the sputtering yield due to implantation of O and to differences in the enhancement of signal strength due to the different equilibrium concentrations of O at the surface. The higher energy O is implanted deeper and takes longer to reach equilibrium, i.e. a thicker layer must be sputtered to reach saturation. A second example (Fig. 7.7b) shows the change in Si^+ yield as a function of partial pressure of O_2 surrounding the Si sample (Wittmaack, 1982b). At least an order of magnitude lower chamber pressure can be used to achieve the same enhancement with an O_2 jet rather than a static presssure. However, the enhancement depends on the composition of the sample, being ~ 100 times greater for Si than GaAs.

b. Bombardment with Neutrals Formed by Beam Interaction with Residual Gas

Focussing, or rastering by electrostatic means is only effective for the charged component of the incident flux. If there is a neutral component, it is not possible to control it electrostatically and bombardment of the central spot with neutrals persists. The neutral beam thus produces a signal irrespective of the rastering. The neutral beam affecting depth profiling most will be that component formed by charge exchange between the incident ions and background gas atoms or molecules after the last beam deflection. The neutrals will be energetic atoms of the beam, or alternatively energetic atoms of the background gas formed by recoil from ions of the incident beam. The background gas could be the gas used to charge the ion source or active gas (e.g. O_2) introduced to produce enhancement of the secondary ion signal. The effects of neutrals formed near the ion source can be reduced by differential pumping. The neutrals formed from and by interaction with the active gas introduced

---

◀ **Fig. 7.7** a. An illustration of the effect of the beam energy on the SIMS measurement of a B implant profile. The differences in measured profile relate to the different concentration profiles of the implanted ion and the effect of that concentration on the SIMS signal (after Wittmaack, 1982b). b. An example of the effect of adsorbate concentration at the surface on the secondary ion emission enhancement. The enhancement due to a molecular beam of adsorbate is compared with adsorption from a background gas consisting of the adsorbate (Wittmaack, 1982b).

for enhancement of the signal are much harder to avoid and it might be better not to use such a method for enhancement. The effect of neutral bombardment as a result of neutralisation along the ion beam path can also be greatly reduced by using a small dog-leg in the beam transport system just before the sample.

c. Variation of Signal Strength due to Impurities in the Beam and Background Gas

Signal enhancement may also be inadvertently introduced by impurities in the beam or in the residual gas environment. In some cases, e.g. profiling for H or O, adsorption of molecules from the background and subsequent implantation or ionisation at the surface may induce high background levels and distorted profiles. Improved residual vacuum can minimise such effects.

7.5.4 SIMS Detection Capabilities

In sputter profiling, the ultimate performance is determined by the ion, electron or photon yields and instrumental limitations such as the current densities available, spot size and detection efficiency. Instrument performance is discussed in Chapter 2 and beam optics are dealt with in more detail in Chapter 10. The effect of these on performance is discussed in Highlight 7.2.

HIGHLIGHT 7.2
DETECTION CAPABILITIES IN SPUTTER PROFILING

Three equations which are important to the capabilities of sputter profiling are:

i. *Available Current Limit*
Chromatic Aberration dominant:

$$I_c = (\pi^2 \beta d^4 / 16 C_c^2) \, (\Delta E/E)^2 \qquad (7.23)$$

Spherical Aberration dominant:

$$I_s = 3\pi^2 \beta d^{8/3} / 16 C_s^{2/3} \qquad (7.24)$$

where β is the source brightness, E is the beam energy, ΔE is the energy spread of the source, C_c and C_s are the chromatic and spherical aberration coefficients of the imaging system.

a.

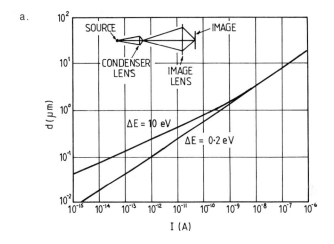

b.

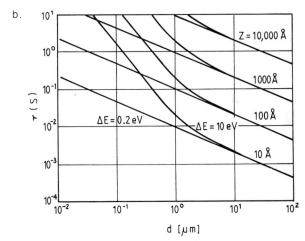

Fig. 7.8 a. Optimum current density as a function of beam diameter in a conventional ion source system. Data are shown for a beam energy spread $\Delta E = 0.2$ eV and $\Delta E = 10$ eV; b. the time τ required for a beam of the given diameter to sputter to the depth indicated on the individual curves (Welkie and Gerlach, 1982).

ii. *Time to Sputter to Depth z*

$$\tau = z\rho/JY \qquad (7.25)$$

where J is the current density, ρ is the bulk atom density and Y is the sputtering yield; τ can be evaluated for the ion source used. Following Welkie and Gerlach (1982), we can use typical values of the quantities in

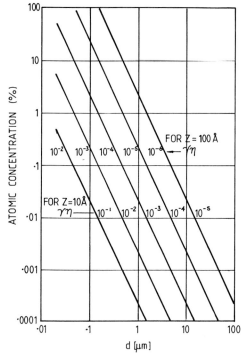

Fig. 7.9 SIMS detection limits for typical values of ion source and rastering parameters (Welkie and Gerlach, 1982).

Equations (7.23) to (7.25) to estimate sputtering times for low energy spread (0.2 eV) such as obtained from electron impact sources or high energy spread (10 eV) as given by plasma or sputter sources. Using the following parameters: $\beta = 100$ Å cm^{-2} s^{-1}; $E = 10$ keV; $C_c = 5$ cm; $C_s = 30$ cm; $\Delta E = 0.2$ or 10 eV; the beam current as a function of diameter has the values plotted in Fig. 7.8a.

If we now assume that $Y = 2$ atom ion^{-1} and values of J from Fig. 7.8a, the sputter time as a function of beam diameter (d) is obtained (Fig. 7.8b). For example, a 200 point depth profile to a depth of 2 μm(beam rastered over an area $10d$ x $10d$) requires a measuring time of 8 x 10^3, 3.6 x 10^5 or 8 s using electron impact, plasma sputter or field emission ion sources respectively. The energy spreads for these sources are 0.2, 10 or 5 eV. These practical limits in depth profiling are important but not often taken into consideration. For example, a duoplasmatron source is often used because of its simplicity and high current. However, because it also has a large energy spread and beam diameter, it is often better to use the lower current but lower energy spread from an electron impact source.

The latter also produces the best depth resolution in sputter profiling. The exceptionally high brightness of the LMIS, even with demagnification of the beam provides the shortest measuring times.

iii. *Detection Limits*

Welkie and Gerlach (1982) define a minimum detectable atomic concentration (c_L) in SIMS as:

$$c_L = 4\,\eta_0/(\pi d^2 z p \gamma \eta) \tag{7.26}$$

where η_0 is the minimum number of counts required for detection, η is the transmission of the SIMS analyser and γ is the ionisation efficiency. Detection limits for typical values of SIMS parameters are shown in Fig. 7.9. Most depth profiling experiments with SIMS use a quadrupole mass spectrometer for which the transmission is typically $\eta = 0.005$. For such a system, with $d = 10\ \mu\text{m}$, $\gamma = 10^{-1}$ (a high value) and the sputtered depth $z = 100$ Å, the minimum detectable concentration is estimated to be less than 1 part in 10^6.

7.5.5 Auger Electron Spectroscopy (AES)

AES is a simple and comparatively well understood technique, which is available in many laboratories. Fig.7.10 indicates schematically the principle of the technique. A relatively high energy electron (5 to 10 keV) creates a hole in an inner level which is filled by an electron from one of the higher energy, less-strongly-bound levels. The electron may undergo a full radiative transition from the higher energy state to fill the hole or alternatively two electrons may couple quantum mechanically, one electron filling the hole, the other, usually from a state of lower binding energy, being ejected with the energy difference as kinetic energy. If this energy difference is corrected for work function and binding energy effects the electron appears with an energy characteristic of the atomic state from which it came and is characteristic of the element. Measurement of the energy spectrum with an electron energy spectrometer allows elemental identification. Even though AES is widely used, it is difficult to obtain quantitative measurements of elemental concentrations. Most results are expressed in relative terms, which for concentration profiling is sufficient. Quantitative analysis will almost certainly involve the use of comparative standards but, even here, only the bulk concentration is known. As indicated earlier, there could be a difference in the concen-

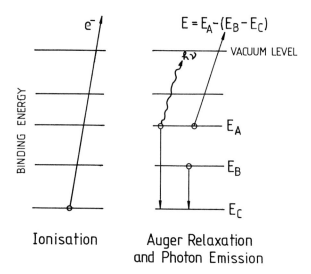

Fig. 7.10 Mechanism of Auger electron emission. A hole in an inner core level of the atom can be filled either by photon emission or by Auger relaxation with electron emission.

tration between the first few surface layers and the bulk of the solid.

The depth resolution of AES is determined by the mean free path of the mono-energetic Auger electrons ejected from atoms in the solid. The mean free path as a function of electron energy is shown in Fig. 7.11 (Brundle, 1975). The best depth resolution is obtained with electrons of about 100 eV. Secondary electrons are also produced with a continuous energy spectrum peaking at a few eV and then falling off in intensity as a power law in E. Their intensity is often much larger than that of Auger electrons and this limits the sensitivity of AES.

The sensitivity of AES is limited by the need to separate the Auger peak from the general secondary electron background. The most popular instrumentation for AES uses electronic differentiation to detect the position of the Auger peak on the energy scale. The differentiation however, makes it difficult to measure intensities accurately because the height of the signal depends very critically on instrumental factors. The new generation of electron spectrometers use pulse counting techniques to measure the energy spectrum. These data are much more amenable to direct quantification. AES has been used widely and many reviews of the technique and its application are available (Seah and Briggs, 1983).

For AES, the analysis volume is fixed by the escape depth of the Auger electrons and the beam diameter. The time constant of the detection system also enters because AES is usually performed with a

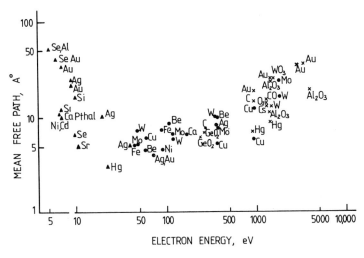

Fig. 7.11 The mean free path of electrons as a function of energy. The values shown are experimental measurements in the solid indicated.

differentiating detection system. Modern systems are tending to use electron counting in the direct energy spectrum. The electron gun is either a thermionic emission or field emission system. A similar analysis to that used with SIMS (Highlight 7.2) suggests that the AES technique is preferable if one is considering only a thin surface layer, but SIMS performs better if one can integrate the accumulated signal over a larger depth. Because of the high secondary electron background, AES is less sensitive in practice than SIMS.

7.6 FACTORS AFFECTING DEPTH RESOLUTION

In the context of sputter profiling, the ideal situation is one in which a uniform beam of ions uniformly removes a surface, layer by layer, as depicted in Fig. 7.12 for an interface between elements A and B. The actual experimental signal strengths will vary as shown in Fig. 7.13. The difference between ideal and experimental profiles arises from instrumental and sputtering effects. The information depth for various analysis methods (Section 7.4) are constant and independent of the depth of surface removed. There are other effects due to the sputtering process itself which result in a depth resolution which is a function of the depth removed. These effects may be on an atomic scale such as recoil mixing, defect formation, preferential sputtering, damage enhanced diffusion, etc., or they may be on a macroscopic scale such as the changes in surface

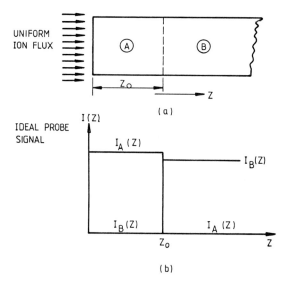

(a)

(b)

Fig. 7.12 A schematic representation (a) of the ideal sectioning using sputter profiling. The signals are shown in (b). The ideal situation assumes layer by layer removal and no matrix effect on the signal strength.

topography mentioned in Section 7.3.2. If these are all random stochastic processes, the interface will be distorted from a sharp step function to an error function and the probability of locating the interface in a region z to $(z + dz)$ where z is the depth parameter, will be:

$$P(z)dz = [1/(2\pi - \sigma)^{1/2}] \exp \{-[(z - z_0)^2/2\sigma^2]\}dz \qquad (7.27)$$

where σ is the standard deviation of the probability distribution and z_0 is the location of the original interface. The two signal strengths become:

$$I_A(t) = I_A(min) + [I_A(max) - I_A(min)] [1 - \exp(z - z_0)]/2\sqrt{2}\sigma \quad (7.28)$$

and

$$I_B(t) = I_B(min) + [I_B(max) - I_B(min)] [1 - \exp(z - z_0)]/2\sqrt{2}\sigma \quad (7.29)$$

Traditionally, the depth resolution δz is defined in terms of the 16% and 84% signal values (i.e. $\pm \sigma$) with the interface defined to be located at the 50% signal points. The interface width is $\delta z = 2\sigma$. If the contributions

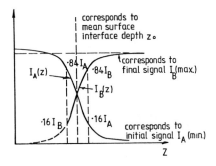

Fig. 7.13 Actual signal strengths observed for sputter profiling across a planar interface. Matrix effects on signal strength are ignored. The various definitions of interface position are shown (Carter *et al*, 1985).

to the depth resolution are independent and can be defined by resolution functions of approximate Gaussian shape then the overall resolution δz is given by:

$$\delta z = \left[\sum_i (\delta z_i)^2 \right]^{1/2} \tag{7.30}$$

This model has its faults, mainly in its presumption that the perturbations to the depth of measurement are random stochastic processes, but provided the depth z is much greater than the half-width, the approximation is reasonable. Further, the signal strengths in practice often show, as in Fig. 7.14, a variation of the maximum and minimum signal strengths in layered structures. Fig. 7.14 shows that the depth resolution is a function of the depth of the probed layer below the surface. This is usually interpreted in terms of an increased contribution from surface topography changes during sputtering. Further discussion of instrumental, topographic and compositional effects on depth resolution is given by Hofmann and Sanz (1984).

A particularly important contribution to the overall depth resolution is due to the rearrangement of target atoms by the bombarding ions. The atoms are mixed by elastic collisions (recoil mixing) or due to migration of irradiation generated defects (radiation enchanced diffusion). All the processes which affect depth resolution, except ion beam mixing, may be circumvented by choice of target and ion species or by correct experimental procedure.

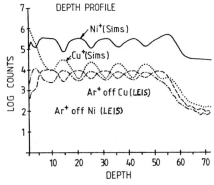

Fig. 7.14 An experimental study of profiling of a multilayer Cu/Ni film using both SIMS and LEIS to monitor surface concentrations. Note the degradation of interface definition with depth (Reed *et al.*, 1983)

7.6.1 Ion Beam Mixing

The signal observed during sputter profiling is always affected by the radiation damage caused by the sputtering beam, since the target atoms are usually displaced by many incident ion cascades before being sputtered. The concentration, $c(x, y)$, of a species at a depth x below the surface after sputtering to a depth y is then different to the initial distribution of the species $c(x, 0)$. The measured depth profile, assumed proportional to the surface concentration, $c(0, y)$, is an integral of the original depth distribution and a function $g(x, y)$ which depends on the amount of ion beam mixing:

$$c(0, y) = \int c(x, 0)\, g(x, y)\, dx \qquad (7.31)$$

In general, y is related to the sputtering time t by:

$$y = \int_0^t U(t')\, dt' \qquad (7.32)$$

where $U(t')$ is the surface erosion rate after a sputtering time t'. The problem which confronts the experimentalist is to decide which features of a sputter depth profile arise from ion beam mixing and which were present in the unirradiated sample. The solution to the problem lies in analysing sputter depth profiles of known samples (thin films or implants). The difference between the measured depth profile and the known species distribution can be described by the depth resolution $\delta z / z_0$

defined in Fig. 7.13 and modelled using ion beam mixing theory. The parameters found by fitting experimental results to the theory can be used to model the depth profiles of unknown samples and deconvolute the effect of mixing from the profiles.

The effect of recoil mixing has been studied in both medium energy (> 100 keV), low fluence RBS and low energy (< 20 keV) high fluence sputter profiling experiments. Theories of recoil mixing which use a transport equation approach (Sigmund and Gras-Marti, 1981; Littmark, 1985) have been developed to describe both these types of experiments. They rely on the calculation of a relocation function, $F(x, z)$, which gives the probability of an atom being transported from a depth x to $(x + z)$ per incoming ion. The relocation function can be split into isotopic and anisotropic contributions of displacements from ion–atom collisions (primary recoil mixing) and atom–atom collisions (cascade mixing). At medium energies the mixing in the low fluence limit is dominated by isotropic cascade mixing (Roosendaal and Sanders, 1983). This is caused by a large number of small relocations and leads to a broadening of a delta function impurity distribution to give a Gaussian depth profile (if sputtering and matrix relocation is ignored) with variance (Sigmund and Gras-Marti, 1981):

$$\sigma^2 = 2Dt = 1/3 \ \Gamma_0 \ \xi_{21} \ F_d(x) \ \Phi \ R_c^2/N \ E_d \qquad (7.33)$$

where Γ_0 is a dimensionless factor, $F_d(x)$ is the energy deposited into collisional processes at depth x, N is the atomic density, E_d is a displacement energy, R_c^2 is a mean square recoil range, ϕ is the ion fluence and ξ_{21} is a mass factor.

In the low fluence limit the contributions of primary recoils and anisotropic cascades are more important (Littmark, 1985; Eckstein and Moller, 1985). The primary recoils consist of single relocations over distances of about 100 Å, which are non-Gaussian in form. However, in high fluence sputtering experiments, an atom may be displaced in 10 to 100 cascades before being sputtered and detected. The relocation function for a large number of recoils is then assumed (by the central limit theorem of statistics) to be described by a diffusion process. In this case the transport equation is rewritten in terms of its first and second moments, $V(x)$ and $D(x)$, given by:

$$V(x) = \quad I \int dz \ z \ F(x, z)$$

$$D(x) = 1/2 \ I \int dz \ z^2 \ F(x, z)$$

$$(7.34)$$

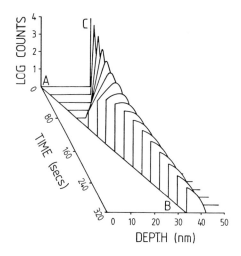

Fig. 7.15 Theoretical variation of the concentration of an impurity marker (line C), originally 20 nm below the surface, as the surface is sputtered at a rate of $U = 0.11$ nm s^{-1}. No preferential sputtering is assumed. The receding surface is represented by the line AB.

where I is the ion flux and $F(x, z)$ represents the relocation of atoms from depth x to depth $(x + z)$. This diffusion approximation is applicable when the mean relocation distance in the low fluence limit is less than one fifth of the initial distribution half width. The concentration of element i at a depth x below the eroded surface after a sputtering time t is given for dilute impurities by, Collins *et al.* (1983):

$$\partial c_i(x, t)/\partial t = (U_i - Y_i)\, \partial c_i/\partial x + \partial(D_i \partial c_i/\partial x)_i\, \partial x \qquad (7.35)$$

where D_i and V_i are the moments for the ith element and the surface erosion rate, $U(t')$, is assumed to be a constant U. Fig 7.15 shows a calculation using Equation (7.35) for the evolution of an impurity layer, originally 20 nm below the surface, as sputtering progresses (King *et al.*, 1983). The signal measured by SIMS or LEIS would be approximately equal to the surface concentration (line AB).

In addition, a term must be included to take account of density changes during bombardment. In the Littmark formulation of mixing, the density is assumed to be constant and the depth scale is transformed to bring the density of each region of the target back to a constant value. In the Collins model, migration currents are assumed to flow away from regions above a certain density. In this case additional fluxes must be incorporated into the transport Equation (7.35):

$$\partial c_i(x, t)/\partial t = (K + U)\, \partial c_i/\partial x + c_i \partial K/\partial x - \partial J_i/\partial x \qquad (7.36)$$

where:$\qquad\qquad J_i = U_i c_i - \partial(D_i c_i)/\partial x$

and K is a reverse collective current to maintain the density.

In general the effect of the reverse collective current is to cancel the anisotropic transport of material given by the terms in V_i. This transport equation formalism has the advantage that it allows the radiation enhanced thermal processes of diffusion and segregation to be easily incorporated into an equation for $c_i(x, t)$ (Lam and Wiedersich, 1982; Rehn *et al.*, 1985).

7.6.2 Computer Simulation

The Monte Carlo based code TRIM (Biersack *et al.*, 1980), uses a binary collision model based upon a screened Coulomb potential for interactions between atoms located randomly in a semi-infinite geometry. The incoming ion and all other recoiling atoms are followed individually until they either pass through the surface (if they have an energy above the surface binding energy E_b) or slow down to an energy below a threshold value, E_f. If more than the binding energy is transferred in a collision a new recoil is generated. Otherwise, energy is dissipated into phonons. Monte Carlo codes have also been developed to directly model sputter profiling. In EVOLVE (Roush *et al.*, 1981, 1983) and TRIDYN 22 (Moller and Eckstein, 1984), the effect of a small fluence (10^{12} ions cm^{-2}) on the target atom concentrations is assumed to be a multiple of the effect of a single ion. For each fluence interval the concentration changes at a certain depth are calculated from the recoil atom redistribution in a single cascade. Since over 10^4 or more trajectories need to be calculated for a typical simulation, relatively long runs on high speed computers are required.

Use of mixing theory to predict experimental sputter profiles is hampered by a lack of reliable values of D for low energy mixing. From Equation (7.33), the quantity $(Dt/\phi F_d)$ depends only on the properties of the target and so provides a convenient parameter for comparing the results of different mixing experiments.

The values of $(Dt/\phi F_d)$ which exist for low energy mixing lie in the range 10 to 200 Å^5 eV^{-1} (Macht and Naundorf, 1986; King *et al.*, 1984). Paine and Averback (1985) and Averback (1986) have reviewed the large number of results for high energy mixing. The range of values for $(Dt/\phi F_d)$ is approximately 1 to 1000 Å^5 eV^{-1}. These values depend on the concentration of the inter-mixed layers and their chemical nature and indicate that low energy processes — replacement sequences, spontaneous defect recombinations and induced defect jumps with a range of one nearest neighbour distance — may dominate the mixing process.

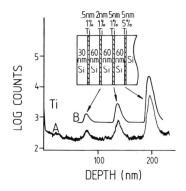

Fig. 7.16 Experimental Ti$^+$ SIMS depth profile (curve A) of a Ti–Si multilayer structure and the depth profile (curve B) calculated from Equation (7.35) using $U = 0.11$ nm s^{-1} and $D_i = 0.3$ nm^2 s^{-1} (King et al., 1984b).

The shape of sputter depth profiles is, however, rather insensitive to values of $(Dt/\phi F_d)$. Fig. 7.16 shows a Ti$^+$ SIMS depth profile of a Ti/Si multilayer. The measured layer thicknesses, 11 nm, are larger than the original widths of the layers and are approximately equal to $(R_p + \Delta R_p)$ where R_p and ΔR_p are the projected range and standard deviation for the sputter ion beam. A computer program based on Equation (7.35) was used to fit the experimental profile (curve B) with a value of $(Dt/\phi F_d) = 14$ Å^5 ev^{-1}. A similar experimental shape and peak width were found for depth profiles of Mo/Si multilayers where $(Dt/\phi F_d)$ was about 200 Å^5 eV^{-1} (King et al., 1984a). As a general rule, the standard deviation in Equation (7.27) is about $(R_p + \Delta R_p)/2$. The peak shape of profiled thin layers shows a rapid rise followed by a more gradual fall. At large depths the decrease is exponential (Wittmaack, 1982c) with a decay depth d_c. For $D(x) = D$, a constant over a range R, d_c is given by:

$$d_c = \quad R \text{ for } D/U \gg R$$

$$d_c = 2D/U \text{ for } D/U \ll R$$

(7.37)

for an erosion rate U.

The effect of implanted ions on the depth profile has been ignored in this simulation. In the absence of chemical or sputter effects, incorporation of ions into the target will cause the apparent magnitude of mixing to decrease (Littmark and Hofer, 1980). This should be the case for self implantation. It has been shown (Menzel and Wittmaack, 1985) that implanted inert gases are detrapped by further irradiation and migrate towards the surface which acts as an unsaturable sink. Since interaction between inert gas and target atoms is small, it is reasonable to disregard the influence of inert gas implantation on measured depth profiles. Reactive gas incorporation, either by implantation or adsorption followed by recoil implantation, causes segregation of impurities (Hues and Williams, 1986) and profile distortion (Vandervorst et al., 1986) and should be considered in the analysis of measured depth profiles.

7.6.3 Deconvolution Procedures

If the resolution function $g(x, y)$ in Equation (7.31) is known then the original depth distribution $c(x, 0)$ can, in principle, be found from the measured depth profile $c(0, y)$. When the initial impurity distribution is concentrated at depths greater than $(R_p + \Delta R_p)$, then $g(x, y)$ can be approximated as $g'(y - x)$ and:

$$c(0, y) = \int c(x, 0)\, g'(y - x)\, dx \qquad (7.38)$$

The resolution function depends on all the contributions to the depth resolution i.e. mixing, escape depth, topography, etc. If the resolution function can be approximated by a Gaussian then the above equation can be solved by Fourier transforms or iterative procedures (Hofmann and Sanz, 1984).

Equation (7.38) has also been solved for the case where the resolution is determined by ion beam mixing alone and $g'(y - x)$ is not assumed to be Gaussian. Carter et al., (1981) expressed the solution to Equation (7.38) in terms of the Laplace transform $c(s)$ of $c(0, t)$, or equivalently $c(0, y)$, where:

$$F(s) = \frac{4 \exp(\lambda/2)\, h(s) \exp(\lambda h(s)/2)}{[h(s) + 1]^2 \exp(\lambda h(s)) - [h(s) - 1]^2} \qquad (7.39)$$

and:

$$h(s) = \sqrt{(1 + 4s)}$$

and $\lambda = UR/D$ for the case of uniform mixing with coefficient D over the ion range R. This was found to approximate to within a few per cent the more general case where D varied with depth (King and Tsong, 1985).

The original depth distribution can then be found as the inverse Laplace transform of $c_0(s)$:

$$c_0(s) = c_\Sigma(s)/G(s) \tag{7.40}$$

where:

$$G(s) = F(s)\{1 + 2k/[\sqrt{1 + 4s} - (2k + 1)]\} \tag{7.41}$$

and k takes into account the effect of preferential sputtering on the profile.

7.7 CONCLUSION

Although sputter profiling is widely used because of its excellent sensitivity and mass resolution as well as the versatility available using a selection of surface analysis probes, great care must be exercised in interpretation, especially when quantitative profiles are sought. It is not sufficient to couple a small ion gun to a system and use an available quadrupole SIMS or an AES unit and hope to obtain meaningful results. A depth profiling system must be a dedicated and sophisticated apparatus with all experimental parameters optimised. To obtain a quantitative profile, due account must be taken of the many factors described in this chapter and elsewhere. However, theoretical developments associated with instrumental effects and ion beam modification of the sample suggest that a reasonable understanding of these processes is emerging. Further development of the model is needed for use in practical applications.

REFERENCES

Andersen, H.H. (1984). *In* 'Ion Implantation and Beam Processing', (J.S. Williams and J.M. Poate, eds.), Academic Press, Sydney, p. 128.

Andersen. H.H. & Bay, H.L. (1975a). *J. Appl. Phys.* **46**. 1919.

Andersen, H.H. and Bay, H.L. (1975b). *J. Appl. Phys.* **18**, 131.

Anderson C. A. and Hinthorne J.R. (1973). *Anal. Chem.* **45**, 1421.

Averback, R.S. (1986). *Nucl. Instrum. Methods* **B15**, 675.

Bayly, A.R. and MacDonald, R.J. (1977). *J. Phys. E: Sci. Instrum.* **10**, 79.

Betz, G. (1980). *Surface Science* **92**, 283.

Biersack, J.P. and Haggmark, L.G. (1980). *Nucl. Instrum. Methods* **174**, 257.

Brundle, C.R. (1975). *Surface Science* **48**, 99.

Bukhanov, V.M., Yurasava, V.E., Sysoev, A.A., Samsonev, G.V. and Nikoloev, B.I. (1970). *Sov. Phys. Solid. State* **12**, 313.

Carter, G., Collins, R. and Thompson, D.A. (1981). *Rad. Effects* **58**, 99

Carter, G., Gras-Marti, A. and Nobes, M.J. (1985). *Rad. Effects* **62**, 119.

Collins, R., Marsh, T. and Jimenez-Rodriguez, J.J. (1983). *Nucl. Instrum. Methods* **209/210**, 147.

Cox. J. (1982). Ph. D. Thesis, Faculty of Military Studies, University of New South Wales, Australia.

Dennis, E. and MacDonald, R.J. (1972). *Rad Effects* **13**, 243.

Dzioba, S., Auciello, O. and Kelly, R. (1980). *Rad. Effects* **45**, 235.

Eckstein, W. and Moller, W. (1985). *Nucl. Instrum. Methods* **B7/8**, 727.

Goto, K., Koshikawa, T., Ishikawa, K. and Shimizu, R. (1978). *J. Vac. Sci. Technol.* **15**, 1695.

Hart. R.R., Dunlop, H.L. and Marsh, O.J. (1975). *J. Appl. Phys.* **46**, 1947

Hofmann, S. and Sanz, J.M. (1984). *In* 'Thin Films and Depth Profile Analysis', (Oechsner, H. ed.), Springer-Verlag, p. 153.

Hues, S.M. and Williams, P. (1986). *Nucl. Instrum. Methods* **B15**, 206.

Husinsky, W. (1985). *J. Vac. Sci. Technol.* **3**, 1546.

King, B.V. and Tsong, I.S.T. (1985). *Nucl. Instrum. Methods* **B7/8**, 793.

King, B.V., Tonn, D.G., Tsong, I.S.T. and Leavitt, J.A. (1984a). *Mat. Res. Soc. Symp. Proc.,* **27**, 103.

King, B.V. and Tsong, I.S.T. (1984b). *J. Vac. Sci. Technol.* **A2**, 1443.

Kolfschoten, A.W., Haring, R.A., Haring, A. and de Vries, A.E. (1984). *J. Appl. Phys.* **55**, 3813.

Lam, N.Q. and Wiedersich, J. (1982). *J. Nucl. Mat.* **103/104**, 433.

Littmark, U. (1985). *Nucl. Instrum. Methods* **B7/8**, 684.

Littmark, U. and Hofer, W.O. (1980). *Nucl. Instrum. Methods,* **168**, 329.

MacDonald, R.J. (1970). *Phil. Mag.* **21**, 519.

MacDonald, R.J. (1974). *Surface Science* **43**, 653.

MacDonald, R.J. and O'Connor, D.J. (1983). *Surface Science* **124**, 423.

Martin, P.J. and MacDonald, R.J. (1977). *Surface Science* **62**, 551.

Matsunami, N., Yamamura, Y., Itikawa, Y., Itoh, N., Kazamata, Y., Miyagowa, S., Morita, K., Shimizu, R. and Tawara, H. (1984). *Atomic Data and Nuclear Data Tables,* **31**.

Macht, M.P. and Naundorf, V. (1986). *Nucl. Instrum. Methods,* **B15**, 189.

Menzel, N. and Wittmaack, K. (1985). *Nucl. Instrum. Methods* **137/8**, 366.

Moller, W. and Eckstein, W. (1984). *Nucl. Instrum. Methods,* **B2**, 814.

Myers, S.M. (1980). *Nucl. Instrum. Methods* **168**, 265.

Newns, D.M., Makoshi, K., Brako, R. and van Wunnick, J.N. (1985). *Physica Scripta,* **T6**, 5.

Oechsner, H. and Stumpe, E. (1977). *Appl. Phys.* **14**, 43.

Okano, J., Ochiat, T. and Nishimura, A. (1985). *Appl. Surf. Science* **22/23**, 72.

Onderlinden, D. (1966). *Appl. Phys. Letts.* **8**, 189.

Onderlinden, D. (1968). *Can. J. Phys.* **46**, 739.

Paine, B.M. and Averback, R.S. (1985). *Nucl. Instrum. Methods* **137/8**, 666.

Pellin, M.J., Wright, R.B. and Gruen, D.M. (1981). *J. Chem. Phys.* **74**, 6448.

Reed, D.A. and Baker, J.E. (1983). *Nucl. Instrum. Methods* **218**, 324.

Rehn, L.E., Lam, H.Q. and Wiedersich, H. (1985). *Nucl. Instrum. Methods* **B7/8**, 764.

Roosendaal, H. and Sanders, J.B. (1983). *Nucl. Instrum. Methods* **218**, 673.

Roush, M.L., Andreadis, T.D., Davarya, F. and Goktepe, O.F. (1981). *Nucl. Instrum. Methods* **191**, 135.

Roush, M.L., Davarya, F., Goktepe, O.F. and Andreadis, T.D. (1983). *Nucl. Instrum. Methods* **209/210**, 67.

Sanders, F.M., Kolfschoten, A.W. and Dielman, J. (1984). *J. Vac. Sci Technol.* **A2**, 487.

Seah, N.P. and Briggs, D. (1983). *In* 'Practical Surface Analysis by Auger and Photoelectron Spectroscopy', (Briggs, D. and Seah, N.P., eds), Wiley and Sons, Chichester, p. 1.

Sigmund, P. (1969). *Phys. Rev.* **184**, 383.

Sigmund, P. and Gras-Marti, A. (1981). *Nucl. Instrum. Methods* **182/3**, 25.

Szymonski, M. and Bhattacharya, R. (1979). *Appl. Phys.* **20**, 207.

Taglauer, E. and Heiland, W. (1978). *Appl. Phys. Lett.* **33**, 950.

Thompson, M.W. (1968). *Phil. Mag.* **18**, 337.

Thompson, M.W. (1981). *Physics Reports* **69**, 337.

Tsong, I.S.T. (1981). 'Inelastic Particle — Surface Collisions', (Taglauer, E. and Heiland, W., eds) *Springer Series In Chem. Phys.* **17**, 258.

Vandervorst, W., Shepherd, F.R., Swanson, M.L., Plattner, H.H., Westcott, O.W. and Mitchell, I.V. (1986). *Nucl. Instrum. Methods* **B15**, 201.

Welkie, D.G. and Gerlach, R.L. (1982). *J. Vac. Sci. Tech.* **20**, 104.

Williams, P., Tsong, I. S. T. and Tsuji, S. (1980). *Nucl. Instrum. Methods* **170**, 591.

Wittmaack, K. (1981). *Appl. Surface Sci.* **9**, 315.

Wittmaack, K. (1982a). *Vacuum* **32**, 65.

Wittmaack, K. (1982b). *Rad. Effects* **63**, 205.

Wittmaack, K. (1982c). *J. Appl. Phys.* **53**, 4817.

Wittmaack, K. (1985a). *J. Appl. Phys.* **A38**, 235.

Wittmaack, K. (1985b). *J. Vac. Sci. Technol* **A3**, 1350.

Zwangobani, E. and MacDonald, R.J. (1973). *Rad. Effects* **20**, 81.

8
Low Energy Ion Scattering from Surfaces

D.J. O'CONNOR
R.J. MACDONALD

Department of Physics, University of Newcastle
Newcastle, Australia

Note

In ion scattering texts the usual notation for incident and scattered ion energies employs the subscripts 0 and 1 respectively. Subscripts 1 and 3 are used here for consistency with the more general notation applicable for nuclear reactions. Subscripts 2 and 4 refer to the target atom and recoil atom respectively.

ION BEAMS FOR
MATERIALS ANALYSIS
ISBN 0 12 099740 1

8.1 INTRODUCTION

Low Energy Ion Scattering (LEIS) can answer three questions:

   i. What is on the surface? i.e. the elemental composition of the surface.
   ii. How much is on the surface? i.e. the relative concentrations of different elements on the surface with accurate quantification for some systems. It is possible to detect less than 1% of a monolayer of many elements on the surface.
   iii. Where is the element on the surface? i.e. the location of surface atoms (e.g. impurity, adsorbate) with respect to the matrix atoms. The crystallographic location of an adsorption site can be determined to an accuracy of the order of 10 pm.

The interpretation is straightforward in some cases while in others it entails the consideration of several different processes and the user should be aware of the potential complications which may arise. Proceedings of two recent conferences (Roosendaal *et al.*, 1984; Biersack and Wittmaack, 1986) indicate that LEIS is becoming a very important analysis tool.

We discuss here the applications of LEIS (at < 10keV) to compositional and structural analysis at surfaces, concentrating on interpretation and quantification of the data.

The main differences between high and low energy ion scattering are:

   i. at high energies RBS detects particles scattered from both the surface layer and from significant depths below the surface; and
   ii. LEIS, using ions formed from inert gases, detects particles scattered only from the uppermost surface layer of the solid. Any inert gas ion penetrating below the surface is neutralised. LEIS uses electrostatic energy analysers and hence the particle detected and measured must be charged. At low energies (< 5 keV) inert gas ions are neutralised if they penetrate below the top two or three layers, in contrast to RBS in which neutralisation processes are negligible. Electrostatic analysers are particularly suited to low energy analysis since they are simple to operate, they provide good energy resolution and their ability to detect only a limited energy window at a time is offset by the significantly greater scattering cross section at low energies.

The general principles governing the application of low and high energy ion scattering to the analysis of the surface are outlined in Chapter

1. The energy of scattered particles (E_3) detected at angle θ and recoil particles (E_4) detected at angle ϕ are independent of the interaction potential:

$$E_3/E_1 = \{[\cos\theta + (\mu^2 - \sin^2\theta)^{1/2}]/[1 + \mu]\}^2 \qquad (8.1)$$

$$E_4/E_1 = 4\mu\cos^2\phi/(1 + \mu)^2 \qquad (8.2)$$

where $\mu = M_2/M_1$. LEIS can also use alkali metal ions with the advantage of a much higher degree of ionisation in the ion–surface interaction, so the same ion yield may be obtained with a very much smaller incident ion beam density. This results in a significant reduction in surface damage but at the expense of surface selectivity since projectiles scattered from below the surface also escape in a charged state.

8.1.1 LEIS Apparatus

A schematic representation of a LEIS system is included in Fig. 7.1. A mass analysed monoenergetic ion beam is directed onto, and rastered over, the target. The requirements for rotation of the target depend on the type of analysis to be performed and will be outlined in later sections. As the channeling angles at low energies are relatively large, angle adjustments of 0.1° are sufficient. In most systems, the energy analyser is an electrostatic device of spherical or cylindrical construction. While high energy resolution is sought in many research analysers, consideration of sensitivity and the existence of a natural energy width of scattered ions (resulting from thermal broadening and kinematic spread) limits the practicable energy resolution of the analyser to about 1%.

LEIS is a surface sensitive technique so it is imperative to ensure that it is performed under clean vacuum conditions. In most research systems, 10^{-10} Torr in the target region is the minimum requirement; however, this may be difficult to achieve for 'industrial' targets and fast interchange of samples. Under poorer vacuum conditions high beam currents can be used to sputter clean the target. Indeed, using LEIS the surface being analysed is constantly being renewed and some of the measured properties of the surface may result from the damage and implantation caused by the incident beam. LEIS is therefore normally a destructive analysis procedure.

8.2 FACTORS AFFECTING THE SCATTERED ION SIGNAL

The yield of ions scattered through an angle θ into a detector of solid angle $d\Omega$ is given by:

$$N^+ = N_i \, . \, N_a \, (d\sigma(E_1, \, \theta)/d\Omega) \, P^+(E) \, . \, T(E) \, . \, d\Omega \qquad (8.3)$$

where N_a is the number of surface atoms of type A per unit area, N_i is the number of incident particles, $[d\sigma(E_1, \, \theta)]/d\Omega$ is the cross-section for scattering of ions of energy E_1 through the angle θ, $T(E)$ is the transmission function of the energy analyser and $P^+(E)$ is the probability that such ions scattered towards the detector will survive the scattering event and subsequent exit trajectory in an ionised state. The main terms affecting the signal strength are the scattering cross section and the probability of neutralisation.

8.2.1 The Scattering Cross-section

The scattering cross-section, $[d\sigma(E_1, \, \theta)]/d\Omega$, depends on the incident ion energy, the scattering angle and the potential governing the interaction between the incident ion and surface atom. The interaction potential is the main factor determining the magnitude of the scattering cross-section. For all LEIS situations considered in this chapter, the repulsive part of the interaction potential is dominant. In general it will be represented by the screened Coulomb potential given in Equation (1.11).

For low energy ion-atom interactions, the Molière representation of the Thomas-Fermi screening function is normally used, where the screening radius is:

$$a = 0.88534 \, a_o/(Z_1^{1/2} + Z_2^{1/2})^{2/3} \qquad (8.4)$$

However, this is often multiplied by a numerical factor to match the Molière to more complex, but more accurate potentials. O'Connor and Biersack (1986) discuss one such fitting factor, Equation (1.46).

Tables of cross-sections as a function of E_1, θ and M_2/M_1 exist, though most laboratories have programs to generate their own data. Although an analytical expression for the cross-section does not exist at the energies associated with LEIS a number of general rules apply. The differential scattering cross-section increases with:

- increasing atomic number of the projectile or target (Z_1, Z_2);

- decreasing energy of the incident projectile (E_1); and
- decreasing scattering angle (θ).

8.2.2 Neutralisation Probabilities

The probability, P^+ (E), that an incident ion remains in a charged state following the scattering event and departure from the surface is given by:

$$P^+(E) = 1 - P_n(E), \qquad (8.5)$$

where $P_n(E)$ is the probability that the incident ion is neutralised as a result of the scattering event, or along the subsequent (or prior) trajectory away from (or towards) the surface. The charge exchange phenomenon is simultaneously one of the greatest strengths of LEIS and one of its greatest weaknesses. The preferential neutralisation of inert gas ions which penetrate below the surface atom layer is responsible for the extreme surface sensitivity of LEIS. The weakness is introduced by the uncertainty in ion yields that result from surface scattering. For example, only 1–10% of the incident inert gas projectiles are scattered off the surface in an ionised state. The lack of a well tested model to predict the ion yield generally hinders quantitative analysis. The use of incident alkali ions reduces the uncertainty associated with the charge exchange process, since a majority of the scattered projectiles remain in an ionised state. However, the resultant energy spectra of these projectiles become more complicated with the appearance of multiple scattering peaks and scattering from subsurface layers.

Although the charge exchange process may be viewed as a hindrance in the proper quantification of LEIS, it can be used to good effect to enhance the sensitivity of surface analysis, particularly when seeking to detect low concentrations of highly electronegative impurities. The probability that incident positively charged inert gas ions scatter off the surface in a negatively charged state is negligible. Thus the energy spectra of the negatively charged ions are dominated by recoiling impurities which readily accept additional electrons (e.g. O, F, Cl). Although the actual improvement in sensitivity has not been quantified, a factor of 100 is possible.

Early theoretical descriptions of electron capture and neutralisation culminated in the time independent perturbation calculations by Hagstrum (1954, 1977). The possible electron exchange mechanisms include resonance exchange or quasi-resonance exchange, and Auger type exchanges. In the resonance processes the energy levels in the solid and the projectile are equal. The electron can transfer from the occupied levels in the solid to an equivalent level in the projectile or from the projectile to the unoccupied levels in the solid above the Fermi level. The quasi-resonance process is similar and involves a near match of energy levels between the projectile and a core atomic level in the target atoms. The degree of mismatch between these levels is less than 5 eV. These suggested mechanisms involve interaction between the states of atoms in the solid and atomic levels of the ion (or atom) near the surface. They take no account of possible quasi-molecular electron exchanges which can occur during the actual elastic collision which forms the basis of LEIS.

According to the calculations of Hagstrum (1954, 1977), the probability that an incident ion survives the interaction to be detected as an ion is given by:

$$P = \exp\left(-A/a_s \cdot v_\perp\right) = \exp\left(-v_c/v_\perp\right) \qquad (8.6)$$

where $v_\perp$ is the component of ion velocity perpendicular to the surface, A is the electron transition rate at the surface and a_s is a length characteristic of the range of the exchange process. These two parameters are commonly combined to give an effective charge exchange velocity represented by v_c. This result was originally developed to describe neutralisation processes affecting ions of very low energy (10 eV or so) and some of the assumptions or approximations used in deriving it are of questionable applicability in the case of LEIS.

More recent calculations have used time dependent perturbation theory, based on the Andersen Hamiltonian (Bloss and Hone, 1978; Muda and Hanawa, 1980; Sebastian et al., 1981). The time dependence is calculated from the classical trajectory of the ion and essentially follows the occupation probability of neutralising states of the incident ion. The calculations are relatively new and the applications complex. The results give a survival probability for the ionised state which resembles Equation (8.6) in form, but without the parameter interpretation of A and a_s.

Experimentally, studies of neutralisation have involved measurement of the scattered ion yield per incident ion as a function of the perpendicular component of the velocity of the scattered ion. This is usually achieved by varying the scattering angle or the incident ion energy, but this changes the collisional parameters (e.g. the impact parameter) which may alter the ion fraction by affecting the collisional charge exchange processes. Experiments have identified four characteristic types of behaviour of the yield as a function of ion energy (Rusch and Erikson, 1977) which are shown schematically in Fig. 8.1. The first is the

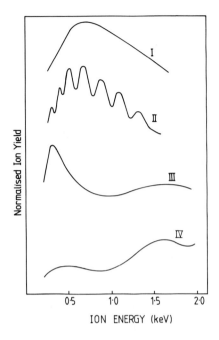

Fig. 8.1 The four different classes of behaviour for ion yield as a function of energy. Class I represents the competing processes of charge exchange and scattering cross-section, while class II behaviour is indicative of a quasi-resonant process. Class III and Class IV behaviour have yet to be identified with particular charge exchange processes.

result of variation in scattering cross-section and survival in an ionised state as the ion energy is varied. Such experiments have been used to determine an experimental v_c. The second, exhibiting characteristic oscillations in the ion yield as a function of incident energy, is associated with a neutralisation event confined to the collision itself and involving an electron exchange between the inner, more tightly bound state of the target atom and the ground state of the neutralised ion. Such oscillations in yield are seen in gas-phase collision processes (Zartner, 1979). No process has yet been identified as responsible for types III and IV behaviour shown in Fig. 8.1.

Experimentally, some researchers have attempted to link their results to behaviour similar to that expected from Equation (8.6) by assuming that one must take account of the incoming and outgoing trajectories through:

$$P = \exp\left[-v_c \cdot (1/v_{\perp\text{in}} + 1/v_{\perp\text{out}})\right] \qquad (8.7)$$

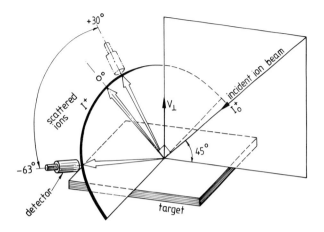

Fig. 8.2 The experimental arrangement used by MacDonald and Martin (1981), and MacDonald and O'Connor (1983) for v_c measurements.

or by adding a component which attempts to correct for exchange processes in the collision itself:

$$P = \exp\left[-b/v - v_c \cdot (1/v_{\perp \text{in}} + 1/v_{\perp \text{out}})\right] \tag{8.8}$$

where v is the relative velocity of the ion and atom during the collision and b is a fitting factor.

Some recent experiments have attempted to study the exchange processes in individual parts of the trajectory rather than integrate over the whole trajectory as in previous experiments. A schematic diagram of the former is shown in Fig. 8.2. By keeping the incident energy and the scattering angle constant, and by moving the analyser in a plane perpendicular to the plane of the incident ion beam and surface normal, the $v_\perp$ of the scattered ion can be varied. The experimental results can be fitted to Equation (8.6), but the value of v_c so obtained is found to be a function of the scattered ion energy (MacDonald and O'Connor, 1983; MacDonald et al., 1984) and lower than that value appropriate to neutralisation integrated over the whole trajectory. Furthermore, it was concluded that the neutralisation occurs approximately equally during the incident trajectory, the collision and the exit trajectory (MacDonald and Martin, 1981).

Attempts have been made to include neutralisation in computer simulation of the scattering event assuming that the rate of neutralisation $R(r)$ is essentially that used by Hagstrum:

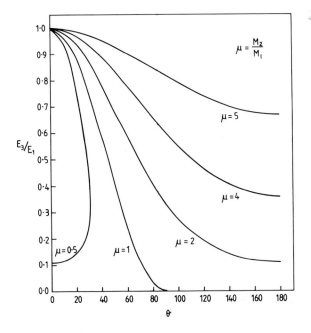

Fig. 8.3 Variation of E_3/E_1 for scattered particles as a function of scattering angle θ, for different values of the mass ratio μ.

$$R(r) = A \exp(-a_s r) \qquad (8.9)$$

where r is the distance between the ion and the atom. This rate equation is integrated along the ion trajectory to determine the overall neutralisation or survival probability. The results of some simulations suggest it is not necessary to include the collisional component of the neutralisation, represented by the first term in the exponential of Equation (8.8), in such models (Woodruff, 1982). Finally, as we indicate in Section 8.4, quantitative LEIS measurements are sometimes hindered by an inability to obtain accurate neutralisation probabilities.

8.2.3 Geometrical Effects

The geometry of the scattering event appears in Equation (8.3) through the term $d\Omega$, the collection solid angle of the analyser which corresponds to an increment $d\theta$ in the scattering angle. Fig. 8.3 shows the variation in E_3/E_1 with θ for various values of μ. One effect is obvious — the energy of ions scattered off light atoms (low μ) is a stronger function of θ than

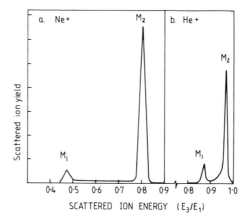

Fig. 8.4 LEIS spectra for He⁺ and Ne⁺ scattered off two masses M_1 and M_2 showing the improved mass resolution and broader peak width achieved with a heavier projectile.

that for similar mass ions scattered off heavy atoms. Furthermore, for scattering off similar mass atoms, light atoms give a smaller E_3/E_1 variation with angle. Thus for a given $d\Omega$ at a constant θ the energy width of the binary (i.e. single ion–atom) scattering peak for a light projectile will be less than for a heavy projectile scattered from the same target atom. This is demonstrated in Fig. 8.4 for Ne⁺ and He⁺ scattered off a target containing masses M_1 and M_2. Geometry is also important in the analysis of single crystal surfaces. The ion yield is strongly dependent on the relationship between the ion beam direction, the detector direction and the major crystal axes of the surface (Section 8.3).

8.2.4 Transmission Function of the Analyser

The most common form of analyser for LEIS is an electrostatic energy analyser which acts as an energy filter. The transmission characteristics depend on the mode of operation. In the constant resolution mode ($\Delta E/E$ = constant) the potentials applied to all elements of the analyser and the transmission factor are proportional to the pass energy. An alternative, referred to as the constant energy window mode (ΔE = constant), involves the acceleration or deceleration of ions to ensure that they pass through the analyser with the same energy at all times. In this case, the focussing lens voltage is a complicated function of the transmission energy. The analysers then have a more complex transmission factor which needs to be determined by computer modelling of the ion trajectories.

After the ions have been transmitted by the analyser they are detected by an electron multiplier with an efficiency which is dependent on the type, charge state and energy of the ions. Since the dependence on ion type and charge state is small, it is usually only necessary to consider the energy dependence which is a rapidly increasing function of energy below 1 keV. Above that energy the response is relatively flat. In most LEIS spectrometers the ion is accelerated after transmission through the analyser by approximately 3 kV to achieve the energies which ensure constant detection response.

8.2.5 Inelastic Energy Loss

In general it is accepted that the stopping power at low energies is proportional to velocity (O'Connor and MacDonald, 1980 a,b; Blume *et al.*, 1982). There have also been reports which indicate that the stopping power may be proportional to energy under some conditions (Blume *et al.*, 1982; Hogberg *et al.*, 1970). The energies of the scattered and recoil ions given by Equations (8.1) and (8.2) assume that there are no inelastic energy losses associated with the scattering event. Although such losses have been observed experimentally in most practical situations they are negligible. Heiland *et al.* (1973) have noted one example of inelastic loss observed as a shift of the energy position of binary scattering, for Ne^+ scattered from Ni. It was suggested that the loss occurred in an ionising collision at the surface. There have been several similar observations involving the detection of two peaks in scattered ion spectra (Shoji and Hanawa, 1984; Souda and Aono, 1986). The higher energy signal is the result of binary scattering while the energy loss associated with the lower energy peak is believed to result from the collisional ionisation of projectiles which were neutralised prior to the collision.

O'Connor and MacDonald (1980a, b) have demonstrated another example of energy loss associated with the extra path length of the scattered ion in the solid when scattering occurs from atoms of the second layer exposed by the appropriate orientation of the incident beam with respect to the regular atomic arrangement in single crystals. This extra path length results in an increased probability for inelastic loss to occur as a result of collision with the electron clouds associated with atoms in the surface layers. The onset of an inelastic loss mechanism has also been observed for Ne^+ scattered from Ag as the distance of closest approach to the surface is reduced. This has been associated with the gradual penetration of the electron selvedge (MacDonald and O'Connor, 1983). Other observations of energy loss, or of shifts in the positions of energy peaks, have been observed and assigned to excitation of target atoms not involved in the collision. An example is the observation of the

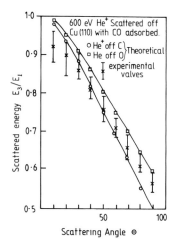

Fig. 8.5 An illustration of the difficulties encountered in target mass identification when the scattered ion peak is relatively broad and significant inelastic energy loss is observed (see text).

shift in the energy of He from TaO which is believed to be associated with atomic excitation processes (Wheeler, 1974; Baun, 1978; McCune *et al.*, 1979).

There have been many experiments showing good agreement with the predictions of Equations (8.1) and (8.2). Part of the reason for not observing inelastic effects in such cases is associated with the resolution of the energy spectrometer. An additional cause is that the so called 'single scattering event' is a reasonably unlikely observation since the ion trajectory almost certainly involves several small angle deviations (with very small energy loss), as a result of large impact parameter collisions with other surface atoms, in addition to a large angle scattering event. As a consequence of such a collision sequence the final energy E_1 can be larger than that predicted by Equation (8.1). This increase in E_1 may tend to compensate for inelastic energy loss and give a measured energy peak inadvertently close to that predicted by Equation (8.1).

Energy loss effects can confuse the interpretation of experimental results as demonstrated for the case of He^+ scattered off a Cu surface covered with an O or CO overlayer (Englert *et al.*, 1985). At a given scattering angle (particularly at small scattering angles) an observed peak associated with scattering from O occurs at the theoretical binary scattering energy, Equation (8.1), for He^+–Cu collisions. This interpretational difficulty was resolved by examining the behaviour of the scattered energy as a function of scattering angle, as shown in Fig. 8.5. The

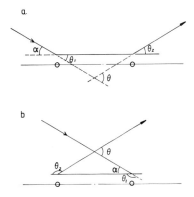

Fig. 8.6 Two possible double scattering sequences leading to the same total scattering angle θ. The second sequence is not observed experimentally as the cross-section for large angle collisions is small.

experimental result is compared with the theoretical variation of the E_3/E_1 value appropriate to scattering from C, O and Cu, as a function of scattering angle, θ. The origin of the inelastic loss has not been identified.

8.2.6 Multiple Collision Events at Surfaces

It is possible for a particle scattered from a surface through an angle θ to reach that scattering angle by a variety of events, of which the single (binary) scattering event is only one example. Two other possible multiple scattering events are demonstrated schematically in Fig. 8.6. These effects are extremely important in LEIS both as a complication in data interpretation and as a possible tool for the analysis of atom positions on surfaces. If an ion undergoes a sequence of binary collisions, the final scattered ion energy is given by:

$$E_3/E_1 = \prod_{i=1}^{n} \{[\cos \theta_i + (\mu^2 - \sin \theta_i)^{1/2}]/(1 + \mu)\}^2 \qquad (8.10)$$

where θ_i is the scattering angle in the ith collision and the collision sequence involves an ion of mass M_1 with a sequence of target atoms each of mass M_2. If the sequence of collisions involved atoms of differing masses, e.g. adsorbates on metals or an alloy target, this expression would be more complex. In general, the results of the collision sequence will involve a numerical calculation and in most cases this has been done with a single chain of atoms. These calculations indicate that the energy

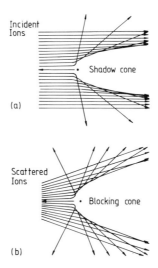

Fig. 8.7 The difference between a shadow cone (a) and a blocking cone (b) is that the angular divergence of a blocking cone is finite and that of a shadow cone is zero at a large distance behind the scattering centre.

distribution is a function of the scattering potential, the lattice constant and the impact parameters. This can be demonstrated for the *double collision*, i.e. a sequence of two collisions. The impact parameter (D_2) in the second collision is a function of the impact parameter in the first collision (D_1), since we can write (for small values of θ_1):

$$D_2 = D_1 + d \cdot \sin(\alpha - \theta_1) \qquad (8.11)$$

where d is the interatomic spacing and α is the angle of incidence. Scattering through the angle $\theta = \theta_1 + \theta_2$ often leads to a peak at a higher energy than that correponding to a single scattering event through the angle θ.

In LEIS of inert gas ions it is generally assumed (and experiment does not contradict) that multiple collisions involving more than two collisions in sequence are not observed, because of the extremely high probability of neutralisation associated with such trajectories. Hence properties of double collisions are worth considering in more detail. For example, the extreme types of double collisions illustrated in Fig. 8.6 have very different properties. In Fig. 8.6a, the magnitudes of the individual scattering angles θ_1 and θ_2 are both less than the total scattering angle $\theta = \theta_1 + \theta_2$, and from Equation (8.10), it can be shown that the energy of the multiply scattered ion is greater than that of an ion singly scattered through the same total scattering angle. This is an

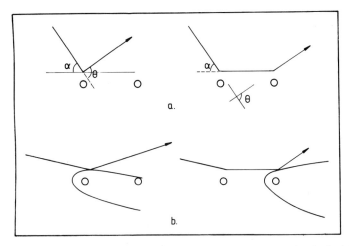

Fig. 8.8a. Single and double scattering sequences at small scattering angles; b. the influence of shadowing and blocking.

important property of small angle multiple scattering and readily allows the identification of multiple scattering peaks in LEIS energy spectra. The other extreme is illustrated in Figure 8.6b where both individual scattering angles are larger than the total scattering angle and hence the final energy of the ion is less than that of an ion singly scattered through the same total scattering angle. This latter sequence is very rare in ion scattering because the cross-sections of each collision are less than the corresponding single scattering cross-section.

Double scattering is also important in atom location experiments. Consider firstly shadow and blocking cones (Chapter 1) which arise from the scattering of a beam of ions of uniform density incident on an isolated atom, as shown in Fig. 8.7. The shadow cone (Fig. 8.7a) occurs for a collimated beam or when the incident ions originate from a source at a large distance from the scattering centre. The angular width of the shadow cone at a large distance behind the scattering centre is negligibly small. The blocking cone (Fig. 8.7b) involves an incident beam emanating from a real or virtual source at a small distance from the scattering centre and has an angular width which is independent of the distance behind a scattering centre. Both the shadow and blocking cones define a forbidden region of interaction and must be taken into account in considering double scattering events.

Single and double scattering events are shown schematically in Fig. 8.8a for a small total scattering angle. Fig. 8.8b shows that the blocking cone formed by the second atom leads to a minimum and maximum

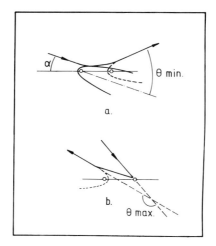

Fig. 8.9 The limits for small and large double scattering angles at a fixed angle of incidence.

angle observed for scattering. These limiting scattering angles are shown in Fig. 8.9. At intermediate scattering angles both single and double collision events are possible but beyond the limits the shadow cones shield the surface in double scattering and no scattering is observable.

The general behaviour of these collision events is worth considering in two ways, namely variation of the angle of incidence at fixed scattering angle or variation of scattering angle at fixed angle of incidence. For constant angle of incidence the result is an E-θ loop as shown in Fig. 8.10a in which the lower part of the loop represents the E_3/E_1 values for single (or quasi-single) collisions and the upper part of the loop represents the values for double scattering sequences. If the scattering angle is kept constant and the angle of incidence to the surface, α, is varied, shadowing and blocking events again impose a similar structure on the variation of scattered energy with angle of incidence to that illustrated in Fig. 8.10a for the E-θ variation. At low angles of incidence the shadow cones interact with the surface and no scattering is observed. At larger angles of incidence (for constant scattering angle) the scattered trajectory, rather than the incident trajectory, can interact with the surface through blocking events and the single and double collision events firstly degenerate, so as to be indistinguishable, and then disappear. Thus the variation of observed scattering energy with angle of incidence will form an E-α loop as shown in Fig. 8.10b where again the lower part of the loop represents the single scattering event and the upper part the double scattering event. In both cases, the small angle limit at which the loop

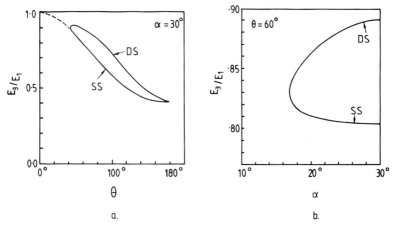

Fig. 8.10 The E-θ(a) and E-α(b) loops for 6 keV Ar scattered off a W $\langle 100 \rangle$ surface chain. The portions of the curve labelled SS and DS refer to the energies for single and double scattering respectively.

closes (called the critical angle) is directly dependent on the interatomic spacing. For small interatomic spacing the critical angle is large and it decreases with increasing interatomic spacing.

Fig. 8.11 shows an experimental result of the energy difference between the double and single scattering peaks as a function of incident angle, and the degeneracy of these peaks at a low incident angle (O'Connor, 1978; MacDonald, 1980). This figure is a collection of portions of energy spectra, (each showing only the energy range incorporating the single and double scattering energies) at angle of incidence increments of $0.75°$ for 6 keV Ar^{+} scattered off a W(110) surface. Ion incidence is along the $\langle 111 \rangle$ direction and the total scattering angle is $60°$. At large angles of incidence (e.g. $30°$) the single and double peaks are the large and small peaks respectively at the right of the figure. The decreasing separation of the single and double scattering peaks with decreasing angle is apparent and the two peaks merge at an angle close to $19°$. From the discussion above, one would not expect to observe any scattered yield for angles of incidence less than this. However a yield having some structure is obviously observed. This yield can result from any disruption of the regular structure of the atomic surface chain, such as thermal vibrations of surface atom, or defects, i.e. missing atoms, or steps in the surface plane. The shadowing and blocking concepts are important in applications of both low and high energy ion scattering to

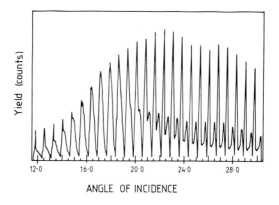

ANGLE OF INCIDENCE

Fig. 8.11 A sequence of energy spectra for 6 keV Ar scattered of a W ⟨111⟩ chain at various angles of incidence and a total scattering angle of 60°. The *E-α* loop should close at 19°, but the existence of yield at angles below that limit is caused by defects and thermal vibrations.

surface crystallography and surface atom location. The methods developed by Aono to simplify the use of the shadow cone in LEIS analysis are discussed in Section 8.5.2.

8.2.7 Computer Simulation

The use of the single scattering Equations (8.1) and (8.2) can lead to identification of the mass of the scattering atom M_2, but it is difficult to obtain quantitative composition information as a result of neutralisation and other factors treated previously in this chapter. It is even more difficult to relate double scattering peaks to definite sequences of atoms on the surface. Much of our understanding of LEIS processes comes from concurrent studies of the collision sequences in computer simulation of the scattering events. A variety of simulation programmes exist, the best known probably being Marlowe (Robinson and Torrens, 1974) or TRIM (Biersack and Haggmark, 1980) but many experiments have been interpreted using much simpler simulations. Sometimes only one-dimensional atom chains are necessary if single and simple multiple collision events need to be considered but often two dimensional arrays provide clearer insight into experimental results if more complex multiple collisions are suspected (as in the case of 'zig-zag' collisions where the second collision partner of a double scattering sequence is in a neighbouring atomic chain) or if alkali ions are used. Three dimensional lattice simulations are complex and require considerable computational time so should only be considered if both *surface* and *subsurface* structure are being investigated.

Detailed comparison of simulation results with experimental results requires careful choice of the interatomic potential in the simulation and the inclusion of such effects as thermal vibrations of the lattice atoms and perhaps inelastic energy loss of the ions. Simulations have been most useful in conjunction with experimental observations to delineate aspects of LEIS, such as the contribution of subsurface planes. Most computer simulations do not realistically include neutralisation processes though Woodruff (1982) has attempted this. An alternative approach is to avoid the need for the inclusion of neutralisation in the simulation by comparing the simulation results with the scattering of low energy alkali ions. The spectra are more difficult to interpret, but the neutralisation is less dependent on the projectile's trajectory than for scattering of inert gas ions. A very good example of this approach is the work of Englert *et al.* (1982a,b).

8.3 SURFACE ATOM IDENTIFICATION

The identification of atoms in the surface layer is the simplest form of compositional analysis using LEIS. The procedure merely involves the acquisition of an energy spectrum of ions (normally inert gas ions) scattered off the surface when the projectile's mass, incident energy and scattering angle are known. The scattered energy peaks are then compared with those given by Equation (8.1) to reveal the target atom mass. This principle works well when there is some prior knowledge of possible surface atoms which needs to be verified. For samples of largely unknown composition due regard must be paid to potential interferences. The advantage of this simple form of LEIS analysis is that often it is sufficient to establish the mere existence of an element at the surface in order to explain some material property. For example, in studies of the composition of spinel surfaces, Shelef *et al.* (1975) showed that $CoAl_2O_4$ and $ZnAl_2O_4$ do not have Co or Zn in the top surface layer while in $NiAl_2O_4$ and $CuAl_2O_4$, Ni and Cu are present. This observation alone is sufficient to explain the observed differences in catalytic activity.

The monotonic relationship between E_1 and M_2, Equation (8.1), means that the energy axis can also be regarded as a non linear mass scale with heavier target atoms corresponding to large values of E_1 and lighter atoms corresponding to lower values of E_1. This relationship applies down to energies of 100 eV or less which are below the energies normally used in LEIS.

For mass identification, the simplest experimental arrangement involves a scattering angle of 90°, which is chosen to take advantage of a simplified form of Equation (8.1):

$$E_3/E_1 = (M_2 - M_1)/(M_2 + M_1) \qquad (8.12)$$

An added advantage is that at angles of 90° or more peaks from recoiling particles are not observed in the energy spectra, and this simplifies the interpretation. Angles in excess of 90° are also chosen to reduce the probability of observing the multiple scattering peaks mentioned in Section 8.2.6. Typical scattered ion energy spectra for He^+ and Ne^+ are shown in Fig. 8.4. The use of angles in excess of 90° limits the range of detectable masses to those which exceed the projectile mass, M_1. For angles, less than 90°, the limiting detectable mass (M_L), is given by:

$$M_L = M_1 \sin \theta \qquad (8.13)$$

As shown in Fig 8.4, the ability to resolve atoms of different mass is a function of the projectile mass and the importance of projectile mass and scattering angle are highlighted by Taglauer and Heiland (1976). In order to optimise the mass resolution of LEIS it is advisable to use a projectile with a mass as near the target atom mass as possible and to employ the largest possible scattering angle. Under these conditions it is possible to detect low levels of contaminants on surfaces. For heavy impurities (e.g. Au on Si) the ultimate sensitivity is 5×10^{-4} monolayers using He^+ ions and 10^{-4} monolayers using Ar^+ ions, while the detection limit for O is between 10^{-2} and 10^{-1} monolayers. Elements lighter than O become increasingly difficult to detect even with ions as light as He^+. This is the result of both the cross section dependence on the target atomic number and on charge exchange processes. As a result, LEIS is insensitive for the detection of low levels of C on surfaces.

A systematic survey of the detection limits of LEIS has not been undertaken as for most elements it is not possible to verify the existence of 10^{-4} of a monolayer accurately by alternative calibration procedures. The detection sensitivity is also a function of the state of the surface. A highly contaminated (e.g. O or CO) or insulating surface will produce an energy spectrum of scattered ions with a relatively high background. Under such circumstances it becomes difficult to detect small signals from low concentration impurities and the detection sensitivity is reduced.

8.3.1 Light Atom Detection

The light atom detection capabilities of LEIS can be enhanced by monitoring the recoiling ion signal which necessitates the use of small angle scattering ($\theta < 90°$). Schneider *et al.* (1983) have shown that elements down to H^+ and D^+ can be observed in the recoil spectrum as well as the heavier components (O^+, C^+, Ni^+). In such positive ion spectra there is always a strong signal from the scattered inert gas ions

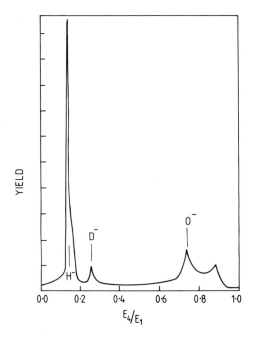

Fig. 8.12 Light atom identification can be achieved by observing recoiling ions. The sensitivity can be further enhanced by detecting negative ions, thus eliminating the background from the positively charged inert gas ions used as projectiles. In this example the projectile is 3 keV Ne^+ and the recoil angle is 30°.

which can mask some of the recoil peaks of low intensity. To overcome this problem, advantage can be taken of the fact that the inert gas ions are not scattered in the negative charge state. Thus an energy spectrum of negative ions is extremely sensitive to recoiling surface ions and is a useful probe of light ion impurities (see Fig. 8.12). This technique is particularly sensitive to strongly electronegative ions (e.g. O^-, F^-) with sensitivities estimated to be between 10^{-4} and 10^{-5} of a monolayer (O'Connor *et al.*, 1986).

8.3.2 Multiple Scattering

At small scattering angles there is a high probability of observing double scattering if the target surface is ordered (crystalline). The double scattering peak (which usually appears at a higher energy than the corresponding single scattering peak) may therefore be mistaken for a heavy impurity on an otherwise clean surface. Such observations rarely arise if the samples analysed are random or polycrystalline or if the scattering angles are large. To test whether a particular peak is a multiple scattering peak or heavy surface impurity the first step is, if possible, to rotate the target about its normal, to preserve the scattering angle, and observe the intensity of the suspect peak. If it does not change appreciably then it is likely to be a heavier mass atom on the surface. If this option is not available then the use of an incident ion of differing mass or study of E_3 as a function of θ (Section 8.3.3) normally clear up the ambiguity, by comparing the dependences with those expected from Equation (8.1).

At low scattering angles ($< 60°$) there is a strong possibility that the projectile will suffer weak collisions with neighbouring target atoms (to the principal scattering atom) on the approach to or departure from the surface. The resultant ion energy will be slightly in excess of that predicted by Equation (8.1), (Niehus and Bauer, 1975), but as such scattering conditions are rarely encountered in practical analysis (except when detecting recoil ions) it is not usually a point to consider.

8.3.3 Further Departures From Single Scattering

In the analysis procedures described so far it has been assumed that the binary collision process has been free of the effects of inelastic energy loss which can result in false identification of surface elements. As indicated in Section 8.2.6, the effects of inelastic processes can be observed, to varying degrees, at any scattering angle and are potentially more serious than multiple scattering. The inelastic loss reduces the scattered ion energy to a value below that predicted by Equation (8.1), and can thus appear to be the result of scattering off a lighter element. The most marked shifts have been observed when O covered surfaces have been analysed (Wheeler, 1974; Baun, 1978; McCune *et al.*, 1979). Although this peak shift has not as yet been identified with a specific inelastic process, it remains one of the possible explanations.

The most reliable technique for the analysis of surface composition when inelastic losses are suspected is outlined in Section 8.2.6. The recommended approach is to monitor E_3 at different scattering angles θ (Smith 1964). This procedure will most reliably identify the mass of the target atoms by comparing with Equation (8.1) as well as allowing differentiation between scattered ion peaks and recoil peaks. However, the flexibility needed to perform E-θ scans is rarely found outside purely research facilities and the measurements can be time consuming. The alternative, though not as satisfactory, is to monitor the peak energy as the incident energy is increased. In most cases the value of E_3/E_1 asymptotically approaches the value predicted by Equation (8.1). This approach can be subject to a reduction in the sensitivity to some impurities as the energy is increased as shown by McCune (1979) for O on Ta. In increasing the incident ion energy from 500 eV to 2500 eV, the ability to detect O on the surface diminished considerably.

8.3.4 Shadowing

It may be possible to miss detection of an element at the surface if it is shadowed by neighbouring or overlying atoms. A particular case, reported by Smith (1964), involves CO adsorption on Ni. In this example the C bonds to the Ni surface and the O is oriented away from the surface. This arrangement results in the O shadowing the underlying C from the ion beam and the yield from C is much lower than expected compared to the O yield. Similar shadowing processes can be observed for adsorbates on other pure surfaces (Niehus and Bauer, 1975; Taglauer and Heiland, 1975) and the increase in adsorbate signal is accompanied by a decrease in the substrate signal. Shadowing effects can be applied for precise location of adsorbates (Section 8.5).

8.4 QUANTITATIVE ASPECTS OF ION SCATTERING

8.4.1 Theory

The ion yields detected in surface analysis can, in theory, be directly related to the composition of surface constituents by Equation (8.3). The cross-section, solid angle and transmission factor have been described previously and are derivable quantities. The main uncertainty is that due to the charged fraction of particles scattered off the atoms under investigation. The theories proposed for estimating the ion yield can be tested under limited conditions for elemental targets but there is no suitable theoretical or experimental information on the charge fraction expected in most practical applications. The following sections describe how to overcome this problem and obtain reliable surface composition information.

Firstly, we consider the role played by the incident ion energy in surface analysis. The ratio of cross-sections for scattering by different elements is a slowly varying function of energy. However, charge exchange and shielding processes are so significant that the apparent relative concentration of elements on the surface, as reflected by the scattered ion yields, can change by more than two orders of magnitude (McCune, 1981) when varying the incident energy over the range 0.15 to 4 keV. A number of reports (Niehus and Bauer, 1975; Nelson, 1978; McCune, 1981) indicate that the scattered ion yield from O increases dramatically with reduced ion energy, relative to the yield from other target constituents (Ta, Nb, Ti, Mg, Al, Si). The opposite behaviour has been observed for scattered ion yields from O compared with yields from Na, Ca, Zn. The charge exchange process plays an important role in these observations (McCune, 1980) but there may also be a shielding effect in that the O sits over the cation and shadows it from the incident beam. The size of the shadow cone increases with decreasing energy in agreement with the observed changes for the first group of target elements. There are reports that these energy-dependent relative ion yields are not confined to oxides with examples involving NaF, MgF_2 and AlF_3 (McCune, 1980) and the Cd/Hg ratio in CdHgTe (Baun, 1981). The conclusion is that a calibration procedure must be carried out at the operational ion energy and extrapolation to other energies, or atomic species, should be avoided.

8.4.2 Relative Concentration Measurements

Many surface problems do not require absolute concentration information. In some cases the existence or non existence of an element, and the relative change as the surface is sputtered away yields sufficient information on which to develop models and draw conclusions about surface properties. An implicit and sometimes dubious assumption is that the ion yield is independent of surface composition. In Section 8.3, we have described an application of this approach by Shelef *et al.* (1975), who related the catalytic activity of various spinels to their relative surface composition. In a study of Ni-Mo-Al_2O_3 catalysts by Jeziorowski *et al.* (1983), LEIS was used to sputter profile the surface layers and relative ion yields were taken as a guide to concentration changes. There is a long list of studies which only follow relative changes in composition using LEIS. Some representative examples are:

- Ni-Mo-Al_2O_3 catalysts (Jeziorowski *et al.*, 1983);
- Iron based metallic glasses (Berghaus *et al.*, 1983);
- $BaTiO_3$ and $Gd_2 (MoO_4)_3$ (Tongson *et al.*, 1980);
- Ba surface enrichment in Tungsten Cathodes (Baun, 1980; Marrian *et al.*, 1985);
- Be and Sn segregation in Cu (Creemers *et al.*, 1981);
- BaO:CaO:Al_2O_3 cathodes (Lampert *et al.*, 1981);
- Pb impurity in AgBr (Tan, 1976);
- Ni catalysts on alumina and silica (Wu and Hercules, 1979);
- Preferential Sputtering of TiC by Ar^+ (Kang *et al.*, 1983);
- Surface composition of a $Pt_{10}Ni_{90}$ (111) alloy (Bertolini *et al.*, 1982); and
- Polymer surface composition (Thomas *et al.*, 1980).

There have also been many successful illustrations of the linearity of ion yield with surface composition for simple surfaces. Two examples, involving the monitoring of ion yields during film deposition, are of Pb on Ni (Verbeek, 1978) and Ti on Mo (Sagara *et al.,* 1980). These showed that the signals from the overlayer and the substrate are complementary with near total extinction of the substrate signal at one monolayer coverage (determined by RBS and thin film monitor, respectively). More chemically active adsorbants have also been monitored as a function of surface coverage, e.g. O-Ni using He^+ (Taglauer and Heiland, 1975); S-Ni using Ne^+ (Taglauer and Heiland, 1974, 1975); and Br-W using He^+ (Brongersma *et al.,* 1981). The existence of a linear complementary change in the yields of the adsorbate and the substrate is regarded as sufficient evidence for the proposition that the ion fraction is independent of surface composition, and, therefore, that ion yield is proportional to composition. In the examples involving Ni, an additional check on the surface composition was obtained by AES measurements.

Despite the supportive evidence, the linear response to coverage cannot be regarded as a universal rule. Niehus and Bauer (1975) reported that the scattered ion yield from Be and O adsorbed on W was not linear with coverage. In fact, the scattered ion yield from Be reached a maximum and then declined with additional coverage beyond 0.6 monolayers. In both cases the substrate signal showed a monotonic decrease with coverage, but it was not linear. The origin of this behaviour has not been identified but it may be due in part to changes in neutralisation rate resulting from a modified surface electronic state.

8.4.3 Quantitative Analysis Using Standards

The most reliable method for quantitative analysis using LEIS involves a comparison with elemental standards to determine concentrations in alloys or chemical compounds. In this technique, as in the previously described method, there is the inherent assumption that the ion fraction does not change significantly between the elemental surface and the alloy surface.

Consider the analysis of an alloy (Flaim, 1975) or compound of two elements, A and B. If the scattered ion yield I_a^o is obtained from a pure A surface (surface concentration of A is N_a^o atoms cm^{-2}) and an ion yield I_a is obtained from the alloy surface, then it is assumed that the number of surface atoms of type A in the alloy (N_a) is given by:

$$N_a = (I_a/I_a^o) \cdot N_a^o \qquad (8.14)$$

and similarly:

$$N_b = (I_b/I_b^o) \cdot N_b^o \qquad (8.15)$$

If the total number of surface sites in the alloy is N_o, (i.e. $N_a + N_b$), then the relative concentration of A is given by:

$$N_a/N_o = S.I_a/(S.I_a + I_b) \qquad (8.16)$$

where:

$$S = (I_b^o/I_a^o) \cdot (N_a^o/N_b^o) = (I_b^o/I_a^o) \cdot (d_b/d_a)^2 \qquad (8.17)$$

where d_a and d_b are the lattice constants in the elemental standards.

In a test of this procedure, a series of Nb-V alloys were analysed using LEIS (Martin *et al.,* 1981). The Nb-V system was chosen since no surface segregation is expected and, as the sputtering yields for each element under He^+, Ne^+ and Ar^+ bombardment are very similar, no preferential sputtering (Chapter 7) is expected. Both He^+ and Ne^+ ions over the energy range 0.5 to 3.0 keV were used and, despite some problems associated with O contamination, the results agreed with the bulk values to within $\pm$ 5% for Ne^+. The results for He^+ were less reliable as illustrated in Fig. 8.13.

A more general approach is described by Swartzfager (1984) in which the effect of different surface roughness between samples can also be included. A sensitivity factor, K, and surface roughness factor, R, are defined for each element by the relation:

$$I_a = I_o \cdot K_a \cdot N_a \cdot R \cdot \qquad (8.18)$$

where I_o is the number of projectiles. A relative sensitivity factor (S) can be defined for an alloy by Equation (8.17). The relative sensitivity factor (K_a/K_b) for Ne scattered off Au and Pd was determined by several different techniques and compared. Firstly, the sensitivity factor from pure elemental surfaces gave a value of 2.49, while that obtained from a range of alloys yielded 2.37. Alternatively, a range of surface compositions were obtained by sputtering away the surface of a Pd–Au (Pd 70%, Au 20%) sample which gradually removes a thermally grown Au rich surface layer, and monitoring the Au and Pd signals. A relative sensitivity factor of 2.67 was obtained. The third method involved the physisorption of Xe on a pure sample of Pd and Au. The sensitivity factor of Au to Pd was determined by taking the ratio of the shadowing/sensitivity factor for Xe to Au and Xe to Pd, yielding a value of 2.49. The overall

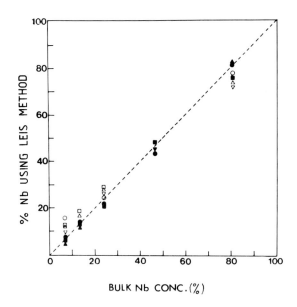

Fig. 8.13 Comparison of the surface concentration of Nb, measured by LEIS using He^+ and Ne^+ projectile, with the known bulk concentration of Nb in Nb/V alloys (Martin *et al.*, 1981).
Key: a) He^+ (keV) 0.5 ○, 1.0 □, 1.5 △, 2.0 ▽. b) Ne^+ (keV) 0.5 ■, 1.0 ▼, 2.0 ●, 3.0 ▲.

agreement is very good, with a relative mean deviation of 3.4%. This level of reproducibility under differing experimental conditions is evidence for the validity that, in the Au–Pd system, the ion fraction is independent of surface conditions. Similar results were obtained for Ag–Au, Cu–Ni, Au–Ni and Cu–Au alloy systems.

One of the most popular applications of LEIS to quantitative surface analysis is in the field of surface segregation and preferential sputtering. In these studies good surface layer selectivity and sensitivity are essential and LEIS is unmatched by alternative analysis techniques. The range of materials studied for surface segregation have been predominantly binary metallic alloys: Cu–Ni, Ag–Au, Cu–Ni and Cu–Pt, Ni–Au, Pd–Au, Pt–Au, Ni_3Sn, Pt_3Sn, Pt–Rh, Pt–Ni (111). Frankenthal and Malm (1976) have also successfully used LEIS to analyse the depth profile of an oxide layer formed on a range of Cr–Fe alloys. A relative sensitivity factor (I_a^o/I_b^o) was determined for Cr–Fe from pure metal standards and from the oxides (Fe_2O_3 and Cr_2O_3), in both cases the sensitivity factor was the same to within the experimental accuracy (i.e. $I_a^o/I_b^o = 1.5 \pm 0.1$). The measurements from the oxides also allowed the determination of the relative sensitivity factor for O of 0.11 for O/Fe, and 0.08 for O/Cr. This low sensitivity highlights the difficulty in detecting light atoms with LEIS. Using the measured sensitivity factors, the concentration profile revealed an enrichment of Cr at or near the surface of Fe–Cr alloys, with the O content diminishing uniformly through the film.

In LEIS measurements of surface segregation, care should be taken to ensure that the composition is not altered by processes associated with the ion beam. The incident ions have the potential to modify both the surface and near-surface regions by sputtering, recoil implantation and radiation enhanced diffusion (Chapters 1 and 7). Their influence on the measurements can be minimised by keeping the ion beam dose low during the measurement and/or by employing a projectile which gives the same sputtering yield for all components of the alloy.

Surface modification and simultaneous surface analysis (by LEIS) have been exploited in the study of preferential sputtering, recoil mixing and radiation enhanced diffusion processes. Typically, the depth of the altered layer produced by an ion beam and the modified surface layer composition can be determined (Taglauer, 1982). In some cases there is disagreement between LEIS and alternative analytical techniques (principally AES) on the nature of the surface composition. One possible reason for this difference is the different sampling depths of the two techniques (Chapter 7). LEIS monitors the composition of the uppermost surface layer while, under optimum conditions, AES samples to depths of 0.5 nm or more and thus tends to average out any strong surface segregation effects. These different sampling depths can be used to advantage to deduce both the surface and sub-surface compositions. For example, in a recent study of a Cu–Ni alloy, the concentration of Ni at the surface was found to be less than that of the bulk, but just below the surface it was much greater than the bulk value (Okutani *et al.,* 1980). Consequently, LEIS and AES in combination promises to be a fruitful approach to near surface (i.e. < 2 nm) composition analysis (Frankenthal and Siconolfi, 1985).

Wheeler (1975) determined the sensitivity factors in a range of oxides directly from the oxide rather than from elemental standards. The sensitivity factor, referred to as a relative cross-section, is given by the expression:

$$S = (I_M/I_O) \cdot (c_O/c_M) \tag{8.19}$$

where I_M and I_O are the scattered ion yields, and c_M and c_O are the atomic concentrations of the metal and O, respectively, in the oxide. It was assumed that the surface composition reflected the bulk composition.

From 11 different oxides a relationship between the relative cross-section and the atomic number of the metal was derived:

$$S = 0.04 \ Z^{1.57} \tag{8.20}$$

The worst departure from this relationship was for Mo which was just over half the value given by Equation (8.20). A limited survey of the effect of the ion energy on S was also reported showing that the exponent in Equation (8.20) increased with increasing ion energy. This energy-dependent behaviour has been investigated more thoroughly by McCune (1980, 1981). The energy dependence of S for oxides and fluorides showed extreme differences for a variety of elements. Much more research is needed on the surface composition of ion bombarded compounds and the charge-exchange processes before relationships for relative sensitivities can be established.

Several attempts to analyse complex surfaces by comparisons with chemical compound standards have been made with varying degrees of success. Bertrand *et al.* (1983) used standard spectra from MoO_3 and $BiPO_4$ to derive surface compositions of each element in a $BiPO_4$–MoO_3 composite catalyst. The sputter profile of each element was monitored but no absolute concentration estimates were made from these spectra. In a later study (Beuken and Bertrand, 1985) the spectra for Co_9S_8, MoS_2 and Al_2O_3 were used to analyse $CoMoS/Al_2O_3$ catalytic surfaces. McCune (1979) has shown that this procedure cannot be used successfully for quantitative analysis of other chemical compounds. This was demonstrated by using MgO and Al_2O_3 as standards for the analysis of $MgAl_2O_4$, and MgO and SiO_2 as standards for Mg_2SiO_4. In these cases the measured surface compositions were best represented by $MgO(Al_2O_3)_2$ and $(MgO)_{1.5}SiO_2$.

The basic assumption made in the use of standards for LEIS composition analysis is that the charge fraction of surface scattered ions is independent of surface composition. While there is some evidence to support this proposition, the study by McCune described above reveals that it cannot be regarded as a general rule. The use of standards may assist in monitoring relative changes in concentration during profiling, but until the charge-exchange process is better characterised it cannot be relied upon to give absolute concentrations.

8.5 SURFACE CRYSTALLOGRAPHY AND STRUCTURE

An area of intense study has been the application of LEIS to surface crystallography. The surface sensitivity of LEIS, coupled with shadowing and blocking effects, offers a powerful method to determine the structure of the outermost atomic layer. The data analysis is greatly simplified by the observation that the interactions with the target atoms are governed by classical mechanics and thus, although useful, computer simulation is not essential to extract structure information. This is in contrast with the best alternative method (LEED) which not only samples the first few atomic layers but also requires sophisticated computer simulations to extract structure information. LEIS is not sensitive to atomic arrangements in second and deeper layers when using inert gas ions, but low energy H^+ can be used (Feijen *et al.,* 1973; Feijen, 1975; Bhattacharya *et al.,* 1979; Schneider and Verbeek, 1981; Saiki *et al.,* 1982) in the same way as higher energy scattering (Turkenburg *et al.,* 1976; Tromp, 1983) to index the surface layer to the structure of deeper layers.

8.5.1 Single and Multiple Scattering

Useful information about the atomic arrangement on a crystal surface can be obtained by monitoring the single scattering peaks while rotating the target (azimuthally) about its normal, with both the analyser and the incident ion beam at high angles with respect to the surface. Under these conditions, the second atomic layer will be visible at some angles of azimuth and shielded at others. Consequently, if the incident energy is sufficiently large, 2 to 5 keV, then an increased ion yield will be observed from the second layer at the appropriate azimuthal angles (Buck *et al.,* 1979). For clean surfaces, the information obtained from this simple scan can usually do little more than identify the surface orientation. More detailed structural information can be deduced by changing to low angles of incidence as we will describe later.

Double scattering is also useful for identifying low index directions in the surface layer and for determining the interatomic spacing. As indicated previously, this event, involving two weaker collisions than the single scattering event, results in a peak at a higher energy. This peak is most prominent when the two scattering centres are 'in plane' with the directions defined by the incident beam and the detector. Under these conditions the scattered ion energy is also a maximum (Higginbottom *et al.,* 1985). The magnitude of the double scattering peak I_{DS}, relative to the single scattering peak I_{SS} is given by:

$$I_{DS}/I_{SS} = (\sigma_1 \, \sigma_2 \, P_D^+)/(\sigma_o \, d^2 \, P_S^+) \qquad (8.21)$$

where σ_1 and σ_2 are the cross-sections for the two collisions in the double scattering event, d the interatomic spacing, σ_0 is the single collision cross-section, and P_D^+ and P_S^+ are the double and single collision charge fractions, respectively. This equation reveals that the double scattering yield is proportional to d^{-2} and is thus most prominent in low index surface directions. When the two atoms are not 'in plane' with the incident beam and the detector, the double scattering event is referred to as a zigzag collision and the energy of the peak is lower than for 'in plane' scattering. The monitoring of multiple collision peaks as a function of the azimuthal angle gives detailed information on structure, low index directions and interatomic spacings. However, the height of the multiple scattering peak is also modified by the charge-exchange process and it cannot be used to determine spacing. It is therefore necessary to use the energy of the peak to determine the surface spacings. In this process the total scattering angle is kept constant and the angle of incidence to the surface is varied over a range of angles (E-α scan). As described in Section 8.2.6, the angle at which the E-α loop closes is characteristic of the interatomic spacing (Begemann, 1972; O'Connor, 1978; MacDonald, 1980).

8.5.2 Shadow Cone

The use of shadowing allows atomic positions to be identified with a minimum of ambiguity arising from the charge-exchange process. While the E-θ loop can be explained on the basis of shadow cones, perhaps the simplest use of the shadow cone has been made by Aono (1984) in what is termed Impact Collision Ion Surface Scattering (ICISS).

In ICISS the distance between two atoms on the surface, or between the surface and second layer, is determined by changing the angle of incidence of the ion beam to the surface. The increasing angle of incidence rotates the shadow cone formed by the surface atom and, at a critical angle, the second atom in the interaction emerges from the shadow cone, resulting in a measured scattered ion yield. This technique can be used to determine the spacing along a surface chain of atoms by a single measurement of the critical angle and a knowledge of the shadow cone dimensions. The shape and size of the shadow cone is a function of the ion energy and the interatomic potential used to describe the scattering. To aid in identifying atomic positions Oen (1983) recently reported a polynomial which accurately describes the shadow cone for the Molière potential. To locate atoms below the surface it is preferable to measure the critical angle along different surface crystallographic directions and then locate the second atom by triangulation (Fig. 8.14). Aono *et al.* (1981) demonstrated this approach by locating the position of the C layer in a TiC(111) surface to be 87 ± 8 pm below the surface.

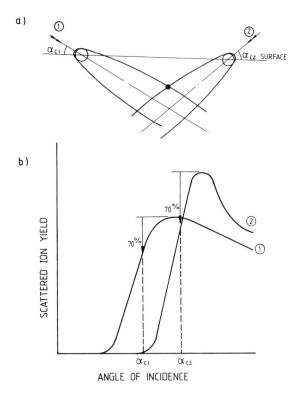

Fig. 8.14 An illustration of ICISS a method of surface structure analysis devised by Aono *et al.* (1981) using shadowing to locate the position of subsurface atoms and the distance between surface atoms.

The geometry used in ICISS emphasises the use of 180° scattering to simplify the analysis. To achieve 180° scattering, the projectile must have a 'head-on' (zero impact parameter) collision which will occur only above a limiting (critical) angle of incidence with respect to the surface. (i.e. when the second atom just intersects the shadow cone, Fig. 8.14a). In practice, such a geometry is difficult to achieve and Yamamura and Takeuchi (1984) have shown that scattering angles as low as 90° can be used if small adjustments to the critical angle are made for the effect of the non-zero impact parameter in the second collision.

The determination of the critical angle is complicated by a number of factors. With a perfectly stationary collection of lattice atoms the critical angle can be clearly defined as the angle at which the yield of scattered ions commences. However, thermal vibrations act to smooth out this dependence on angle of incidence. Generally, a sharply increasing yield

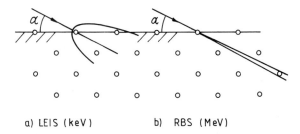

a) LEIS (keV) b) RBS (MeV)

Fig. 8.15 A comparison of the role played by the shadow cone dimensions in the application of LEIS and RBS to surface structure analysis (Aono, 1984).

with increasing angle of incidence is observed with a peak yield at an angle just above the critical angle. Fauster and Metzner (1986) have recently shown that the shape of the yield curve depends on the projectile energy leading to the conclusion that charge exchange may also play a role. In their first article on the ICISS technique, Aono *et al.,* (1981) recommended that the critical angle be determined from the onset of yield as a function of angle of incidence, though in a more recent article Aono (1984) recommends that the critical angle be determined at 70% of the maximum yield (Fig. 8.14b). The criterion used by Niehus and Comsa (1984) is the angle at which the yield is 50% of the maximum yield while Overbury and Huntley (1985) have concluded, on the basis of computer simulations, that the criterion should be 90%. While such an uncertainty in determining the critical angle may seem to limit the accuracy of this approach, it is still capable of determining the relative position of two atoms to within 10 pm.

Aono (1984) has demonstrated important differences between the uses of shadow cones in the high energy and low energy cases (Fig. 8.15). At high energies the shadow cone radius is small while at low energies it is large, though we would never want it to be larger than half the atomic spacing at the surface. The small radius at high energies means that, by choice of suitable geometry, the shadow cone (or blocking cone) can be made to interact with atoms in layers different to the one containing the atom which is the origin of the cone. However, the angle of the incident beam to the surface would have to be prohibitively small for the shadow cone to be able to interact with ions in the same layer as the atom at the origin of the cone. In contrast, at low incident ion energies, the larger cone radius means it is very reasonable to use a geometry involving interactions with coplanar surface atoms. Furthermore, neutralisation effects associated with ions which go below the surface layer at low energies, means that the ion scattering yield from subsurface atoms is negligible.

8.5.3 Adsorbates

Adsorbates can be located either by the effect they have on single and double scattering yields from the substrate atoms or by changes in adsorbate scattering yield with orientation. In the former case the adsorbate may block a double scattering process, it may shield some of the surface atoms from the incident ion or it may block some of the scattered ions from the analyser direction. Quantitative information for such shadowing and blocking has been used to identify probable surface sites of the adsorbate. To illustrate this point Equation (8.3) can be modified to describe the ion yield from a substrate surface:

$$N^+ = N_i (N_a - \beta \cdot N_b) \cdot P_a^+ \cdot T \cdot d\sigma/d\Omega \cdot d\Omega \qquad (8.22)$$

where N_b is the number of adsorbate atoms on the surface and β is a shadowing factor (i.e. the number of substrate atoms which are shadowed or for which scattered ions from the substrate are blocked by each adsorbate atom). To obtain β, the ion yield scattered from the substrate is monitored as the adsorbate coverage is increased. The magnitude of β so determined can then be used to identify probable adsorption sites. For values of β greater than unity the adsorption site is on a low index row, while low values (< 1) can be associated with positions between rows. A number of studies which have used this technique are:

- O on Ni(110) (Heiland and Taglauer, 1972);
- O on Ag(110) (Heiland *et al.*, 1975);
- O on W(100) (Prigge *et al.*, 1977);
- O on Ni (100) (Brongersma and Theeten, 1976);
- N_2 on Ni(110) (Moller *et al.*, 1984);
- O on W (Niehus and Bauer, 1975); and
- O on Pb (Honig and Harrison, 1973).

The value of these measurements is greatly enhanced if they are made along different crystallographic directions. This approach is only applicable if the adsorption site is above or in the surface layer. To locate subsurface adsorbates it is necessary to adopt the shadow cone procedure of Aono (Section 8.5.2).

8.6 COMPARISON WITH OTHER TECHNIQUES

The comparison of LEIS with other techniques should not be viewed in a competitive way, with the aim of determining which is 'the best' technique overall. In each application some particular technique will stand out as the most appropriate but this usually does not exclude the use of alternatives to supply supportive or complementary evidence. In some of the investigations to verify that the ion yield in LEIS is linearly related to the surface composition of adsorbates, the 'proof' was taken as the linear relationship between the LEIS signal and the AES signal (Taglauer and Heiland, 1974). Such a comparison is valid as long as the adsorbed atoms sit on the surface and are not incorporated into the surface region by diffusion or reconstruction. The combination of LEIS and AES also allows the surface and subsurface composition to be elucidated (Okutani *et al.*, 1980). Composite studies involving LEIS and XPS have been useful for determining changes in Ta_2O_5 surfaces under ion bombardment (Nelson, 1978) and the surface composition of Au–Pd alloys (Varga and Hetzendorf, 1985). LEIS, UPS and AES have been applied to the surface composition of metallic glasses (Berghaus *et al.*, 1983). The combination of techniques supplies complementary information on the surface layers, including the possibility of chemical bonding information from the XPS and UPS techniques.

Although more research needs to be concentrated on the characterisation of the neutralisation processes associated with LEIS, it has been shown that the relative concentrations determined from LEIS more closely reflect the true situation than those determined by SIMS. SIMS is extremely sensitive to low concentrations of alkali ions which have a low ionisation potential resulting in a large ion fraction. The alkali ion fraction has been measured for low energy scattered ions to approach 100% in some cases at low energies (Algra *et al.*, 1982) and it would be expected to have similar values for sputtered material. Thus the ion yield from alkali ions measured by SIMS will be disproportionate to the true concentration of the element on the sample under investigation.

A comparison between SIMS and LEIS (Grundner *et al.*, 1974) revealed basic agreement between these techniques in the analysis of the surface composition of Cu and stainless steel but it also highlighted the extreme sensitivity of SIMS to specific elements and contaminants. During the cleanup of Cu by Ar^+ bombardment, the Cu^+ SIMS signal dropped by a factor of 100 (Fig. 8.16) while the LEIS signal for He^+ scattered off Cu rose by only 25%. In the same study, after stainless steel was sputter cleaned by He^+ bombardment, SIMS indicated a 29% surface concentration of Al while the same analysis with LEIS using He^+ ions resulted in an estimate of 0.01%.

In structure studies, LEED should be used when available to characterise the surface being investigated by LEIS (Englert *et al.*, 1973; Heiland and Taglauer, 1972). Usually, a high quality LEED pattern indicates a "clean" surface suitable for structure determination by LEIS.

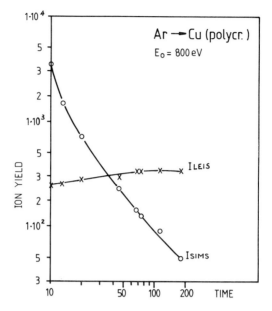

Fig. 8.16 An illustration of the relative importance of the 'matrix' effect for LEIS and SIMS. The Cu SIMS signal decreases by a factor of 100 during cleanup by Ar^+, but the LEIS signal increases by only approximately 15% (Grundner *et al.*, 1974).

Surface studies which apply several techniques are infrequent but two cases involving surface segregation investigated with AES, XPS, LEIS and CO adsorption are good examples of their potential. In Table 8.1, the composition of a $Pt_{10}Ni_{90}$ (111) surface has been listed (Bertolini *et al.*, 1982). Detailed agreement under such circumstances cannot be expected but the overall agreement on a significant enrichment of the surface in Pt over its bulk composition is obvious. In a separate study of Pt_3Sn and Ni_3Sn, differences between techniques were attributed to the different sampling depths of the analysis procedures (Biloen *et al.*, 1979) as shown in Table 8.2.

TABLE 8.1
Composition of a $Pt_{10}Ni_{90}$ (111) surface as determined by various techniques.

Technique	%Pt	%Ni
AES	30 ± 3	70 ± 3
XPS	34 ± 2	66 ± 2
ISS (Ne^+)	27.6	66.4
ISS (He^+)	30.8	67.1
CO adsorption	40	60

TABLE 8.2
Comparision of techniques in studies of Pt_3Sn and Ni_3Sn

Technique	Pt₃Sn		Ni₃Sn	
	% Pt	% Sn	% Ni	% Sn
XPS	68	32	–	–
AES	60	40	22	78
AES/XPS	50	50	10	90
LEIS	43	57	27	73
CO adsorption	42	58	15	85

In conclusion, LEIS is a surface sensitive analysis technique which can probe the outermost one or two atomic layers. It can be used to elucidate the surface structure and adsorption sites. Analysis can be made quantitative in many cases by comparison to standards, providing the ability to detect surface contaminants with a sensitivity limit ranging from 10^{-1} for C to 10^{-4} for Au. LEIS appears to be far less sensitive to the 'matrix' effects which makes the interpretation of SIMS so difficult when absolute or relative concentration evaluations need to be made. The spatial resolution is limited by ion beam size which is typically 500 μm, though it can be reduced to 100 μm or less if necessary by use of appropriate ion optics (Chapters 7 and 10). LEIS should be used, whenever possible, in conjunction with the many available complementary surface analysis techniques to maximise the compositional and structural information obtained.

REFERENCES

Algra, A.J., van Loenen, E., Suurmeijer, E.P.Th.M. and Boers, A.L. (1982). *Rad. Effects* **60**, 173.
Aono, M, (1984). *Nucl. Instrum. Methods* **B2**, 374.
Aono, M., Oshima, C., Zaima, S., Otani, S. and Ishizawa, Y. (1981). *Japanese Journal of Applied Physics* **20**, L829.
Baun, W.L. (1978). *Phys. Rev.* **A17**, 849.
Baun, W.L. (1980). *Appl. Surf. Sci.* **4**, 374.
Baun, W.L. (1981). *Appl. Surf. Sci.* **7**, 46.
Begemann, S.H.A. (1972). Thesis, University of Groningen.
Berghaus, Th., Neddermeyer, H., Radlik, W. and Rogge, V. (1983). *Physica Scripta* **T4**, 194.
Bertolini J.C., Massarddier, J., Delichere, P., Tardy, B., Imojik, B., Jugnet, Y., Tran Minh Duc, Temmerman, L., Creemers, C., Van Hove, H. and Neyens, A. (1982). *Surf. Sci.* **119**, 95.
Bertrand, P., Beuken, J.–M. and Delvaux, M. (1983). *Nucl. Instrum. Methods* **218**, 249.
Beuken, J.–M. and Bertrand, P. (1985). *Surf. Sci.* **162**, 329.
Bhattacharya, R.S, Eckstein, W. and Verbeek, H. (1979). *Nucl. Materials* **79**, 420.
Biersack, J.P. and Haggmark, L.G. (1980). *Nucl. Instrum. Methods* **174**, 257.

Biersack, J.P. and Wittmaack, K., eds. (1986). *Nucl. Instrum. Methods* **B15**.

Biloen, P., Bouwman, R., Van Santen, R.A. and Brongersma, H.H. (1979). *Appl. Surf. Sci.* **2**, 532.

Bloss, W. and Hone, D. (1978). *Surf. Sci.* **72**, 277.

Blume, R., Eckstein, W., Verbeek, H. and Reichelt, K. (1982). *Nucl. Instrum. Methods* **194**, 67.

Brongersma, H.H. and Theeten, J.B. (1976). *Surf. Sci.* **54**, 19.

Brongersma, H.H., Van Der Ligt, G.C.J. and Rouweler, G., (1981). *Philips J. of Res.* **36**, 1.

Buck, T.M., Wheatley, G.H. and Verheij, L.K. (1979). *Surf. Sci.* **90**, 635.

Creemers, C., Van Hove, H. and Neyeens, A. (1981). *Appl. Surf. Sci.* **7**, 402.

Englert, W., Heiland, W. and Taglauer, E. (1973). *Surf. Sci.* **83**, 243.

Englert, W., Taglauer, E. and Heiland, W. (1982a). *Nucl. Instrum. Methods* **194**, 663.

Englert, W., Taglauer, E. and Heiland, W. (1982b). *Surf. Sci.* **117**, 124.

Englert, W., MacDonald, R.J. and Taglauer, E. (1985). *Verhandl. DPG (VI)* **20**, 904.

Fauster, Th. and Metzner, M.H. (1986). *Surf. Sci.* **166**, 29.

Flaim, T.A. (1975). Research Publication GMR-1942, Research Laboratories, General Motors, Warren, Michigan.

Feijen, H.H.W. (1975). Thesis, University of Groningen.

Feijen, H.H.W., Verhey, L.K. and Boers, A.L. (1973). *Phys. Lett.* **45A**, 31.

Frankenthal, R.P. and Malm, D.L. (1976). *J. Electrochem. Soc.* **123**, 186.

Frankenthal, R.P. and Siconolfi, D.J. (1985). *Surf. Interface Anal.* **7**, 223.

Grundner, M., Heiland, W. and Taglauer, E. (1974). *Appl. Phys.* **4**, 243.

Hagstrum, H.D. (1954). *Phys. Rev.* **96**, 336.

Hagstrum, H.D. (1977). *In* 'Inelastic Ion-Surface Collisions', (Tolk, N.H., Tully, J.C., Heiland, W. and White, C.W., eds) Academic Press, New York.

Heiland, W. and Taglauer, E. (1972). *J. Vac. Sci. Technol.* **9**, 620.

Heiland, W., Schaffler, H.G. and Taglauer, E. (1973). *Surf. Sci.* **35**, 381.

Heiland, W., Iberl, F., Taglauer, E. and Menzel, D. (1975). *Surf. Sci.* **53**, 383.

Higginbottom, P.R., Homer, J., O'Connor, D.J. and MacDonald, R.J. (1985). *Appl. Surf. Sci.* **22/23**, 100.

Hogberg, G., Norden, H. and Skoog, R. (1970). *Phys. Stat. Sol.* **42**, 441.

Honig, R.E. and Harrison, W.L. (1973). *Thin Solid Films* **19**, 43.

Jeziorowski, H., Knozinger, H., Taglauer, E. and Vogdt, C. (1983). *J. Catalysis* **80**, 286.

Kang, H.J., Matsuda, Y. and Shimizu, R. (1983). *Surf. Sci.* **134**, L500.

Lampert, W.V., Baun, W.L., Lamartine, B.C. and Haas, T.W. (1981). *Appl. Surf. Sci.* **9**, 165.

McCune, R.C. (1979). *Anal. Chem.* **51**, 1249.

McCune, R.C., Chelgren, J.E. and Wheeler, M.A. (1979). *Surf. Sci.* **84**, L515.

McCune, R.C. (1980). *Appl. Surf. Sci.,* **5**, 275.

McCune, R.C. (1981). *J. Vac. Sci. Technol.* **18**, 700.

MacDonald, R.J. (1980). *Aust. J. Phys.* **33**, 1.

MacDonald, R.J. and Martin, P.J. (1981). *Surf. Sci.* **111**, L739.

MacDonald, R.J. and O'Connor D.J. (1983). *Surf. Sci.* **124**, 423.

MacDonald, R.J., O'Connor, D.J. and Higginbottom, P.R. (1984). *Nucl. Instrum. Methods* **B2**, 418.

Marrian, C.R.K., Shih, A. and Haas, G.A. (1985). *Appl. Surf. Sci.* **24**, 372.

Martin, P.J., Loxton, C.M., Garrett, R.F., MacDonald, R.J. and Hofer, W.O. (1981). *Nucl. Instrum. Methods* **191**, 275.

Moller, J., Heiland, W. and Unertl, W. (1984). *Nucl. Instrum. Methods* **B2**, 396.

Muda, Y. and Hanawa, T. (1980). *Surf. Sci.* **97**, 283.

Nelson, G.C. (1978). *J. Vac. Sci. Technol.* **15**, 702.

Niehus, H. and Bauer, E. (1975). *Surf. Sci.* **47**, 222.

Niehus, H. and Comsa, G. (1984). *Surf. Sci.* **140**, 18.

O'Connor, D.J. (1978). Thesis, Australian National University.

O'Connor, D.J. and MacDonald, R.J. (1980a). *Rad. Effects* **45**, 205.

O'Connor, D.J. and MacDonald, R.J. (1980b). *Nucl. Instrum. Methods* **170**, 495.

O'Connor, D.J. and Biersack, J.P. (1986). *Nucl. Instrum. Methods* **B15**, 14.

O'Connor, D.J., MacDonald, R.J., Eckstein, W., and Higginbottom, P.R. (1986). *Nucl. Instrum. Methods* **B13**, 235.

Oen, O.S. (1983). *Surf. Sci.* **131**, L407.

Okutani, T., Shikata, M. and Shimizu, R. (1980). *Surf. Sci.* **99**, L410.

Overbury, S. and Huntley, D. (1985). *Phys. Rev.* **B32**, 6278.

Prigge, S., Niehus, H. and Bauer, E. (1977). *Surf. Sci.* **65**, 141.

Robinson, M.T. and Torrens, I.M. (1974). *Phys. Rev.* **B9**, 5008.

Roosendaal, H.E., Lutz, H.O., Heiland, W. and Andra, H.J. (eds) (1984). *Nucl. Instrum. Methods* **B2**

Rusch. T.W. and Erikson, R.L. (1977). *In* 'Inelastic Ion-Surface Collisions', (Tolk, N.H., Tully, J.C., Heiland, W. and White, C.W., eds) Academic Press, New York.

Sagara, A., Akaishi, K., Kamada, K. and Miyahara, A. (1980). *J. Nucl. Materials* **93/94**, 847.

Saiki, K., Tanaka, S. and Koma, A. (1982). *Applied Phys.* **A27**, 263.

Schneider, P.J. and Verbeek, H. (1981). *J. Nucl. Materials* **97**, 319.

Schneider, P.J., Eckstein, W. and Verbeek, H. (1983). *Nucl. Instrum. Methods* **218**, 713.

Sebastian, K.L., Jyothi Bhasu, V.C. and Grimley, T.B. (1981). *Surf. Sci.* **110**, L571.

Shelef, M., Wheeler, M.A.Z. and Yao, H.C. (1975). *Surf Sci.* **47**, 697.

Shoji, F. and Hanawa, T. (1984). *Nucl. Instrum. Methods* **B2**, 401.

Smith, D.P. (1964). *J. Appl. Phys.* **38**, 340.

Souda, R. and Aono, M. (1986). *Nucl. Instrum. Methods* **B15**, 114.

Swartzfager, D.G. (1984). *Anal. Chem.* **56**, 55.

Taglauer, E. and Heiland, W. (1974). *Appl. Phys. Lett.* **24**, 437.

Taglauer, E. and Heiland, W. (1975). *Surf. Sci.* **47**, 234.

Taglauer, E. and Heiland, W. (1976). *Appl. Phys.* **9**, 26.

Taglauer, E. (1982). *Appl. Surf. Sci.* **13**, 80.

Tan, Y.T. (1976). *Surf. Sci.* **61**, 1.

Thomas, G.E., Van Der Ligt, G.C.J., Lippits, G.J.M. and Van Der Hei, G.M.M. (1980). *Appl. Surf. Sci.* **6**, 204.

Tongson, L.L., Bhaila, A.S., Cross, I.E. and Knox, B.E. (1980). *Appl. Surf. Sci.* **4**, 263.

Tromp, R.M. (1983). *J. Vac. Sci. Technol.* **A1**, 1047.

Turkenburg, W.C., Soska, W., Saris, F.W., Kersten, H.H. and Colenbrader, B.G. (1976). *Nucl. Instrum. Methods* **132**, 587.

Varga, P. and Hetzendorf, G. (1985). *Surf. Sci.* **162**, 544.

Verbeek, H. (1978). *In* 'Materials Characterisations Using Ion Beams', (Thomas, J.P. and Cachard, A., eds), Plenum Press, New York, 303.

Wheeler, M.A. (1974). Third Annual 3M ISS Users Conference.

Wheeler, M.A. (1975). *Anal. Chem.* **47**, 146.

Woodruff, D.P. (1982). *Surf. Sci.* **116**, L219.

Wu, M. and Hercules, D.M. (1979). *J. Phys. Chem.* **83**, 2003.

Yamamura, Y. and Takeuchi, W. (1984). *Rad. Effects* **83**, 73.

Zartner, A. (1979). 'Oscillatory Ion Yields of He$^+$ Scattered from Atomic and Solid Pb Targets', Max-Planck-Institut fur Plasmaphysik Report, IPP 9/31.

9

Ion Scattering from Surfaces and Interfaces

L.C. FELDMAN

AT & T Bell Laboratories
Murray Hill, USA

This chapter is concerned with interactions between a high energy ion beam and surface atoms of an aligned single crystal. Such interactions are useful in studying the structure of the surface region of solids and solid/solid interfaces. Here high energy refers to the velocity region greater than the Bohr velocity (2.2×10^8 cm s^{-1}; i.e. 25 kcV H$^+$ or 100 keV He$^+$) where the basic particle–solid interactions are well understood. The scattering that takes place in the first few monolayers of a single crystal solid can be measured with precision giving detailed information on the arrangement of atoms at a single crystal surface or interface. Ion

ION BEAMS FOR
MATERIALS ANALYSIS
ISBN 0 12 099740 1

beam/surface studies have been applied to the structure determination of atomically clean surfaces, adsorbate/surface interactions, ultra-thin epitaxial layers and internal surfaces such as solid/solid interfaces.

9.1 BASIC PRINCIPLES

The concepts of ion scattering and channeling are introduced in Chapter 1 and discussed in detail in Chapter 3 and 6. Correlated scattering, which is responsible for channeling, also plays a role at the surface or within the first few monolayers of a crystalline arrangement of atoms. The correlation effect results in an interaction between the incident beam and the first layer of the crystalline solid plus any near-surface atoms displaced from regular lattice sites. In backscattering this surface interaction gives rise to a surface peak (SP) which provides a measure of the scattering yield from the surface. This chapter reviews the current state of quantitative understanding of the SP and describes its use in surface and interface studies.

9.1.1 Shadowing

The simplest example of correlated small angle scattering is illustrated in Figure 9.1. Ions incident at the smallest impact parameter undergo large angle scattering; those at large impact parameter suffer small impact deflection which determines the flux distribution of ions near the second atom. In the absence of thermal vibrations this flux has a distance of closest approach to the second atom, R_c, within which there are no ions. For Coulomb scattering:

$$R_c = 2(Z_1 Z_2 e^2 d/E)^{1/2} \tag{9.1}$$

where d is the atomic spacing, E is the incident ion energy and Z_1, Z_2 the atomic numbers of the incident ion and target atom respectively. For 1.0 MeV He^+ incident along the $\langle 100 \rangle$ direction of W, $R_c = 0.16$ Å. Since the bulk value of ρ, the two dimensional thermal vibrational amplitude, is 0.06 Å (at room temperature), the shadowing is still substantial, even in a nonstatic case.

In the Coulomb approximation the flux distribution at a distance r from the second atom, $f_c(r)$, is given by (Lindhard, 1965)

$$f_c = \begin{cases} 0 & r < R_c \\ \dfrac{1}{2}\left[\left[1 - \dfrac{R_c^2}{r^2}\right]^{1/2} + \left[1 - \dfrac{R_c^2}{r^2}\right]^{-1/2}\right], & r > R_c \end{cases} \tag{9.2}$$

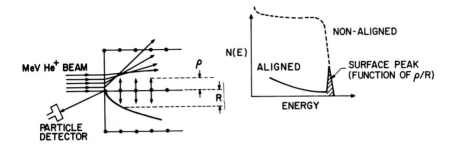

Fig. 9.1 Schematic showing interactions at the surface of an aligned single crystal and the formation of a shadow cone of radius R_c. Typical room temperature thermal vibrations are shown (amplitude ρ). Energy spectra for the aligned and nonaligned case are shown at the right.

This distribution is extremely sharp for a static lattice. To provide the scattering yield from the second atom, the flux distribution, $f(r)$, must be convoluted with the positive distribution of the second atom, $P(r)$, which is governed by thermal motion and is given by:

$$P(r) = [\exp(-r^2/\rho^2)]/\pi\rho^2 \tag{9.3}$$

The scattering yield from the second atom is then given by:

$$P_2 = \int_0^\infty f(r)\, P(r)\, 2\pi r dr \tag{9.4}$$

When the flux distribution in the Coulomb approximation, Equation (9.2), is folded with the Gaussian (thermal) position distribution of first and second atoms, an estimate of the two atom surface peak intensity is obtained as (Feldman *et al.*, 1977):

$$I_c = 1 + (1 + R_c^2/2\rho^2)\exp(-R_c^2/2\rho^2) \tag{9.5}$$

The first term represents the unit contribution from the first atom in the string, and the second term represents the variable contribution from the second atom. In most cases this two-atom Coulomb approximation is not adequate for detailed comparison with experiment; however, it is useful for its suggestion that the SP intensity is a function of one parameter, ρ/R.

This use of the pure Coulomb potential and the two-atom model allows us to illustrate many of the features of the shadowing concept. These assumptions, however, are not sufficiently accurate or complete and must be refined for comparison with actual experiment. The impact parameters important in determing the flux distribution are of order 0.1 Å and are large compared to the radius of the inner electron orbits of the scattering atoms. This means that the bare Coulomb approximation does not adequately describe the flux distribution of the beam. A variety of screened Coulomb potentials, which take into account the shielding of the nucleus by the atomic electrons, have been derived. The most accurate are based on the Thomas-Fermi model of the atomic potential and the one most used is the Molière numerical approximation, $V(\check{r})$, to the exact Thomas-Fermi potential, given by

$$V(\check{r}) = \frac{Z_1 Z_2 e^2}{\check{r}} (0.1 e^{-6\check{r}/a} + 0.35 e^{-0.3\check{r}/a} + 0.55 e^{-1.2\check{r}/a}) \qquad (9.6)$$

where $\check{r}$ is the distance from the nucleus in spherical coordinates. In this equation a is the Thomas-Fermi screening distance given by:

$$a = 0.8853 a_0 (Z_1^{1/2} + Z_2^{1/2})^{-2/3} \qquad (9.7)$$

where a_0 is the Bohr radius, 0.528 Å. (Commonly used alternate definitions of the screening distance are given in Equation (1.4).) Analogous to the derivation of the Coulomb shadow cone radius, one can derive the Molière shadow cone radius R_M, corresponding to the use of the screened potential for any particular case of interest. R_M is expected to be quite accurate, whereas R_c remains a useful approximation.

9.1.2 Practical Considerations and Computer Simulation

In cases of practical interest the surface peak yield corresponds to scattering from a few atoms per string. Since a screened Coulomb potential is necessary to describe the interactions, analytical approaches are formidable and maybe even impossible. The most direct approach is computer simulation in which a large number of incident ion trajectories are computed and the large angle scattering probability calculated for each layer, as outlined by Barrett (1971). An additional advantage of the simulation technique is that particular surface structure models are easily included for comparison with experiments.

The assumptions in the simulation centre on two aspects of the problem: i. the scattering of the projectile, and ii. the model of thermal vibrations used to govern the relative positions of the atoms in the crystal. The Molière approximation is used in most calculations. Results obtained using other reasonable potentials deviate insignificantly from those using the Molière potential (Stensgaard *et al.*, 1978). Once the potential is specified, the scattering problem is reduced to the simplest and most accurate method of solving the scattering integral associated with a binary collision, Equation (9.4), and summing over all ion–atom collisions (see Feldman *et al.*, 1982).

While the scattering properties of energetic ions are well understood, the least established part of the simulation concerns the model of thermal vibrations of the atoms. An MeV ion passes a lattice spacing in a time which is short compared to the characteristic phonon times in a solid. Thus the solid can be viewed as an ordered arrangement of atoms, with the atoms statically displaced from their equilibrium positions in accord with the chosen model of thermal vibrations. We now consider three aspects of these thermal vibration models:

i. In most cases a harmonic approximation is used in which the thermal position distribution of atoms is a Gaussian, Equation (9.3), whose width (ρ) is related to a measured Debye temperature.

ii. Although correlations in the thermal vibration amplitudes of atoms are expected in almost any dynamic model of a solid, computer simulation techniques often treat the lattice as an array of independent vibrators. The effect of correlations is to decrease the relative displacement of nearby atoms (as compared to an independent vibrator model of the same amplitude) and thus increase the shadowing. The magnitude of the correlation is usually expressed in terms of the two-body correlation coefficient, c, which is the ratio of the thermal average of the amplitude, $u_{i,j}$, of two adjacent atoms, i, j, to the average mean square amplitude, i.e. $c = \langle u_i u_j \rangle / \langle u_i^2 \rangle$. In the Debye model it can be shown that the two-body correlation coefficient is ~ 0.13 in the low temperature limit ($T \ll \Theta_{Debye}$) and ~ 0.4 in the high temperature limit ($T \gg \Theta_{Debye}$).

iii. Enhanced vibration amplitudes of surface atoms have long been considered a possibility and represent the largest uncertainty in SP estimates. As we will show, the ambiguity in the interpretation of an experiment due to this uncertainty in vibration amplitude can be reduced or even eliminated by measuring at low temperature.

A typical simulation result is shown in Fig. 9.2 (Stensgaard *et al.*, 1978). The scattering intensity from each atom along the string is represented as a histogram. Since a backscattering experiment will typically have ~ 40 Å depth resolution, the measured spectrum is

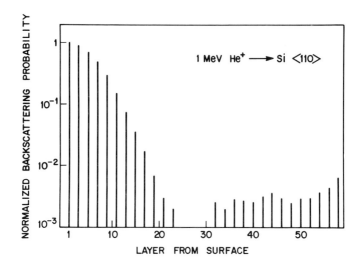

Fig. 9.2 Calculated normalized backscattering probability from successive layers of a single crystal for 1.0 MeV He incident on Si at 300°K. (Stensgaard *et al.,* 1978)

represented by the fold of the simulation result with a Gaussian depth resolution function of the appropriate width. The measured SP intensity in this case would be the sum of all the contributions from the first ~ 20 layers. Contributions from greater depths come from particles which have 'crossed-over' from one string of atoms to an adjacent string and then make a large angle scattering.

The main purpose of this section has been to consider the interaction of an aligned ion beam with the ions in the surface monolayers of an ideal single crystal. The principal feature of this geometry is that shadowing effects exist, the extent of shadowing depending on the parameter ρ/R_M. The generality of ρ/R_M parameterisation is illustrated in Highlight 9.1. The comparison of data with calculations indicates that our understanding of SP intensities in ideal structures is quite good and gives a measure of confidence for using SP yields to give practical surface structure information. Highlight 9.2 summarises the concepts of application of SP yields to practical surface structure determinations. Practical examples of many of these concepts are given in later sections.

HIGHLIGH 9.1
SURFACE PEAK PARAMETERISATION

The results of SP calculations for a large number of cases are shown in Fig. 9.3a. The parameterisation of the results is still valid in terms of a

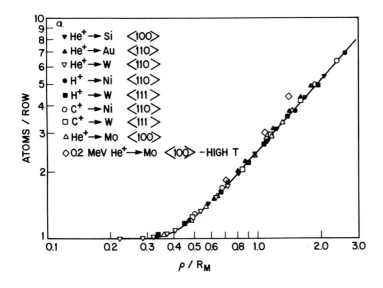

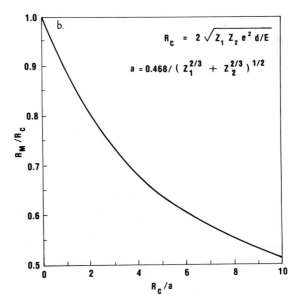

Fig. 9.3a. Calculated values of the close encounter probability or SP in units of atoms/row versus ρ/R_m where ρ is the two dimensional thermal vibration amplitude and R_m is the Moliere shadow cone radius. b. Ratio of the shadow cone radius from the Moliere potential to the unscreened (Coulomb) potential as a function of R_c/a. (From Stensgaard *et al.*, 1978)

420   *L.C. Feldman*

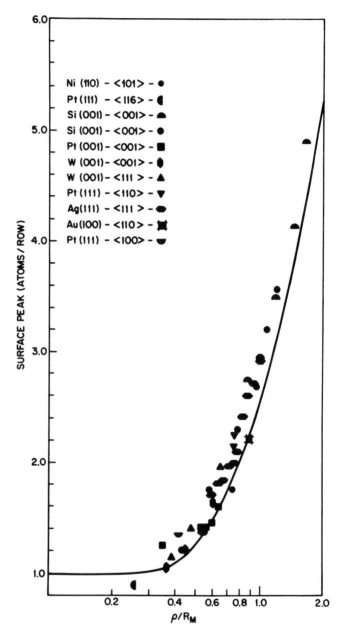

Fig. 9.4 Comparison of the 'universal' curve from Fig. 9.3a, with experimental values from backscattering measurements for a number of different 'bulk-like' surfaces. The notation Pt(111) — ⟨116⟩ indicates a Pt crystal with a (111) surface plane and the backscattering measurement in the ⟨116⟩ axial direction. (From Feldman, 1981b)

ρ/R scaling; R_M corresponds to the shadow cone radius at the second atom using the Molière approximation to the Thomas-Fermi potential. R_M is often close to R_c and can be calculated exactly from R_c (Fig. 9.3b) (Stensgaard *et al.*, 1978). Using Figs 9.3a and 9.3b one can estimate the SP for a particular scattering situation. In an ideal single crystal, the intensity of this peak is governed by the ratio of the two-dimensional thermal vibration amplitude, ρ, to the shadow cone radius, R_M, at the second atom. The shadow cone radius describes the flux distribution and the fact that particles do not approach closer than R_M to the equilibrium position on the second atom. ρ is a measure of the 'disorder' in the static string due to thermal vibration and measures the extent to which the atoms in the string can extend beyond the 'shadow cone'. The scaling of the SP intensity in terms of single parameter ρ/R_M was first suggested by Barrett (1971) and later shown to be true over a broad range of parameters by Stensgaard *et al.* (1978). It is remarkable that this one simple parameter describes the SP so well.

The first 'observations' of the ρ/R_M scaling were based primarily on computer simulation of ideal crystal structures. There have been a considerable number of SP measurements reported which allow an experimental test of this scaling. Fig. 9.4 shows the 'universal' SP curve as a function of ρ/R_M and a listing of experimental points. In each case the ion scattering measurement is on an atomically clean crystal which shows a (1×1) low energy electron diffraction (LEED) pattern. Such a LEED pattern suggests strongly that the surface atomic structure is arranged in a 'bulk-like' fashion and thus comparison with computer simulations assuming ideal, 'bulk-like' structures is valid. The small difference between simulation and experiment is discussed in Section 9.4.1.

HIGHLIGHT 9.2
SURFACE STRUCTURE DETERMINATION

One of the main advantages of the use of ion scattering studies is the simple, geometrical pictures that can be employed in the design of experiments. In Fig. 9.5 we show schematically the three most common methods in which SP measurements are used to probe surface structures. Fig. 9.5a shows the ideal SP from a 'bulk-like' structure. Surface reconstruction (atomic translations in the plane of the surface) is indicated by a SP increase for a beam incidental normal to the surface (Fig. 9.5b). Surface relaxation is shown in Fig. 9.5c. In this case reordering of the surface occurs by a change of spacing between the first and second atomic layers; atomic spacings in the plane of the surface

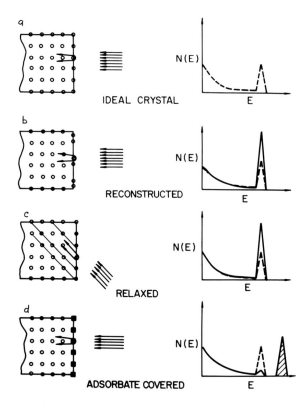

Fig. 9.5 Representations of different surface structures on a simple cubic crystal. The backscattering spectra shown on the right-hand side represent the expected signal from the different structures. The dashed line represents the signal from the bulk-like-crystal. (From Feldman *et al.*, 1982)

remain the same. A measurement of the SP in a non-normal direction yields an increase above the expected value. Relaxation is usually explored by measurement of the SP as a function of small angular variations about the off-normal direction. Asymmetry in this angular scan reveals the sign of the lattice constant change (contraction or expansion) and the magnitude.

Adsorbate site determination can be explored by SP measurements as a function of adsorbate coverage (Fig. 9.5d). An adsorbate will form a shadow cone which will reduce the substrate SP in certain cases. Obviously a positive effect can only be expected if the absorbate is aligned to an atomic string of the substrate. This limitation makes this technique most useful in cases where the LEED pattern suggests registry or in epitaxial crystal growth investigations. An elegant and general method for absorbate site determination is given by the channeling/ blocking technique described by van der Veen *et al.* (1979). More detailed discussion of these effects can be found in Feldman *et al.* (1982) and in Feldman (1981a, b), Jackson and Barrett (1979) and van der Veen *et al.* (1979).

9.2 EXPERIMENTAL CONSIDERATIONS

A typical surface scattering system consists of a 45 cm diameter stainless steel UHV chamber coupled via a differentially pumped beamline to an ion accelerator. Ion and Ti sublimation pumps maintain the chamber in the low 10^{-10} torr range. The beamline, connected to the chamber by a 1 mm diameter aperture, is held to 10^{-8} torr by a turbomolecular pump. At the upstream end of the beamline (2.5 m away), a variable collimator provides further isolation from the accelerator and beam switching magnet, which are typically in the 10^{-6} torr range. The two collimators define the divergence of the beam which must be $\leq 0.06°$ for channeling applications. The accelerator should produce microamps of ions of any gas at energies from 0.1 to 2.0 MeV.

As shown in Fig. 9.6, the chamber should contain LEED optics (also used with a glancing incidence electron gun for retarding field Auger spectroscopy) for monitoring the condition of the sample surface, sputter ion and electron bombardment guns for sample cleaning, and a residual gas analyzer. A leak valve manifold permits the admission of a variety of gases for sputtering or adsorption studies; this relatively high-pressure gas may then be removed rapidly by the turbopump through an auxiliary exhaust line which bypasses the chamber's beam entrance aperture.

The sample manipulator used in these experiments is based on a UHV concentric rotary and linear motion feedthrough mounted on a movable stage. It permits three orthogonal translations, a tilt, and two rotations; one about an axis in the sample surface ($\pm 180°$) and one about an axis normal to the sample ($\pm 90°$). Both rotations have $\sim 0.1°$ precision, making crystal alignment for ion channeling relatively easy. The compact design of the sample holder allows a clear view of the LEED screen, while the open center makes transmission experiments through thin crystals possible. Samples may be heated by a resistive heating element or by electron bombardment from either side, and temperature is usually monitored externally with infrared and optical pyrometers. For accurate ion beam charge integration, the sample is biased at $+300$ V to suppress secondary electron emission; when thin samples are used, a Faraday cup may be moved behind the holder to collect the ion current.

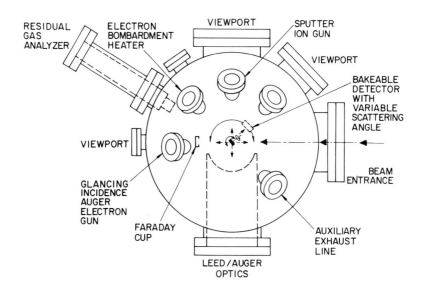

Fig. 9.6 Schematic of a typical UHV scattering chamber for ion scattering studies. (From Feldman, 1981a.)

Also held on the manipulator, above the sample position, is a scattering 'standard' consisting of a calibrated concentration of a heavy element implanted at low energy into Si. By comparing the scattering from the sample of interest, an absolute determination of the areal atomic density in the sample can be made without specific knowledge of the solid angle subtended by the detector or the exact scattering angle.

The energy of scattered ions is measured with a special implanted surface barrier detector made to withstand the 200°C chamber bakeout without deterioration. Cooling is provided for improving the energy resolution (Chapters 2 and 3). The detector is mounted on an arm which can be rotated around the chamber's central axis over a range 0 to 160° using an external control. In this way, a grazing exit-angle geometry can be obtained for all incident beam directions, resulting in improved depth resolution and a reduced background level under the surface scattering peak in the energy spectrum (Chapter 6). A slit over the detector which limits its acceptance angle to ~ 1° can be changed *in vacuo* to a 10 μm thick Al foil for elastic recoil scattering detection of hydrogen.

Additional features are (i) sample cooling to liquid helium temperatures, (ii) detector motion in both the vertical as well as the horizontal plan for blocking experiments, and (iii) the use of electrostatic or magnetic analyzers which provide better energy resolution than solid state detectors, although at the expense of efficiency.

9.3 ANALYSIS OF SPECTRA

The most critical aspect of the analysis of data is the accurate conversion of the intensity of the SP to an absolute number of atoms/row or monolayers. As in most backscattering experiments the intensity of the peak is measured directly in atoms cm^{-2} through the use of the equation:

$$Y = N_s \, (d\sigma/d\Omega) \, \Delta\Omega \, N_1 \qquad (9.8)$$

where Y is the measured yield, $d\sigma/d\Omega$ is a known scattering cross-section at angle θ, $\Delta\Omega$ is the detector solid angle and N_1 is the number of incident ions. If the cross-section is given in units of $cm^2 \, sr^{-1}$, N_s, the number of scattering centers contributing to the SP, is in units of atoms cm^{-2}. The unit 'atoms/row' is convenient for comparison to ion scattering theory. For the surface scientist a more useful unit for the SP intensity may be in monolayers. The conversions between these different units for the SP is straightforward and is based on the orientation of the crystal, the lattice constant and other known geometrical factors. In this section we concentrate on the analysis of the basic spectrum and the representation of the SP intensity in useful units.

9.3.1 Background Subtraction

An ideal backscattering/channeling energy spectrum, measured with perfect depth resolution, can be represented by the histogram-like figure shown in Fig. 9.2. Each monolayer contributes a finite yield ranging from 1.0 for the first layer to very small values for monolayers within the crystal. The SP is essentially the sum between the first layer and the first minimum. With finite detector resolution the observed depth distribution of scattering intensity is as shown in Fig. 9.7.

As illustrated in Fig. 9.8 and discussed by Stensgaard *et al.* (1978), one simple method of background subtraction is the straight-line construction. This background correction is determined by a triangular area specified by a straight line from the origin (peak position of the SP) to the minimum in the yield directly behind (at lower energy) the SP (dashed curves in Fig. 9.7). The justification for this subtraction procedure is determined from computer simulations, which 'fold-in' the detector resolution and hence are accurate simulations of the real data. Fig. 9.8 shows a practical example of background subtraction.

9.3.2 Conversion to Absolute Areal Density

The most convenient method to convert from raw counts to atoms cm^{-2} is to use a standard sample containing an implanted impurity of a known

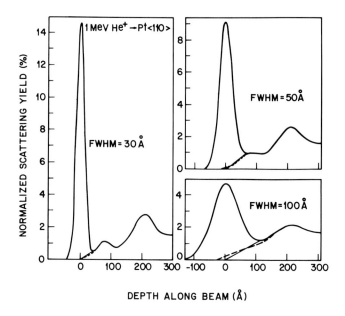

Fig. 9.7 Calculated backscattering/channeling spectra showing the effect of the finite detector resolution on the shape of the SP. (From Stensgaard *et al.*, 1978.)

number of atoms cm^{-2}. A commonly used standard is the '*Bi* implanted into *Si*' series originally produced at Harwell (L'Ecuyer *et al.*, 1979). This method of conversion is straight forward and merely requires the scaling parameters associated with Rutherford scattering. The small 'non-Rutherford' correction described by L'Ecuyer *et al.* (1979) must also be applied for completeness.

Another method of establishing the areal density of the surface peak is to normalise with a 'random' spectrum as shown in Fig. 9.8. Here a random spectrum refers to the equivalent backscattering spectrum measured in an amorphous solid. Usually an amorphous solid is not available and a random spectrum is approximated by beam incidence far from major channeling directions or by employing a 'rotating random'. This latter case refers to rotating the sample during the time of spectrum acquisition so as to average over incident directions. Both of these methods are discussed in Chapter 6.

9.3.3 Conversion to Monolayers or Atoms/Row

The number of atoms cm^{-2} associated with a given crystallographic direction, *hkl*, is given by Nd_{hkl} where N is the volume density of atoms in

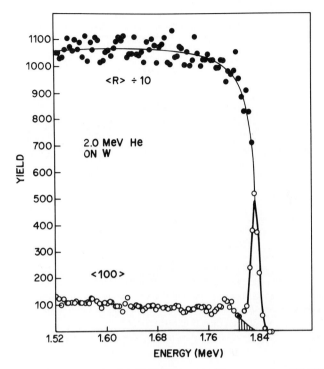

Fig. 9.8 Backscattering spectra for 2.0 MeV He ions incident on a clean W (100) surface along the normal [100] axis (open circles) and for the beam in a non-aligned (closed circles) direction. The method of background subtraction is illustrated by the cross-hatched area under the SP. (From Feldman, 1981a.)

the crystal and d_{hkl} is the spacing of atoms along the string. (This is the same factor that is used in the calculation of the minimum yield, χ_{min}, discussed in Chapter 6.) In the case of an uneven spacing of atoms, d is the average spacing. For example, in the $\langle 111 \rangle$ direction of a silicon-like lattice with lattice constant a there are two atom spacings of $\sqrt{3}a/4$ and $3\sqrt{3}a/4$; the appropriate factor d is given by $\sqrt{3}a/2$. The ratio of the measured value of N_s (atoms cm^{-2}) to Nd_{hkl} is the measured number of atoms/row. The minimum number of atoms/row is always unity.

The number of monolayers visible to the beam depends on the cut of the crystal as well as the channeling direction. The number of monolayers visible to the beam, n_{hkl}, in an axial direction hkl at an angle θ to the normal direction is given by

$$n_{hkl} = n_n \cos \theta \, d_{hkl}/d_n \tag{9.9}$$

where n_n is the number of layers visible to the beam in the normal

TABLE 9.1
Number of monolayers visible to the beam—body centered cubic crystals

Surface 'cut'	n_n	Incident Direction		d_{hkl}	d_n	n_{hkl}
		Crystal Axis	Angle θ to normal			
(100)	2	$\langle 100 \rangle$	0°	1	1	2
(100)	2	$\langle 110 \rangle$	45.0°	$\sqrt{2}$	1	2
(100)	2	$\langle 111 \rangle$	54.7°	$\sqrt{3}/2$	1	1
(110)	2	$\langle 110 \rangle$	0°	$\sqrt{2}$	$\sqrt{2}$	2
(110)	2	$\langle 111 \rangle$	35.3°	$\sqrt{3}/2$	$\sqrt{2}$	1
(110)	2	$\langle 100 \rangle$	45.0°	1	$\sqrt{2}$	1
(110)	2	$\langle 110 \rangle$	60.0°	$\sqrt{2}$	$\sqrt{2}$	1
(111)	3	$\langle 111 \rangle$	0°	$\sqrt{3}/2$	$\sqrt{3}/2$	3
(111)	3	$\langle 110 \rangle$	35.3°	$\sqrt{2}$	$\sqrt{3}/2$	4
(111)	3	$\langle 100 \rangle$	54.7°	1	$\sqrt{3}/2$	2
(111)	3	$\langle 111 \rangle$	70.5°	$\sqrt{3}/2$	$\sqrt{3}/2$	1

Note. Tables 9.1–3 give the values of n_{hkl}, the number of monolayers visible to the beam, for a variety of crystal types and surface orientations. In this table, n_n is the number of monolayers visible in the normal direction, θ refers to the angle (from the normal) of the off-normal axial direction, d_{hkl} and d_n are the atom spacings in units of the lattice constant in the [hkl] direction and normal direction respectively.

TABLE 9.2
Number of monolayers visible to the beam—face centered cubic crystal

Surface 'cut'	n_n	Incident Direction		d_{hkl}	d_n	n_{hkl}
		Crystal Axis	Angle θ to normal			
(100)	2	$\langle 100 \rangle$	0°	1	1	2
(100)	2	$\langle 110 \rangle$	45.0°	$\sqrt{2}/2$	1	1
(100)	2	$\langle 111 \rangle$	54.7°	$\sqrt{3}$	1	2
(110)	2	$\langle 110 \rangle$	0°	$\sqrt{2}/2$	$\sqrt{2}/2$	2
(110)	2	$\langle 111 \rangle$	35.3°	$\sqrt{3}$	$\sqrt{2}/2$	4
(110)	2	$\langle 110 \rangle$	45.0°	1	$\sqrt{2}/2$	2
(110)	2	$\langle 110 \rangle$	60.0°	$\sqrt{2}/2$	$\sqrt{2}/2$	1
(111)	3	$\langle 111 \rangle$	0°	$\sqrt{3}$	$\sqrt{3}$	3
(111)	3	$\langle 110 \rangle$	35.3°	$\sqrt{3}/2$	$\sqrt{3}$	1
(111)	3	$\langle 100 \rangle$	54.7°	1	$\sqrt{3}$	1
(111)	3	$\langle 111 \rangle$	70.5°	$\sqrt{3}$	$\sqrt{3}$	1

See Table 9.1 for definition of symbols.

TABLE 9.3
Number of monolayers visible to the beam—diamond structure (Si) crystal

Surface 'cut'	n_n	Incident Direction		d_{hkl}	d_n	n_{hkl}
		Crystal Axis	Angle θ to normal			
(100)	4	$\langle 100 \rangle$	0°	1	1	4
(100)	4	$\langle 110 \rangle$	45.0°	$\sqrt{2}/2$	1	2
(100)	4	$\langle 111 \rangle$	54.7°	$\sqrt{3}/2$	1	2
(110)	4	$\langle 110 \rangle$	0°	$\sqrt{2}/2$	$\sqrt{2}/2$	4
(110)	4	$\langle 111 \rangle$	35.3°	$\sqrt{3}/2$	$\sqrt{2}/2$	4
(110)	4	$\langle 100 \rangle$	45.0°	1	$\sqrt{2}/2$	4
(110)	4	$\langle 110 \rangle$	60.0°	$\sqrt{2}/2$	$\sqrt{2}/2$	2
(111)	3	$\langle 111 \rangle$	0°	$\sqrt{3}/2$	$\sqrt{3}/2$	3
(111)	3	$\langle 110 \rangle$	35.3°	$\sqrt{2}/2$	$\sqrt{3}/2$	2
(111)	3	$\langle 100 \rangle$	54.7°	1	$\sqrt{3}/2$	2
(111)	3	$\langle 111 \rangle$	70.5°	$\sqrt{3}/2$	$\sqrt{3}/2$	1

See Table 9.1 for definitions of symbols. In this table d_{111} is taken as $\sqrt{3}a/2$ which is actually the average value, $\left(\dfrac{\sqrt{3}a/4 + 3\sqrt{3}a/4}{2}\right)$, of the spacing in the (111) direction.

direction and d_n is the average lattice spacing in the normal direction. Tables 9.1–3 summarize the values of n_{hkl} for a variety of commonly studied crystals with various surface cuts and corresponding to measurements along various incident directions. The measured value of the SP in monolayers is given by $N_s n_{hkl}/N d_{hkl}$.

9.4 DATA COMPILATIONS

9.4.1 Comparison of SP Measurements with Theory

A comparison of an array of SP measurements with theory (universal curve) was shown in Fig. 9.4. This curve (Feldman, 1981b) represents all SP measurements known to that time. In each measurement the sample consisted of a clean surface (by Auger spectroscopy standards) and displayed a (1 × 1) LEED pattern. The (1 × 1) LEED pattern is taken as an indication of a bulk-like surface so that comparison with a bulk-like calculation is valid. (This is strong evidence for, but not necessarily proof of, the existence of a bulk-like surface.) Nevertheless one can see that the overall trend is well established and the systematics of the SP are well explained by the universal theory.

In establishing this curve, bulk thermal vibrations were used

TABLE 9.4
Summary of recent SP measurements

Semiconductors	
Surface	Reference
Si (100)	Stensgaard et al., 1981; Tromp et al., 1981
Si (111)	Culbertson et al., 1980; Narusawa and Gibson, 1981; Tromp et al., 1983
GaAs (100)	Narusawa et al., 1985
GaAs (110)	Gossmann and Gibson, 1984 (a,b)
GaSb (110)	Smit et al., 1984
Ge (100)	Culbertson et al., 1986
Ge (111)	Culbertson et al., 1986

Metals	
Surface	Reference
Ni (100)	Smeenk et al., 1981a, Frenken et al., 1983
Ni (110)	Smeenk et al., 1981b
Ni (111)	Narusawa et al., 1981, 1985
Cu (100)	Culbertson, et al., 1985, Alkemade et al., 1985
Cu (110)	Adams et al., 1982; Feidenhans'l and Stensgaard, 1983
Pd (111)	Kuk et al., 1983a
Ag (110)	Kuk and Feldman, 1984
Ag (111)	Culbertson et al., 1981
W (100)	Feldman et al., 1979, Stensgaard et al., 1979
Pt (100)	Norton et al., 1979, 1981; Davies et al., 1981
Pt (110)	Jackman et al., 1982; Jackson et al., 1983
Pt (111)	Bogh and Stensgaard, 1978; Davies et al., 1978
Au (110)	Jackson et al., 1983; Kuk et al., 1984
Au (100)	Zehner et al., 1975
Pb (110)	Frenken and van der Veen, 1985

throughout, i.e. for calculation and experimental plotting. This assumption is almost certainly invalid at some level, as surface thermal vibrations are not expected to be the same as bulk values. In general, surface vibration amplitudes are not known for a specific material and surface. Theoretical estimates suggest that the root-mean-square thermal vibration amplitude can be up to a factor of $\sqrt{2}$ larger than the bulk value. The overall deviation in Fig. 9.4 between theory and experiment can be explained by an overall enhancement in amplitude of a factor of ~ 1.2.

9.4.2 Compilation of SP Measurements

Table 9.4 lists SP measurements on various crystals. In many cases, the study of a surface requires a long and laborious effort which often results in a number of publications relating to a particular surface. In those cases, only the last publication is cited.

9.5. APPLICATIONS

An understanding of the SP is useful in a variety of surface, materials science and thin film applications. The purpose of this section is to briefly described the use of SP measurements and discuss some of the assumptions implicit in the measurements.

9.5.1 Surface Structure: Reconstruction and Relaxation

A major use of SP measurements has been in elucidating the surface structure of clean surfaces. The basic concepts underlying these measurements were discussed in Section 9.1.5. Fig. 9.5 shows how the SP intensity is sensitive to different kinds of surface modifications. In practice each study of this type consists of much more than a single SP measurement. For relaxation measurements one requires the SP as a function of small angular variations from the bulk crystalline directions. Asymmetry in this angular scan reveals the sign of the surface lattice constant change, contraction or expansion, and the magnitude. By choosing appropriate directions it has been possible to establish multi-layer relaxation in which the change in lattice constant in the first few monolayers of the crystal is extracted (Adams *et al.*, 1982; Feidenhans'l and Stensgaard, 1983; Kuk and Feldman, 1984; Frenken and van der Veen, 1985). The largest relaxation measured to date is that for Pb(110); ~ 15% (Frenken and van der Veen, 1985). An example of this type of measurement is shown in Fig. 9.9 for the case of the Ag(110) surface (Kuk and Feldman, 1984). Surface peaks are measured as a function of angle for small deviations from the ⟨101⟩ off-normal channeling direction using 0.4 MeV He$^+$ with the sample at room temperature. The data clearly show the asymmetry indicating relaxation. The different lines correspond to fits to the data using two-layer relaxation models and indicate the necessity for a two-layer variation to fit the data. A systematic analysis of this type of measurement along a few prominent channeling directions resulted in a surface model consisting of: 7.8 ± 2.5% contraction for the first interlayer spacing and a 4.3 ± 2.5% expansion in the second.

In reconstruction studies one is usually interested in the magnitude of the in-plane displacement. For elemental semiconductors ion scattering has established that the surface may be grossly reconstructed with

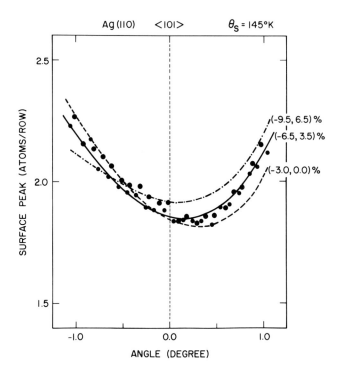

Fig. 9.9 The SP intensity as a function of angle measured near the off-normal ⟨101⟩ direction on a Ag(110) surface. The scattering parameters correspond to 0.4 MeV He⁺ on a surface at 293° K. The lines correspond to 'two-layer' relaxation fits, a contraction of the first layer and a possible expansion of the second layer. The computer simulations include not only the relaxation but an enhanced surface Debye temperature ($\Theta_s = 145°K$) in the first layer, all other vibration amplitudes correspond to the accepted bulk Debye temperature of 215°K. (From Kuk and Feldman, 1984.)

displacements of > 0.1 Å in the first few layers (Culbertson *et al.*, 1980; Stensgaard *et al.*, 1981; Tromp *et al.*, 1981., Culbertson *et al.*, 1986). Since the shadow cone varies with energy, the energy dependence of the SP is useful in estimating the magnitude of the displacements (Feldman *et al.*, 1979; Stensgaard *et al.*, 1979). This type of measurement is illustrated in Fig. 9.10a, b which shows the SP intensity as a function of energy for He incident on a clean Au(110) surface at 293 K and 100 K. The LEED pattern of this surface is (2×1) indicating that the surface is not 'bulk-like', but reconstructed. The relatively poor agreement with the full and dashed lines (Fig. 9.10a), derived via computer simulation assuming a bulk-like model, support this conclusion. Fig. 9.10b shows the fit of the data to a model which includes surface displacments, Δ, in the plane of the surface in a full monolayer of atoms of 0.12 Å and 0.18 Å. Both the energy dependence and temperature dependence were useful in establishing the magnitude of the surface displacement.

9.5.2 Epitaxial Growth

Measurement of the SP has been useful in studies of the initial stages of epitaxial growth. In the ideal case one observes the decrease of the substrate SP as a function of coverage of an epitaxial overlayer (see Fig. 9.5d). Analysis of the resulting curve can yield information on the registry of the overlayer with respect to the substrate, islanding and the quality of the epitaxial layer. An example of this kind of study is shown in Fig. 9.11 in which Au has been deposited on the clean surfaces of Ag(111) and Pd(111). The decrease of the substrate SP is a measure of the crystal registry between the crystalline Au overlayer and the substrate. The full curve indicates the theoretical expectation for perfect registry. The data indicate that the Au/Ag system with a lattice constant differences of 0.2% forms a better epitaxial system than Au/Pd with a lattice constant difference of 4.7% (Kuk *et al.*, 1983b). It is possible to study both heteroepitaxy and homoepitaxy. An example of the latter case is Si growth on Si in which it is shown that epitaxy is influenced by the nature of the substrate reconstruction (Gossmann and Feldman, 1985a, b).

9.5.3 Thin Film Formation: Silicides

The object of these experiments is to understand the initial phases of reactive thin film formation under UHV conditions. The experiment usually consists of the measurement of a substrate SP as a function of the thickness of a deposited overlayer (Fig. 9.12). For example, the deposition of Ti on clean Si results in the formation of a silicide, $Ti_x Si_y$ which is assumed non-epitaxial. In this case the reaction rate of the deposited material with the clean substrate is measured. An accurate measure of the variation of the SP with coverage yields the stoichiometry of the initial phase that is formed as shown in Fig. 9.13 (van Loenen *et al.*, 1985). It is important to note that the substrate (Si) SP, which is dashed in Figs 9.12c and d, must be properly taken into account when analysing thin film stoichiometry in a channeling direction. At relatively high energy (~ 2.0 MeV) the intrinsic (substrate) SP can correspond to scattering from ~ 20 to 30 Å of the substrate material, thus the intrinsic surface peak must be subtracted from the thin film yield for accurate stoichiometry determination.

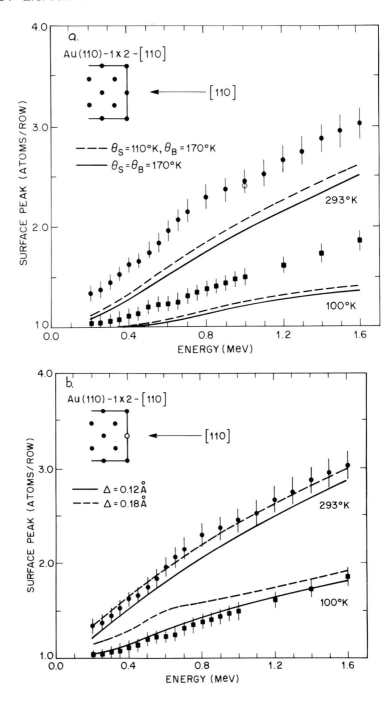

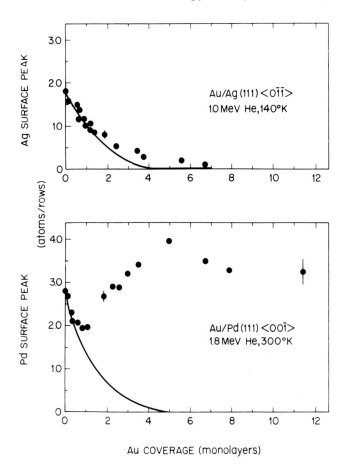

Fig. 9.11 The decrease of the substrate SP for a) Ag or b) Pd, as a function of Au coverage. The full line indicates the expected dependence for perfect crystallinity and registry of the Au overlayer. The data indicate close agreement for the Au/Ag (111) system with a lattice match of 0.2%, but not for the case for Au/Pd with a lattice match of 4.7%. (From Kuk *et al.*, 1983b).

Fig. 9.10 The SP intensity as a function of energy and temperature for He$^+$ scattering from the Au(110)(1 x 2) surface along the normal, ⟨110⟩, channeling direction. The open symbol in a). is from Jackson *et al.* (1983) and indicates the good agreement in SP intensities measured by different laboratories a. Comparison of the data to models using a 'bulk-like' surface and including a possible enhanced surface vibration (surface Debye, Θ_s = 110°K) in the first monolayer of the solid. The bulk Debye temperature is 170°K. b. The same data fit to a model which includes surface displacements of 0.12 Å or 0.18 Å in one monolayer of the solid (from Kuk *et al.*, 1984).

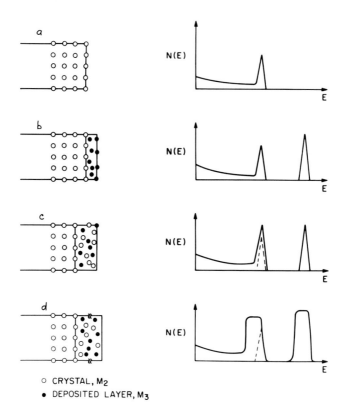

○ CRYSTAL, M$_2$
● DEPOSITED LAYER, M$_3$

Fig. 9.12 Schematic ion scattering spectra for amorphous films on single crystal substrates (from Feldman *et al.,* 1982). a. The uncovered crystal with a bulk-like surface . b. A passive (non-reacting) overlayer. c. A thin reactive film with an increase of atoms off-lattice sets. d. A thick reactive film.

9.5.4 Analysis of Amorphous Film/Single Crystal Interfaces

Here the major interest is in characterising the crystal structure at the interface of an amorphous layer and a single crystal. Such interfaces are common in technology as in the Si/SiO$_2$ interface or single crystal/metal oxide structures. The principal measurement is the substrate SP as a function of amorphous layer thickness, similar to the thin film formation studies. The amorphous layer/interface peak gives information on possible reconstructions (or relaxation of reconstructions) at the buried interface. The most extensive analysis of this type has been carried out in the Si/SiO$_2$ system (Haight and Feldman, 1982).

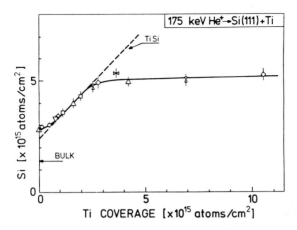

Fig. 9.13 Increase of the Si substrate SP as a function of Ti deposition. The slope of the line indicates the initial stoichiometry, TiSi. (From van Loenen *et al.*, 1985).

9.5.5 Analysis of Crystal/Crystal Interfaces

The interface of crystalline heterostructures continues to be an important subject particularly in the electronics industry where semiconductor/heterostructure research is an active area. Bulk channeling techniques (Chapter 6) can be applied to many of these problems although one aspect of the growth is particularly suitable for SP measurement. In general semiconductor surfaces are reconstructed, so that SP measurements tend to yield values above that expected for the bulk-like structure. Deposition of another semiconductor may reorder the reconstruction resulting in a reduction of the substrate SP. This process of surface reordering is shown schematically in Fig. 9.14. It has been studied most extensively in the Ge/Si and Si/Si systems (Gossmann and Feldman, 1985a, b).

9.5.6 Other Surface Sensitive Geometries

Since the SP is a general feature of all backscattering/channeling spectra it is important to demonstrate that the SP is both qualitatively and quantitatively understood. In this section we describe three other surface sensitive ion scattering geometries which are less commonly encountered.

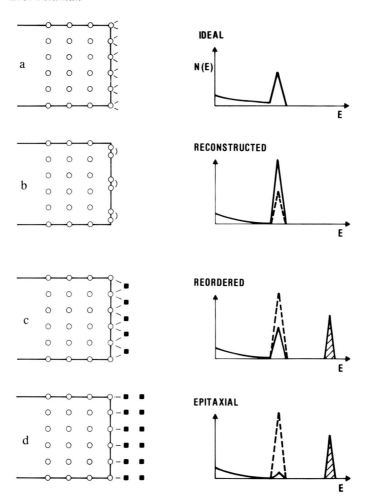

Fig. 9.14 Schematic of the decrease in the SP for adsorbate induced surface reordering. a. The SP from a 'bulk-like' surface. b. The increased SP from a reconstructed surface. c. A decrease in the substrate SP due to an adsorbate induced reordering. d. A further decrease in the SP due to overlayer epitaxy. In some cases, such as Ge on Si, the SP simply acquires the bulk-like value indicating a non-epitaxial overlayer, but a reordered substrate.

a. Double Alignment

In a double alignment geometry one makes use of channeling and blocking to determine atom positions. The basic concept is shown in Fig. 9.15 (a through c). The principal idea is to use single alignment to confine backscattering to the first few exposed layers. Backscattering particles may be blocked on their outward path. The blocking pattern is then used to yield surface co-ordinates. This double alignment technique has been

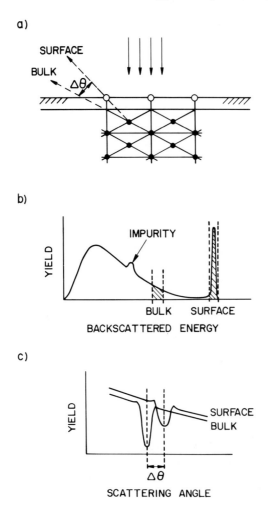

Fig. 9.15 Geometry for the double alignment experiment.

pioneered by van der Veen *et al.* (1979). A clever variation by Boulliard *et al.* (1984) has allowed channeling and blocking to be used for the study of surface steps. The double alignment technique has been particularly fruitful, providing information over and above a single alignment experiment.

b. Thin Crystal Transmission

In this geometry adsorbate surface location and thin films are studied on the back side of a thin single crystal substrate. The major advantage in such thin film studies is that the substrate SP makes essentially no contribution to the scattering yield. Interpretation of ultra-thin film stoichiometry is then simplified, since the substrate SP need not be considered (Cheung *et al.*, 1979). The technique is valuable because essentially no calculations are required in adsorbate site determination. This technique has been used in studies of the Si/SiO$_2$ interface (Cheung *et al.*, 1979) and for impurity site determination on Ni(100) (Stensgaard and Jakobsen, 1985).

c. Grazing Angle Channeling

Here the beam is incident at angles almost parallel (within $\sim 1°$) to the crystal surface. The close encounter probability is measured via an ion-induced Auger process, a nuclear reaction or a backscattering event. Studies of the azimuthal dependence of the yield are then used to determine adsorbate locations or other geometric features on the surfaces (Schuster and Varelas, 1983; Sailer and Varelas, 1984).

REFERENCES

Adams, D.L., Nielsen, H.B., Andersen, J.N., Stensgaard, I., Feidenhans'l, R., and Sorensen., J.E. (1982). *Phys. Rev. Lett.* **49**, 669.

Alkemade, P.F.A., Turkenburg, W.C. and van der Weg. W.F. (1985). *Nucl. Instrum. Methods* **B15**, 126.

Barrett, J.H. (1971). *Phys. Rev.* **B3**, 1527.

Bogh, E. and Stensgaard, I. (1978). *Phys. Lett.* **65A**, 357

Boulliard, J.C., Cohen, C., Bomange, J.L., Drigo, A.V., L'Hoir, A., Moulin, J. and Sotto, M. (1984). *Phys. Rev.* **B30**, 2470.

Cheung, N.W., Feldman, L.C., Silverman, P.J. and Stensgaard, I. (1979). *Appl. Phys. Lett.* **35**, 859.

Culbertson, R.J., Feldman, L.C. and Silverman, P.J. (1980). *Phys. Rev. Lett.* **45**, 2043.

Culbertson, R.J., Feldman, L.C., Silverman, P.J. and Boehm, H. (1981). *Phys. Rev. Lett.* **47**, 657.

Culbertson, R.J., Kuk, Y. and Feldman, L.C. (1985). *Mat. Res. Soc. Symp. Proc.* **41**, 187.

Culbertson, R.J., Kuk, Y. and Feldman, L.C. (1986). *Surf. Sci.* **167**, 127.

Davies, J.A., Jackson, D.P., Matsunami, N., Norton, P.R. and Anderson, J.U. (1978). *Surf. Sci.* **78**, 274

Davies, J.A., Jackman, T.E., Jackson, D.P. and Norton, P.R. (1981). *Surf. Sci.* **109**, 20.

L'Ecuyer, J., Davies, J.A. and Matsunami, N. (1979). *Nucl. Instrum. Methods* **160**, 337.

Feidenhans'l, R. and Stensgaard, I. (1983). *Surf. Sci.* **133**, 453.

Feldman, L.C. (1981a). *Crit. Rev. in Solid State and Material Science* **2**, 143.

Feldman, L.C. (1981b). *Nucl. Instrum. Methods* **191**, 211.

Feldman, L.C., Silverman, P.J. and Stensgaard, I. (1979). *Surf. Sci.* **87**, 410.

Feldman, L.C., Mayer, J.W. and Picraux, S.T. (1982). 'Materials Analysis by Ion Channeling'. Academic Press, New York.

Frenken, J.W.M., van der Veen, J.F. (1985). *Phys. Rev. Lett.* **54**, 134.

Frenken, J.W.M., van der Veen, J.F. and Allen, G. (1983). *Phys. Rev. Lett.* **51**, 1876

Gossmann, H.J. and Feldman, L.C. (1985a). *Phys. Rev.* **B32**, 6.

Gossmann, H.J. and Feldman, L.C. (1985b). *Surf. Sci.* **155**, 413.

Gossmann, H.J. and Gibson, W.M. (1984a). *J. Vac. Sci. Tech.* **B2**, 343.

Gossmann, H.J. and Gibson, W.M. (1984b). *Surf. Sci.* **139**, 239.

Haight, R. and Feldman, L.C. (1982). *J. Appl. Phys.* **53**, 4884.

Jackman, T.E., Davies, J.A., Jackson, D.P., Unertl, W.N. and Norton, P.R. (1982). *Surf. Sci.* **120**, 389.

Jackson, D.P. and Barrett, J.H. (1979). *Phys. Lett.* **71A**, 359.

Jackson, D.P., Jackman, T.E., Davies, J.A., Unertl, W.N. and Norton, P.R. (1983). *Surf. Sci.* **126**, 226.

Kuk, Y. and Feldman, L.C. (1984). *Phys. Rev.* **B30**, 5811.

Kuk, Y., Feldman, L.C. and Silverman, P.J. (1983a). *Phys. Rev. Lett.* **50**, 511.

Kuk, Y., Feldman, L.C. and Silverman, P.J. (1983b). *J. Vac. Sci. Tech.* **A1**, 1060.

Kuk, Y., Feldman, L.C., Robinson, I.K. (1984). *Surf. Sci.* **138**, L168.

Lindhard, J. (1965). *Mat. Fys. Medd.*, Selsk, Kgl. Dan Vidensk **34**, 14.

Narusawa, T. and Gibson, W.M. (1981). *Phys. Rev. Lett.* **47**, 1459.

Narusawa, T., Gibson, W.M. and Tornquist, E. (1981). *Phys. Rev. Lett.* **47**, 417.

Narusawa, T., Kobayashi, K.L.I. and Nakashima, H. (1985). *Jap. J. Appl. Phys.* **24**, L98.

Norton, P.R., Davies, J.A., Jackson, D.P. and Matsunami, N. (1979). *Surf. Sci.* **85**, 269.

Norton, P.R., Davies, J.A., Creber, D.K., Sitter, C.W. and Jackman, T.E. (1981). *Surf. Sci.* **105**, 205.

Sailer, E. and Varelas, C. (1984). *Nucl. Instrum. Methods* **B2**, 326.

Schuster, M. and Varelas, C. (1983). *Surf. Sci.* **134**, 195.

Smeenk, R.G., Tromp, R.M., Frenken, J.W.M. and Saris, F.W. (1981a). *Surf. Sci.* **112**, 261.

Smeenk, R.G., Tromp, R.M. and Saris, F.W. (1981b), *Surf. Sci.* **107**, 429.

Smit, L., Tromp, R.M. and van der Veen, J.F. (1984). *Phys. Rev.* **B29**, 4814.

Stensgaard, I. and Jakobsen, F. (1985). *Phys. Rev. Lett.* **54**, 711.

Stensgaard, I., Feldman, L.C. and Silverman, P.J. (1978). *Surf. Sci.* **77**, 513.

Stensgaard, I., Feldman, L.C. and Silverman, P.J. (1979). *Phys. Rev. Lett.* **42**, 247.

Stensgaard, I., Feldman, L.C. and Silverman, P.J. (1981). *Surf. Sci.* **102**, 1.

Tromp, R.M., Smeenk, R.G. and Saris, F.W. (1981). *Surf. Sci.* **104**, 13.

Tromp, R.M., van Loenen, E.J., Iwami, M. and Saris, F.W. (1983). *Sol. State Comm.* **44**, 971.

van der Veen, J.F., Smeenk, R.G., Tromp, R.M. and Saris, F.W. (1979). *Surf. Sci.* **79**, 212.

van Loenen, E.J., Fischer, A.E.M.J. and van der Veen, J.F. (1985). *Surf. Sci.* **155**, 65.

Zehner, D.M., Appleton, B.R., Noggle, T.S., Miller, J.W., Barrett, J.H., Jenkins, L.H. and Schow, O.E. (1975). *J. Vac. Sci. Tech.* **12**, 454.

10
Microprobe Analysis

G.J.F. LEGGE

School of Physics, University of Melbourne,
Parkville, Australia

ION BEAMS FOR
MATERIALS ANALYSIS
ISBN 0 12 099740 1

10.1 INTRODUCTION

10.1.1 Microanalysis

With many specimens, significant inhomogeneities occur over spatial intervals of a few microns or less. There is thus a need for high resolution information on the spatial distribution of elements or isotopes and for highly localised distributions in depth. This is the area of microanalysis using instruments referred to as microprobes. In this chapter the design and performance of these instruments will be discussed and illustrated with appropriate applications.

The history of elemental microanalysis dates from the development of the first electron microprobe (EMP) by Castaing and Guinier (1950). The general principles of quantitative analysis using EIXE were worked out by Castaing (1951). His microprobe had an electron beam size of less than 1 μm and was marketed by Cameca in 1956 as an electron probe microanalyser. Cosslett and Duncumb (1956) introduced the concept of beam scanning to generate maps of elemental distributions in the first commercial Scanning Electron Microprobe (SEMP) marketed in 1960 by Cambridge Instrument Corporation. The SEMP is the subject of several excellent books (Andersen, 1973; Reed, 1975; Chandler, 1977; Goldstein *et al.*, 1981).

The Sputter Ion Microscope (IMS), which was developed by Castaing *et al.* (1960) and later improved (Castaing and Hughes, 1962; Castaing and Slodzian, 1962; Slodzian, 1964), utilises mass selected secondary ions sputtered off the specimen by a fixed beam of heavy ions to form an image of the irradiated specimen area. The Scanning Ion Microprobe (SIMP) was developed by Liebl (1967) and marketed in 1968 by Applied Research Laboratories. It utilises a well focussed beam of primary ions to scan the specimen and is analogous in its primary ion optical system to the SEMP. The SIMP has been described by Andersen (1973), Robinson (1973) and Colby (1975).

The Scanning Proton Microprobe (SPMP), or nuclear microprobe, utilises a focussed beam of high energy protons or other light ions to scan a specimen and is similar in its operation to a SEMP. It uses PIXE, RBS or RFS and NRA for elemental or isotopic analysis. Simultaneous depth profiling is also possible. The first fixed beam trials were made by Pierce *et al.* (1966, 1968) and by Bird and associates (Mak *et al.*, 1966; Price and Bird, 1969) using collimated beams of protons. The first scanning probe was developed by Cookson *et al.* (1972). The SPMP is described in review articles (Cookson, 1979, 1981; Legge, 1980; Maggiore, 1980; Martin, 1980), conference proceedings (Demortier, 1982; Grime and Watt, 1988), a useful discussion mostly on non-scanning ion opitcs (Martin and Nobiling, 1980) and a detailed discussion of quadrupole lens optics (Grime and Watt, 1984). Principles and applications have been presented by Watt and Grime (1987).

10.1.2 Basic Principles of a Focussed Microprobe

The basic principles of the ion optical system of the three microprobes (electron, ion and proton) are shown in Highlight 10.1. Space limitations preclude a full description of the ion optical system of a secondary ion microscope. This is primarily an imaging instrument, though quantitative analyses may be made by selecting points of the image. Much of what is said of the SIMP will apply also to the IMS and reference should be made to the literature (Robinson, 1973) for comparisons. The basic principles used in all three probes for scanning system analysis are described in Highlight 10.2. Some modifications to these principles, which give much greater quantitative power and methods of handling line scans are discussed in Section 10.3.

HIGHLIGHT 10.1
ION OPTICS OF ELECTRON, ION AND
PROTON MICROPROBES

Schematic diagrams of each probe type are shown in Fig. 10.1. In each case an electron or ion source (positive or negative) and low energy lens produces a beam which is focussed successively by two further lenses to form a spot of about one micron diameter or less on the specimen. The aberrations of the final or probe lens are, because of its demagnification, much more important than those of previous lenses and are controlled by limiting the entering beam divergence with an aperture diaphragm. The focussed spot can be scanned across the specimen by means of magnetic scan coils or electrostatic plates placed either before or after the probe lens. In the SEMP and SIMP, this spot is essentially an image of the beam emitting area in the source.

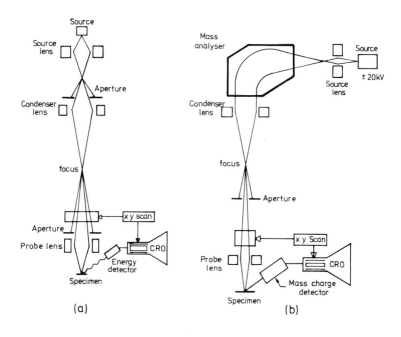

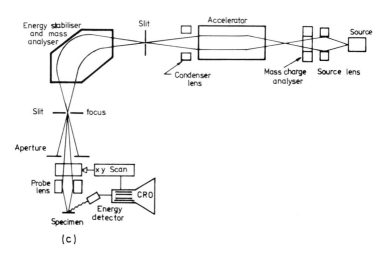

Fig. 10.1 Electron optical or ion optical systems of (a) SEMP, (b) SIMP and (c) SPMP, showing major components.

With positive ion sources, some form of charge-mass analysis is generally required to separate and thereby eliminate undesired charge states or mass components from the ion beam. The SPMP is further distinguished by its operating voltage of several million volts. This necessitates the incorporation of a long acceleration section, generally stabilised by feedback from the exit slits of an energy analysing magnet (Chapter 2). Such a magnet acts as a lens with fixed object and image points where the entrance and exit slits are positioned. Consequently the intermediate lens of the SPMP acts only as a condenser lens — the only effects of reducing its strength, or even turning it off, are to decrease the beam current accepted by the analysing magnet and to decrease the divergence of this beam, which may thereby be made more acceptable to the final lens aperture. The final focussed spot is an image of an object diaphragm (or slit), placed shortly after the exit slits of the analysing magnet unless an extra lens is employed to focus between these two points. Thus the final spot size and hence the spatial resolution of the SPMP are determined solely by the size of the object diaphragm and final lens aberrations; the object and image distances and hence the demagnification factor of the final lens are generally fixed. In contrast, the SEMP and SIMP have no fixed diaphragm or slit between their intermediate and final lenses, apart from a divergence limiting aperture associated with the final lens. It is therefore possible, by weakening the intermediate lens current, to throw the intermediate focus forward and thus reduce the demagnification of both lenses. This increases both the spot size and the beam current, because the divergence angles are simultaneously reduced and more beam is accepted by the final lens aperture.

HIGHLIGHT 10.2
BASIC SCANNING ANALYSIS

In order to measure distributions of elements along a line or to map elemental distributions over an area, the focussed beam spot must be scanned (Fig. 10.2) and the detector signal recorded as a function of the displacement of the beam from its normal position. In basic systems, mapping is usually done for one element at a time, by selecting the appropriate range of detector signals and using these signals to brighten the beam of an oscilloscope whose X and Y deflections are made proportional to the deflections of the microprobe beam. In this manner a pattern is recorded on the oscilloscope screen, depicting the distribution of the selected element. The magnification of the system is the ratio of the scan amplitudes for oscilloscope and microprobe. The pattern can be

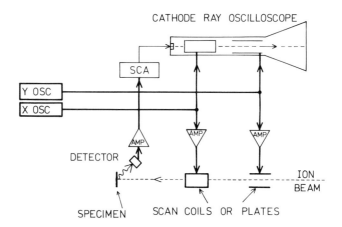

Fig. 10.2 Basic principles of scanning analysis by an electron or ion beam. The oscillators which sweep the focussed spot of the beam over the surface of the specimen also sweep the electron beam of a cathode ray oscilloscope across its screen. A signal from the detector which is characteristic of some element is selected by the single channel analyser (SCA) and used to brighten the oscilloscope display. The resulting display of bright spots on the oscilloscope screen forms a map showing the distribution of the selected element within the scanned area.

either viewed directly by means of a long persistence or storage screen, or recorded photographically as it is built up.

10.1.3 Principal Signal Channels

The major signals used for imaging and microanalysis in a charged particle microprobe are summarised in Table 10.1. However, only those in bold type yield quantitative data on the composition of the specimen. With a thick specimen, the first three signals are, of course, unavailable. In specific instances, other signals may be used, e.g. cathodoluminescence, speciment current, magnetic contrast, voltage contrast and channeling contrast. For the most part these give only qualitative information of value only in imaging a limited range of specimen types. Although applicable to the SPMP, they have mostly been employed with the SEM (Goldstein and Yakowitz, 1975; Goldstein *et al.*, 1981). Channeling is an increasingly important quantitative tool and is discussed in Section 10.4.1c. Signals which are not discussed in other chapters are described in the following Sections.

TABLE 10.1
Principal signal channels available in charged particle microprobes

	Thin Specimens					Thick Specimens			
Signal	US	IS	FS	SE	E	X	BS	γ	ION
PROBE									
EMP	STEM(b)	**EELS**	STEM(d)		SE, **AES**	**EIXE**	BSE		
IMP					SE, **AES**				SIMS
IMS					SE				SIMS
PMP			STIM	**RFS, ERA**	SE		**PIXE**	**RBS, CH**	**PIGME**	NRA

The techniques in bold type give quantitative information. US=Unscattered, IS=Inelastic Scattering, FS=Forward Scattering, BS=Backscattering, SE=Secondary Electrons, (b)=Bright field, (d)=Dark field; other symbols – see Preface, Table 1.

a. Transmission Imaging and Analysis by Inelastic Scattering

When a beam of charged particles passes through a thin specimen, the beam transmitted in the forward direction includes some particles that scattered elastically off atomic nuclei or lost energy due to interaction with atomic electrons (inelastically scattered) as well as those particles that suffered no interaction (unscattered). By analogy with light optics, an image formed with this forward transmitted beam is referred to as a bright field image. In a conventional transmission microscope such as a TEM, this imaging is done with an objective lens and the energy degradation of the inelastically scattered particles limits use of the TEM to very thin specimens. Mere uniformity of thickness does not give uniformity of energy in the transmitted beam, because of energy straggling. However, in a scanning transmission instrument, this energy degradation does not affect spatial resolution so thicker specimens may be used. With protons or heavy ions, the mean free path between ionising events is generally much less than the specimen thickness and hence multiple inelastic scattering occurs. The energy loss spectrum is then a measure of specimen thickness rather than elemental content. The specimen thickness exceeds the range of the slow heavy ions in an IMP (Chapter 7) and transmission signals are not available. However, in a PMP, where the energy is typically 3 MeV, the proton range may be some hundreds of microns and the mean free path between inelastic collisions will be less than 100 nm. It is therefore possible to use specimens several microns in thickness for which the energy loss of transmitted protons is readily measured but the energy straggling is relatively small, though not necessarily insignificant. The major energy spread is then the detector resolution if a solid state detector is used. This direct measurement of specimen thickness with STIM is extremely efficient if individual particles are recorded, because nearly all beam particles contribute to this signal. For very thin specimens magnetic or electrostatic energy analysis enables the use of low energy (100 keV) protons (Levi-Setti and Fox, 1980).

b. Elastic Scattering

Some particles will be scattered out of the beam path and may be used to produce a dark field image of the specimen — essentially an image of thickness or density variation. With an electron beam, whatever the angle of scattering, there is no specific elemental information in this signal, apart from the fact that the scattering cross-section is a function of atomic number. But with an ion beam, energy exchange results in a mass spectrum whose mass resolution increases with angle of scatter (Chapter 3). The cross-section is high at forward angles and, although other elemental peaks are not well separated in the mass spectrum, accurate quantitative measurements of H can be made. Other elements are best resolved at backward angles. RFS and RBS are used in the SPMP for light element analysis with thin or semithin specimens ($< 10 \mu$m) as EELS is used in the SEMP with very thin specimens (< 50 nm).

c. Secondary Electron Emission

Secondary electron emission is the one interaction which is common to all three types of microprobe. It is not a technique for elemental analysis, being sensitive to surface contours and local potential, as well as composition; but, because of the short range of secondary electrons, secondary emission is only a surface phenomenon and hence is insensitive to beam divergence (loss of spatial resolution due to scattering) in a thick specimen. Secondary electron emission then provides the most useful signal for imaging. The basic principles of secondary electron detection and signal handling are the same for all instruments and are covered in most books on scanning electron microscopy. Its application to proton microscopy was first described by Younger and Cookson (1979) and the Z contrast obtainable was investigated by Traxel and Mandel (1984). Its application to high resolution ion microscopy is described by Levi-Setti *et al.* (1985a).

10.1.4 A Comparison of Microprobes

To some extent the three instruments discussed so far provide complementary information. Table 10.2 gives a rough guide to spatial resolution, elemental sensitivity, useful specimen thickness, depth resolution,

TABLE 10.2
Performance of charged particle microprobes in microanalysis

	EMP	IMS	IMP	PMP
Spatial resolution (μm)	0.001–10	0.1	0.1[†]	1[†]
Depth resolution (nm)	–	0.2	0.2	1–10
Elemental sensitivity (ppm)	1000	0.1–10	0.1–10	1–10
Quantitative accuracy	fair	–	poor	good
Isotopic ratio measurement	–	good	good	sometimes
Elemental mapping	fair	good	poor	good
Secondary electron imaging	yes	no	yes	yes
Channeling	yes	no	no	yes

* depending on specimen thickness
[†] at 100 pA beam current

and so on, but these figures depend on many conditions. The SEMP cannot be used for depth profiling and it is much less sensitive in terms of mass fraction (mg g^{-1}) than the other two instruments. However it can achieve much better spatial resolution in thin specimens than the other instruments. The specimen thickness needs to be comparable with or less than the desired resolution, which may further limit sensitivity. Much valuable work is done with this instrument by making a moderate resolution analysis of an area previously identified by means of a high resolution surface image. Most SEMPs are not equipped with EELS and therefore cannot easily measure the light elements (Z < 11). If EELS is available, the specimens must be very thin (< 50 nm) but the efficiency is high. The SPMP is particularly suited to high sensitivity multielement analysis of both thick and thin specimens, with spatial resolution at the moment limited to about a micron but expected to improve. It has available the greatest number of techniques for elemental analysis and will generally supply the greatest quantitative precision. Depth profiling is possible with the SPMP but greater depth resolution can be achieved with the SIMP. The sensitivity of the SIMP is superior to that of the SPMP for some elements and inferior for others. The SIMP is less quantitative, because interaction rates are variable and the measurements are more difficult to interpret. The continuous etching away of the specimen surface may be an advantage in some analyses and intolerable in others. For imaging purposes, when much lower currents are needed, big improvements in resolution are possible for both the SIMP (Levi-Setti *et al.*, 1985a) and the SPMP (Sealock *et al.*, 1983).

The SEMP or SEM with X-ray attachment is likely to remain the workhorse of microanalysis. It is commercially available from many suppliers and its excellent spatial resolution of surface features coupled with its ease of operation will ensure its continued use for many problems. The SIMP is a very specialised and more sophisticated instrument, having great strengths and limitations in its applicability. It requires an experienced interpreter to handle the data if quantitative results are required. The SPMP is the most recently developed instrument. It is the most versatile and accurate for quantitative work and is generally the most sophisticated in operation and in data processing.

10.2   REQUIREMENTS OF A FOCUSSED PROTON MICROPROBE

10.2.1   Aberrations and Requirements for Good Spatial Resolution

In this section we consider the factors determining the spatial resolution achievable with proton microprobes, discuss the means of handling these factors and the progress achieved, and indicate the procedures necessary to set up and ensure optimum operation of such a microprobe.

a. Brightness — Ion Source and Accelerator

The reduced brightness of an ion beam is defined as:

$$\beta_r = I/A\Omega E \tag{10.1}$$

where I/A is the current per unit area, Ω is the solid angle of divergence or convergence of the beam, E is the kinetic energy of the beam and the units are $A\,\mathrm{m}^{-2}\,\mathrm{rad}^{-2}\,\mathrm{eV}^{-1}$. The kinetic energy is included to allow for the decrease in angular spread when the velocity is increased. All parameters must be measured at the same cross-section of the beam where there is a waist. If axial symmetry applies at some point considered as a source, then the reduced brightness is usually defined equivalently as:

$$\beta_r = I/\pi^2 r^2 \alpha^2 E \tag{10.2}$$

where r is the effective radius of the source and α is the semiangle of divergence at the source. It is a fundamental principle of ion optics (Liouville's theorem) that this brightness cannot be increased by any focussing system, though it can readily be decreased by aberrations of the ion optical system. Consequently it is important for microprobe work that the ion source produce a beam of the maximum possible brightness and that this brightness be maintained so far as possible during transmission of the beam through the accelerator and any associated beam optical equipment. Because of the complexity of this path and the doubtful quality of the optics, it is customary to start with a large beam current at the ion source and then to select the core of the transmitted beam as it enters the mircroprobe section. Aberrations in the accelerator section are then not so critical; but even so brightness is lost in some accelerators.

With a single ended Van der Graaff or Cockroft Walton type of electrostatic accelerator, a RF ion source is commonly used because of its simplicity. It is suitable for production of beams of ^{1}H, ^{2}H, ^{3}He and ^{4}He singly charged positive ions and fortunately is a relatively bright source. With such a source coupled to an Electrostatics International Incorporated 5 MV single ended Pelletron, this brightness can be maintained in the accelerator and analysing magnet, the reduced brightness being 13 A m^{-2} rad^{-2} eV^{-1} measured inside the entrance of an attached microprobe. Tandem accelerators are more commonly equipped with a duoplasmatron source for negative ion production. The brightness is comparable with that of the RF source but the beam quality is impaired by the necessary transmission through a stripper in the centre terminal of the accelerator (Chapter 2). A reduced brightness of about 0.8 A m^{-2} rad^{-2} eV^{-1} has been achieved in a microprobe attached to a High Voltage Engineering Corporation EN tandem accelerator using direct extraction off-axis duoplasmatron proton injection and a gas stripper (Watt *et al.*, 1982). This brightness was reduced by an order of magnitude when a foil stripper was used.

It should be possible to achieve greater brightness in the beam from a tandem accelerator as better ion sources are developed. These will probably be high power sources from which much of the beam is thrown away after acceleration. In a single ended accelerator, there is less space and power available for the ion source but the ion optics are better. Field ionisation sources are much brighter than RF sources, but the proton currents as yet have been too low for microprobe analysis, though adequate for microscopy (Levi-Setti and Fox, 1980). Recent progress (Allan *et al.*, 1985a) indicates that it may be possible to extract sufficient proton current from such a source, 100 pA to 1 nA, to use it for microprobe analysis, but a parallel improvement in the accelerator ion optics is needed in order to take advantage of the brightness offered. Reliability and long life are also important characteristics of an ion source. The energy spread is important in so far as it sets a limit to the energy spread of any accelerator beam and leads to chromatic aberrations which arise throughout the ion optical system.

b. Chromatic Aberrations — Accelerator Stability

Charged particle beams are focussed by means of magnetic or electrostatic lenses. The focal length of such a lens depends on the energy of the particles and on the excitation of the lens. In Fig. 10.3 a beam of half divergence angle α' is focussed by a lens to a half convergence angle α, where $D = \alpha/\alpha'$ is the demagnification factor and also the ratio of object and image distances. For a thin lens and large object distance, α' is small and the image distance is approximately equal to the focal length f and to the working distance W. A variation in energy of the incoming beam results in a variation in the focal length. Consequently, as illustrated in Fig. 10.3, at the correct focus a point object will be imaged as a circular disk of diameter:

$$d_c = C_c\,\alpha\,\Delta E/E \tag{10.3}$$

where ΔE is the full range of energy and C_c is the chromatic aberration coefficient of the lens. It is readily shown that for magnetic deflection in a thin lens, $C_c \simeq f$ and hence the dimensionless quantity $C_c/f \simeq 1$. This quantity C_c/f is an important property of the lens. Likewise it is readily shown that a short electrostatic lens should have $C_c/f \simeq 2$. We shall see in Section 10.2.1d to what extent the thin lens approximations are valid. Note that the aberration depends to the first order on α but does not depend directly on α', provided that $\alpha' \ll \alpha$. It is possible to use an aberration coefficient C_c' expressed in terms of α':

$$d_c = C_c'\,\alpha'\,\Delta E/E \tag{10.4}$$

but then $C_c' = D\,C_{c'}$ so that $C_c'/f = D\,C_c/f$ is not merely a property of the lens but depends on the choice of object and image distances. Since equations and coefficients in image coordinates, Equation (10.3), and in object coordinates, Equation (10.4), are to be found in the literature without primes, care must be taken in interpreting these.

Equations (10.2) and (10.3) may be used to predict the spatial resolution achievable by a microprobe which is limited mainly by chromatic aberration. Thus, assuming that the chromatic aberration d_c will add to the primary image of diameter d_0 in quadrature to produce a resultant focussed spot of diameter:

$$d = (d_0^2 + d_c^2)^{1/2} \tag{10.5}$$

and assuming axial symmetry with $A = \pi d_0^2/4$, we find from Equations

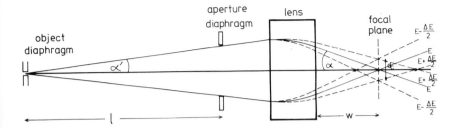

Fig. 10.3 Ion optics of a microprobe with chromatic aberration. A point source is focussed to a chromatic aberration disk of diameter d_c at the working distance W from the back of the lens.

(10.2), (10.3) and (10.5) that maximum current is achieved for a given resolution d when $d_c = d_0$, and the resolution is then:

$$d = 2^{1/2}d_0 = 2(C_c/f)^{1/2} f^{1/2} (\Delta E/E)^{1/2} (I/\pi^2 \beta_r E)^{1/4} \qquad (10.6)$$

The available current is:

$$I = \pi^2 \beta_r E \, d^4/16(C_c/f)^2 f^2 (\Delta E/E)^2 \qquad (10.7)$$

for a half convergence angle:

$$\alpha = \left(\frac{1}{2}\right)^{1/2} d/(C_c/f) f (\Delta E/E) \qquad (10.8)$$

Therefore, for the best spatial resolution, we need to seek the minimum achievable value of C_c/f, use a short focal length and reduce to a minimum the energy instabilities in the accelerator. Brightness is also important, but any improvements in brightness must be very large in order to give a significant improvement in the resolution for a given current. Alternatively, great sacrifices in current are demanded in order to improve resolution. For such a microprobe, in which the only significant aberration is chromatic, the object diaphragm diameter (or slit width), d_{object}, and the aperture diaphragm diameter (or slit width), $d_{aperture}$, should be selected with a fixed ratio, for maximum current at all resolutions, that ratio being:

$$d_{aperture}/d_{object} = (2\ell/f) D^{-2} (C_c/f)^{-1} (\Delta E/E)^{-1} \qquad (10.9)$$

where ℓ is the spacing between these two diaphragms (or slit systems), as

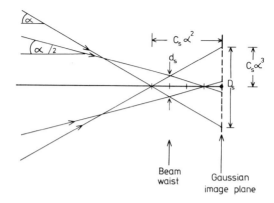

Fig. 10.4 Ion optics of a microprobe with spherical aberration. A point source is focussed at different points for rays of different convergence angle. A beam waist of diameter d_s is formed in front of the paraxial ray focal plane (Gaussian image plane).

shown in Fig. 10.3, and:

$$d_{\text{object}} = D \, d_0 \tag{10.10}$$

c. Spherical Aberrations

Even a monoenergetic beam will not produce a single point image of a point object with a lens of finite aperture. Rays of successively greater convergence angle will focus at points of successively shorter focal distance and hence produce a spherical aberration disk of diameter D_s in the focal plane of the paraxial rays (the Gaussian image plane). The aberration or displacement of extreme rays in this plane $D_s/2$ has a third order dependence on the convergence half angle α. So we define the spherical aberration coefficient C_s by writing the aberration:

$$D_s/2 = C_s \, \alpha^3 \tag{10.11}$$

As indicated in Fig. 10.4, the beam envelope is composed of many rays and it is readily shown that a waist is formed at the intersection of the above extreme rays with those of half the maximum convergence angle. This waist has a diameter:

$$d_s = \frac{1}{2} C_s \, \alpha^3 \tag{10.12}$$

It is known as the *circle of least confusion* and is formed at a distance 3/4 $C_s \alpha^2$ in front of the Gaussian image plane as shown in Fig. 10.4.

Analogous to the treatment of chromatic aberration, Equations (10.2) and (10.12) may be used to predict the spatial resolution of a microprobe which is limited mainly by spherical aberration. Again we assume that the primary image diameter, d_0, and the spherical aberration waist diameter, d_s, add in quadrature to give a resultant focussed spot of diameter:

$$d = (d_0^2 + d_s^2)^{1/2} \tag{10.13}$$

with axial symmetry ($A = \pi/4\, d_0^2$). If we further assume that primary and aberration contributions are equal:

$$d_0 = d_s \tag{10.14a}$$

then

$$d = 2^{1/2}\, d_0 = 2\, C_s^{1/4}\, (I/\pi^2 \beta_r E)^{3/8} \tag{10.15a}$$

or

$$I = \pi^2 \beta_r E\, C_s^{-2/3}\, (d/2)^{8/3} \tag{10.16a}$$

$$d_{\text{object}} = D\, d_0 = (1/2)^{1/2}\, D\, d \tag{10.17a}$$

and

$$d_{\text{aperture}} = 2^{1/6}\, 2(\ell/D)\, (d/C_s)^{1/3} \tag{10.18a}$$

However, Equations (10.2), (10.12) and (10.13) predict a maximum current for a given resolution if we use a slightly larger object and smaller aperture so that:

$$d_0 = 3^{1/2}\, d_s \tag{10.14b}$$

$$d = (2/3)^{1/2}\, 2^{1/2}\, d_0 = (16/27)^{1/8}\, 2\, C_s^{1/4} (I/\pi^2 \beta_r E)^{3/8} \tag{10.15b}$$

or

$$I = (27/16)^{1/3}\, \pi^2 \beta_r E\, C_s^{-2/3}\, (d/2)^{8/3} \tag{10.16b}$$

$$d_{\text{object}} = D\,d_0 = (3/4)^{1/2}\,D\,d \qquad (10.17b)$$

and

$$d_{\text{aperture}} = 2(\ell/D)\,(d/C_s)^{1/3} \qquad (10.18b)$$

These optimum conditions however show only a 20% gain in current when equations (10.16a) and (10.16b) are compared. Comparison of Equations (10.17) and (10.18) shows that we need to keep the ratio $d_{\text{aperture}}^3/d_{\text{object}}$ constant if the resolution is limited mainly by spherical aberration, in which case only minor changes in the aperture size will be required. We also need to keep in mind the significant longitudinal movement of the waist as the aperture is changed.

The above discussion applies strictly to lenses with cylindrical symmetry. With a quadrupole lens, the aberration coefficients are functions of the azimuthal angle and, although it is possible to shape the aperture (that is, to make α a function of azimuthal angle) so as to make the spherical aberration $1/2\ C_s\alpha^3$ independent of azimuthal angle (Jamieson, 1985), it is not possible simultaneously to make $3/4\ C_s\alpha^2$ independent of azimuth. Thus the position of the minimum waist will be dependent on the azimuth angle and the *circle of least confusion* is not readily defined. This will tend to slightly worsen lens performance.

d. Lens Configurations and Performance

The majority of proton microprobes have a compound lens comprising two, three or four magnetic quadrupole lenses. The first combination to be used (Cookson *et al.*, 1972) was the *Russian quadruplet*, following the investigation of its aberrations as an electron lens (Dymnikov and Yavor, 1964). This, the most popular combination, is antisymmetric, the lens currents being in order designated as $+$ A, $-$ B, $+$ B, $-$ A, and only two independent current supplies are needed. It also has the advantage of being orthomorphic; that is, when a stigmatic image is formed, the demagnification in both planes of focussing is the same, so that a drilled hole can be used as an object diaphragm to produce a circular focussed spot on the target. The doublet is a simpler combination and can achieve greater demagnification in a shorter length (Bosch, *et al.*, 1978) but the demagnification factors for the two planes of focussing may differ by a factor of five or more. This requires that two sets of independently controlled slits be used in place of an object diaphragm. Aperture slits may also be required. The low excitation triplet is similar to the antisymmetric quadruplet but has a lower demagnification factor and requires the use of three lens current supplies. The high excitation triplet can achieve large demagnification by forming a prefocus (crossover) within the lens for one plane of focussing (Watt *et al.*, 1982), but this combination also has unequal demagnification factors. Asymmetric and high excitation quadruplet combinations are also possible. For further discussion see Legge (1982) and the references therein and Grime and Watt (1984).

The strong focussing of magnetic quadrupole lenses was used for proton microprobes because the weak focussing of cylindrical lenses, as used in most electron instruments, would require fields well above the saturation limits of any iron or current densities well above the thermal limits of any normal iron free coils. These limits may be overcome if the necessary field is obtained with a superconducting solenoid coil. Two such lenses have been developed, a long solenoid at MIT (Koyama-Ito and Grodzins, 1980) and a combination pancake winding at Los Alamos (Maggiore, 1981). These lenses are basically symmetrical. The Los Alamos design gives a field distribution approximating to a Glaser field (Glaser, 1941) and as such may be treated analytically. The MIT design gives a field which is less extensive than a Glaser field and must be handled by ray tracing. However the theoretical performance of this lens is also similar to that of a Glaser field.

In estimating the potential performance of any lens combination, we must first look at the aberration coefficients. First order aberration coefficients may be calculated by rapid matrix computer programs. More complex matrix programs have also been developed to handle third order aberrations (Heck and Kasseckert, 1976; Heck, 1976). Alternatively third and higher order aberrations can be calculated by tracing individual rays through the system to form the image. An excellent discussion of some available programs and the techniques adopted has been given by Grime *et al.* (1982) and a detailed coverage of the beam optics of quadrupole probe forming systems by Grime and Watt (1984). The most important aberration coefficients have been evaluated in object space (α') for several magnetic quadrupole systems (Grime *et al.*, 1982), enabling the performance of each to be predicted. (Although the focussed spot sizes arising from some aberrations are incorrectly plotted in Fig. 6.11 of this paper, giving an incorrect view of the relative performance of various systems, the coefficients used to generate these computer plots and listed in the tables are correct.) For the moment we are concerned only with the chromatic and spherical aberration coefficients. In Table 10.3, these coefficients are given in image space (α) and then normalised to the focal lengths, so that the different systems may be compared in principle, despite the large difference in scale.

TABLE 10.3

Major parameters for several proton microprobe configurations and their associated chromatic and spherical aberration coefficients

	Heidelberg Doublet[a]	Karlsruhe Spaced[a]	Doublet Close[c]	Oxford Triplet[a]	Harwell Triplet[a]	Harwell Quadruplet[a]	Melbourne Short WD[b]	Quadruplet Long WD[b]	MIT Solenoid[d]	Los Alomos Solenoid[d]	
Length of microprobe	2.030	3.115	3.298	6.450	4.590	4.590	8.605	8.605	4.21	1.25	m
Length of lens system (coils)	0.120	0.480	0.315	0.446	0.900	0.900	0.314	0.314	0.100	0.050	m
Diameter of lens bore (usable)	0.005	0.050	0.050	0.030	0.038	0.038	0.013	0.013	0.016	0.019	m
Demagnifications											
D_x	−26.4	−29.0	−28.5	67.8	32.1	−5.3	−20.4	−15.6	−25.0	−10.0	
D_y	−4.6	−2.3	−3.48	−15.2	−10.2	−5.3	−20.4	−15.6	−25.0	−10.0	
Focal Lengths											
f_x	0.071	0.083	0.103	−0.091	−0.069	0.642	0.403	0.508	0.150	0.100	m
f_y	0.257	0.380	0.439	0.566	0.361	0.642	0.403	0.508	0.150	0.100	m
$f = (f_x f_y)^{1/2}$	0.135	0.178	0.213	−0.277	−0.158	0.642	0.403	0.508	0.150	−0.100	m
Chromatic aberration coefficients											
$\langle x\,|\,\theta\delta\rangle$	0.07	0.10	0.108	−0.19	−0.23	0.60	0.36	0.49	0.150	0.093	m
$\langle y\,|\,\phi\delta\rangle$	0.43	1.53	1.002	2.58	2.16	0.98	0.44	0.54	0.150	0.093	m
$C_c = (|\langle x\,|\,\theta\delta\rangle\langle y\,|\,\phi\delta\rangle|)^{1/2}$	0.173	0.391	0.329	0.700	0.705	0.767	0.398	0.514	0.150	0.093	m
C_c/f	1.285	2.197	1.544	3.084	4.46	1.19	0.988	1.013	1.00	0.93	
Spherical aberration coefficients											
$\langle x\,|\,\theta^3\rangle$	−0.17	−0.16	−0.231	0.28	0.43	−7.4	−17.2	−37.3	0.79	0.14	m
$\langle x\,|\,\theta\phi^2\rangle$	−2.9	−16.9	−2.68	7.1	7.4	−39.0	−58.4	−117	–	–	m
$\langle y\,|\,\phi^3\rangle$	−11.3	−160	−49.7	−125	−78.3	−30.9	−25.3	−55.2	0.79	0.14	m
$\langle y\,|\,\theta^2\phi\rangle$	−2.5	−5.0	−2.68	−6.1	−8.0	−37.6	−50.7	−121	–	0.14	m
C_s	5.5	18.3	8.6	21	22	29	38	83	0.79	0.14	m
C_s/f^3	2230	3288	897	1763	5467	110	580	632	234	140	m^{-2}
Working distance (W.D.)	0.110	0.125	0.129	0.170	0.199	0.199	0.236	0.350	0.110	0.070	m

Convergence angles (θ in the xz plane and ϕ in the yz plane) are expressed in radians and energy spread, δ, is expressed as a fraction of the full energy. In this nomenclature, for example, the aberration coefficient $\langle x\,|\,\theta\delta\rangle$ measures the x-displacement in the Gaussian image plane per unit convergence angle, θ, per unit energy spread, δ.

[a] Data from calculations on quadrupole systems by Grime et al. (1982) converted to image coordinates; confirmed by independent calculations.

[b] Calculations by D.N. Jamieson (priv. com.).

[c] Calculations by Heck (priv. com.) for modified system.

[d] Data from calculations on solenoid systems by Koyama-Ito and Grodzins (1980) and Maggiore (1981); confirmed by independent calculations. Calculations by M. Dix (priv. com.).

The quadrupole lens systems have parameters which differ for their two planes of focussing (x and y) and their associated half convergence angles (θ and ϕ). The symmetric quadruplet systems, like the solenoid lenses, are assumed to employ circular apertures giving circular chromatic aberration patterns. The asymmetric systems are assumed to employ rectangular apertures giving square chromatic aberration patterns. The effective chromatic aberration coefficient C_c is that of a symmetric lens giving a circular aberration disc of the same width when used with a circular aperture of equal area. The effective focal length f is the geometric mean of the values f_x and f_y for the two planes of focussing. The *Russian quadruplet* (Harwell and Melbourne) and the doublet, if compact (Heidelberg), have values of C_c/f which are close to the value predicted for a thin magnetic lens. (For asymmetric quadrupole multiplet lenses, the parameters have different values for the two planes of focussing. However, when predicting the current available for a given resolution, we always need the product of such parameter pairs and this is comparable with the equivalent parameter for a symmetrical lens. Thus the notation $(C_c/f)^2$ for a quadrupole lens here denotes $(C_c/f)_x(C_c/f)_y$. When the spacing is increased (Karlsruhe) or when a triplet with an internal crossover is employed (Oxford and Harwell), the value of C_c/f increases significantly above the thin lens prediction. The two superconducting lenses also have values of C_c/f approximately equal to the thin lens value. The equivalent figure of merit expressed in object coordinates is C_c'/L_o, where L_o is the object distance.

The situation with the spherical aberration coefficients is more complicated. There are now four aberration coefficients for quadrupole lenses, two associated with the two planes of focussing and two cross terms. We assume that the aperture dimensions for each system are those chosen above to give square chromatic aberration patterns. The cross terms then dominate the spherical aberration of the quadrupole systems and the major dimension of the aberration disc is used in the calculation of an effective spherical aberration coefficient, again for a symmetric lens used with a circular aperture of the same area. If the major aberration is spherical, an improvement would be obtained with a tilted square (diamond) aperture on the quadruplet lens and an elliptical aperture on the asymmetric lenses if that were practical.

The superconducting solenoid lenses achieve, at least in theory, the best values of C_s. For a solenoid lens $C_s \simeq f^3/a^2$, where a is the half width of the field. Electron microscope lenses commonly operate in the region where $f \simeq a$ and so $C_s \simeq f$, but for proton microprobe lenses f is much larger and less dependent on a (Dix, 1983). The dominant influence of focal length is shown by the greatly reduced spread in the values of C_s/f^3 for all lens systems. The overall length, L, of the microprobe system is not significant in a discussion of lenses, greater length serving to increase demagnification and reduce the significance of object scattering but having little effect on aberration coeficients. However the choice of focal length must be a compromise, for the diminished aberrations of a shorter focal length are generally gained at the expense of a shorter working distance and this distance is of great practical importance, defining the space available for detectors, microscopes and scanning coils or plates between the lens and the specimen.

The performance of proton microprobes is at present limited mostly by chromatic aberrations. In such circumstances, the superconducting lenses have no great advantage, and they are much more expensive to operate because of their liquid He consumption, about $1.5\,l\,h^{-1}$ per hour for the Los Alamos lens (Maggiore, 1981). The development of alloys which allow superconductive operation at liquid N_2 temperature or higher will greatly reduced this problem. Table 10.4 shows that the magnetic quadrupole lens multiplets have a limiting chromatic performance similar to that of the solenoid lenses. Electrostatic quadrupole lenses have sometimes been used because of their simplicity of construction (Augustyniak *et al.*, 1978), but these have twice the chromatic aberration limits of magnetic lenses. Because their focal length is independent of ion mass, electrostatic lenses have been proposed for heavy ion beams; but such beams are brightest for high charge states and can be handled by suitably designed magnetic lenses. Martin and Goloskie (1982) constructed an achromatic doublet consisting of two lenses, each having superimposed electric and magnetic quadrupoles. With this combination a reduction of the chromatic aberration by a factor of 16 over that of the magnetic doublet alone was achieved. With poorly stabilised accelerators such a lens design may be the best choice, as long as the complexity of the design does not lead to unacceptable values of the other aberration coefficients. However the energy stability of many single ended accelerators, typically about one part in 10^3 which leads to the present dominance of chromatic aberrations, could be improved by at least an order of magnitude by the application of known techniques (Chapter 2.3.1). The spherical aberration would then be of major importance to the attainment of better spatial resolution. Recent attempts to cancel the high order aberrations of magnetic quadrupole lenses by the addition of magnetic octupole lenses have met with some success (Jamieson and Legge, 1987; 1988) but the task is complex and is still under investigation. On an accelerator of very high energy stability, the superconducting solenoid lenses may ultimately provide the best resolution, but only if parasitic aberrations, to date unmeasured, can be controlled in the construction. Even then, the best parameter values that are currently achievable $(C_s = 1.4 \text{ m and } \beta_r = 13 \text{ A m}^{-2} \text{ rad}^{-2} \text{ eV}^{-1})$ would give a spherical aberration limit of 0.1 μm at 100 pA and 3 MeV. Because of magnetic field and space limitations, the value of C_s is not likely to be sufficiently further reduced to significantly improve the resolution as given by Equation (10.15). Consequently the useful resolution limit set by the lens itself will be about 0.1 μm until brighter sources are available.

The present resolution limit of about 1 μm at 100 pA and 3 MeV has been achieved only with magnetic quadrupole multiplet lenses. The Melbourne quadrupole lens of 12 mm bore diameter utilises a long object distance in order to achieve large demagnification for a quadruplet together with a conveniently large working distance and orthomorphic imaging. The latter gives uniform convergence angles at the specimen, which is desirable for channeling applications. The Oxford triplet lens of 30 mm bore diameter, also with a large object distance, achieves very large demagnification, operation in a high excitation mode producing a crossover within the lens for one plane of focussing. Such high excitation modes of operation are possible for all quadrupole multiplet lenses other than the doublet, but the demagnification, although high, differs greatly in the two planes of focussing and the large value of C_c requires a fairly bright source and very nonuniform convergence angle at the specimen. The Heidelberg doublet lens of 5 mm bore diameter almost attains the above performance. It has been designed on a small scale to minimise the values of C_c and C_s but the object distance of this system also will be increased in order to increase the demagnification. Several other magnetic quadrupole multiplet systems have achieved almost equal performances. The superconducting solenoid lenses are limited by accelerator performance; but the one at Los Alamos of 20 mm bore diameter has attained a resolution of 5 μm with 4 nA and therefore should be capable of 2 μm with 100 pA. The radial self electric field of a trapped electron column may be used to form a focussing lens for ions which traverse it. Although such plasma lenses were rejected for electron microscopy because of their complexity, a modern version has been developed for positive ion microprobe work (Lefevre *et al.*, 1983). The aberrations are not yet completely understood and stability is still a problem, but a spatial resolution of 5 μm has been obtained and this lens is worthy of further study.

e. Parasitic Aberrations

There are other aberrations, known as parasitic aberrations, which may be associated with defects of manufacture, misalignment or excitation instability. In theory these can be eliminated and in practice, with sufficient care in the design and alignment stages, they can be sufficiently controlled so as to not limit the spatial resolution. The major aberrations of this type are excitation, rotation, translation and mechanical aberrations.

Excitation aberrations are similar in effect to the chromatic aberration, but may arise in each lens independently, in which case they are normally assumed to add in quadrature. We are here concerned with both long and short term true instabilities and with ripple. In order to minimise excitation aberrations, whenever lens components are to be excited equally but with opposite polarity, such as occurs with the first and last components of a Russian quadruplet and also with the two central components of such a system, they should be powered by the one supply, so that the excitation aberrations of opposing components tend to cancel. These aberrations may be incorporated into Equation (10.3) for chromatic aberration of a single or compound lens by writing the latter as (for one electrostatic lens with voltage V):

$$d_c = C_c \alpha \left[(\Delta E/E)^2 + (\Delta V/V)^2 \right]^{1/2} \qquad (10.19)$$

where $C_c \alpha \Delta V/V$ is the excitation aberration of the electrostatic lens, or, for one magnetic lens with current I:

$$d_c = C_c \alpha \left[(\Delta E/E)^2 + 4(\Delta I/I)^2 \right]^{1/2} \qquad (10.20)$$

where $C_c \alpha 2\Delta I/I$ is the excitation aberration of the magnetic lens. However if more than one excitation supply is needed, the chromatic aberration coefficients of the different excitation groups must be known and the excitation contributions added in quadrature. In general the lens supplies need to be an order of magnitude more stable and ripple free than the accelerator energy and, in view of the relative costs involved, it is worthwhile making their performance and reliability even better than this, so that their contribution to the total aberration is negligible.

Any departure from symmetry in a lens may result in aberrations. In particular, a slight relative rotation of the components of a quadrupole multiplet can introduce severe first order aberrations, because of the departure from a pure quadrupole field. Therefore in setting up such a compound lens, provision must be made for the accurate relative rotational alignment of the various component lenses. Fortunately, to first order, the effects of slight rotational displacements of the various components add algebraically so that a final adjustment may be made with any one component lens. An accuracy of 0.01 mrad may be required (Grime et al., 1982), but the correction may be made electrically with a weak quadrupole field at 45° provided by a small correcting lens (Prins and Hoffman, 1981). Slight rotation of a lens about an axis perpendicular to the beam axis does not produce any serious aberrations.

In axially symmetric lenses, transverse displacements of the lens can produce astigmatism. In axially asymmetric lenses, this is not a problem, because a stigmatic image is always obtainable with the two or more independent lens controls available. However any lens whose axis is displaced from the beam axis will tend to pull the beam onto its own axis, if it is a converging lens, or push it further off axis, if it is a diverging lens. Observation and elimination of resultant *steering* of a beam by an off-axis lens as the lens strength is changed should be used to centre all lenses independently, and for this purpose a fine adjustment should be provided on the horizontal and vertical positioning of each lens. Displacement of the axis by a distance comparable with the aperture dimensions would increase chromatic and spherical aberrations. A slight displacement of the axis may not affect the obtainable spatial resolution but the resultant steering whenever the focus is adjusted makes any fine work very tedious.

Mechanical aberrations in magnetic lenses may arise from inhomogeneities in magnetic materials, from imperfect machining or assembly of the parts or from subsequent changes in magnetic properties or mechanical placements. Any lack of symmetry in magnetic reluctance around the yoke paths between pole tips or in uniformity of coil winding may upset the field of a quadrupole lens; but the most critical dimension is the distance between adjacent pole tips. It is strongly recommended that this be set with non-magnetic spacers which may be left in place if desired (Legge *et al.*, 1982a). Otherwise the pole tips should be pinned in place (Watt *et al.*, 1982) before removal of the spacers. Magnetic materials should have low hysteresis and high saturation fields; they should be aged and, after completion of major machining, should preferably be stress relieved by annealing with subsequent work restricted to fine grinding (Nobiling, 1982). Knies *et al.* (1982) have quantitatively demonstrated the effects of these machining stresses in the absence of permanent spacers. The interelectrode spacing of electrostatic quadrupoles is obviously of similar critical importance and uniformity of winding is important in a solenoid lens. The precision of currently available fabrication techniques is adequate to ensure that these aberrations are small but may set the limit on further improvements in spatial resolution.

f. External Limitations

The achievement of good spatial resolution requires a good basic ion optical design, high precision in fabrication, careful alignment and careful attention to sources of interference. Inside a metallic section of beam tubing electric fields can arise from insulating surfaces or ungrounded conductors that collect charge from the beam. They do not have to be in the direct path of the beam for this to occur. The major problem however, usually arises from stray magnetic fields. Over a flight path of several metres, the earth's magnetic field can deflect a beam of 3 MeV protons by several mm. As a result the beam alignment will differ from any optical alignment. There will also be a change in alignment whenever the beam energy is changed. Misalignment will result in steering of the beam by the lens. A more serious problem is that stray AC magnetic fields will scan the beam spot and thus effectively enlarge it. The entire line must therefore be shielded from the magnetic fields set up by neighbouring motors, solenoids, transformers and other electrical equipment.

Another serious problem arises from vibrations. Several systems have been built, like that at Heidelberg (Bosch *et al.*, 1978), on a heavy concrete base. The Melbourne system is built on a single box girder supported on rubber pads at two points. The major sources of vibration — turbomolecular pumps, other motors, water cooling turbulence, air streams, traffic, local movements and the experimenter — must be isolated from the system as far as possible, remembering that movements at the specimen or lens are more serious than those at the object, since the latter will be diminished in their effect by the demagnification factor of the system. This is fortunate because the one source of vibrations which can never be completely removed is the necessary section of vacuum tubing (bellows) which couples the microprobe line to the accelerator system. It is recommended that an accelerometer be used to measure vibrational levels, though the finger is capable of sensing vibrational amplitudes of less than a micron. For further discussion of the requirements in developing a microprobe, see Legge (1982) and other papers in Demortier (1982).

10.2.2 Scattering and Requirements for Low Background

a. Gas Scattering — Vacuum Requirements

The focussed spot represents a demagnified image of the object diaphragm or slits broadened by aberrations of the lens system, plus a halo of particles which have undergone some scattering en route from the object. The likelihood of scattering from residual gas molecules is directly proportional to the pressure and the length of the path. However, the effect of scattering on the object side of the lens is reduced by the focussing

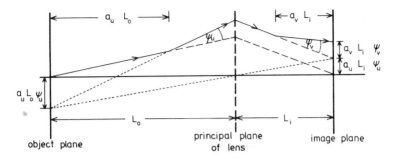

Fig. 10.5 Path of a particle deflected once before the lens (by an angle ψ_u) and once after the lens (by an angle ψ_v). The points of deflection are here defined in terms of object and image distance, L_o and L_i. Gas scattering involves a series of such deflections and scanning involves a single deflection. Prelens deflection is related to the shift $a_u L_o \psi_o$ of a virtual object and this in turn to an image shift $a_u L_i \psi_u$. Postlens deflection is related directly to a further image shift $a_v L_i \psi_v$.

action of the lens. Fig. 10.5 shows a deflection on each side of the lens; in practice there may be a number of such deflections. If ψ_u is a deflection on the object side of the lens at a distance $a_u L_o$ from the object ($0 \leqslant a_u \leqslant 1$) and ψ_v is a deflection on the image side of the lens at a distance $a_v L_i$ from the image ($0 \leqslant a_v \leqslant 1$), we may write the net deflection of a particle at the image as the vector sum:

$$r = \left| \frac{1}{D} \sum_u a_u L_o \, \underset{\sim}{\psi_u} + \sum_v a_v L_i \, \underset{\sim}{\psi_v} \right| \tag{10.21}$$

$$= L_i \left| \sum_u a_u \, \underset{\sim}{\psi_u} + \sum_v a_v \, \underset{\sim}{\psi_v} \right| \qquad \text{(since } L_o/D = L_i) \tag{10.22}$$

$$= L_i \left| \sum_w a_w \, \underset{\sim}{\psi_w} \right| \qquad \text{with } w = u \cup v \tag{10.23}$$

where $\sum\limits_w a_w \psi_w$ is the weighted vector sum of all deflection angles (on both sides of the lens); $a_w = 0$ for a scattering event at the object or image and $a_w = 1$ at the principal plane of the lens, here treated as thin. It is obviously important to maintain a low residual gas pressure in the vicinity of the lens. However, if the pressure preceeding the lens is P_o and that following the lens is P_i, we may regard this as equivalent to a pressure

$(L_o/L_i) P_o + P_i$ in the latter section alone. As a limit we may treat this as a scatterer at the lens (all $a_w = 1$) of thickness (mass per unit area):

$$t = (L_o P_o + L_i P_i) M/N_0 kT \qquad (10.24)$$

where M is the molecular weight of the gas, N_0 is Avogadro's number, k is Boltzmann's constant and T is the absolute temperature. Thus an overall path length of 10 m at a pressure of 10^{-6} Torr of N would be equivalent to a thickness of approximately 0.8 pg cm^{-2}. The effect of such a scatterer is not simple to calculate. We are concerned with very small angle scattering, for which the potential of the scattering nucleus is heavily screened. Multiple scattering in such a thin target is relatively improbable and the screening would be even greater for individual smaller deflection angles. The energy transfer in a small angle scattering event is negligible, so the scattered beam particles are indistinguishable from the unscattered particles. Nobiling *et al.* (1975) used the breakup of H_3^+ molecular beam particles to measure independently the proportion of particles scattered from gas molecules and from slit surfaces. Their results showed that when a highly collimated beam of H_3^+ ions travelled 1 m through a vacuum of 2×10^{-5} Torr, there was, around the beam, a uniform background of ~ 1 gas scattered particle per μm^2 for every 10^9 ions in the collimated beam. As expected, this background was directly proportional to the pressure. Whilst the gas scattered particles were mostly from break-up of H_3^+ ions and the uniformity of the background may not hold for simple scattering of protons or other ions, this estimate is a useful guide to the intensity of halo that can be expected.

b. Slit Scattering

It is important that no slits or diaphragms be placed in the vicinity of the specimen where they might act as a source of background radiation or scattered particles. The lens aperture diaphragm or slit should be sufficiently thin, upstream and where the beam density is low, so that few beam particles will be scattered onto the specimen from its surface. The beam in a microprobe system can be classified as very clean. However, the object diaphragm or slit is a major potential source of scattered particles. Because an image of the object is formed at the specimen by the lens, the problem is not one of deflection, but of energy loss by a beam particle at the surface of the object. The energy loss arises chiefly from ionisation losses in this thick target. It is serious because of the chromatic aberration of the lens.

The proportion of scattered to unscattered beam transmitted through the object will depend on the areas of useful scattering surface that the beam can see, compared to the open area of the object diaphragm. Therefore it may be argued that the object diaphragm should consist of a smooth parallel sided hole through a Ta or Pt foil that is just thick enough to stop the beam (Legge *et al.*, 1979). In order to further reduce the effective area of the walls that are *visible* to the incoming beam, the convergence angle of this beam should be limited in some way at an earlier stage, rather than bombard the object with beam that without scattering would not be acceptable to the lens aperture. Such an object diaphragm is appropriate for an orthomorphic system and a set of such holes in an *object strip* can be made to order by suppliers of electron microscope accessories. If the demagnifications differ or if a rectangular image is desired, then a similar parallel sided rectangular object hole can be formed with thin slits.

A very different approach was taken by Nobiling *et al.* (1975) who constructed very thick slits with deliberately angled surfaces (a long 15° chamfered entrance followed by a short 4° chamfered exit). Such a slit surface will obviously produce much more scattering than a smooth non-angled surface; but the shape is designed to prohibit single scattering directly into the lens. It might be argued that the inevitable thin region of transparency at the neck of this slit is a major weakness and that a second thin (antiscattering) slit would be more efficient in removing scattered particles than the second scattering surface deliberately provided in this slit design; but it is claimed that the smooth polishing obtainable on such thick slits yields a better result than a so-called paralled sided hole with rougher surface finish. It is not possible in this brief discussion to do justice to the considerations involved in the above slit design. A thorough comparison needs to be made, preliminary results existing only above 20 MeV (Al-Ghazi and McKee, 1982). Certainly, whatever the design, the aperture or slit surfaces should be as smooth as possible. They should be cooled as efficiently as possible to prevent beam damage and be protected from the unusable fringe beam by a beam monitoring precollimator. A clean all-metal high vacuum system with ion or turbomolecular pumps will help to ensure that carbon deposits do not contaminate the smooth surfaces. The measurements of Nobiling *et al.* (1975) suggest that gas scattering will dominate over slit scattering above pressures of 10^{-6} Torr, though this will depend strongly on the system geometry.

10.2.3 The Scanning System

Both electrostatic plates and iron free electromagnetic coils have been used for sweeping a focussed beam spot across the surface of the specimen. Plates can be positioned even within the field of a magnetic lens, but this requires the lens to have a large bore. Coils can be readily superimposed and it is easier to move or modify coils. In either case the field should be made as uniform as possible. The coils or plates can be placed before or after the lens. They will have a maximum effect if placed at the principal plane of the lens and obviously no effect if placed at the object or image. From the geometry of Fig. 10.5 with only one deflection and a_u (or a_v) = 1, it can be seen that the effect of principal plane deflection by an angle ψ is to displace the apparent position of the object by a distance $L_o\psi$ off-axis as the image moves a distance $L_i\psi$ off-axis, L_o and L_i being the object and image distances which are related by the demagnification factor $D = L_o/L_i$.

There may be aberrations associated with any scanning system, even if it is placed after the lens, but additional aberrations will be introduced by the passage of the deflected beam through the lens. Fig. 10.5 shows that deflection in the principal plane should be equivalent to a tilt of the lens and this does not introduce serious aberrations. Heck (1978) has obtained a similar effect by double deflection before the lens so that the beam follows a doglegged path for the paraxial ray to pass through the centre of the principal plane, as in a scanning electron microscope. This can only be arranged for one plane of focussing in a quadrupole doublet and Heck (1982) has calculated the aberrations also for a quadruplet with such an arrangement. If single deflection is used in some plane preceding the principal plane, serious additional chromatic and spherical aberrations can be introduced as the beam is driven further off the lens axis, a second reason for placing such beam deflectors as close as possible to the lens. The aberration of a single prelens deflector would be reduced by placing the aperture diaphragm after the deflector in the principal plane of the lens; but, apart from the inconvenience of this arrangement, if the principal plane lies within the lens, it would most likely lead to non-uniform distribution of beam current over the scanned area of the specimen.

Opposite pairs of poles in a magnetic quadrupole lens may be used to provide additional dipole fields for relatively slow scanning (Grime et al., 1984) and an electrostatic quadrupole lens could be used in like manner to higher frequencies. For slow scanning the alternative of moving the specimen under a fixed beam (Maggiore, 1980; den Ouden et al., 1981) has the advantage that the probe spot then remains fixed with respect to any detectors and the field of view of an optical microscope. Movement of the specimen can also be carried out over large distances with no problems of scanning aberrations, nonlinearities or amplitude limitations. The choice between low and high frequency scanning rates is discussed in Section 10.3.2. Here we describe the techniques for high frequency scanning. The required waveform is usually either triangular or sawtooth and, if distortions at the edges of the scan are to be avoided, the frequency of response of the entire system must be much higher than the maximum scanning frequency to be used. Scanning frequencies up to 20 kHz are possible with iron-free scanning coils of low inductance wrapped around small diameter beam tubing. The deflection may be measured by the voltage across a stable resistor in series with the coils. This gives an accurate measure of the field produced by the coils but ignores the fields produced by any eddy currents in metallic beam tubing. The latter fields produce an effective delay in the scanning field with additional distortion near maximum amplitude. To avoid the problem of eddy currents, the scanning coils can be wound on an insulating section of beam tubing, coated on the inside with a continuous thin film of C to prevent the build up of charge. The same resistance to eddy current induction can be obtained if the coils are wound on a thin stainless steel bellows. These precautions are not necessary for low (0 to 100 Hz) scanning frequencies, for which the eddy currents are small and the timing is less critical. At high frequencies, the effect of any uncompensated delay in the scanning field with respect to the detector signal is to shift any display (line scan or two dimensional image). This will invalidate the edges of the display and multi-imaging will occur unless unidirectional scanning is employed.

10.3 SCANNING ANALYSIS

The design of a specimen chamber requires much deliberation and some compromises which are described in Highlight 10.3.

HIGHLIGHT 10.3
THE SPECIMEN CHAMBER

Table 10.1 shows that an important attribute of the proton microprobe is the large number of signal channels that are available to provide information on the specimen. For maximum efficiency, versatility and analysing power, several of these channels should be available simultaneously. The space in front of the specimen is inevitably congested, with many detectors requiring a view of the front surface, and this problem is compounded by the requirement that the beam lens have a short focal length, for maximum demagnification with minimum aberration. In practice, focal length is limited by the necessity to provide an adequate working distance between the lens and the specimen. There are

TABLE 10.4
Specimen chamber requirements for a proton microprobe

1.	Adequate pumping system with low vibration — preferably ion pump.
2.	Specimen support locatable in both transverse and axial directions to about 1 μm and stable to greater accuracy.
3.	Microscopes for transmission and reflective viewing of specimen and any associated mirrors and sources of illumination.
4.	Faraday cup with electron suppression for measuring currents from 1 pA to 100 nA.
5.	Si (Li) detector for efficient detection of X-rays — mounted as close as possible to the specimen.
6.	Surface barrier detectors for backward and forward scattering and reactions.
7.	Large observation window.
8.	Secondary electron detector.
9.	Two-axis goniometer for channeling work.
10.	Facilities for transmission imaging.
11.	Vacuum lock for rapid changing of specimens.
12.	Facilities for selecting different specimens within chamber.
13.	Facilities for movement of several detectors with accurate positioning.
14.	Facility for changing filters on Si(Li) detector.
15.	Facilities for changing magnification of internal objective lens.
16.	Wavelength dispersive detector for optimum energy resolution of X-rays.
17.	Ge(Li) or NaI(Tℓ) crystal for detection of γ-rays.
18.	Facility for maintaining a specimen at liquid nitrogen temperature.

many additional items needed for the purposes of accurate observation and manipulation of the specimen. Table 10.4 lists those accessories or facilities that are most likely to be required in a specimen chamber. The list is not exhaustive. Fig. 10.6 shows a relatively simple octagonal chamber. The material is stainless steel, all gaskets are metal and the eight small flanges are 70 mm diameter to accept standard UHV fittings. It is pumped on the far side by a $120 \, l \, s^{-1}$ ion pump, after roughing to 10^{-7} Torr by a $100 \, l \, s^{-1}$ turbomolecular pump. This chamber fulfils the first seven requirements of Table 10.4. A secondary electron detector was later added and a rough goniometer for channeling work. Transmission imaging was also developed in this chamber and such a chamber is

Fig. 10.6 Simple SPMP specimen chamber which meets the first seven requirements of ▶ Table 10.4. Beam enters from the left and is focussed on the specimen at centre. This is mounted on an external micrometer stage below. RBS and X-ray detectors are mounted in lower left and upper left ports respectively. The mechanism at right moves a RFS detector mount and a Faraday cup or a microscope objective lens behind the specimen. The microscope eyepiece is outside the chamber at right. An internal mirror and external stereozoom microscope are used for front viewing of opaque specimens. (Legge *et al.*, 1979)

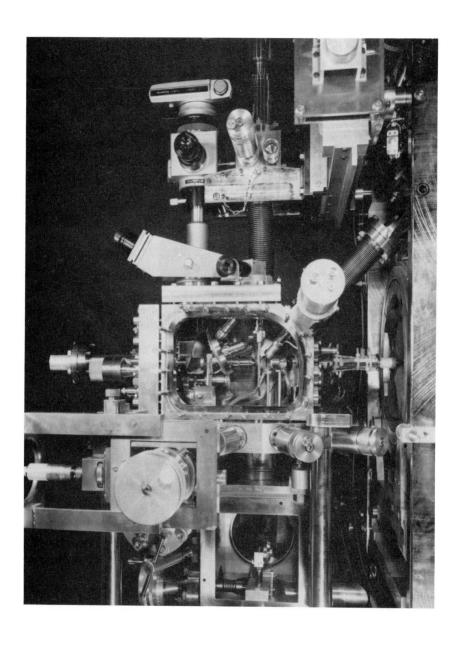

adequate for most work. A much more elaborate chamber with provision for all the requirements of Table 10.4 is shown in Fig. 10.7. The instrumental complexity is justified by the time saved in setting up and a feature identified in a scan may also be rapidly centred.

The specimen chamber should be as large as possible within the constraints, the severest of which is the limited working distance. With many detectors or other items of instrumentation movable within a confined space, it is important to provide a clear view of all parts of the chamber. Both for preliminary focussing and for setting up a new specimen, there is also no satisfactory substitute for a good optical microscope. A short depth of focus ×20 microscope objective within the chamber can be used to view a thin microscope cover glass on the other side of which the beam produces fluorescence. It is possible to focus the beam to a diameter of about 1 μm, if care is taken to focus on the beam side of the glass. However to make an accurate measurment of spot size or spatial resolution, the beam must be scanned across an edge. X-rays may be used for this measurement or, for greater speed and efficiency, electrons. A secondary electron detector will collect sufficient electrons to produce a real-time image of the specimen on an oscilloscope. This may be used for focussing but its greatest use is in precisely locating a specimen for study and it is an essential component of the specimen chamber for high resolution work. When optimum resolution is needed to locate and identify a specimen of low contrast or low secondary electron emission, the low beam current available may result in inadequate electron yields for a satisfactory real-time image . Under these conditions a variable decay time oscilloscope should be used for greatly improved imaging detail and image contrast. Correct identification and localisation of a sample are just as important as correct data collection and analysis.

◀ **Fig. 10.7** SPMP specimen chamber which can meet all the requirements of Table 10.4. The beam enters from the right and the specimen is supported from a goniometer and microprocessor controlled micrometer stage above the chamber. The X-ray detector is mounted above right, and its snout is partly obscured by the SE detector. RBS, RFS or NRA detectors, mirrors and so on are mounted on six externally controlled rings in the base of the chamber. A vacuum lock with automatic specimen changer is below the chamber. An externally controlled turret of objective lenses is mounted behind the specimen. The STIM detector and the Faraday cup are outside the chamber at left. Near and far side of chamber can be used for illumination, front viewing microscopy or large detectors. (Legge *et al.*, 1985)

10.3.1 General Principles of Scanning Analysis

When a beam of ions scans an area of the specimen, the emitted radiation carries information in three degrees of freedom — the two scanning dimensions and the energy. In the case of backscattering, the energy represents depth of penetration into the specimen. In a PIXE spectrum it contains only elemental information. In either case, recording information directly on a two-dimensional storage device (such as the screen of a cathode ray oscilloscope) requires selection of the values to be accepted for the third dimension. Thus an energy window is set up to filter the spectrum. This mode of operation has several serious disadvantages:

   i. all data outside the energy window are ignored and repetition for different energies is time consuming and may cause specimen damage or elemental loss;
  ii. the maps collected on a storage oscilloscope or photographic film are not quantitative;
 iii. a complete energy spectrum can be collected but does not contain any spatial information; and
  iv. it is not possible to perform elemental analysis for an extended structure of irregular shape except by repeated measurements with a fixed beam at many selected points.

In theory, these problems can be solved by direct storage of the collected data in allotted locations within a computer memory. However, for high resolution in all three dimensions, the number of memory locations would be prohibitively large. The memory space requirements may be reduced by preselection of energy windows, allotting some of these to background measurements, but the complete spectrum is then no longer available and background corrections will be less reliable. Full three-dimensional recording is discussed in Highlight 10.4.

HIGHLIGHT 10.4
PRINCIPLES OF TOTAL QUANTITATIVE SCANNING ANALYSIS

Three-dimensional data may be retained for analysis in their entirety by collecting and storing them in *event by event* mode on a *hard memory* device — magnetic tape or disc (Legge and Hammond, 1979). Each three-dimensional event is stored as three words — two spatial coordinates and one energy coordinate — initially in a computer buffer memory and, when the buffer is filled, then on tape or disk, but still in the original time sequence of events. If several detectors are collecting data, the particular one associated with each event may be indicated by setting one or more bits in these words. The minimum requirements for such a data storage system are a small laboratory computer, a number of ADCs (Section 10.3.2), an interface to handle coincident data, a magnetic disk for programs and a magnetic tape for data storage. Data collection programs of the required type for storing multiparameter events in sequence are available in many nuclear physics laboratories.

It is possible to monitor the spectra of the three dimensions and one selected map as the data are stored, but most of the data will be available only after sorting the events. This can be performed at some later time by a large off-line computer, immediately after completion of data collection by the same laboratory computer or simultaneously with data collection by a large dedicated or linked computer running a high speed real time SORT program. Availability of the sorted data to the laboratory computer immediately or shortly after completion of data collection greatly increases the analytical capabilities of the system. If the laboratory computer itself is to perform the task of sorting the three-dimensional data, then a second or larger magnetic disk will be required (to provide *virtual* memory space) and as much core memory as possible (for efficiency), though the first such system performed the task with a memory of only 48 kB (Legge and Hammond, 1979). The sorted data are stored for rapid access on magnetic disk.

The techniques available for data analysis are illustrated in Fig. 10.8. The sample is a botanical one — a 40 μm thick transverse section of dessicator dried mature wheat seed supported between thin ($< 1 \mu$m) nylon foils. The section was irradiated with 300 pA of 3 MeV protons for 90 mm. The beam was focussed to a spot of 12 μm diameter and scanned over 1 mm square (Mazzolini *et al.*, 1981). The three-dimensional block of sorted data is depicted with the map of P distribution at the front. This shows the concentration of P in the embryo of the seed. The energy spectrum for the entire scanned area is shown top left. An energy window set on the Mn X-ray peak produced the Mn distribution map, shown as an energy-slice pulled out of the three-dimensional block in Fig. 10.8. At this pregermination stage of the seed, Mn is apparently concentrated in the coleorhiza surrounding the primary and two lateral roots. The area of the coleorhiza has been outlined on the computer display bottom right and specified as a *region of interest*. The energy spectrum for this *region of interest* was extracted from the total block of data and is shown top right. Only a small fraction of the available data is shown in Fig. 10.8. RBS and RFS data were also collected and yielded information on light elements and hence matrix thickness. Charge counts were simultaneously recorded and used to monitor the stability of both light and heavy elements (Section 10.3.3). Maps of all elements in the block of data, line

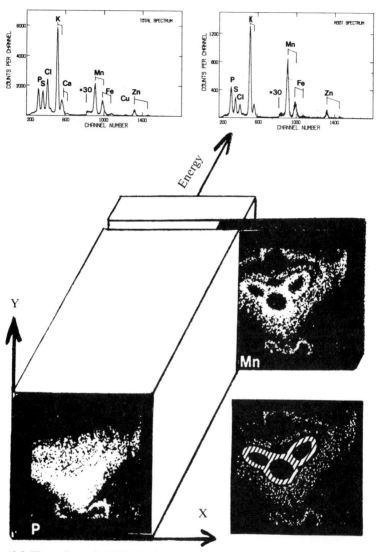

Fig. 10.8 Three-dimensional block of sorted PIXE data from a 1 mm square area of wheat seed, transversely sectioned and scanned with a 12 μm focussed spot of 3 MeV protons. (Mazzolini *et al.*, 1981). The data may be treated as a stack of elemental maps, any of which may be extracted by windowing the appropriate element in the total spectrum (top left). *Regions of interest* may be defined (bottom right). The projection through the block of such a region yields its elemental spectrum (top right).

scans and spectra for all *regions of interest* may be extracted. In this way spectra may be corrected for background and quantitative values may be obtained for the elemental composition. In this form of interactive programming, it is common for the analyst to move from spectral displays to maps and back to spectra as a feature discovered in one map is then defined as a *region of interest* and analysed for all its elements.

The advantage of such a total data handling system may be summarised as follows:

i. no data are lost and there is less probability of damage or elemental loss (Sections 10.3.2 and 10.3.3);
ii. elements not visible in the total spectrum but present as some localised concentration may show up unexpectedly when the spectrum of some feature is inspected;
iii. all forms of data analysis are possible;
iv. the elemental content of any shaped feature can be obtained by defining a *region of interest*;
v. the data are permanently recorded, and the map of any element or the spectrum of any feature can be studied at any later date;
vi. all data are quantitative and background corrections can be made reliably;
vii. because data on all elements are collected simultaneously, the elemental maps are directly comparable with regard to beam charge, specimen thickness and movement or change of beam conditions should that occur;
viii. the availability of the analysed data in digital form provides additional advantages for further analysis; and
ix. if the data are stored in time sequence, then a history of the analysis during irradiation is recorded (Section 10.3.3).

10.3.2 The Scanning System and its Control

There have been two approaches to scanning:

- rapid continuous scanning, obtains quantitative information by the digital storage of all information from a continuous scanning measurement; and
- slow sequential scanning, obtains spatial information from a series of measurements combined with a digitally controlled positioning of the spot.

There are significant differences in the operation and applicability of the two methods.

a. Rapid Continuous Scanning

In this method (Legge, 1978; Legge and Hammond, 1979) the beam spot is continuously moving, driven by two independent oscillators (the x and y oscillators) of comparable but unrelated frequencies — related frequencies would cause the beam to follow a closed Lissajous path. Both waveforms are linear triangular waves, so that the beam zig-zags rapidly from top to bottom and side to side of the rectangular scanned area. There are a number of consequences:

   i. The rapid movement of the beam over the entire scanned area minimises any possible heating effects.

   ii. An image of the specimen can be rapidly accumulated and the details filled in during the scan.

   iii. The beam charge to any given point, or *pixel*, within the scan cannot be directly measured; but every point is scanned many times so that beam current fluctuations average out. (Effectively each point is scanned n times in a period t, where $n \simeq 2\ dft/D$, d is the effective resolution of the beam, D is a scan dimension and f an oscillator frequency*.) The beam current to the entire scanned area is monitored. In the case of a thick specimen, the beam charge to a given point may often be inferred from some matrix yield. In the case of a thin specimen, variations in specimen thickness are crucially important and the ratio of an elemental yield to the matrix yield is all that matters.

   iv. With high frequency scanning, the timing circuit must be accurately set up, because any delay in registering the x and y deflections corresponding to a detected event will lead to an apparent displacement of that event on a display. With triangular wave scanning from both oscillators, this displacement occurs in four alternative directions, so that every point is imaged quadruply. The recording delays must be compensated to within a period $\Delta t \leqslant d/4Df$, where d is the desired spatial resolution, D is the dimension of the scan and f is the frequency of scanning in that dimension.

A simplified schematic diagram of the system is shown in Fig. 10.9 and should be compared with the more basic scanning system of Fig. 10.2. The detection of any energy signal above some threshold level will trigger the three ADCs to sample the energy signal level and the

---

* More precisely, for scan frequencies f_x and f_y, scan dimensions D_x by D_y and pixels of area $d_x d_y$, $n = 2\,[(d_x^2 + d_y^2)\,(D_x^2 f_x^2 + D_y^2 f_y^2)]^{1/2}\, t/D_x D_y$.

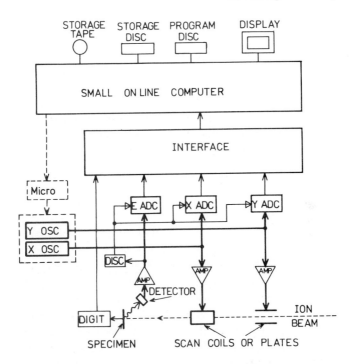

Fig. 10.9 Schematic diagram of a system for rapid continous scanning analysis with quantitative collection, storage, sorting and display of all data. The circuitry shown in heavy lines is that of the basic qualitative scanning system of Fig. 10.2. Additions to this are shown in light lines. Unessential controlling circuitry is shown dashed.

simultaneous values of the x and y scanning displacements. These provide the three words to be written on magnetic tape for that event. Several detectors may be used with several ADCs or with a mixer, in which case additional bits may be set in the energy word to indicate the detector which produced the signal. For example, the data of Fig. 10.8 were collected with a system having an addressable data block of 8000 energy channels, 1000 x channels and 1000 y channels. The 8000 energy channels could be shared between the three detectors — 4000 for X-rays, 2000 for backscattered particles and 2000 for forward scattered particles.

b. Slow Sequential Scanning

In this method (Bonani *et al.*, 1978) the beam is stationary for each point in turn whilst data are collected. The deflection amplifiers are controlled by DACs with registers set by scalers. One scaler counts up and down to sweep the beam back and forth linearly, whilst the other is incremented at the end of each line to give the required displacement between lines. Thus the beam traces a continuous series of parallel lines in alternate directions to completely scan the specimen. The clock pulses to set up the beam are generated by beam charge counts (corrected for detector amplifier dead time); so the system measures a complete spectrum at each point of the specimen (or pixel of the subsequent display), with each spectrum corresponding to the same predetermined beam charge.

The advantages of sequential point measurements are the beam charge normalisation already mentioned and the natural ordering of the addresses for successive spectra, so that the amount of sorting required is reduced. The disadvantages are concerned with specimen damage. Consider a scan of a 1 mm × 1 mm area which is to be digitised to 1000 × 1000 points. Typically the beam exposure might be about 1000 s; so the time per point is 1 ms. Thus in 1 ms the spot will move a distance of 1 μm. In contrast, a beam scanning continuously at a frequency of 500 Hz would move the full 1 mm in 1 ms. Kirby (1983) has estimated that a thin (3 μm) biological specimen will reach equilibrium temperature in about 1 ms for a microprobe fixed spot irradiation. Thermal damage is minimised by rapid continuous scanning, and can be serious for many specimens (Sections 10.3.3 and 10.4).

The use of digitised scan generators and registers means that only one ADC is required for each signal channel or, if a mixer is used, for all signal channels. For fast counting with several detectors, dead time is minimised if each signal channel has its own ADC and also its own pair of x and y registers so that the counting circuits for the various signals are independent. These features however are not restricted to slow sequential scanning. If up-down scalers are fed by fast free running clocks in place of those controlled by the beam charge, the circuit is appropriate to rapid continuous scanning. Thus the one circuit can easily be adapted to run in either mode.

c. Computer Control

Although the data handling computer should not be burdened with the task of generating fast scanning signals, there are many advantages if this computer controls the scan in so far as it sets the scanning amplitudes and the centre of the scan (by movement of the beam or the specimen) as indicated in Fig. 10.9 with dashed lines. In particular, specimen areas identified in the data from previous scans can be automatically centred and rescanned. Computer control of the scan can also be used to generate a scan of complex shape over some *region of interest* (Höfert *et al.*, 1984) as an alternative to extracting the required data from an encompassing rectangular scan as discussed in Highlight 10.4. This would be worthwhile when the *region of interest* formed only a small fraction of the encompassing rectangle, the areas outside the *region of interest* were of no interest even as background and the complex shape of the *region of interest* was known beforehand or could be readily identified. These conditions are not often satisfied, but they are more likely to be found with ordered samples such as those involving semiconductors.

10.3.3 Temporal Analysis and Beam Damage

Changes to the specimen that might result from beam bombardment are:

  i. Discolouration — generally not in itself a problem and often a convenience in identifying a probed area.
 ii. Physical deformation at the microscopic level — generally only a problem with sensitive specimens under intense bombardment by a fixed beam.
iii. Atom relocation due to recoil or implantation — generally only a problem for depth profiling studies and channeling where submicroscopic crystalline damage can occur rapidly, as is well known from electron diffraction studies.
 iv. Loss or movement of any element of interest due to destabilisation of the specimen matrix by the beam.
  v. Mass loss — loss of matrix elements from the specimen causing changes in specimen thickness or surface concentrations of important elements, most likely for biological specimens under intense bombardment by a fixed beam.

Discolouration is usually due to ionisation whereas physical deformation is caused by overheating. Elemental loss and mass loss may arise from either or both of these causes. All effects are minimised by keeping irradiation to a minimum, that is by efficient collection of data over as large an area as possible. The manner in which the scanning is performed (fast or slow) may have a large bearing on temperature effects, but cannot influence purely ionization effects. However, if the data are stored in time sequence, this historical record may be inspected at a later date. The circuit of Fig. 10.9 shows the digitised beam charge fed to the data handling comptuer interface. These beam charge pulses are employed to flip additional bits in the event words to record accumulated charge on the data tape. A simple program can later sort the data tape and plot the yield of each element recorded between successive charge pulses (Legge and Mazzolini, 1980). An example of such elemental monitoring is shown in Fig. 10.10 (Mazzolini *et al.*, 1981) where a fixed beam (250 pA

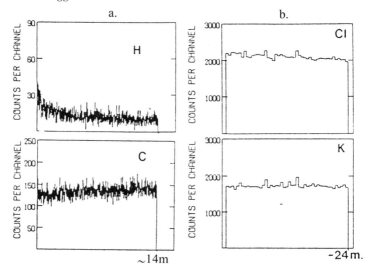

Fig. 10.10 (a) Yield of two matrix elements, H and C, measured by RFS when a 10 µm thick cross-section of wheat seed was irradiated for 14 min with a 250 pA fixed beam of 3 MeV protons focussed to a spot 6 µm × 10 µm. With a fixed charge per channel, the scale is effectively 1.2 s per channel and shows rapid initial loss of H. (b) Yield of Cl and K measured by PIXE for the same sample as that above with a 120 pA beam scanned over an area of 1.5 mm × 1.5 mm for 24 min (30 s per channel). There is a slight but measurable loss of Cl over this period. (Mazzolini *et al.*, 1981)

in a 6 µm × 10 µm spot) of 3 MeV protons on a specimen of wheat seed led to no loss of the matrix element C but rapid loss of the matrix element H (both recorded with an RFS detector). The first H loss was noticeable after a charge of > 10^{-9} C in only a few seconds. Yields of heavier elements, monitored with PIXE during continual scanning of a 120 pA beam over an area of 1.5 mm × 1.5 mm, showed only a slow but steady loss of Cl for which a correction could be made if necessary. Further discussion may be found in Section 10.4. Similar monitoring may be performed for any other recorded spectral information such as depth profiles or channelling spectra. A further advantage of such recording of data in time sequence is that, should damage or any disturbance of the specimen or apparatus have occurred during the analysis procedure, those sections of the data record which preceded and followed such an event may be separately analysed.

10.4 APPLICATIONS

Early applications of the proton microprobe were reviewed by Cookson (1979), those involving PIXE analysis have been reviewed by Cookson (1981) and Legge (1984) and numerous authors have discussed the work of their own groups (Demortier, 1982; Martin, 1984; Biersack and Wittmaack, 1986). Many references are given in a review of multidimensional analysis by Doyle (1986) and a detailed coverage of applications is given by Watt and Grime (1987). We shall present examples which illustrate different facets of microprobe work and the considerations involved. It is convenient to divide discussion between physical and biological sciences. This generally corresponds to a division between hard, thick, inert specimens and soft, thin, unstable ones.

10.4.1 Physical Sciences

a. Metallurgy

Metals provide ideal specimens for microanalysis by high velocity electron or positive ion beams. Their high thermal conductivity minimises thermal damage and their high electrical conductivity removes any possibility of specimen charging giving rise to beam deflection or to discharging with accompanying emission of continuum X-rays. Changing elemental concentrations across or along grain boundaries are probably best measured with an electron instrument (Reed, 1975) with its high spatial resolution, as long as the sensitivity is adequate. However, the light elements, O, N and C, which can drastically affect the physical properties of metals, are not readily measured by electron instruments. The proton microprobe, with its ability to detect these elements by NRA with high quantitative accuracy, has therefore been in heavy demand for such work since its first development (Cookson *et al.*, 1979) and all light elements, including H, have been successfully measured with the Harwell microprobe. Table 10.5 from McMillan *et al.* (1982) lists the reactions used and the sensitivities for the isotopes of various elements detected, mostly in metallic specimens. The use of a microbeam adds no complication apart from an associated decrease in the available beam current. This is a serious consideration, because the cross-sections for nuclear reactions are low compared to those for PIXE or RBS. Consequently high resolution two-dimensional elemental mapping of trace elements is not a practical possibility. The measurements made are usually spot measurements or line scans and, where permissable, the current is maximised by opening the microprobe slits in the transverse direction. The reactions of Table 10.5 are not characterised by sharp resonances suitable for depth profiling. The relationship that does exist between reaction depth in the specimen and energy of any detected particle has been used primarily to distinguish between surface (possibly contaminant) effects and bulk effects. Depth profiling has been performed by sectioning and polishing, and examining the resultant edge

TABLE 10.5
Nuclear reactions and their sensitivities for light element microanalysis[a]

Element	Isotope	Reaction	Matrix	Sensitivity (wt.ppm)
H	^1H	$(^{11}B, \alpha)2\alpha$	metals	1000
		$(^3He, {}^3He)^1H$	catalyst pellets	< 1000
	^2H	$(^3He, p)\alpha$	steels, zircaloy, carbon	< 10
	^3H	$(d, n)\alpha$	zircaloy	< 1
Li	^7Li	$(p, \alpha)\alpha$	steels, B_4C	100
Be	^9Be	$(\alpha, n_0)^{12}C$	Be diffusion in Cu	200
		$(d, p_0)^{10}Be$		200
B	^{10}B	$(d, p)^{11}B$	steels, nimonics, welds	50
	^{10}B	$(\alpha, p)^{13}C$	borided steel	100
	^{11}B	$(\alpha, p)^{14}C$	borided steel	100
	^{11}B	$(p, \alpha)2\alpha$	steel	< 10
C	^{12}C	$(d, p_0)^{13}C$	metals, oxides, welds, catalysts, wear particles, biological, nuclear fuel	1
	^{13}C	$(d, p_0)^{14}C$	nimonics	40
N	^{14}N	$(d, p_0)^{15}N$	steels, nimonics, TZM, nitrided steels, biological	20
	^{14}N	$(d, \alpha)^{12}C$	nitrided metals	100
	^{15}N	$(p, \alpha\gamma)^{12}C$	biological	100
	^{16}O	$(d, p_1)^{17}O*$	steel, nimonics, V, zircalogy, welds, oxides, nuclear fuel	50
	^{16}O	$(d, p_{1\gamma})^{17}O*$	oxidised Si_3N_4, metals, nuclear fuel	1000
	^{18}O	$(p, \alpha_0)^{15}N$	corroded metals	100
F	^{19}F	$(p, \alpha_0)^{16}O$	composites	10
Si	^{28}Si	$(d, p_0)^{29}Si$	steels, nimonics, TZM	200

[a] McMillan *et al.* (1982)

with the microbeam. Excellent agreement has been demonstrated (Pierce *et al.*, 1973) between profiles found by this direct technique and those found by the tedious metallographic technique of repeated examination following stepwise removal of surface material by polishing. Non-destructive depth profiling of metals with NRA has been performed with resonant reactions in the Albuquerque microprobe by Doyle and Wing (1983) and by Heck (1985).

A good example of the application of these NRA techniques is presented by McMillan and Pummery (1980) who measured the corrosion of two alloys, PE-16 (Cr 16.75%, Ni 44.1%, Mn 0.10%, Si 0.25%, Al 1.20%, Mo 3.3%, Ti 1.15%, C 0.06%, B 0.005%, remainder Fe); and Zircaloy-2 (Sn 1.48%, Fe 1.1%, Cr 0.09%, Ni 0.05%, O 0.113%, remainder Zr). They give an excellent discussion of the reasons for choosing the $^{16}O(d, p)^{17}O$ reaction, the possibility of interfering reac-

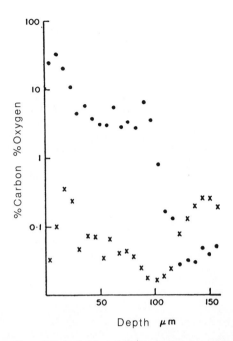

Fig. 10.11 Depth profiles of O (points) and C (crosses) in a metal surface obtained by sectioning the oxidised alloy PE–16 and scanning the exposed polished surface with a 2D microbeam. Elements were distinguished by NRA. (McMillan and Pummery, 1980)

tions, the choice of deuteron beam energy (1.1 MeV) and range of proton energies, and the determination of an O calibration curve. Samples and standards were sectioned, mounted in Woods metal and polished with diamond or alumina. Line scans were made as a series of points of dimensions $70\,\mu m \times 10\,\mu m$ and some maps were made with probe dimensions of $5\,\mu m \times 5\,\mu m$. For the Zircaloy-2, a complete O profile, consisting of 40 points at $10\,\mu m$ spacing, was measured in 1 h. For PE-16, extension of the beam spot to $70\,\mu m$ parallel to the sample edge was insufficient to average over local variations and averages were taken therefore of at least four such profile measurements. Fig. 10.11 shows such an O concentration profile and with it a profile for C measured with the $^{12}C(d, p)^{13}C$ reaction at a beam energy of 1.3 MeV. There is a strong anticorrelation between the two concentrations. Strong local variations in the distributions of O, C, Cr and Ti were observed in maps of these elements. For these maps, O and C were again identified by the energies of their associated proton groups, but Cr and Ti were identified by their deuteron induced X-rays. Such measurements are very difficult to make by any other technique.

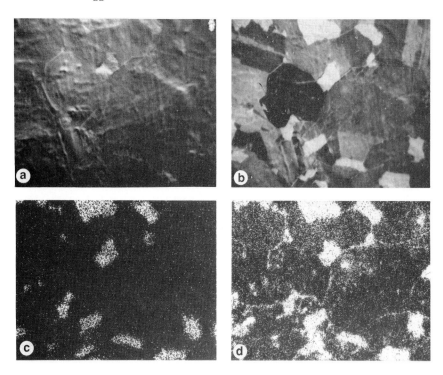

Fig. 10.12 Surface analysis with a SIMP — the specimen was a weld edge in stainless steel and the instrument a 50 nm probe of Ga ions from a MIG 100 metal ion gun, VG Ionex Ltd (Bayly and Waugh, 1983). (a) Initial secondary electron image; (b) clearer image obtained after sputter cleaning; (c) $^{16}O^-$; and (d) $^{56}Fe^+$ maps.

As an alternative to the PMP, both the IMP and the LMP have equal or even better sensitivity. They can also detect light elements and their spatial resolution is comparable if not better. Apart from their destructivity, the major problem is their poor quantitative accuracy which stems mostly from matrix effects. Fig. 10.12 illustrates the strengths and weaknesses of the IMP. A MIG 100 *microfocus* metal ion gun from VG Ionex, producing a $0.05\,\mu m$ probe of Ga ions at $10\,kV$, was used to examine the surface of a weld edge in stainless steel (Bayly and Waugh, 1983). The scan dimensions were approximately $100\,\mu m \times 100\,\mu m$. Fig. 10.12a is an initial secondary electron image of the oxidised surface. Fig. 10.12b shows a later image after initial sputtering has removed most of the oxide to clearly reveal the sharp grain boundaries. Fig. 10.12c is an $^{16}O^-$ map and shows the concentration of O at grain boundaries and on the surface of certain grains. Fig. 10.12d is a $^{56}Fe^+$ map and shows how the ion yield is affected strongly by the O content or other features of the matrix grain structure, making quantitation difficult. In a situation such as this, the PMP and IMP provide complementary information.

b. Isotopic Measurements and Tracer Techniques

NRA, like SIMS and laser analysis, provides a measurement of individual isotopes. The PMP has been used to measure stable isotopic tracers and also isotopic ratios, in which case two or more different nuclear reactions must be employed. Such work, on biological as well as metallic specimens, was reviewed by McMillan *et al.* (1982) and promises to be very useful in situations where a non-destructive technique is required. When specimen damage is not a problem and good depth resolution is required or possibly extreme sensitivity combined with good spatial resolution, the IMP is more suitable for measuring isotopic ratios, since matrix effects should not affect the ratio measurement. However, in situations where it is not safe to assume that the common isotope is uniformly distributed, the absolute concentration of the tracer must be measured and, in an inhomogeneous matrix, only the PMP will have the required accuracy.

The various microprobes can not normally compete in sensitivity with autoradiography, since the latter technique can in theory detect nearly every atom of the radioactive isotope present. However good spatial resolution in autoradiography can only be obtained with very low energy emitters, which implies very long half-lives and hence very long irradiation times. The microprobes also provide information on many elements but, unless a time series of measurements is made, do not distinguish between elemental components in a fixed state and those in a state of flux. Thus, although measurements made with a microprobe may suggest others to be made with autoradiography and vice versa, the two techniques seldom compete.

c. Semiconductors

The spatial resolution of the PMP is ideally matched to that required for the examination of microcircuitry. Although some probes can achieve a resolution of 1 μm with a current of 100 pA, RBS measurements of trace elements normally require higher beam currents. From Equation (10.7), the achievable beam current, when limited by chromatic aberration, should be proportional to the fourth power of the achievable beam spot diameter, so the above probes should achieve a resolution of 3 μm with a beam current of 10 nA.

However, for channeling, the half angle of convergence at the

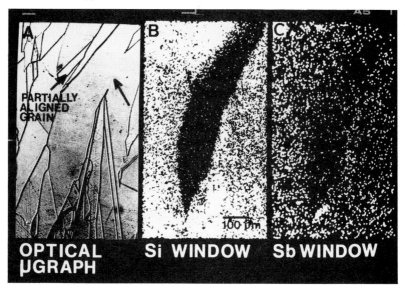

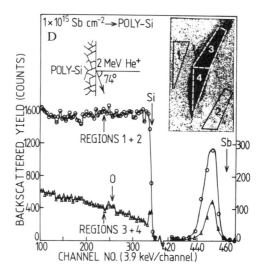

Fig. 10.13 Channeling contrast microscopy with a microbeam of 2 MeV α-particles and a specimen of polished, Sb implanted and annealed polycrystalline Si (McCallum *et al.,* 1986). (a) Optical micrograph showing surface crystal structure; (b) Si map extracted from total RBS data; (c) Sb map from same data set; and (d) RBS geometry and inset map of Si, showing channeling (very low RBS yield) occurring for the central crystal. The RBS spectra are from nonchanneling regions (1 and 2) and channeling regions (3 and 4).

specimen must be restricted in both planes of focussing ($\leqslant 3$ mrad for 2 MeV He ions channeled along the major axis of Si). For optimum resolution, the lens should have similar aberration coefficients in the two planes. The probe current may only be increased by increasing the object size and hence for large currents will be proportional to the square of the spatial resolution. Accelerator beam brightness is crucial and the $10 \, \text{A m}^{-2} \, \text{rad}^{-2} \, \text{eV}^{-1}$ achievable with a single stage accelerator gives a resolution of $7 \, \mu\text{m}$ for a 10 nA beam of 2 MeV and a half angle focus cone of $2°$. Scanning may increase the angular deviation of the beam but, since aberrations are of less concern for large object sizes, this effect may be reduced by prelens scanning and even eliminated by placing the scanning coils or plates in the first focal plane of the lens.

Because of ideal beam optics, this work has nearly all come from the SPMP at Melbourne. McKenzie, Williams and coworkers demonstrated microbeam channeling with α-particles (Ingarfield *et al.*, 1981) and employed grazing-exit-angle geometry with solid state detectors and RBS to obtain a depth resolution of 3 nm (McCallum *et al.*, 1983). *Channelling contrast* microscopy has proved to be useful in investigations of the annealing behaviour of individual grains in ion implanted polycrystalline Si (McCallum *et al.*, 1986). An example of this work is given in Fig. 10.13 which shows (a) an optical micrograph of a polycrystalline sample which had been polished, etched, implanted with 10^{-15} ions cm^{-3} of 80 keV Sb$^+$ ions, furnace annealed at $650°$C for 30 minutes and then scanned with a 1 nA 6 μm diameter beam of 2 MeV He$^+$ ions over an area of $570 \times 960 \, \mu\text{m}^2$. The channeling half angle was maintained at less than 3 mrad by the use of prelens scanning for these large deflections. Also shown (b) is a channeling contrast Si map which shows channeling alignment within the large central crystal and (c) a channeling contrast Sb map, both maps being extracted from the one block of sorted RBS data. Fig. 10.13d shows the glancing exit angle RBS geometry and two RBS spectra — one for the unaligned regions 1 and 2 of the inset channeling map and the other for the aligned regions 3 and 4. Levi-Setti *et al.* (1985b) used low energy ion channeling contrast to observe flat microcrystallites in pyrolitic graphite surfaces.

d. Geology

Geological samples usually contain a large number of elements, both major, minor and trace. This situation does not normally call for depth profiling but does call for elemental mapping or at least fixed probe measurements with good spatial resolution and reliable quantitative accuracy. This inhomogeneous mixture of elements in various states of

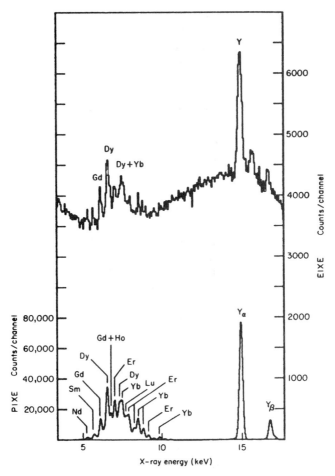

Fig. 10.14 Direct comparison of EIXE and PIXE data for a geological specimen (xenotime) rich in rare earths. The PMP is here about three orders of magnitude more sensitive because of the much lower background. (Legge *et al.*, 1979)

chemical composition makes quantitative work with the IMP questionable although it has often been called on because of its sensitivity for some elements. The EMP has been more generally available and is more quantitatively reliable although much less sensitive. Deflection of the electron beam due to charging of the insulating specimen can generally be avoided by evaporation of a surface layer of C or Al. However, the scattering of electrons in the sample leads to a spatial resolution comparable with their effective range (the range at which the electrons have insufficient energy to excite the X-rays of interest). If a wide spread of X-ray energies is to be excited by the EMP, the spatial resolution for the lower energy X-rays may be several μm (Goldstein and Yakowitz, 1975) and worse than that of a PMP.

Many PMPs have been utilised in mineralogical studies and one at Sydney (Sie and Ryan, 1986) was installed specifically for work in this area. The advantage of the PMP over the EMP in the quantitative determination of rare earth elements is shown in Fig. 10.14, a direct comparison by McKenzie of spectra generated in a JEOL electron microscope and the Melbourne PMP, using Si(Li) detectors of similar energy resolution (Legge *et al.*, 1979). The thick sample of the mineral Xenotime was exposed to 4×10^{-10} A of 35 keV electrons for 10 min and 10^{-9} A of 3 MeV protons for 25 min. The scales have been normalised to show approximately the same peak heights for characteristic radiation and are plotted without correction on the same baseline. The intense background of bremsstrahlung radiation which underlies the EIXE spectrum, limiting the statistical accuracy, is invisible on the scale for the PIXE spectrum, the actual ratio of background intensities being approximately 10^3. These spectra illustrate another problem with rare earth detection. Although many more elements can be identified in the PIXE spectrum because of the absence of a significant background, there is still a severe problem of peak overlap due to the restricted energy resolution of the Si(Li) detector. This can be overcome by use of a wavelength dispersive spectrometer but the detection efficiency is then greatly reduced. Nevertheless the specimens can generally withstand the increased irradiation required and such a spectrometer is frequently used with an EMP where it also improves the peak to background ratio by an order of magnitude. Similar work should be possible with the SPMP but the available beam current is limited. There is one feature of the PMP which puts it at a disadvantage and that is the great penetration depth of its beam. Because it is impractical in most cases to make very thin geological or mineralogical specimens, there is a high probability of penetrating through a surface feature to excite the matrix or some unknown feature beneath and it is very difficult to correct for this. It is most likely to occur in high resolution studies of grains with small lateral dimensions and hence probably small depth. A recommended precaution is either to collect quantitative data in scanning mode or to collect preliminary spectra at sample points over the grain of interest and then to select the region of the grain giving maximum signal levels for the major elements attributable to that grain (Rogers *et al.*, 1984).

An interesting comparison of quantitative results for analyses of the rare-earth-rich mineral Durango apatite by PIXE, nuclear activation,

TABLE 10.6

Comparison of concentrations (ppm) measured by several techniques in the mineral Durango Apatite.

	PIXE[a]	SD	NAA[b]	SD	XRF and AAS[c]	OES[c]
La	3610	130	3660	84	3330	4200–6400
Ce	4400	150	5310	210	4150	4700–7000
Pr	320	130	–	–	540	800
Nd	1410	100	1330	190	1230	2000
Sm	–	–	158.3	8.1	–	200–300
Eu	–	–	18.0	1.9	17	< 100
Gd	320	60	–	–	< 110	200–300
Tb	353	62	12.8	1.3	–	100
Dy	294	52	88.6	6.5	90	100–150
Ho	< 34	–	–	–	10	30–40
Er	–	–	–	–	35	100
Tm	–	–	–	–	7	10–20
Yb	–	–	39.0	2.7	23	50–80
Lu	–	–	–	–	–	< 100
V	30	39	33.3	2.0	–	30–50
Cr	40	23	88.0	6.1	–	< 1
Mn	117	17	94.6	9.0	100	100–130
Fe	402	16	530	160	350	400
Co	12.0	9.4	–	–	–	–
Ni	26	10	–	–	–	< 30
Cu	36.4	6.4	< 61	–	–	< 1
Zn	14.9	5.2	–	–	10	< 200
Ga	< 5	–	< 44	–	–	–
Hf	31	17	–	–	–	–
Tl	16.3	6.1	–	–	–	–
Th	143	12	196.3	6.1	202[d]	–
U	20	10	10.43	0.29	11[d]	–

[a] Rogers *et al.* (1984)
[b] Garcia (1984)
[c] Young *et al.* (1969)
[d] Data from isotope dilution (Young *et al.*, 1969)

XRF, atomic absorption and optical spectroscopy is given in Table 10.6 from Rogers *et al.* (1984), who used the Los Alamos PMP. In this case the PIXE spectrum was analysed by a process of peak deconvolution. This gave good agreement with the analysis by bulk techniques for most elements. On the other hand, peak deconvolution techniques are only applicable to data with good counting statistics and, if these are achievable, it may sometimes pay to sacrifice them in order to utilise the spectral resolution of a wavelength dispersive spectrometer. Much interest has centred on the examination of lunar rocks and the quantitative accuracy and trace elemental sensitivity of the PMP have been utilised at Heidelberg by Blank, El Goresy and coworkers to examine the partitioning of trace elements (Zr and Nb) in various coexisting minerals. Spot (6 μm) measurements were made within large ($\sim$ 50 μm) grains and corrections made for the thick (infinite) specimen. Zr concentrations as low as 4 ppm atomic were measured and both major and trace elements exhibit zoning (Blank *et al.*, 1984). Vis *et al.* (1987) have used a SPMP at Amsterdam to detect many trace elements in meteorites and in cosmic dust particles.

e. Archaeology, Difficult Specimens, External Beams and Collimated Probes

It is possible to extract a microbeam from the vacuum system, but it will suffer considerable scattering from the air or gas and from any exit foil. Nevertheless such an external microbeam can be very useful for low resolution studies. Several groups currently use this technique to good effect (Horowitz and Grodzins, 1975; Van Patter *et al.*, 1981; MacArthur *et al.*, 1981; Trail, 1982; Shroy *et al.*, 1978; Swann and Fleming, 1987). Generally the beam is reduced to submillimetre dimensions by means of an exit collimator, rather than by focussing. The specimen must be placed as close as possible to this collimator. If the collimator is small enough, differential pumping may be used to dispense with the foil. MacArthur *et al.* (1981) employed such a system to obtain a spatial resolution of 10 μm. A detailed discussion of collimator design, scattering by the collimator and gas, beam halo and beam profile measurements for this system is given by Barfoot *et al.* (1982). Two magnetic quadrupole doublets were used to concentrate the beam onto the collimator but the demagnification factor of six is fairly low, so that beam divergence from the collimator is not unworkably high. Even so, the specimen must be 0.5 mm from the collimator exit to achieve optimum resolution. External beams have naturally found use with bulky specimens such as archaeological objects, works of art, historical records or forensic specimens (Demortier and Hackens, 1982; Swann, 1982; and Demortier *et al.*, 1984). In biology they have been used on live specimens, such as the shells of living oysters (Carriker *et al.*, 1982) but there is little advantage in using an external microprobe beam on non-living soft biological tissue which could be sectioned and examined with the superior resolution and detection geometry of an internal microprobe beam. It is true that such tissue should not be directly inserted into a vacuum system where it would rapidly dry; but drying would occur rapidly in dry air also and, even if some form of wet cell were employed, the intensely ionising beam would give rise to strong chemical changes in the cells of the unfixed specimen.

10.4.2 Biological Sciences

When biological or medical samples are to be examined with a PMP, a number of constraints and problems are introduced, associated with specimen preparation and integrity, microprobe operation, data collection and handling. These were reviewed by Legge and Mazzolini (1980), Lindh (1981), Legge *et al.* (1982b) and Malmqvist (1986). Further discussion of specific examples is to be found in the third PIXE conference proceedings (Martin, 1984) and in (Vis, 1985). It is also advisable for anyone contemplating biological measurements with a PMP to become familiar with the extensive body of literature relating to EMP analysis of biological specimens (e.g. Hall *et al.*, 1974; Chandler, 1977; Erasmus, 1978; Hall and Gupta, 1983). The problems to be overcome are similar and many of the techniques of the EMP have been carried over to the newer instrument.

a. The Nature of the Specimen

Soft biological tissues and cells are largely composed of H, C, O, N, a few per cent of some of the light elements from Na to Ca and generally only trace amounts of some heavier elements which may occur at any level between 1 and 1000 ppm. For example, Watt *et al.* (1984) obtained high contrast maps and good spectra of 5 μm diameter Cu accumulations in primary biliary cirrhosis liver tissue. Such specimens are ideally suited to PIXE analysis in a PMP, since the light elements are easily seen above the background of bremsstrahlung and this is insignificant beneath the heavy trace elements. In contrast, the EMP, being at least two orders of magnitude less sensitive, must either be restricted to relatively abundant elements or be used with a wavelength dispersive spectrometer and consequently much higher beam charge deposition. Even then it cannot approach the 1 ppm sensitivity of the PMP. It is generally not practical to use NRA with soft biological specimens though F has been depth profiled in bone by means of the reaction $^{19}F(p, \alpha\gamma)^{16}O$ (Coote and Sparks, 1981; Coote *et al.*, 1982). The reactions $^{14}N(d, p_0)^{15}N$ and $^{15}N(p, \alpha\gamma)^{12}C$ have been successively used in a line scan to determine the uptake of N by barley roots (McMillan *et al.*, 1982), using the stable isotope ^{15}N as a tracer. The IMP is not ideally suited to the examination of biological specimens and their complex chemical structure, because of the difficulties in quantification although it has good sensitivity, and some biological applications have been reviewed by Burns (1982, 1986). The preferred instrument for high spatial resolution is the EMP (with thin samples) and for sensitivity and quantitative accuracy the PMP with PIXE analysis. Generally the beams required will be at least 100 pA of protons and the energy about 3 MeV.

b. The Preparation of Tissues

As soft biological specimens are very unstable outside their normal environment, it is necessary to remove a specimen rapidly from that environment, fix it in order to stabilise all elements against chemical reactions or movement, cut sections of suitable thickness, mount them and stabilise the water content, before introducing the sample to the vacuum chamber. In normal circumstances, chemical fixatives, embedding compounds, stains and drying agents can not be used; so cryofixation is used and also freeze drying.

Cryofixation is generally performed by plunging the specimen beneath the surface of some cryoliquid which has been cooled to a temperature well below its boiling point. Nitrogen slush, made by evaporative cooling of nitrogen liquid is a common choice; others are isopentane and propane (fire hazards), Freon 22 and Freon 12 (Bald, 1984); for hazards see Ryan and Liddicoat (1987). Even with these precautions, the cooling rates within the tissue block will be too slow to completely prevent ice crystal formation with consequent structural damage and movement of elements ahead of the ice front. Only sections cut close to the surface of the frozen block will be relatively free of ice crystal artifacts and this must be considered when rapidly dissecting the tissue before immersion in the cryoliquid. Cryoprotectants are sometimes used but they themselves endanger the elemental integrity of the specimen. Sectioning must be carried out in the frozen state with a cryomicrotome. For the highest resolution, good sections may need to be thinner than 1 μm; but usually they can be several μm. Obviously when an organelle of micron dimensions is to be examined the sections should not be appreciably thicker. The average histological cryomicrotome is inadequate for work at the micron scale, offering insufficient control over the temperature of the specimen and knife which need to be maintained independently at constant temperatures of about $-80°C$ or lower, depending on the particular tissue. The surroundings should be maintained at a much lower temperature. It is essential that the specimen does not thaw until freeze drying has been completed. Therefore the frozen sections must be transferred onto precooled plastic foils within the microtome cryochamber and kept at liquid nitrogen temperature whilst they are transferred to the freeze dryer. Several materials are used as foils,

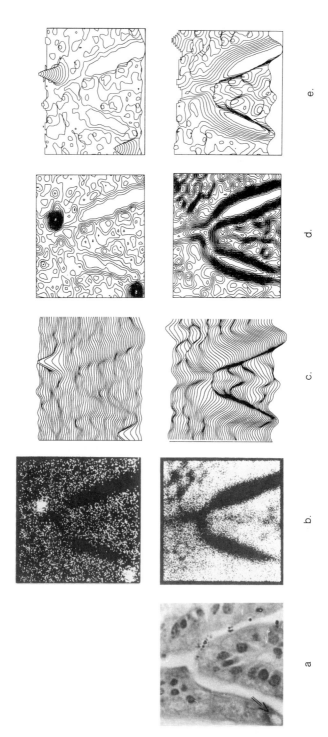

e.g. stretched polypropylene, Formvar, collodium, but the toughest and one of the cleanest is Nylon which can be made to a thickness of 0.1 μm or greater by the method developed by Hall (Echlin and Moreton, 1974).

These techniques are examplified in work of Kirby *et al.* (1988). The tissue was ileum of 10-day old *brindled mouse.* The cryofixative was liquid propane cooled in liquid nitrogen. The 4 μm thick cryosection was mounted without foil (self-supporting) over a 0.5 mm diameter hole in the Hostaphan target frame, freeze dried and examined in the SPMP with a 400 pA beam of 3 MeV protons in a 2.5 μm diameter spot scanned over an area of 60 μm $\times$ 60 μm for a period of 7h. The adjacent 4 μm cryosection was stained with eosin and haemotoxylin to show cell boundaries and nuclei. In Fig. 10.15, the stained section on the left shows the region scanned — a tongue shaped villus tip separated from neighbouring villi by gaps of 3 to 5 μm. Cell nuclei are darkly stained. Of the many elements extracted from the data block, P maps are shown in the top row and Ca maps in the bottom row. These are presented, from left to right, as dot maps, as intensity scans (y-modulation), as two-dimensional contour maps and as three-dimensional contour maps. P is seen to be fairly uniformly distributed within the villi with some concentration at the edges. Ca however is sparsely distributed except in two regions of heavy concentration. These are probably associated with goblet cells and one is marked with an arrow in the adjacent stained section — the lower left unstained tissue (Legge *et al.*, 1988a).

The surest way to maintain specimen integrity is to keep the specimen in the frozen hydrated state after sectioning and during examination in the microprobe. This avoids shrinkage and consequent distortion during freeze drying and the tendency of all lumen contents to collapse onto the nearest membrane. However it is a technique so far used by relatively few EMP groups. As well as being a difficult technique, it results in higher bremsstrahlung backgrounds and also gives poor secondary electron images unless surface thawing is permitted while the specimen is on the chamber cryostage. All these techniques are covered in the review by Hall and Gupta (1983).

◀ **Fig. 10.15** Elemental mapping in soft biological tissue. a. Villi cells and their nuclei are seen in stained 4 μm cryosection of mouse ileum at left. Adjacent section was freeze dried and scanned by 400 pA, 2.5 μm spot of 3 MeV protons over 60 μm $\times$ 60 μm for 7.5 h. P (top) and Ca (bottom) are displayed (left to right) as b. PIXE dot maps, c. intensity scans, d. two-dimensional contour maps and e. three-dimensional contour maps. (Legge *et al.*, 1988a).

c. Quantitative Measurements

With very thin specimens, corrections for beam energy loss and X-ray attenuation are negligible and the relative concentrations of different elements may be derived from an X-ray spectrum measured with a calibrated detector (Chapter 5). The spectra from two samples of the same matrix and thickness may be compared without need for correction or accurate beam charge measurement provided the beam is scanned across both samples in the one operation. This technique was used to compare the concentrations and distributions of several elements in green and chlorotic eucalyptus leaves (Mazzolini *et al.*, 1982). If absolute concentrations are required then, whether or not beam energy loss or X-ray attenuation corrections are needed, the specimen thickness and beam charge must be accurately measured.

Thin sections cut with a cryomicrotome may not have a reliably known thickness. Moreover the thickness of such sections, even if uniformly cut, may vary after freeze drying and have local density fluctuations. It is therefore desirable to measure the specimen thickness at every point to be analysed. Hall and Werba (1971) used the level of continuum bremsstrahlung radiation in an EMP X-ray spectrum as a measurement of specimen thickness. In a dry biological specimen this radiation is mostly associated with the C of the matrix. The same technique is applied to PIXE spectra from a PMP, although the continuum statistics are much lower than for the EMP. However, with a PMP, the matrix elements C, N and O may be measured individually by RBS and even H may be measured by RFS. Thus a direct measurement of matrix composition and hence thickness (areal density) is obtained simultaneously with the collection of a PIXE spectrum if back scattered and forward scattered protons are collected together with X-rays (Legge *et al.*, 1979). Heck and Rokita (1984) used this technique to normalise and quantify each point of a line scan across the posterior part of a rat eye. In areal scans the thickness normalisation should be applied to each selected region of interest as a whole, after extraction of the relevant data, and this was done by Mazzolini in producing a table of absolute concentrations (dry weight) for nine elements in seven principal anatomical features of a wheat seed — a total of 63 quantitative microanalytical results from a single scanning operation (Mazzolini, 1983, Mazzolini *et al.*, 1985). Because the elemental yields measured by PIXE are normalised to the thickness of R areal density measured by RBS and RFS at the same time, neither beam charge nor dead time are needed as parameters.

d. The Bombardment of Tissues — Elemental Loss and Mass Loss

Some elements, such as Br and Cl, may be lost from the specimen under severe bombardment conditions making quantitative measurements unreliable for these elements. If a major matrix element is lost however, it may affect quantisation of all elements. This we refer to as *mass loss*. Such mass loss has been measured for electron bombardment by Hall and Gupta (1974) who monitored the bremsstrahlung yield as a function of time with a constant electron beam current. The beam current in a PMP is less steady, but if beam charge pulses are recorded on tape together with PIXE, RBS and RFS events, the levels of all elements can be measured with a time resolution of about 1 s (Legge and Mazzolini, 1980; Mazzolini *et al.*, 1981). Of the matrix elements, C is generally stable but the exponential loss of H is rapid for an unscanned beam. The rapid loss of Cl also observed makes this a difficult element to quantify in spot analysis, though with the above system of data recording the initial level of Cl may be inferred. it is advisable to study beam sensitive materials in rapid scanning mode rather than spot or slow scanning mode. Although the net ionisation damage is not affected, the net thermal damage is reduced by rapid scanning. However scanning in general greatly reduces the problem of beam damage simply because the number of damage events is spread over a much larger number of molecules. Thus it is safer to scan an entire cell and extract from the accumulated data the spectrum of the cytoplasm than to conduct a high intensity spot measurement on the cytoplasm with the same beam charge. Because of the poor contrast available in unstained biological tissue, such a scan may be advisable in any case for purposes of identification.

e. Cytology

The handling and examination of isolated cells (cytological microanalysis) is an important subsection of soft tissue biological microanalysis and is covered in the aforementioned reviews. It is, however, distinguished by peculiar difficulties of sample preparation techniques. In general the examination of a cell requires that its normal environment or growth medium be first replaced by some medium which will not interfere with the elemental analysis. The techniques involved must not introduce any contaminant element or permit the loss of any element. Considering the potential of the cell membrane for passive or active diffusion and its known sensitivity to level of pH, tonicity and individual elemental levels, washing is always a problematical exercise and the literature on available

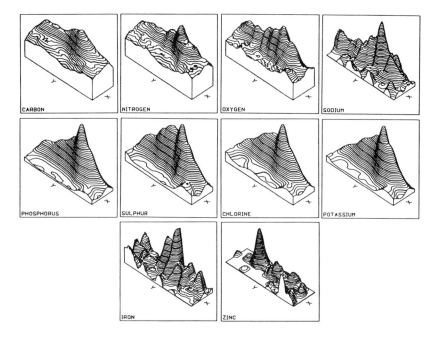

Fig. 10.16 Three-dimensional contour maps of freeze dried skin fibroblast cell on Nylon foil — Na and lighter elements from RBS, heavier elements from simultaneous PIXE. P peaks at cell nucleus. 1 nA beam of 3 MeV protons in 3 μm spot scanned over 24 μm $\times$ 144 μm for 150 min. (Allan *et al.*, 1985b).

washing buffers is vast. The high sensitivity of the PMP to a wide range of elements makes this problem more acute and the only reassurance as to the integrity of the cellular content currently available is the repeatability of results with a variety of buffers. For subcellular work it may be necessary to section sells in a cryomicrotome, in which case the normal environmental background of the cell may be removed without resort to washing techniques. Because of the spatial resolution required, such work to date has been restricted to the EMP and the thin sections required restrict the sensitivity obtainable within a reasonable time limit. In PMP work, a resolution of 1.5 μm and 150 pA beam of 3 MeV protons was needed to show the structure of erythrocytes (O'Brien and Legge, 1987) but Allan *et al.* (1985b) were able to use a resolution of 3 μm and hence currents of 1 nA to map the gross structure of skin fibroblast cells. The erythrocytes were left in their blood plasma, but the fibroblasts were first washed in ammonium acetate buffer. In each case the cells were plated on thin Nylon foils, snap frozen and freeze dried. Fig. 10.16 shows

a.

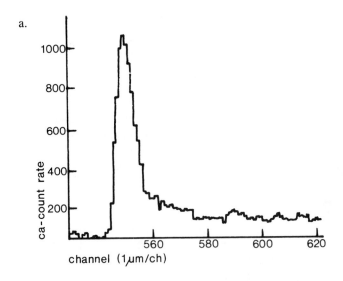

b.

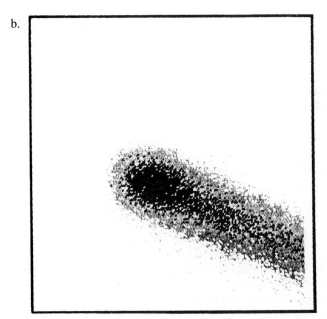

Fig. 10.17 Concentration of Ca ions in growing tip cell of pollen tube. Cell was fixed in glutaraldehyde, dried and scanned with SPMP. (a) Line scan showing strong concentration at tip of cell. Heidelberg SPMP (Reiss *et al.*, 1983). (b) Area scan showing map of strong Ca concentration. Oxford SPMP (Reiss *et al.*, 1985).

elemental distributions for one skin fibroblast cell as three-dimensional contour maps. The C, N, O and Na data were from RBS (Na including a 20% contribution from Mg). The other data were from PIXE. The P concentration peaks at the cell nucleus. The background underlying the C, N and O of the cell is associated with the Nylon foil.

O'Brien *et al.* (1982) employed the sensitivity of the PMP to map the distribution of heavy metals in normal and tumorous macrophage cells, a field in which the PMP is likely to make important contributions. Lindh has studied the elemental profiles of several blood cells in the normal state (Lindh *et al.*, 1984) and in the blood from leukaemic and arthritic patients (Johansson and Lindh, 1984; Lindh *et al.*, 1985). Another area of great significance to cytology is the movement of Ca ions. They are known to play a major role in the regulation of some cell functions and their part in regulating the tip growth of pollen tubes and moss protonema, as demonstrated qualitatively by the fluorescence studies of Reiss and Herth (1978, 1979) was well quantified by PMP studies (Reiss *et al.*, 1983). Fig. 10.17 shows from this work a Ca line scan along a growing tip cell of the pollen tube *Lilium longifolium* after glutaraldehyde fixation. Chemical fixation is expected to result in the loss of unbound ions; so Fig. 10.17a demonstrates that there is a gradient of bound Ca peaked strongly at the growing tip of the cell. This was confirmed in an independent study (Reiss *et al.*, 1985) in which the 40 μm long tip area of a pollen tube was scanned. Strong concentrations at the tip were seen in the maps of bound Ca (Fig. 10.17b) and of the trace element Zn. Further work by Reiss and Traxel (1987) suggests a polar distribution of Ca channels in the plasma membrane of pollen tubes.

10.5 SPATIAL INFORMATION AND TRANSMISSION IMAGING — TOWARDS HIGHER RESOLUTION

The elemental analysis of any specimen will be meaningful only to the extent that the structure of the specimen is known. We have already discussed the uncertain thickness of small grains in geological samples. Mazzolini *et al.* (1983) have discussed the effects observed with semithick cylindrical specimens, in this case rye grass root, if the elemental distributions studied are predominantly surface concentrated or axis concentrated. The correction of data for depth distribution effects is necessary for any quantitative work with thick or semithick specimens.

The interpretation of an elemental map or correct positioning of the beam spot on the specimen for a fixed spot analysis may require the availability of an image of the specimen having greater detail and possibly spatial resolution superior to that of the elemental maps. Since the microprobes considered are all working close to or below the spatial resolution of simple optical microscopes, such an image must be generated by the probe itself. In order to obtain the required detail, advantage may be taken of the high efficiency for secondary electron emission and for inelastic scattering. As the efficiency for the production and detection of a signal increases, we can afford to stop down both the object and aperture diaphragms, so decreasing both the first order spot diameter and its aberrations. With bright field transmission imaging (Overley *et al.*, 1983; Sealock *et al.*, 1983) in which the unscattered transmitted beam is run directly into a detector, it may be necessary to restrict the beam current to about 10^4 particles per second or about 1 fA. Fig. 10.18 shows the resulting improvement in spatial resolution when a PMP giving 1 μm spatial resolution for microanalysis with 100 pA beams of protons or alpha particles was operated by Sealock with reduced diaphragms (Legge *et al.*, 1985). In this case the beam was 1.5 fA of 2 MeV He$^+$ ions, the specimen was a diatom on a thin (0.1 μm) Nylon foil, the beam was scanned over an area of 16 μm $\times$ 14 μm and the signal was collected from a detector at 0° giving a bright field image. The images shown in Fig. 10.18a are energy slices of the *three-dimensional* sorted data block (Highlight 10.4). Since the different energy losses of the collected beam particles represent different thicknesses or densities of the specimen, the data block is truly three-dimensional and may be presented in a single three-dimensional contour map which is also shown. The parameter contoured here is the average energy loss for each pixel of the bright field image and this parameter is smoothed in both transverse directions over 0.1 μm. The image is then tilted and rotated to show the maximum information on the peripheral structure. It must be remembered that only thickness of the specimen can be depicted and not the true morphology. The spatial resolution achieved was about 0.3 μm. Only one particle per pixel is needed to measure the energy loss, but contrast is greatly improved if a mean or median of several measurements is made (Overley *et al.*, 1983). STIM images may be *bright field* or *dark field* and in each case scattering contrast is an alternative to energy loss contrast. A quantitative treatment of both is given by Sealock *et al.* (1987). Lefevre *et al.* (1987) have used STIM data to normalise data for each pixel of PIXE elemental maps. High resolution microbeams may also be used for etching, as demonstrated by Fischer (1988) with single event pulsing of a heavy ion microprobe.

In conclusion we look at the future development of microprobes. There is a continual demand for higher resolution. The SEMP is limited by scattering and the IMP by mixing but the PMP is mostly limited by accelerator performance (particularly of ion sources). Fig. 10.19 shows two

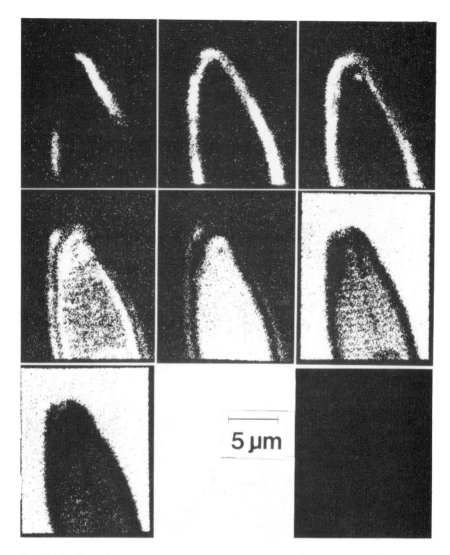

Fig. 10.18a. Scanning transmission ion microscopy at a resolution of 0.3 μm with a beam of 2 MeV α-particles. The specimen is a diatom examined in brightfield mode with energy loss contrast. The images represent slices of different energy loss and hence different density. (Legge *et al.*, 1985)

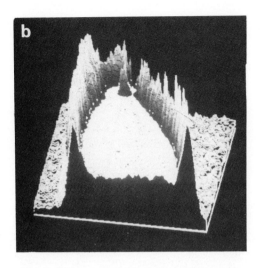

Fig. 10.18b. The three-dimensional contour map shows the total data block presented in one average energy loss image. (Legge *et al.,* 1988b)

successive bright field STIM three-dimensional contour images of an imperfection on a SiO_2 grating, measured by Bench (Legge *et al.*, 1988b). The thickness ranges from 30 nm to 250 nm and the scan area is 1.9 μm × 2.5 μm. Before contouring, the data were energy averaged for each pixel and smoothed in both transverse directions with a Gaussian of FWHM 60 nm. The 2 MeV beam of He^+ had a resolution of about 100 nm and the current used was 0.06 fA. Each image was collected over 18 min and over the 36 min period mechanical plus electrical drift was 40 nm. Though more current was available, useful elemental analysis is certainly not yet possible at this resolution. However the measurement demonstrates that the mechanical and electromagnetic stability required to work at resolutions of 100 nm or better is achievable in a SPMP and therefore further development of accelerators and their ion sources will bring immediate benefits.

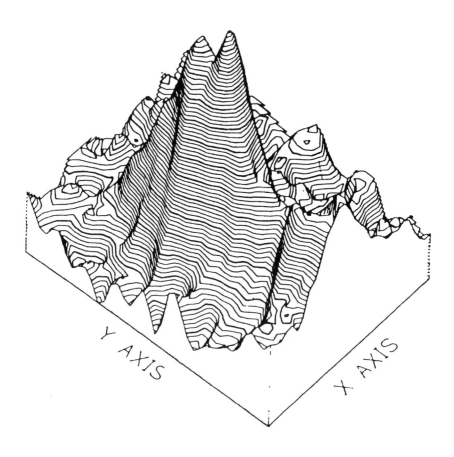

Fig. 10.19 High resolution (100 nm) STIM bright field energy loss image of imperfection on SiO_2 grating. Scan area is 1.9 μm $\times$ 2.5 μm. Three-dimensional contour image obtained 100 nm depth resolution with 2 MeV He^+ beam. Mechanical and electrical drift was 40 nm over 36 min. Melbourne SPMP (Legge *et al.*, 1988b).

REFERENCES

Al-Ghazi, M.S.A.L. and McKee, J.S.C. (1982). *Nucl. Instrum. Methods* **197**, 117.

Allan, G.L., Zhu, J. and Legge, G.J.F. (1985a). Proc. 4th Aust. Conf. Nucl. Tech. Analy., Lucas Heights, *ISSN 0811–9422, 49.*

Allan, G.L., Camakaris, J., Zhu, J. and Legge, G.J.F. (1985b). Proc. 4th Aust. Conf. Nucl. Tech. Analy., Lucas Heights, *ISSN 0811–9422,* 105.

Andersen, C.A., ed. (1973). 'Microprobe Analysis'. Wiley, New York.

Augustyniak, W.M., Betteridge, D. and Brown, W.L. (1978). *Nucl. Instrum. Methods* **149**, 669.

Bald, W.B. (1984). *J.Microsc.* **134**, 261

Barfoot, K.M., MacArthur, J.D. and Vargas-Aburto, C. (1982). *Nucl. Instrum. Methods* **197**, 121.

Bayly, A.R. and Waugh, A.R. (1983). Private communication with Ryder, G.C.

Biersack, J.P. and Wittmaack, K. eds (1986) *Nucl. Instrum. Methods* **B15**, 1.

Blank, H., El-Goresy, A., Janicke, J., Nobiling, R. and Traxel, K. (1984). *Nucl. Instrum. Methods* **B3**, 681.

Bonani, G., Suter, M., Jung, H., Stoller, Ch. and Wölfli, W. (1978). *Nucl. Instrum. Methods* **157**, 55.

Bosch. F., El-Goresy, A., Martin, B., Povb, B., Nobiling, R., Schwalm, D. and Traxel, K. (1978). *Nucl. Instrum. Methods* **149**, 665.

Burns, M. (1982). *J. Microsc.* **127**, 237.

Burns, M. (1986). *In* 'Microbeam Analysis — 1986' (Romig, A.D. and Chambers, W.F., eds) San Francisco Press, San Francisco, 193.

Carriker, M.R., Swann, C.P. and Ewart, J.W. (1982). *Marine Biol.* **69**, 235.

Castaing, R. and Guinier, A. (1950). 'Proc. Int. Conf. Electron Microsc.' Delft (1949), p. 60.

Castaing, R. (1951). PhD Thesis, University of Paris.

Castaing, R. and Hughes, L. (1962). *C. R. Acad. Sci.* **255**, 76

Castaing, R. and Slodzian, G. (1962). *C. R. Acad. Sci.* **255**, 1893

Castaing, R., Jouffrey, B. and Slodzian, G. (1960). *C. R. Acad. Sci.* **251**, 1010.

Chandler, J.A. (1977). *In* 'Practical Methods in Electron Microscopy', (Glauert, A.M., ed.) North Holland Elsevier/North-Holland Biomedical Press, Amsterdam, Vol. 5, 317.

Colby, J.W. (1975). *In* 'Practical Scanning Electron Microscopy'. (Goldstein, J.I. and Yakowitz, H. eds) Plenum Press, New York, 327.

Cookson, J.A. (1979). *Nucl. Instrum. Methods* **165**, 477.

Cookson, J.A. (1981). *Nucl. Instrum. Methods* **181**, 115.

Cookson, J.A., Ferguson, A.T.G. and Pilling, F.D. (1972). *J. Radioanal. Chem.* **12**, 39.

Cookson, J.A., McMillan, J.W. and Pierce, T.B. (1979). *J. Radioanal. Chem.* **48**, 337.

Coote, G.E. and Sparks, R.J. (1981). *N.Z. J. Archaeol.* **3**, 21.

Coote, G.E., Sparks, R.J. and Blattner, P. (1982). *Nucl. Instrum. Methods* **197**, 213.

Cosslett, V.E. and Duncumb, P. (1956). *Nature* **177**, 1172.

Demortier, G. ed. (1982). *Nucl. Instrum. Methods* **197**, 1.

Demortier, G. and Hackens, T. (1982). *Nucl. Instrum. Methods* **197**, 223.

Demortier, G., Van Oystaeyen, B. and Boullar, A. (1984). *Nucl. Instrum. Methods* **B3**, 399.

den Ouden, J.C., Bos, A.J.J., Vis, R.D. and Verheul, H. (1981). *Nucl. Instrum. Methods* **181**, 131.

Dix, M.R. (1983). University of Melbourne 4th Year Report (unpub.).

Doyle, B.L. and Wing, N.D. (1983). *IEEE Trans. Nucl. Sci.* **NS-30**, 1214.

Doyle, B.L. (1986). *Nucl. Instrum. Methods* **B15**, 654.

Dymnikov, A.D. and Yavor, S. Ya. (1964). *Sov. Phys. Tech. Phys.* **8**, 639.

510   *G.J.F. Legge*

Echlin, P. and Moreton, R. (1974). *In* 'Microprobe Analysis as Applied to Cells and Tissues' (Hall, T.A., Echlin, P. and Kaufmann, R. eds) Academic Press, London, 159.
Erasmus, D.A. ed. (1978). 'Electron Probe Microanalysis in Biology', Chapman and Hall, London.
Fischer, B.E. (1988). *Nucl. Instrum. Methods* **B30**, 284.
Garcia, S.R. (1984). Los Alamos National Laboratory, personal communication to Rogers *et al.*
Glaser W. (1941). *Z. Physik* **117**, 285.
Goldstein, J.I. and Yakowitz, H. (1975). 'Practical Scanning Electron Microscopy', Plenum Press, New York.
Goldstein, J.I., Newbury, D.E., Echlin, P., Joy, C., Fiori, C. and Lifshin, E. (1981). 'Scanning Electron Microscopy and X-ray Microanalysis'. Plenum Press, New York.
Grime, G.W. and Watt, F. (1984). 'Beam Optics of Quadrupole Probe-forming Systems', Adam Hilger, Bristol.
Grime, G.W. and Watt, F. (1988). *Nucl. Instrum. Methods* **B30**.
Grime, G.W., Watt, F., Blower, G.D. and Takacs, J. (1982). *Nucl. Instrum. Methods* **197**, 97.
Grime, G.W., Takacs, J. and Watt, F. (1984). *Nucl. Instrum. Methods* **B3**, 589.
Hall, T.A. and Gupta, B.L. (1974). *J. Microsc.* **100**, 177.
Hall, T.A. and Gupta, B.L. (1983). *Quart. Rev. Biophys.* **16**, 279.
Hall, T.A. and Werba, P. (1971). *In* 'Electronic Microscopy and Analysis'. (Nixon, W.C. ed.) Inst. of Physics, London, 146.
Hall, T.A., Echlin, P. and Kaufmann, R. eds. (1974). 'Microprobe Analysis as Applied to Cells and Tissues', Academic Press, London.
Heck, D. and Kasseckert, E. (1976). Kernforschungszentrum Karlruhe Report KFK 2379, 130.
Heck, D. (1976). Kernforschungszentrum Karlsruhe Report KFK 2288.
Heck, D. (1978). Kernforschungszentrum Karlsruhe Report KFK 2734.
Heck, D. (1982). *Nucl. Instrum. Methods* **197**, 91.
Heck, D. (1985). *Atomkernenergie Kerntechnik,* **46**, 187.
Heck, D. and Rokita, E. (1984). *Nucl. Instrum. Methods* **B3**, 259.
Höfert, M., Bischof, W., Stratmann, A., Raith, B. and Gonsior, B. (1984). *Nucl. Instrum. Methods* **B3**, 572.
Horowitz, P. and Grodzins, L. (1975). *Science* **189**, 795.
Ingarfield, S.A., McKenzie, C.D., Short, K.T. and Williams, J.S. (1981). *Nucl. Instrum. Methods* **191**, 521.
Jamieson, D.N. (1985). 'The Correction of Spherical Aberration in a Proton Microprobe', PhD Thesis, University of Melbourne (unpub.)
Jamieson, D.N. and Legge, G.J.F. (1987). *Nucl. Instrum. Methods* **B29**, 544.
Jamieson, D.N. and Legge, G.J.F. (1988). *Nucl. Instrum. Methods* **B30**, 235.
Johansson, E. and Lindh, U. (1984). *Nucl. Instrum. Methods* **B3**, 637.
Kirby, B.J. (1983). 'Radiation Damage in Biological Tissue During Ion Beam Microanalysis', 4th Year Report, University of Melbourne (unpub.)
Kirby, B.J., McArdle, H., Danks, D. and Legge, G.J.F. (1988). Priv. Comm. See also Legge *et al.* (1988a).
Knies, H., Martin, B., Nobiling, R., Povb, B. and Traxel, K. (1982). *Nucl. Instrum. Methods* **197**, 79.
Koyamo-Ito and Grodzins, L. (1980). *Nucl. Instrum. Methods* **174**, 331.
Lefevre, H.W., Connolly, R.C., Sieger, G. and Overley, J.C. (1983). *Nucl. Instrum. Methods* **218**, 39.

Lefevre, H.W., Schofield, R.M.S., Overley, J.C. and Macdonald, J.D. (1987). *Scan. Microsc.* **1**, 879.

Legge, G.J.F. (1978). Proc. 2nd Aust. Conf. Nucl. Tech. Anal., Lucas Heights *ISN O 642 59991 2*, 18.

Legge, G.J.F. (1980). *In* 'Microbeam Analysis — 1980' (Wittry, D.B. ed.) San Francisco Press, San Francisco, 70.

Legge, G.J.F. (1982). *Nucl. Instrum. Methods* **197**, 243.

Legge, G.J.F. (1984). *Nucl. Instrum. Methods* **B3**, 561.

Legge, G.J.F. and Hammond, I. (1979). *J. Microsc.* **117**, 201.

Legge, G.J.F. and Mazzolini, A.P. (1980). *Nucl. Instrum. Methods* **168**, 563.

Legge, G.J.F., McKenzie, C.D. and Mazzolini, A.P. (1979). *J. Microsc.* **117**, 185.

Legge, G.J.F., Jamieson, D.N., O'Brien, P.M.J. and Mazzolini, A.P. (1982a). *Nucl. Instrum. Methods* **197**, 85.

Legge, G.J.F., Mazzolini, A.P., Roczniok, A.F. and O'Brien, M. (1982b). *Nucl. Instrum. Methods* **197**, 191

Legge, G.J.F., McKenzie, C.D., Mazzolini, A.P., Sealock, R.M., Jamieson, D.N., O'Brien, P.M., McCallum, J.C., Allan, G.L., Brown, R.A., Colman, R.A., Kirby, B.J., Lucas, M.A., Zhu, J. and Cerini, J. (1985). *Nucl. Instrum. Methods* **B15**, 669.

Legge, G.J.F., O'Brien, P.M., Kirby, B.J. and Allan, G.L. (1988a) *Ultramicrosc.* **24**, 283.

Legge, G.J.F., O'Brien, P.M., Sealock, R.M., Allan, G.L., Bench, G., Moloney, G., Jamieson, D.N. and Mazzolini, A.P. (1988b). *Nucl. Instrum. Methods* **B30**, 252.

Levi-Setti, R. and Fox, T.R. (1980). *Nucl. Instrum. Methods* **168**, 139.

Levi-Setti, R., Crow, G., Wang, Y.L., Parker, N.W., Mittleman, R. and Hwang, D.M. (1985a). *Phys. Rev. Lett.* **54**, 2615

Levi-Setti, R., Crow, G. and Wang, Y.L. (1985b). *In* 'SEM/1985/II' (O. Johari, ed.) IIT Research Institute, Chicago, 535.

Liebl, H.S. (1967). *J. Appl. Phys.* **38**, 5277.

Lindh, U. (1981). *Nucl. Instrum. Methods* **181**, 171.

Lindh, U., Johansson, E., Hällgren, R. and Jameson, S. (1985). *Nutr. Res.* **1**, 451.

Lindh, U., Johansson, E. and Gille, L. (1984). *Nucl. Instrum. Methods* **B3**, 631.

MacArthur, J.D., Amm, D., Barfoot, K.M. and Sayer, M. (1981). *Nucl. Instrum. Methods* **191**, 204

Maggiore, C.J. (1980). *In* 'SEM/1980/I' (O. Johari, ed.) IIT Research Institute, Chicago, 439.

Maggiore, C.J. (1981). *Nucl. Instrum. Methods* **191**, 199.

Mak, B.K., Bird, J.R. and Sabine, T.M. (1966). *Nature* **211**, 738.

Malmqvist, K.G. (1986). *Scanning Electron Microsc. 1986.* **111**, 821.

Martin, B. (1980). *In* 'SEM/1980/I' (O. Johari, ed.) IIT Research Institute, Chicago, 419.

Martin, B. ed. (1984). *Nucl. Instrum. Methods* **B3**, 1.

Martin, B. and Nobiling, R. (1980). *In* 'Applied Charged Particle Optics' (A. Septier, ed.) Academic Press, New York, 321.

Martin, F.W. and Goloskie, R. (1982). *Appl. Phys. Lett.* **40**, 191.

Mazzolini, A.P. (1983). 'The Scanning Proton Microprobe and its Application to Elemental Analysis in Botany', PhD Thesis, University of Melbourne, (unpub.).

Mazzolini, A.P., Legge, G.J.F. and Pallaghy, C.K. (1981). *Nucl. Instrum. Methods* **191**, 583.

Mazzolini, A.P., Anderson, C.A., Ladiges, P.Y. and Legge G.J.F. (1982). *Aust. J. Plant. Physiol.* **9**, 261.

Mazzolini, A.P., Jeffrey, J.J., Uren, N.C. and Legge, G.J.F. (1983). *Aust. J. Plant Physiol.* **10**, 153.

Mazzolini, A.P., Pallaghy, C.K. and Legge, G.J.F. (1985). *New Phytol.* **100**, 483.

McCallum, J.C., McKenzie, C.D., Lucas, M.A., Rossiter, K.G., Short, K.T. and Williams, J.S. (1983). *Appl. Phys. Lett* **42**, 827.

McCallum, J.C., Brown, R.A., Nygren, E., Williams, J.S. and Olson, G.L. (1986). *Mat. Res. Soc. Symp. Proc.* **69**, 305.

McMillan, J.W. and Pummery, F.C.W. (1980). *Corrosion Sci.* **20**, 41.

McMillan, J.W., Pummery, F.C.W. and Pollard, P.M. (1982). *Nucl. Instrum. Methods* **197**, 171.

Nobiling, R. (1982). Quoted in Legge (1982).

Nobiling, R., Civeleköglu, Y., Povh, B., Schwalm, D. and Traxel, K. (1975). *Nucl. Instrum. Methods* **130**, 325.

O'Brien, P.M. and Legge, G.J.F. (1987). *Biological Trace Element Research* **13**, 159.

O'Brien, P.M., Legge, G.J.F., Bradley, T.R. and Hodgson, G.S. (1982). *Aust. Phys. Eng. Sci. Med.* **5**, 30.

Overley, J.C., Connolly, R.C., Sieger, G.E., MacDonald, J.D. and Lefevre, H.W. (1983). *Nucl. Instrum. Methods* **218**, 43.

Pierce, T.B., Peck, P.F. and Cuff, D.R.A. (1966). *Nature* **211**, 66.

Pierce, T.B., Peck, P.F. and Cuff, D.R.A. (1968). *Nucl. Instrum. Methods* **67**, 1.

Pierce, T.B., McMillan, J.W., Peck, P.F. and Jones, I.G. (1973). *Radiochem. Radioanal. Lett.* **14/5-6**, 357.

Price, P.B. and Bird, J.R. (1969). *Nucl. Instrum. Methods* **69**, 277.

Prins, M. and Hoffman, L.J.B. (1981). *Nucl. Instrum. Methods* **181**, 125.

Reed, S.J.B. (1975). 'Electron Microprobe Analysis'. Cambridge University Press, Cambridge.

Reiss, H-D. and Herth, W. (1978). *Protoplasma* **97**, 373.

Reiss, H-D. and Herth, W. (1979). *Planta* **145**, 225.

Reiss, H-D. and Traxel, K. (1987). Biological Trace Element Research, **13**.

Reiss, H-D., Herth, W., Schnepf, E. and Nobiling, R. (1983). *Protoplasma* **115**, 153.

Reiss, H-D., Grime, G.W., Li, M.Q., Takacs, J. and Watt, F. (1985). *Protoplasma* **126**, 147.

Robinson, C.F. (1973). *In* 'Microprobe Analysis'. (Andersen, C.A. ed.) Wiley, New York, 507.

Rogers, P.S.Z., Duffy, C.J., Benjamin, T.H. and Maggiore, C.J. (1984). *Nucl. Instrum. Methods* **B3**, 671.

Ryan, K.P. and Liddicoat, M.I. (1987). *J. Microsc.* **147**, 337.

Sealock, R.M., Mazzolini, A.P. and Legge, G.J.F. (1983). *Nucl. Instrum. Methods* **218**, 217.

Sealock, R.M., Jamieson, D.N. and Legge, G.J.F. (1987). *Nucl Instrum. Methods* **B29**, 557.

Shroy, R.E., Kraner, H.W. and Jones, K.W. (1978). *Nucl. Instrum. Methods* **157**, 163.

Sie, S.H. and Ryan, C.G. (1986). *Nucl. Instrum. Methods* **B15**, 664

Slodzian, G. (1964). *Ann. Phys.* **9**, 591

Swann, C.P. (1982). *Nucl. Instrum. Methods* **197**, 237.

Swann, C.P. and Fleming, S.J. (1987). *Nucl. Instrum. Methods* **B22**, 407.

Trail, C.C. (1982). *Phys. Rev. Lett. (USA)* **48**, 308.

Traxel, K. and Mandel, A. (1984). *Nucl. Instrum. Methods* **B3**, 594

Van Patter, D.M., Swann, C.P. and Glass, B.P. (1981). *Geochim et Cosmochim Acta* **45**, 229.

Vis, R.D., (1985). 'The proton microprobe: Applications in the biomedical field', CRC Press, New York.

Vis, R.D., Van der Stap, C.C.A.H. and Heymann, D. (1987). *Nucl. Instrum. Methods* **B22**, 380.

Watt, F. and Grime, G.W. eds. (1987). 'Principles and Applications of High-Energy Ion Beams'. Adam Hilger, Bristol.

Watt, F., Grime, G.W., Blower, G.D. and Takacs, J. (1982). *Nucl. Instrum. Methods* **197**, 65.

Watt, F., Grime, G.W., Takacs, J. and Vaux, D.J.T. (1984). *Nucl. Instrum. Methods* **B3**, 599.

Young, E.J., Myers, A.T., Munson, E.L. and Conkling, N.M. (1969). U.S.G.S. Prof. Paper 650-D, D84-D93.

Younger, P.A. and Cookson, J.A. (1979). *Nucl. Instrum. Methods* **158**, 193.

11
Critical Assessment of Analysis Capabilities

J.R. BIRD

ANSTO Lucas Heights Research Laboratories
Menai, Australia

J.S. WILLIAMS

Microelectronics and Materials Technology Centre, RMIT
Melbourne, Australia

ION BEAMS FOR
MATERIALS ANALYSIS
ISBN 0 12 099740 1

In this chapter, the strengths and weaknesses of ion beam analysis techniques are summarised and compared with other techniques used for similar or complementary measurements. Finally, the role and contribution of IBA to various fields of science and technology are summarised.

11.1 GUIDE TO ION BEAM TECHNIQUES

The many types of ion-atom interaction and the Z versus E regimes exploited for sample analysis are illustrated in the Figure in the Preface and Fig. 1.16.

11.1.1 Scattering

Scattering is the most commonly used ion–atom interaction and has a number of advantages for analytical work (Highlight 11.1) — especially for the study of details of structure or depth profiles in samples of known composition or for the study of the location of surface atoms of known mass.

HIGHLIGHT 11.1
SCATTERING

Advantages:

- high yield and hence rapid measurements (< 15 min measuring time and often < 1 min) and good sensitivity for heavy elements in a light element matrix;
- simple equipment requirements, a choice of geometries allows optimisation for specific problems;
- the energies of scattered ions and recoil atoms are dependent on the target mass thus providing mass analysis;
- the results are depth dependent allowing non-destructive depth profiling; and
- the yield is dependent on atomic arrangements with respect to the beam and detector directions and hence is sensitive to both surface and bulk atomic structure.

Limitations in the use of scattering techniques arise from:

- the ambiguity in scattered energy between a high energy ion scattered from a light atom and a low energy ion scattered from a heavy atom;
- the mass resolution decreases as target mass increases being up to 10 amu at mass 200;
- the depth resolution deteriorates with increasing scattering depth as a result of energy straggling and multiple scattering;
- the possibility of multiple scattering leading to ambiguous results, particularly for low energy ions;
- the effects of surface topography, voids, cavities or pinholes on the depth versus energy relationship; and
- beam-induced changes in atomic positions and compositions, particularly in insulators.

---

a. LEIS

At low energies (< 5 keV), LEIS is well suited to the study of surface structure, segregation, reconstruction, adsorbates and topography (Chapter 8.6) through the determination of the mass and location of atomic species present at the surface. When using rare gas ions, neutralisation processes ensure that there is no yield from sub-surface layers and all spectrum features convey information concerning the state of surface atoms. Alkali ions are less prone to neutralisation and can be used to probe to depths of several monolayers.

b. MEIS

At medium energies (10 to 200 keV), ion scattering is mainly used with channeling and blocking to provide atom location to moderate depths (20 nm). High, near-surface depth resolution is possible with electrostatic analysers but quantification is difficult. This is not a frequently used regime but it is a viable application of high energy ion implanters (100 to 400 keV).

c. RBS and Channeling

The advantages of RBS and channeling are summarised in Highlight 11.2. RBS and channeling are *de rigueur* for the development of semiconductor technology and justify the cost of a small dedicated accelerator installation.

HIGHLIGHT 11.2
RBS and CHANNELING

Advantages:

- a high and analytically calculable cross-section allows quantitative analysis without standards;
- the smooth dependence of yield on Z^2 gives excellent sensitivity (< 0.01 monolayer) for heavy elements in a light substrate;
- non-destructive depth capability ideal for profiling trace elements and for studying interfaces, thin films and multi-layer structures with thickness from 10 to 1000 nm;
- the surface peak, when channeling in single crystals, provides information on reconstruction and relaxation of surfaces, epitaxial growth, topography and other surface features;
- sensitivity to the regular arrangement of atoms in crystals, when combined with channeling techniques, allows location of substitutional or interstitial impurities and studies of damage and annealing of crystalline materials and thin film structures; and
- non-Rutherford cross-sections, especially for incident protons or for light element scattering, can be exploited for enhanced sensitivity in light element determination and profiling.

---

d. ERA

ERA is complementary to RBS since light atoms receive the greatest recoil energy and can be observed in the presence of heavy substrate atoms. The advantages of ERA include:

- high count rates in the depth profiling of H or He;
- Heavier isotopes can be profiled — for example up to O using heavy incident ions (Cl, Ar, etc.) at high energies; and
- the absolute Rutherford cross-section applies in some cases, but for the important case of H determination, this is not always true.

ERA is limited to forward geometries ($\theta < 40°$) to permit observation of recoiling atoms. This limits the penetration depth but allows good depth resolution to be achieved.

11.1.2   Sputtering

In sputter profiling, the composition of a freshly exposed surface is studied using positive or negative ions, neutral atoms or deexcitation photons associated with the primary sputtering process or using an independent ion, electron or photon beam to probe the surface. The general features of sputter profiling are summarised in Highlight 11.3.

HIGHLIGHT 11.3
SPUTTER PROFILING

Attractive features of Sputter Profiling include:

- for keV ion energies, sputter rates vary from 1 to 20 atoms per ion and sometimes higher; surface removal rates up to 1 Å s^{-1} can therefore be obtained by controlling the ion current (typically up to the order of 1 μA);
- the depth resolution is controlled by the depth from which interaction products are detected and the effects of ion beam mixing and microroughness (typically > 1 to 10Å);
- reasonable depth resolution can be maintained to depths of tens of micrometres;
- a finely focussed primary beam (< 1 μm) can be used to obtain excellent spatial resolution; and
- a variety of ion, electron or photon excitation methods can be used to obtain depth or spatial distributions of isotopic, elemental or chemical species.

Problems which must be taken into account in the use of sputter profiling include:

i. Primary Beam Effects

- implantation of ions in the primary beam can alter the target composition;
- neutralisation of the primary beam and residual gas contamination can dramatically alter the sputter rate;
- beam induced atomic mixing, diffusion and segregation processes can change the composition of the surface layers. Apparent composition changes often take many minutes to reach equilibrium values representative of the bulk composition;
- sputter-induced roughness increases with sputtering time and compromises the depth resolution;
- preferential sputtering of some atomic species causes surface enrichment of others; and
- the best sensitivity is achieved with a large analysed volume, i.e. with poor spatial and depth resolution.

Some of these effects can be minimised by the use of reactive gas ions (e.g. O) or a residual partial pressure of the reactive gas whereas others require minimisation of residual gas contamination.

ii. Sample Effects

- original surface roughness or topography can reduce the depth resolution;
- polycrystalline or multi-phase samples, or the presence of dislocations also cause the development of topography although this can be minimised by rotating the sample; and
- if the primary beam is at an angle to the sample surface, this causes an asymmetric sputter pit unless the sample is rotated.

iii. Instrumental Effects

- signals generated while the walls of the sputter pit are being irradiated must be rejected by rastering the primary beam and accepting signals only while it is clear of the walls;
- redeposition of sputtered atoms (e.g. from crater walls) can modify the depth profile;
- neutral atoms in the primary beam cause errors in dose determination; and
- non-uniform beam intensity during rastering causes errors in spatial distributions.

a. SIMS

SIMS is the most widely used detection method for sputter profiling and is often carried out with low mass resolution QMA devices having low transmission. Many special instruments have been developed to achieve high mass resolution (> 5000, for molecular ion studies), multiple mass detection and high transmission (to reduce the sample destruction rate). Commercial instruments give good performance in sputter profiling but special-purpose UHV systems are often used for the study of surface physics. The general features of SIMS are listed in Highlight 11.4.

HIGHLIGHT 11.4
SIMS

The major attractions of the SIMS technique are:

- reactive gas ions (e.g. O) have advantages of high and relatively constant yield of positive ions. A Cs beam is preferable for the production of high yields of negative ions of electronegative elements or molecules (e.g. for studying dopants in Si);
- excellent sensitivity (< 1 ppm) and wide dynamic range is available for many elements;
- the ability to select positive or negative secondary ions increases the versatility;
- isotopic sensitivity (depending on mass resolution) which can be exploited for oxygen isotope ratio, lead isotope dating and other such applications;
- molecular ions are observed and can be used to obtain information on chemical bonding of surface atoms;
- suitability for semi-quantitative analysis of many materials; quantitative when using suitable standards; and
- with SIMP, a spatial resolution of 0.3 μm or less is possible but the sensitivity is reduced in proportion to the resolution area.

Problems additional to those listed in Highlight 11.3, include:

- many orders of magnitude variation in sensitivity can occur for different elements in one substrate;
- at least an order of magnitude variation in sensitivity can occur for one element in different substrates;
- the secondary ion yield is sensitive to surface roughness and contamination which may change the relation between ion yield and element concentration;

- in many SIMS analysers, only one mass peak can be measured at a time and cycling of the analyser is necessary to compare different masses; this can make quantitative comparison of profiles difficult and time-consuming; and
- radiation induced changes in the surface layer composition and instrumental effects make quantitative composition measurements very difficult except when a suitable calibration sample is available.

---

b. SNMS

Neutral atoms ejected during sputter profiling can be ionised by laser or electron beams, by excitation in various forms of plasma or discharge cell or by surface ionisation. Neutral atom yields are high, especially for low energy incident ions and SNMS sensitivies are similar for most elements, being unaffected by large changes in secondary ion yields for different elements.

c. SIPS

The detection of photons emitted by sputtered atoms or ions has advantages over the direct use of the ions — particularly for insulating samples since surface charging does not affect photon emission. Photon yields vary over many orders of magnitude being highest for alkalis, alkaline earths and other elements such as Ag (Table 14.28). SIPS sensitivities are of the order of $1~\mu g~g^{-1}$ for elements such as F, P and Zn and $\sim 10~\mu g~g^{-1}$ for many other elements. Photon yields are strongly dependent on chemical binding and SIPS is useful for studying surface bonds.

d. SIMP

The ion microprobe is one of the most sensitive methods for in-situ element or isotope distribution analysis. Spatial resolution down to $0.1~\mu m$ can be achieved and can also be used with imaging techniques (ion microscope) but the sensitivity depends inversely on the beam diameter. Its use for light elements is particularly valuable. Difficulties which must be taken into account include:

- charging of insulating components of a complex sample;
- the sputter removal of some components which prevents repeat investigations; and

- contrast may arise from chemical or topographic rather than isotopic differences.

11.1.3 X-rays

a. PIXE

PIXE is a multi-elemental analysis technique with a number of special advantages (Highlight 11.5). It is widely used in routine analysis of thin samples such as air filters, other environmental and biological samples and for general purpose thick sample analysis. Such analyses can be readily automated and carried out on a commercial basis with cost recovery meeting the cost of an accelerator installation.

HIGHLIGHT 11.5
PIXE

Special features of PIXE include:

- high yield and hence rapid measurements ($<$ 15 min measuring time for many applications);
- simple absolute measurements on thin samples requiring $< 10^{-14}$ g of material for elements with good sensitivity;
- absolute measurements on thick samples using sophisticated computer codes for spectrum fitting and yield calculation;
- multielement analysis (from Na to U) but performance depends on matrix;
- excellent sensitivity ($< 10~\mu g~g^{-1}$) for a range of elements ($20 < Z < 40$ for K X-rays and $60 < Z < 90$ using L X-rays and 2.5 MeV protons);
- location of impurity atoms is possible using channeling and this is especially useful for light species in a heavier matrix;
- in-vacuum or external beam capabilities give great versatility for non-destructive analysis of almost any type of sample; and
- measurements can be made on large areas ($> 1~cm^2$) using beam sweeping or on very small spots (1 μm or less) including spatial distributions using the proton microprobe.

Limitations of the PIXE technique include:

- depth profiling is not readily achievable;
- complex spectra require fitting of multiple X-ray lines, sum and escape peaks and complex background shape;
- interference between L X-rays of high Z elements with K X-rays of low Z elements;

- thickness of layer analysed is limited by attenuation of X-rays from low to medium Z elements;
- light element analysis is very sensitive to self-absorption changes introduced by roughness or topography and to particle size and inhomogeneities for small beam sizes; and
- difficulty in measurement of trace elements in samples with major elements in the region of high sensitivity (e.g. metals) requires the use of special filters but even then the sensitivity is compromised.

b. HIXE

Heavy ion excitation of X-rays has not been widely used but can be expected to become more common for a number of applications:

- to obtain higher yields for low Z-elements which are favoured because of the lower velocity of heavy ions;
- to exploit the shorter ion range for analysis of thin layers (with reduced self-absorption) — a common example being semiconductor studies; and
- to exploit selective excitation and emission of specific X-ray lines or to produce spectrum structure which is sensitive to chemical binding.

c. PIXRF

The problem of high PIXE intensities from major elements in the medium to high Z range can be overcome by using X-rays from a suitable PIXE target as the primary beam for XRF measurements. The primary target is selected to produce X-rays with energies below the absorption edges of the major elements in the sample but high enough to excite trace elements of interest. For example, medium Z elements can be determined in Au alloys. The method has the disadvantages typical of XRF (e.g. high background) and is only attractive as a way to exploit existing PIXE equipment to solve special problems.

11.1.4 Nuclear Reactions

The diversity of nuclear reactions and the performance figures when using these for sample analysis make it difficult to summarise the overall advantages and disadvantages and so different types of reaction are considered separately.

a. Ion–Ion Reactions

Used only for light elements, ion–ion reactions require the same equipment as RBS (usually plus an absorber foil) and a careful choice of incident ion type and energy. They are therefore valuable for supplementing the information obtained from RBS which is usually insensitive to light isotopes. Special features are listed in Highlight 11.6.

HIGHLIGHT 11.6
ION–ION REACTIONS

Special features include:

- preferential sensitivity to light elements which are often difficult to determine by other methods; isotopes of H, C, N, O and Si are the most commonly analysed but suitable reactions are available for most light isotopes (Chapter 4.6 and Table 14.12);
- isotopic specificity which can be exploited in stable isotope tracer techniques;
- non-destructive depth profiling of light isotopes (including H which is of special interest) with a depth resolution of 10 to 100 nm and a maximum depth of $< 1 \ \mu$m;
- absolute determinations provided that cross-sections, solid angle and detector efficiency are calibrated for specific experimental conditions using standard samples;
- use with a microbeam for lateral imaging especially of light isotopes; and
- use with channeling for structure studies such as the location of substitutional or interstitial light element impurities.

Limitations include:

- cross-sections are relatively low but large solid angle and 100% detector efficiency give reasonable count rates; even so the sensitivity is limited except in a few favourable cases; and
- different experimental conditions including specific ion energies are often required for each application with careful calibration and good stability also being important — this generally implies a need for considerable experience in accelerator and equipment operation.

---

b. PIGME

This is the most versatile of the NRA techniques and, because of its multi-element capability, it is an important supplement to other IBA techniques (especially PIXE). The strengths and weaknesses of PIGME are listed in Highlight 11.7.

HIGHLIGHT 11.7
PIGME

The main advantages of PIGME include:

- absolute multi-element analysis but favouring light nuclides; sensitivities $\sim 1\ \mu g\ g^{-1}$ for Li, Be, F and Na;
- used with PIXE, a total major element (except H, C, N, and O) analysis is usually obtained;
- excellent depth resolution (1 to 10 nm) for non-destructive depth profiling of selected light isotopes (H, ^{15}N, ^{19}F, ^{23}Na, ^{27}Al);
- immunity from attenuation problems which plague light element analysis using X-ray techniques; not sensitive to surface topography;
- in-vacuum or external beam applications to samples of any size.

Limitations include:

- need for large volume, expensive detector; and
- suitability only for selected isotopes rather than wide range use.

---

c. Ion–Neutron Reactions

Although seldom used, except by those experienced in measuring neutrons, ion–neutron reactions have useful performance for many light isotopes. Some parameters are particularly attractive — such as the ability to obtain depth profiles to depths of 10 μm or greater. Even so, no important problems have arisen which have demanded this technique instead of being solved in other ways.

d. Ion-induced Neutron Beam Applications

In contrast to the previous topic, this is one with widespread use ranging from low-energy neutron generators for fast neutron activation analysis (FNAA) to 50 or more MeV cyclotrons used for FNAA, neutron therapy, in vivo analysis and other applications. Many of the accelerators used for

IBA are suitable for neutron-beam applications and a number are used for both purposes. However, this possibility is overlooked in many cases.

e. PAA

Ion irradiation of a sample produces radionuclides which are usually different to those from neutron activation. Ions are therefore widely used for both radioisotope production and for analysis. The ion energies required generally exceed 3 to 5 MeV and are often above 10 MeV. PAA is thus restricted to higher energy accelerators than those most commonly used for other IBA techniques. Advantages and disadvantages of PAA are summarised in Highlight 11.8.

HIGHLIGHT 11.8
PAA

The advantages of PAA include:

- sensitivities down to 1 ng g^{-1} can be obtained for a different suite of elements to those for which NAA is most sensitive;
- use with autoradiography gives a spatial resolution as good as with SIMP or SPMP but with much higher sensitivity for favoured elements;
- the available choice of ion type and energy provides versatility for optimising sensitivity and selectivity to suit particular problems;
- radioactivity is only produced in a surface layer, the thickness of which depends on the ion energy and the variation of reaction cross-section with energy; the total activity produced is therefore low;
- incident particle fluxes can be very high which offsets relatively low cross-sections.

Disadvantages include:

- sample heating, when a high beam current is used, may cause changes or damage to the sample;
- counting times of 10 to 100 hours are necessary to achieve high sensitivity and may need to be repeated several times to determine half-lives; this reduces the number of samples that can be analysed when compared with most of the other IBA methods; and
- ionisation energy loss prevents the irradiation of large volumes so that this is less of a bulk analysis technique than NAA.

11.1.5   Choice of Facility

One attraction of ion beams for analytical work is that a suite of techniques can be applied with the same basic equipment. Low energy facilities are the best choice for the study of surface layers and surface phenomena and can be used to study depths of many micrometres by sputter profiling. A simple ion gun and mass analyser can be used in conjunction with an SEM or other UHV chamber for sensitive and semi-quantitative assessment of surface composition. A more sophisticated SIMS analyser (mass filtered incident beam, well designed extraction optics and low resolution mass analyser) is preferable and can be used with electron based techniques, such as AES, LEED, for versatile surface analysis. The SIMP or ion microscope are powerful techniques for studying spatial distributions at the surface, qualitative depth profiles and for obtaining molecular information. Such equipment is limited to semi-quantitative analysis and does not give information on atomic structure below a surface and does not have the range of absolute analytical capabilities provided by high energy IBA. The latter is thus the best choice for pseudo-bulk analysis (little affected by surface contamination or modification) and for the non-destructive study of composition and structure of sub-surface layers. Medium energies are not usually a good choice for either type of analysis although useful information can be obtained if existing facilities (such as an ion implanter) are used. The choice of accelerator for high energy IBA depends on three factors: the maximum ion energy, maximum beam current and choice of ion types. These are discussed in Chapter 2. There is no single accelerator specification which is likely to provide optimum performance in all applications.

11.2 COMPARISON OF TECHNIQUES

The key features of IBA and other instrumental analytical techniques based on various kinds of probing radiation (see the figure in the Preface) are summarised in Table 11.1. The performance figures are illustrative rather than definitive since they may vary from element to element or from matrix to matrix and are very dependent on equipment choices. Even so, Table 11.1 shows the major points of similarity and difference between the various techniques as they affect four aspects of materials analysis:

- composition analysis;
- depth profiling;

- microprobe analysis and imaging; and
- atomic structure studies.

No one technique can adequately cover all these aspects but each has particular strengths which can be assessed from the Table. An extensive comparison is given in Whan *et al.* (1986). The various IBA methods between them give good performance in most aspects but they should be considered as complementary to each other and to non ion beam techniques.

11.2.1 Element or Isotope Analysis

a. Bulk Analysis

Except for PAA, in which the analysed volume may approach 1 cm³, IBA methods do not provide a bulk analysis unless the volume analysed is a true average sample of the original material. This is also the case for other irradiation based methods such as XRF, AES, XPS and EMP. The only instrumental method for true bulk analysis without sampling or homogenisation is NAA but even this is subject to error if heterogeneities are present. If an ion beam has reasonably large diameter (> 5 mm) and energy (> 2 MeV), the volume probed is > 5 mm³ which may be sufficient to average out small scale variations in composition. Otherwise, it is necessary to homogenise the sample before analysis in which case IBA becomes a destructive technique. Electron and photon based techniques can often analyse a greater volume provided that attenuation of emerging photons is not significant.

The speed, and hence cost, of analysis is often important and optical spectroscopy with new forms of excitation (e.g. ICP) can give information on 20 elements in as many seconds. However, the preliminary dissolution of solid samples is much more time consuming. The collection of PIXE data for as many elements can also be done in less than a minute if major and minor elements are of interest and, in this case, data processing may well take longer than the measurement. Likewise, MS can produce a record of a mass spectrum more rapidly than it can be analysed.

Detection limits define a number of overlapping analytical regimes (Table 11.1). Techniques such as AES, EMP, XRF and light mass RBS provide major and minor element analysis down to ~ 1 mg g⁻¹. Other techniques with lower detection limits can usually be used for major/ minor element analysis if necessary. For example, XRF or PIXE are often used for major cation analysis but they need to be supplemented,

TABLE 11.1 Comparison of Performance of Analytical Techniques

Method	Characteristics					Composition			
	Des.[1]	Volume[2]	Sp.[3]	, Vacuum[4]	Prep.[5]	Species[6]	, Meas.[7]	Multi[8]	Sensitivity[9]
Ion Beam Techniques									
AMS	D	B	Me	HV	SPR	I,Se	Q,St	S	UT(1E-3)
ERA	-	NS	F	HV	-	I;(<M$_1$)	Q,St	F	T,MA(1)
FIM	D	S	F	HV	SPR	-	-	-	MA
LEIS	-	S	Me	UHV	SPR	I;(A>M$_1$)	Q,St	F	T(0.1-1ML)
MEIS	-	S,NS	Me	UHV,HV	SPR	I;(A>M$_1$)	Q,A	F	MA,T(1)
NRA-II	-	NS	Me	HV	-	I,Se	Q,St	F	Ma,T(1)
-IG	-	NS	Me	HV,N	-	I;(Z<16)	Q,A	M	MA,T(1)
-IN	-	NS	Me	HV	-	I,Se	Q,St	F	MA,T(1)
PAA	-	NS,B	Sl	HV	-	I,Se	Q,A	M	T(1E-3)
PIXE	-	NS,B	Me	HV,N	-	E;(Z>12)	Q,A	M	T(0.1)
RBS	-	NS,S	F	HV	SPR^c	I;(A>M$_1$)	Q,A	F	MA,T(1)
SIMS	D^{sp}	S,NS^{sp}	Me	UHV	-	I,All,Cb	Q,St	S	UT(1E-3)
SIPS	D^{sp}	S,NS^{sp}	Me	UHV	-	E,Se	Q,St	M	T(1)
SNMS	D^{sp}	S,NS^{sp}	Me	UHV	-	I,All	Q,St	S	UT(1E-3)
Other Techniques									
AAS	D	B	F	N	SPR	E,MostZ	Q,St	S	UT(1E-3)
AES	D^{sp}	S,NS^{sp}	Me	UHV	-	E;(Z>12),Cb	Q,St	S	MA(1E3)
EELS	D^t	NS	Sl	HV	SPR	I,Se	Q,St	F	MA(1E4)
EIXE,SEM	-	NS	F	HV	-	E;(Z>12)	Q,St	M	T(1E2)
EPR,ESR	-	B	Sl	N	-	-	-	-	-
ESCA	D^{sp}	S,NS^{sp}	Sl	UHV	SPR	E,M,Se,Cb	Ql	S	MA(1E4)
EXAFS	-	NS,B	Sl	N	-	E,M,Se,Cb	Ql	S´	T(1E2)
FTIR	D	NS,B	F	N	-	Mo,Se,Cb	Ql	S	MA(1E3)
LEED	-	S	Sl	UHV	SPR	-	-	-	-
MS	D	B	F	HV	SPR	I,All	Q,St	M	UT(1E-3)
NAA	-	B	Sl	N	-	I,Se	Q,A	M	UT(1E-3)
ND	-	B	Sl	N	-	Mo,Se	-	-	-
NMR	D	B	Sl	N	SPR	Mo,Se,Cb	Ql	S	MA(1E5)
OES-ICP	D	B	F	N	SPR	E,Se,Cb	Q,St	M	UT(1E-3)
RIS	D^{sp}	S,NS^{sp}	F	N	-	I,SE	Q,St	F	UT(1E-2)
RS	-	NS,B		N	-	Mo,Se	-	-	-
STM	-	S	Sl	N,UHV	-	Cb	-	-	-
TEM	D^t	S,NS,B	Sl	HV	SPR	Mo,E;(Z>12)	Ql	F	MA(1E3)
XRD	-	NS,B	Sl	N	-	Mo,Se	Ql	F	MA(1E3)
XRF	-	NS,B	Me	N	-	E;(Z>12)	Q,St	M	T(10)

[1]Description: D=Destructive;
[2]Volume Analysed: S=Surface; NS=Near Surface; B=Bulk
[3]Speed of Analysis: Sl=Slow(> 1 h); Me=Medium(< 1 h); F=Fast(few minutes)
[4]Vacuum used: UHV=Ultra-high (<10^{-8}Torr); HV=High (10^{-5} to 10^{-7} Torr); NV=Non-vacuum
[5]Preparation: SPR=Sample Preparation Required;
[6]Species Analysed: I=Isotope;E=Element; Mo=Molecule; Cb=Chemical bonding; Se=Selected cases only
[7]Measurement: Q=Quantitative; A=Absolute; Ql=Qualitative; St=with standards;   (Z or A range);
[8]Multi: S=Single; F=Few; M=Multi-isotope or element
[9]Sensitivity: MA=Major (>0.1%); T=Trace (ppm); UT=Ultra-trace (<0.1 ppm); Best detection limit (μg/g or
^{sp} = when used with sputtering; ^{t} = with thin samples only (< 0.2 μm).   ML=Monolayers);

TABLE 11.1 (*cont.*)

Method	Depth				Microprobe		Structure	Comments
	Meth[10]	Cal.[11]	Res.[12]	Limit[13]	Method[14]	Res.[15]	Information[16]	
Ion Beam Techniques								
AMS	-	-	-	-	-	-	-	Radioisotopes
ERA	ES	A	30	0.2	-	-	-	Forward angles
FIM	-	-	-	-	I	1E-5	PD,FA,LA	High m.p. samples
LEIS	ES	R	0.1	0.01	-	-	LA,FA	Single crystals
MEIS	ES	A,R	2	0.01	-	-	PD,LA,FA,ED^c	
NRA-II	ES	A	10	10	Sc	0.5	FA^c	Selected isotopes
-IG	YC	A	1	10	Sc	0.5	-	Light isotopes
-IN	EY	A	10	50	-	-	-	" "
PAA	-	-	-	R	-	-	"	"
PIXE	-	-	-	-	Sc	0.3	FA^c	Multielement
RBS	ES	A	2	3	Sc	0.3	PD,ED,LA,FA^c	Heavy elements
SIMS	SP	R	0.5	>100	Sc,I	0.1	-	Profiling
SIPS	SP	R	0.5	>100	-	-	-	-
SNMS	SP	R	0.5	>100	-	-	-	-
Other Techniques								
AAS	-	-	-	-	-	-	-	Single element
AES	SP	R	2	>100	Sc	0.1	-	Multielement
EELS	-	-	-	-	Sc	1	-	Light elements
EIXE,SEM	-	-	-	-	Sc,I	5E-3	-	Microstructure
EPR,ESR	-	-	-	-	-	-	PD,FA	Bulk defects
ESCA	SP	R	2	>100	-	-	PD,FA	Surface structure
EXAFS	-	-	-	-	-	-	PD,FA	Bulk bonding
FTIR	-	-	-	-	-	-	PD	Light Isotopes
LEED	-	-	-	-	-	-	LA,FA	Modelling
MS	-	-	-	-	-	-	-	ICP,SS,GC
NAA	-	-	-	-	-	-	-	Bulk analysis
ND	-	-	-	-	-	-	PI,LA	Bulk structure
NMR	-	-	-	-	-	-	PD	Bulk defects
OES-ICP	-	-	-	-	-	-	PI	" "
RIS	SP	R	-	>1	-	-	-	Developmental
RS	-	-	-	-	-	-	LA,PI	
STM	-	-	-	-	SP,I	2E-6	LA,PD,FA	Topography
TEM	-	-	-	-	Sc,I	1E-5	PI,LA,PD,ED	Crystal structure
XRD	-	-	-	-	I	-	LA,ED	Crystal structure
XRF	-	-	-	-	-	-	-	Multielement

[10]Method: SP=Sputter Profile; ES=Energy Spectrum; YC=Yield Curve; EY=ES & YC
[11]Calibration: A=Absolute; R=Relative;
[12]Resolution (nm);
[13]Depth Limit (μm);
[14]Method: Sc=Scan; I=Image; R=Radiograph; SP=Spot;
[15]Lateral Resolution (μm);
[16]Information: PI=Phase identification; PD=Point defect; ED=Extended defect; FA=Foreign Atom location; LA=Lattice arrangement; ICP=Inductively Coupled Plasma; SS=Spark Source; GC=Gas Chromatograph

^{c} = when used with channeling;

for example by PIGME, MS or AAS, for the determination of light elements. Geochemistry and related investigations require that a sum of oxides (including H_2O) should be close to 100% for a satisfactory analysis. IBA comes close to meeting this requirement although H, C, N, and O are more difficult to determine than other elements.

b. Trace Analysis

Trace element analysis (typically 10^{-7} to 10^{-3} g g^{-1}) is the major application of SIMS, SIPS, NAA, PAA, PIGME, MS, PIXE, XRF, etc. — in fact most irradiation based analytical methods. Each method is usually used for the determination of specific elements or suites of elements for which their sensitivity is most favourable (Table 11.1). For example, SIMS and SIPS are not suitable for noble gases and a number of other elements that have low ion yields (Fig. 14.17); NAA is excellent for rare earths and other elements that have high cross-sections; PIGME is best for light isotopes; PIXE and XRF are best for medium to heavy elements; MS can be used for all isotopes if molecular interference permits. The use of various ion sources in MS gives great versatility, for example LAMMA is excellent for solid samples and MS-ICP is excellent for samples in solution.

c. Ultra-trace Analysis

NAA, LAMMA, as well as new methods such as AMS and RIS, extend the limits to 1 pg g^{-1} or less. Such ultra-trace analysis is important in fields such as semiconductor development, environmental tracing and health/pollution studies. Ion beam methods are proving valuable in this work and, as usual, they provide capabilities which are complementary to other available techniques. Laser excitation (RIS) has been applied to counting single atoms of noble gases such as ^{81}Kr by repeatedly exciting and observing the deexcitation of the one atom (Hurst and Morgan, 1986).

d. Absolute Sensitivity

A separate aspect of sensitivity is the analysis of micro-volumes such as minute quantities of pollutants trapped on air filters. PIXE, MS and SXRF are able to detect the order of 10^{-12}g or less of material. Multi-element analysis is possible with MS and PIXE whereas 'atom counting' techniques have so far been developed for individual isotopes. This is also true of AMS (Gove *et al.*, 1987). PIXE has the attraction that absolute measurements can be made in this regime and this has provided the justification for dedicated accelerator installations for PIXE analysis.

e. Precision and Accuracy

A most important aspect of composition analysis is the reliability (precision accuracy) of each technique. The usual approach to assessing these factors is the 'round-robin'. An example is given in Table 10.6 of a comparison of PIXE with four non-ion beam techniques for mineral analysis. The results show fair agreement but each technique is better for some elements than for others. Another intercomparison of XRF, PIXE, OES and PAA for analysis of solutions, aerosols and powdered rock samples showed satisfactory performance by PIXE and XRF, with XRF having a little lower standard deviations (Cahill, 1986). The optical methods also agreed for solutions but showed wide variations for solid samples — indicating problems with the necessary dissolution step. More recently, a comparison of different spectrum fitting methods in PIXE gave < 4% standard deviations from the overall means for biological, environmental and geological specimens (Campbell *et al.*, 1986). Using one fitting technique, less than 2% variation can be expected. An absolute precision of 20% for many elements with concentrations over 5 decades has been reported for measurements on a variety of reference materials (Fig. 5.17). The sensitivity and precision of RBS analysis has been demonstrated by inter-laboratory comparisons of implanted ion concentrations (Davies *et al.*, 1986). A 4% difference between samples having the same nominal implantation dose was readily observed in measurements having 1% precision indicating that RBS gives good absolute precision when appropriate attention is given to methods and calibration. With suitable attention to detail, high energy ion beam methods give very good precision and have the attraction that they usually also have good absolute accuracy.

f. Surface and Near-surface Analysis

The thickness of sample which can be analysed with various techniques is illustrated in Fig. 11.1. The depth scale is shown at the left and the regions of importance in adsorption, contamination, oxidation, corrosion and other surface modification processes are shown schematically beside the depth scale. The figure shows the emphasis on the surface layer of atoms in low energy ion beam techniques. Sputtering can be used to allow these to be applied as a function of sputtering depth. On the other hand, the high energy ion beam techniques can probe depths to 1 μm or

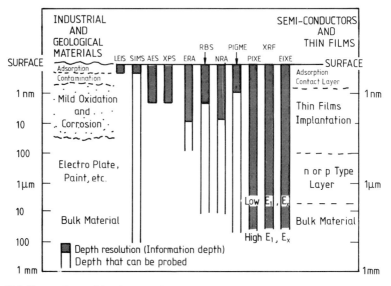

Fig. 11.1 Comparison of depth resolution (shaded areas) and maximum depth for various analytical techniques in comparison with typical thicknesses in materials of interest.

greater non-destructively. The depth limit is shown for 2 to 3 MeV incident ions but, for some types of measurement, the limit can be extended to 100 μm or even more by the use of higher energy beams. In the case of EIXE, XRF and PIXE analyses of low-Z elements, the depth sampled is limited to < 10 μm by X-ray attenuation and much less for high-Z matrix material.

11.2.2 Depth Profiling

The important parameters in depth profiling are the resolution and the depth limit, the latter being similar to the thickness of sample which can be analysed (Fig. 11.1). For composition and trace element depth profiling of the near-surface, IBA techniques have unique capabilities. There is a clear distinction between surface analysis techniques (which can only be applied to depth profiling using sputter profiling or other surface removal techniques) and the non-destructive depth profiling which is essentially unique to high energy IBA and derives from ion energy loss during penetration. A third category of techniques which have no intrinsic depth resolution (e.g. AAS, OES) are not included in Fig. 11.1. The intrinsic depth resolution (observed within a short distance of the surface) is different for each technique (Table 11.1). Low energy sputter profiling has a depth resolution typical of the escape depth of the radiation observed (0.1 to 1 nm). With high energy IBA, the depth resolution is ~ 1 to 10 nm at the surface and deteriorates to larger values at large depths (> 1 μm) because of the dominating influence of ion energy straggling and multiple scattering. The use of glancing angles with such techniques can improve the near-surface depth resolution by a sine factor with an optimum usually being reached for angles from 5 to 10° (see Chapter 12). Non-destructive measurements with high energy IBA are of great value for following changes in profiles produced by heat, radiation damage, chemical interactions, poisoning and many other time dependent processes. These can also be studied by sputter profiling with greater speed and sensitivity but not as quantitatively or non-destructively. Profile measurements are also used to a limited extent for artefact dating.

11.2.3 Microprobe Analysis and Imaging

The lateral resolution available with various electron and ion techniques range from atomic dimensions (1 Å) to fractions of a mm at which optical imaging and microanalysis are adequate (Figure 11.2). This covers the range from atomic and molecular structure to crystal phases in metals and minerals and the cellular structure of biological materials. It is therefore a range of great interest in modern science and technology which have been dependent for many years on the electron microscope and microprobe. The role of other instruments depends on their having superior performance in some aspect of importance to specific problems. For example, the SIMP is very useful because of its high sensitivity for many elements, including light elements, while the SPMP brings to bear all the advantages of high energy IBA — both having a spatial resolution of 1 μm or better (Table 10.2) (Watt and Grime, 1987). MS and LAMMA also provide micron size spot analyses but are not exploited in microscopy or scanning microprobes. Developments under way should bring the unique performance of AMS and possibly RIS to the IMP by the use of the microprobe ion source in a tandem accelerator. More specialised examples include the Field Ion Microscope (FIM) which gives better than atomic resolution in the study of high melting point materials (Tsong, 1983) and the Scanning Tunneling Microscope (STM) which can map topographic and electron distributions at the surface of a sample, providing information on atomic arrangements, defects and chemical bonding at the surface.

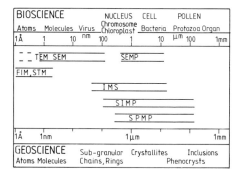

Fig. 11.2 Comparison of lateral resolution of various microprobes with radii of typical features of interest in bioscience and geoscience.

11.2.4 Structure Analysis

Neutron and X-ray diffraction techniques generally provide information on long-range order in bulk materials whereas low energy electron techniques such as LEED sense surface structure or nearest-neighbour bond lengths. Such information is complementary to that obtained with ion beam techniques which provide additional information on localised and depth dependence of structure. LEIS can be used, exploiting the shadowing and blocking effects, to determine the location of surface atoms at concentrations down to 10^{-3} parts per monolayer and to investigate the effects of topography (steps, ledges and missing atoms), adsorption and other surface changes (Chapter 8). Such measurements often benefit from a comparison with LEED and RHEED results. LEIS with alkali ions and the surface peak observed in RBS-C at higher energies are also useful for extending atom location studies to the first few monolayers (Chapter 9). Channeling with RBS, NRA or PIXE gives structure information as a function of depth (Chapter 6) (Feldman *et al.*, 1982). More precise atom location is possible because of the much smaller shadow cone radius ($\sim$ 5 pm) compared to that for LEIS ($\sim$ 500 pm). Sensitivities down to 10^{-2} monolayers can be achieved in favourable cases in atom location experiments. SEM and TEM are usually needed to aid full interpretation of the ion beam results.

11.3 FIELDS OF APPLICATION

The following sections summarise the significance of the contributions of IBA to various areas of science and technology but without details of performance which have been described in earlier sections. Also listed are the section numbers where examples can be found in earlier Chapters.

11.3.1 Materials Science

Adhesion, Adsorption, Annealing, Atom Location, Crystal Structure, Defects, Diffusion
Displaced Atoms, Epitaxy, Interfaces
Ion Beam Mixing, Ion Implantation, Ion–Solid Interactions, Phase Transformations, Precipitation, Radiation Damage, Relaxation Reconstruction, Segregation, Solid State Reactions,
Strain, Sub-surface Layers,
Surface Physics, Thin Film Physics,

a. Surface Physics

Examples: 1.4.1, 1.4.2, 6.3.5, 7.5.4, 7.5.5, 8.2.5, 8.2.6, 8.3.3, 8.3.4, 8.4.2, 8.4.3, 8.5.2, 8.5.3, 8.6, 9.3, 9.4, 9.5.

The surface of a solid can be quite different in composition and structure to the underlying material. Processes such as adsorption, contamination, oxidation and leaching may unintentionally modify the surface composition while thin film deposition and ion irradiation are used to intentionally do so. Changes also occur during crystal growth and surface removal (e.g. by sputtering). The arrangements of surface atoms is different from that in the bulk and is also affected by the presence of steps, missing atoms or foreign atoms. The study of these and other surface physics effects requires techniques which probe one, or at most a few monolayers. Low energy IBA techniques are very sensitive to surface composition and atomic structure and are widely used in surface physics studies (Datz, 1983). Ion-induced AES and the surface peak in RBS-C can also be used but the signal is usually averaged over several monolayers. ESCA techniques are needed to obtain additional chemical information, for example in studies of chemi- and physi-sorption. IBA gives more direct information, averaged over $\sim 10^{10}$ atoms, than LEED or RHEED for atom location and the study of surface relaxation and reconstruction but the STM and FIM are the most direct methods for achieving atomic resolution for a wide range of both light and heavy elements with sensitivities less than 10^{-3} parts per monolayer. They are far superior to AES used in conjunction with IBA for surface composition analysis and to other IBA techniques for studying surface defects and topography.

b. Subsurface Crystal Structure Information

Examples: 1.4.3, 6.2.2, 6.2.3, 6.3.1, 6.3.3, 6.3.4, 6.3.5, 6.4.2, 6.4.3.

High energy ion channeling has unique capabilities for providing information on several crystallographic features as a function of depth. When combined with RBS, it is the main method for depth profiling of both point defects, particularly those induced by radiation damage, and extended defects such as dislocations, twins and stacking faults. However, detailed defect characterisation is best achieved by electron microscopy and diffraction (or X-ray techniques) which are more sensitive to strain fields around defects than is RBS-C. However, RBS-C has an advantage in high strain situations such as the detection of phase transformations and strained multilayers where unique information can be obtained on phase transition temperatures and layer misorientation. For low strain studies, X-ray techniques have much higher sensitivity. The non-destructive feature of channeling measurements has particular advantages for analysis of sequential annealing studies in defective crystals and for crystal growth measurements. In such cases, it is useful to combine channeling with TEM or XRD on selected samples to give a comprehensive understanding of microstructure and defect characterisation.

Precise lattice location of both displaced host atoms and foreign atoms is also readily obtained using channeling. RBS, NRA and PIXE can be employed to detect both light and heavy atoms, although the latter technique cannot readily give depth information. Such comprehensive atom location possibilities are not matched by other techniques although emission spectroscopies such as Mossbauer and EXAFS can provide either higher sensitivity for specific elements or limited chemical information. Combining ion channeling with a microbeam (Chapter 10) can also enable polycrystals and other small scale structures to be studied.

c. Layered Structures and Interdiffusion

Examples: 3.4.1, 3.4.2, 6.2.2, 6.3.4, 7.5

For the study of thin film solid state reactions, elemental interdiffusion and trace element profiling in multilayer thin films, particularly during annealing, RBS and SIMS are the IBA techniques with major applicability. These techniques are often used in conjunction with microstructural techniques such as TEM and grazing angle XRD to provide complementary compound phase information. SIMP and SPMP can also be used for laterally non-uniform films. The structure of large area single crystal films can be studied using RBS-C and misorientation in strained layers, misfit dislocations and other crystal growth defects can readily be measured. Cross-section TEM analyses can provide complementary structural information in such cases.

d. Ion Beam Modification

Examples: 1.1.3, 3.3.2, 3.4.2, 3.4.3, 3.5.2, 4.2.4, 4.3.3, 4.4.1, 4.5.3, 5.5.3, 5.5.5, 6.2, 6.3, 6.4.2, 6.4.3, 7.5, 8.4.2, 8.4.3, 8.6, 10.4.1

Ion beams (at keV energies) can be used to modify the near-surfaces of solids (Chapter 1). They introduce radiation damage, produce structural changes and new phases by ion implantation and cause atomic mixing of thin films with underlying substrates. The radiation damage and atomic mixing processes are particularly important in low energy ion beam measurements, especially in sputter profiling, and are consequently studied to better understand these analysis techniques. However, ion beam modifications can also produce desirable property changes (e.g. electrical conductivity, corrosion resistance, improved hardness) (Williams, 1986). It is interesting that ion beam induced structural and composition changes themselves are often best studied by ion beam analysis. For example, ion implanted profiles and ion mixing processes are readily analysed by RBS, NRA and SIMS whereas radiation damage and lattice location of implanted impurity atoms are well suited to analysis by RBS-C. As with composition and structural analyses, complementary techniques such as TEM, XRD and AES are routinely used in combination with IBA.

e. Ion–Solid Interactions

The study of ion–solid interactions has been a major interest for several decades (e.g. the International Conference Series on Atomic Collisions in Solids). Both fundamental studies and those stimulated by the needs of modification or analysis applications are involved. For low energy IBA, ion–solid interactions are important to provide sputtering data and to understand atom removal, ionisation and neutralisation mechanisms. High energy interactions are better understood and catalogued but additional data is still needed on ion ranges, stopping powers and interaction cross-sections (particularly for NRA and HIXE).

11.3.2 Materials Technology

Alloys, Catalysis, Ceramics, Coatings, Corrosion, Electrochemistry, Glass, Glassy Carbon, Metals, Metallic Glasses, Magnetic Devices, Optical Devices, Oxidation, Polymers, Semiconductors, Superconductors, Wear,

a. Catalysis

Examples: 8.3, 8.4.2, 8.4.3.

Catalysis is a surface phenomenon so that the surface composition rather than the bulk composition is important to performance. SIMS is particularly suitable for trace element detection having higher sensitivity than EIXE or AES for surface atom analysis. LEIS is excellent for studying the surface position of atoms of key elements, changes in atom locations and adsorbed atoms affecting catalytic activity. Techniques which provide chemical information, such as ESCA, or structural information, such as TEM, LEED and RHEED, are complementary to IBA techniques. IBA also has potential for analysis of catalytic reactions — preferably when combined with electron diffraction and microscopy. LEIS and SIMS are valuable for the development of new catalysts and for diagnosing problems in their performance. Other IBA techniques such as H determination by ERA or NRA and heavy element determination by RBS and PIXE are also useful for general analysis and depth profiling in catalysts.

b. Corrosion, Oxidation, Electrochemistry

Examples: 3.4.2, 4.2.2, 4.2.3, 4.3.2, 4.5.3, 8.4.2, 8.6.

These topics involve surface layer formation for which IBA is ideally suited. The low energy ion techniques such as SIMS and LEIS have very high sensitivity for the identification of surface contaminants at the very early stages of reaction. High energy IBA, on the other hand, is valuable for depth profiling and lateral scanning. O and other light elements are best measured by SIMS, ERA and NRA. The latter technique allows the use of stable tracer isotopes such as ^{15}N or ^{18}O. Heavier matrix elements can be determined by SIMS, LEIS, RBS, PIXE, etc. Layer thickness and composition, reaction kinetics, diffusion coefficients and other parameters can be determined non-destructively by RBS and NRA so that repeated measurements can be made during the life of a working component. However, great care is needed in the interpretation of the results if oxidation or corrosion roughen the surface (see Chapter 3). Sputter profiling with SIMS and AES is also widely applicable for depth profiling but such techniques are destructive. SIMP and sputter profiling can give three-dimensional composition profiles; SPMP also has significant analysis potential but has not been widely used in these fields. Non-IBA techniques such as ESCA, AES and SEM are complementary to IBA but are most useful when combined with some form of depth profiling.

c. Glass, Ceramics, Polymers

Examples: 3.2.2, 3.4.1, 3.5.1, 4.4.1, 5.5.5, 6.3.5, 8.4.2, 8 6

Provided precautions are taken to limit beam-induced heating and charging, much useful information can be obtained from applications of IBA to insulating materials. SIMS is very sensitive to both surface and semi-bulk contaminants when used with sputter profiling. NRA and ERA are valuable for determining light element concentrations and together with PIXE provide a general-purpose analysis. Of particular value, are the depth profiling techniques (SIMS, RBS, NRA, ERA, etc.) for studying near-surface modification (diffusion, leaching, hydration, weathering, etc.) but care is required to avoid changes induced by the probing ion beam. Polymers are extremely sensitive to damage and large area beams and low currents must be used or the beam should be frequently moved to a fresh spot. Channeling in crystalline insulators is useful for atom location and damage measurements.

d. Metals, Alloys

Examples: 3.4.2, 3.4.3, 3.5.2, 4.2.4, 4.3.3, 4.4.1, 4.5.3, 5.5.3, 5.5.5, 6.4.2, 6.4.3, 8.4.2, 8.6, 9.3, 9.4, 9.5.1, 10.4.1.

Many techniques are available for determining major, minor and trace elements in metals (Whan *et al.,* 1986). For the study of crystal structure and damage effects, the SEMP and XRD are the most widely used methods since they have excellent performance for this purpose. However, metals and conductors in general, are also particularly suitable for IBA because high beam currents can be used with minimum problems from sample heating or charging. This favours the use of NRA for light element analysis, depth profiling and spatial scanning — all of which provide information which is difficult to obtain with other techniques. Little background contribution arises from the metal substrate in NRA measurements. RBS, ERA and PIXE are also useful while low energy IBA can be used for very sensitive detection (SIMS) of surface contamination and for the study of surface composition and structure (LEIS). However, problems may arise with SIMS because of variations in sputter rates from grain to grain. Defect and atom location studies in single crystals can be achieved using channeling combined with RBS, NRA and PIXE. Structural information from TEM, XRD is needed to complement IBA measurements. Wear can be measured with high sensitivity, for example by PAA and the detection of activated products in lubricant or coolant. The effects of ion implantation to enhance wear and corrosion resistance can also be readily monitored using IBA. The general area of composition, structure and modification of real metal surfaces (as distinct from the polished surfaces often used in other techniques) and the role of light elements in the study of segregation, precipitation, diffusion, etc. are the main uses for IBA in metallurgy. Microbeam techniques (SIMP, SPMP, SEMP and scanning AES) are all used widely to obtain information on lateral changes in composition in metals.

e. Semiconductors, Microelectronics and Optoelectronics

Examples: 3.2.1, 3.2.2, 3.5.1, 4.2.2, 4.2.5, 5.5.3, 6.2.2, 6.2.3, 6.3.1, 6.3.4, 6.4.3, 9.5.5, 10.4.1.

In these fields, the interest lies in the composition and structure of at most the first few microns and usually the first few hundred nm. In such cases, IBA techniques are ideal, offering unique capabilities. Usually, depth profiling requirements (e.g. known element impurity distributions) dictate that RBS and SIMS rather than PIXE be used. SIMS has the necessary sensitivity of depth profiling impurities at very low concentrations (Williams, 1983) but RBS, being non-destructive, is preferable for repeated measurements of implant profiles. For light element profiling, NRA is also useful and sputter AES is appropriate for determination of major elements in multilayered samples. Structural analysis is also vital and LEIS or RBS-C are ideally suited to determination of surface or sub-surface structure respectively. Damage profiles, resulting from ion implantation which is widely used for device manufacture, and annealing of damage can be readily monitored with RBS, TEM, XRD and electrical measurements. Exact lattice location of impurities and defects can be achieved with channeling. EXAFS, EPR and ESR techniques are also used in this work. Microbeams have had limited application in this field but the situation may change as circuit elements become smaller. In summary, IBA is a vital tool in the development of this area of high technology.

f. Thin Films and Coating Technology

Examples: 3.2.3, 3.3.4, 4.2.2, 4.4.2, 6.2.2, 6.3.6, 6.4.2, 7.6, 8.4.2, 9.5.2, 9.5.3.

In these fields, the discussion of the previous section is relevant. Composition and contaminants in multilayers and surface films can be readily monitored with IBA. Crystalline films can be examined by channeling to determine strain and defect levels while surface structure and composition is readily accessed by SIMS and LEIS. Topography is often important in such studies so that STM, SEM, TEM and grazing angle XRD plus metallographic techniques have an important role for microstructure studies.

11.3.3 Energy Technology

Coal, Fission Energy, Fossil Energy, Fusion Energy,
Oil Shale, Solar Absorbers, Solar Cells, Solar Energy.

Much of the support for development of IBA methods, including microprobes, originated from energy technology requirements. High energy IBA is used for many studies of fission reactor materials including the effects of oxidation, welding, leaching and diffusion in fuel, cladding, moderator or control systems. It is also used for analogous studies of fusion reactor materials and an understanding of low energy ion interactions is vital for this work. Many analytical problems arising in fossil fuel technology are studied with IBA and contributions have also been made to the development of solar cells and other solar energy materials (see Sections 11.3.2 e and f).

11.3.4 Earth Science

Geochemistry, Meteorites, Mineral Exploration, Mineralogy,
Lunar Samples.

Examples: 4.3.1, 4.3.2, 4.4.1, 5.1, 5.4.2, 5.4.3, 5.4.4, 5.5.1, 10.4.1.

A wide range of analytical techniques is used in these fields but some of the special capabilities of IBA can add to the information available. At least one accelerator is being used chiefly for high energy IBA in mineral studies. For example, XRF and the SEMP are widely used for major, minor and some trace element studies but PIXE and the SPMP give greater sensitivity for trace elements while NRA and ERA are useful for H and other light element studies including isotope ratios. For PIXE, the detection limits are very dependent on the major element composition, being very good for coal and very element dependent in a matrix containing medium atomic weight elements. Heavy-ion PIXE may also prove useful for preferential observation of specific heavy elements in a lighter substrate although a wavelength dispersive detector may be needed to resolve the X-rays of interest. Of special interest is the isotope ratio capability of SIMP (and IMS) provided that a high resolution mass analyser is used to identify molecular interference such as that from hydride molecules. Isotope ratios are useful for dating terrestrial and extraterrestrial materials and processes (Williams, 1983). NAA has unique performance for studying rare earth elements and this has been exploited for studying geochemical processes. Other elements can also be determined by activation analysis including PAA which is useful for H, C, N, O, Pb, etc. which cannot be determined by NAA.

An important problem in the study of element distributions is the varying grain size in most geological samples. Changes in X-ray intensity may arise from changes in depth distribution as much as from lateral variations. The shorter range of ions and much smaller scattering probability, compared to electrons, favours IBA in such cases but care is still necessary. Once available, the versatility of IBA methods can also satisfy many of the more routine requirements for analysis. By automation, large numbers of samples can be analysed for most elements (except elements such as the rare earths at trace levels). Sample variability raises many problems requiring careful attention to sampling techniques or homogenisation before analysis. A combination of PIXE with PIGME, NRA or ERA often allows a total analysis to be carried out with results summing to 100%.

11.3.5 Environment

Atmosphere, Aerosols, Contamination, Environment, Pollution.

The determination of absolute concentrations of trace elements has become an important task for PIXE in these fields with dedicated accelerators used for measurements on large numbers of samples. The assessment of correlations between trace element concentrations and particle size is very important but only very small quantities of materials are usually available for this purpose. PIXE is an ideal analytical technique for such studies because of its very high absolute sensitivity and its non-destructive, multielement capability. Separate or simultaneous PIGME measurements provide information on light elements (e.g. F) and RBS is useful for identifying the major and minor element composition of individual particles (using the microprobe). Although techniques such as AAS or OES are much used for monitoring specific elements in occupational health or pollution studies, IBA methods give much more information which may reveal the presence of unexpected elements and correlations. When combined with total mass, particle size and reflectivity measurements, a comprehensive understanding can be obtained of the production and distribution of pollutants. The use of several IBA techniques to determine all significant elements in air filters, fluids, tissues, etc. gives information beyond the capabilities of other methods with the added advantages that the measurements are non-destructive and can be repeated if necessary.

11.3.6 Medicine and Life Sciences

Agriculture, Biochemistry, Biology, Botany, Cytology
Medicine, Occupational Health.

Examples: 4.2.3, 4.3.2, 4.3.3, 5.2.2, 5.3.7, 5.4.4, 5.5.1, 10.3.1, 10.3, 10.3.3, 10.4.2, 10.4.3.

The dominant interest in the life sciences is in the many techniques for organic analysis and specific biological testing. IBA is of little assistance in such studies although some use has been made of molecular spectrometry in SIMS. However, trace elements are important through both beneficial and deleterious effects on the function of cell and organism. At high levels most elements are toxic, whereas at low levels (< 10 μg g^{-1}), metals, etc. may be toxic or essential depending on their concentration. Techniques such as AAS and OES have the necessary sensitivity and are ideally suited for use with small quantities of liquids (IAEA, 1980). Techniques such as NAA are useful for some elements but require larger samples. Notwithstanding problems in sample preparation (especially for use in vacuum) and the possibility of beam-induced damage, PIXE and the SPMP are being used in an increasing number of biomedical studies (Van Rinsvelt *et al.*, 1987; Watt and Grime, 1987). The organic matrix makes no contribution to X-ray spectra and the very high absolute sensitivity, together with multi-element capability, make PIXE a powerful technique. PIGME can also be used for light elements such as F and can even be used for *in vivo* studies in teeth. Multi-element measurements on non-invasive samples, such as hair, urine, blood, are particularly useful in occupational health surveys. The SPMP is well suited for element mapping in small biological and botanical specimens, including thin sections, where the resolution of 1 to 10 μm is useful. An improvement in resolution to 0.1 to 1 μm should make possible many more studies of trace element distribution within single cells. The SIMP has had some use, for example, exploiting the high sensitivity and mass resolution for following stable isotope tracers.

Many pilot studies have been carried out, demonstrating that large data bases can be quickly built up (element composition and spatial distributions) provided that great care is taken with sample preparation, handling and irradiation. Many interesting results have been obtained on normal levels of trace elements in plants, animals and humans, on changes arising during occupational exposure to hazardous materials and on anomalous concentrations occurring in conjunction with symptoms of disorder. The technology is thus well established for IBA to make vital contributions to the life sciences. However, this will require long-term projects with close cooperation between physicists, botanists, biologists, physicians, and others, in order to establish conclusions which are scientifically and statistically fully justified.

11.3.7 Archaeology, Art and Museum Science

Archaeology, Art, Authentication, Bone, Characterisation, Conservation, Dating, Glass, Glazes, Jewelry, Metals, Numismatics, Obsidian, Paper, Pigments, Porcelain, Pottery,

Examples: 4.3.2, 4.3.3, 10.4.1.

Most methods of analysis and imaging find application in these topics (Mairinger and Schreiner, 1982). They have also been a growth area for IBA in recent years (Bird *et al.*, 1983; Lahanier, *et al.*, 1986) including an imaginative project to instal a 2 MV tandem accelerator as part of the 'Grand Louvre' museum laboratory in Paris. Major programs using IBA (chiefly PIXE, PIGME and PAA) have been carried out on materials such as ceramics, obsidian, paper and precious metals and exploratory work has involved a wide range of materials and most of the ion beam techniques. In archaeology, the speed of multi-element analyses is the main attraction for provenancing large collections of artefacts – provided that inhomogeneities and surface modification are taken into account. Hydration of obsidian is a useful dating technique which can be aided by NRA or ERA profiling of H; some work has also been done on fluorine depth profiles in bone and teeth with a view to dating these materials.

Museum science is object oriented and non-destructive major and trace element analysis, including microbeam scanning, is the major attraction of IBA. External beam measurements are particularly useful for studying large and valuable objects. Contributions have been made to authentication and studies of ancient technology and this type of work can be expected to grow. The surface and depth sensitivity of IBA techniques is ideally suited to the study of surface layers (pigments, glazes, coatings, ink, etc.) as well as of aging and deterioration which is a largely unexplored IBA application with important potential.

REFERENCES

Bird, J.R., Duerden, P. and Wilson, D.J. (1983). *Nucl. Sci. Applications* **1**, 357.

Cahill, T.A. (1986). *In* Whan *et al.,* (1986), 102

Campbell, J.L., Maenhaut, W., Bombelka, E., Clayton, E., Malmqvist, K., Maxwell, J.A., Pallon, J. and Vandenhaute, J. (1986). *Nucl. Instrum. Methods* **B14**, 204.

Datz, S. (ed.) (1983). 'Applied Atomic Collision Physics', Vol. 4. Condensed Matter, Academic Press, Orlando.

Davies, J.A., Jackman, T.E., Eschbach, H.L., Dobma, W., Watjen, U., and Chivers. D. (1986). *Nucl. Instrum. Methods* **B15**, 238.

Feldman, L.C., Mayer, J.W. and Picraux, S.T. (1982). 'Materials Analysis by Ion Channeling', Academic Press, New York.

Gove, H.E., Litherland, A.E. and Elmore, D. (eds). (1987). *Nucl. Instrum. Methods* **B29**.

Hurst, G.S. and Morgan, C.G. (eds) (1986). 'Resonance Ionisation Spectroscopy 1986'. *Inst. Phys. Conf. Ser.* **84**, Inst. Phys., Bristol.

IAEA (1980). 'Elemental Analysis of Biological Materials', International Atomic Agency, Vienna.

Lahanier, C., Amsel, G., Heitz, C., Menu, M. and Andersen, H.H. (eds) (1986). *Nucl. Instrum. Methods* **B14**.

Mairinger, F. and Schreiner, M. (1982). 'Science and Technology in the Service of Conservation', Int. Inst. for Conservation, London, 5.

Tsong, T.T. (1983). *In* Datz (1983), 380.

Van Rinsvelt, H., Bauman, S., Nelson, J.W. and Winchester, J.W. (1987). *Nucl. Instrum. Methods* **B22**.

Watt, F. and Grime, G.W. (eds) (1987). 'Principles and Applications of High-Energy Ion Microbeams', Adam Hilger, Bristol.

Whan, R.E., Mills, K., Davis, J.R., Destefani, J.D., Dieterich, D.A., Crankovic, G.M. and Frissell, H.J. (1986). 'Metals Handbook, 9th Ed., Vol 10, Materials Characterisation', American Soc. for Metals, Ohio.

Williams, J.S. (1986). *Rep. Progr. Phys.* **49**, 491.

Williams, P. (1983). *In* Datz (1983), 327.

Section 2
General Analytical Methods

12
General Methods

J.R. BIRD

ANSTO Lucas Heights Research Laboratories
Menai, Australia

J.S. WILLIAMS

Microelectronics and Materials Technology Centre, RMIT
Melbourne, Australia

ION BEAMS FOR
MATERIALS ANALYSIS
ISBN 0 12 099740 1

12.1 DETERMINATION OF ELEMENT OR ISOTOPE CONCENTRATIONS

As explained in Chapter 1.1.1g, the yield of interaction products is directly dependent on the cross-section (which defines the probability of a specific type of interaction) and the density of atoms of the kind to be determined, Equation (1.20). This simple situation only applies for extremely thin samples (e.g. a monolayer). The possibility of energy loss by the incident ions before the interaction of interest occurs and interactions by the product radiation must be taken into account in most practical situations. Although the nature of the significant interactions is different for each analytical technique discussed in Chapters 3 to 9, the following treatment provides equations and methods which are common to most aspects of ion beam analysis.

12.1.1 Thin Sample Analysis

If there is no interaction by the product radiation, the number of counts (Y_3) observed in a particle or photon detector having efficiency (e) and subtending a solid angle $d\Omega$ during an irradiation of a sample by N_1 incident ions is given by:

$$Y_3 = N_1 \, e \, d\Omega \, d\sigma \, dx \qquad (12.1)$$

where $d\sigma$ is the differential cross-section (barn atom^{-1} sr^{-1}) for a specific interaction (the notation $d\sigma$ is used instead of the full notation $d\sigma/d\Omega$); and dx is the ion path length in the sample (atoms cm^{-2}). If the path length is expressed as dd g cm^{-2} then we have:

$$dd = m_2 \, dx \qquad (12.2)$$

where m_2 is the mass of an atom of type 2 (g atom^{-1}) ($m_2 = M_2/N_0$), M_2 is the mass number, N_0 is Avogadro's number (6.028 10^{23} atoms per gram atom). Likewise, if the path length is expressed as dt cm we have:

$$dt = m_2 \, dx/\rho_2 \qquad (12.3)$$

where ρ_2 is the sample density (g cm^{-3}).

a. Mixture

The yield of observed counts (Y_i) from interacting atoms of type 2, which are present with an atom fraction f_i and weight fraction c_i in a compound or mixture, is:

$$Y_i = f_i N_1 e \, d\Omega \, d\sigma \, dx \qquad (12.4)$$

$$= c_i N_1 e \, d\Omega \, d\sigma \, dd/m_i \qquad (12.5)$$

Inverting these equations gives expressions for the composition of a thin sample.

If a comparison is made with results (Y_s) from a standard sample of known composition using identical beam exposure and detection efficiency factors, then:

$$f_i = f_s Y_i \, dx_s / Y_s \, dx \qquad (12.6)$$

$$c_i = c_s Y_i \, dd_s / Y_s \, dd \qquad (12.7)$$

These equations apply only to simple two-body interactions. At low energies, multiple interactions are common and require the use of a product of cross-section terms, Equation (8.7).

12.1.2 Thick Samples — Energy Loss

The most frequent interaction at high energies is ionisation energy loss and it becomes necessary to consider the energy dependence of the cross-section for major interactions. This may involve a simple analytical relation, as is the case for RBS (Equation 3.4), or empirical fits as for X-ray production (Fig. 5.7). Many different energy dependences are possible for nuclear reactions (Chapter 4). A special case is a narrow resonance in the cross-section for which the resonance area (integrated cross-section) can be used instead of the differential cross-section in the above equations (see also Equation 4.5). For so-called thick samples (thickness greater than the range of the incident ion), and provided that the product radiation is not significantly attenuated by travelling through this still quite thin layer, the observed yield obeys the following relations:

$$Y_3 = (N_1 e \, d\Omega f_i) \int_o^R d\sigma \, dx$$

$$= (N_1 e \, d\Omega f_i) \int_{E_1}^o [d\sigma / \varepsilon_m(E)] dE \qquad (12.8)$$

$$Y_3 = (N_1 \, e \, d\Omega \, c_i/m_i) \int_o^R d\sigma \, dd$$

$$= (N_1 \, e \, d\Omega \, c_i/m_i) \int_{E_1}^o [d\sigma/S_m(E)] dE \qquad (12.9)$$

where $\varepsilon_m(E)$ is the stopping cross-section and $S_m(E)$ is the stopping power of the sample with $\varepsilon_m(E) = \sum_i f_i \, \varepsilon_i$ and $S_m(E) = \sum_i c_i \, S_i(E)$ for a mixture or compound.

The ratio of yields from standard and unknown samples now involves the ratio of integrals. At first sight, this is a major complication for sample analysis — particularly since accurate data on the cross-section are not always available and the major element composition of each sample is needed to calculate the stopping powers. If necessary, ion beam analysis methods such as RBS and PIXE can be used but iteration may be necessary to obtain self-consistent values of composition and stopping power.

a. Analytical Expressions

If the cross-section can be described by a suitable analytical function, then the Bethe-Bloch relation for the stopping power or a polynomial fitted to stopping power data can be used to evaluate Equations (12.8) and (12.9). In most cases the cross-section is small at low energies and the effects of inadequate fitting of stopping powers at these energies does not cause problems. The integration can be carried out analytically or numerically using tabulated cross-sections. The power of this approach is best demonstrated for RBS since here the reliability of analytical expressions for the cross-section is so accurate that absolute concentrations can be derived directly from the observed intensities provided that the detection efficiency and solid angle are accurately known. PIXE and nuclear reaction analysis are equally good when the cross-section has been measured with sufficient accuracy at all energies up to the incident ion energy. Low energy techniques are, at this stage, subject to experimental factors which are often difficult to define quantitatively, but rapid strides are being made in this direction.

b. Approximations

The stopping power changes relatively slowly with energy and, provided that the cross-section is also a relatively smooth function of energy, then the integral in the yield equation can be approximated with sufficient accuracy in a number of ways.

i. *Average Cross-section.*   Ricci and Hahn (1965) defined an average cross-section ($\bar{\sigma}$) such that:

$$\int (d\sigma/S)\,dE = \bar{\sigma} \int (1/S)\,dE = \bar{\sigma}\,R \qquad (12.10)$$

where R is the range of the incident particle in the sample. The concentration is then given approximately by:

$$f_i = Y_i/N_1\,\bar{\sigma}\,R \quad ; \quad c_i = Y_i\,m_i/N_1\,\bar{\sigma}\,R \qquad (12.11)$$

The ratio of concentrations from standard and unknown samples is also approximately:

$$f_i = Y_i f_s\,R_s/Y_s\,R \quad ; \quad c_i = Y_i\,c_s\,R_s/Y_s\,R \qquad (12.12)$$

The assumption that the average cross-section is the same for both materials can introduce systematic errors and, if the analytical expression for stopping power (Equation 14.2) is used to simplify the determination of $\bar{\sigma}$, errors of up to 6% can be introduced (Ishii *et al.*, 1978a).

ii. *Equivalent Thickness Method.*   An equivalent thickness (R') can be defined as:

$$R' = (1/\sigma(E_1)) \int_0^R [d\sigma\,(E)/d\Omega]\,dx \qquad (12.13)$$

and the product $\sigma(E_1)\,R'$ can be used instead of $\bar{\sigma}\,R$ in Equation (12.11) (Engelmann, 1981). This method has similar limitations to those of the average cross-section method.

iii. *Average Stopping Power.*   Ishii *et al.* (1978a) proposed the use of an average stopping power defined as the stopping power at an energy $\bar{E}$ such that:

$$\bar{E} = \int E\,d\sigma\,dE/\int d\sigma\,dE \qquad (12.14)$$

The ratio of concentrations from standard and unknown samples is now:

$$f_i = f_s\,(Y_i/Y_s)\,[\varepsilon\,(\bar{E})/\varepsilon_s\,(\bar{E})]$$
$$c_i = c_s\,(Y_i/Y_s)\,[S(\bar{E})/S_s(\bar{E})] \qquad (12.15)$$

This approach gives deviations from the true yield by 0.5% or less in most cases but it involves an integration to determine the 'mean energy' which is only marginally simpler than the full calculation (not requiring full tabulated stopping powers for the integration). However, accurate stopping powers must be used at the mean energy in order to obtain accurate results.

iv. *Stopping Power Ratios.* Other choices of energy for the stopping power ratio in Equation (12.15) have been tested with considerable success. The half-yield energy $(E_{1/2})$ is the energy at which the yield from a thick target is one-half of that at the full incident ion energy (Deconninck, 1978). The use of stopping powers at this energy for standard and unknown has been shown to give an accuracy of better than 1% (Kenny *et al.*, 1980). Even the incident ion energy itself can be used to obtain a stopping power correction which is accurate to 1% for a matrix with effective atomic number up to 50 or 60 in many cases.

v. *Double Ratio Method.* The form of the stopping power dependence on energy is similar for all elements and can be expressed as:

$$S_m(E) = \sum_i c_i \, a_i \, S_s(E) \, [1 + b_i(E)] \qquad (12.16)$$

where a_i and $b_i(E)$ are a constant and energy dependent function respectively which describe the differences between the stopping powers of unknown and standard samples. If we ignore the energy dependent term $b_i(E)$, then Equation (12.16) can be used in equation (12.9) to obtain an approximate expression for the yield:

$$Y_i = (N_1 \, e \, d\Omega \, c_i/m_i \, \Sigma_i \, c_i \, a_i) \int_{E_1}^0 d\sigma \, dE \qquad (12.17)$$

The ratio of concentration from standard and unknown samples is now:

$$c_i = Y_i \, c_s \, (\Sigma c_i \, a_i)/Y_s \, (\Sigma c_i \, a_i)_s \qquad (12.18)$$

The ratio of concentration weighted stopping powers in the two samples can be calculated or, alternatively, it can be measured by a second experiment. For example, if two radiation products are detected during measurements on standard and unknown samples (using the same beam energy and dose) then we can obtain a double ratio of results for products i and j in samples s and u (Boulton and Ewan, 1977):

$$(Y_i/Y_j)_u/(Y_i/Y_j)_s = (c_i/c_j)_u/(c_i/c_j)_s \qquad (12.19)$$

For example, two gamma-rays or ions at different energies can be used provided that the energy dependence of the stopping powers of the two samples is the same. The same approach can even be applied to the observation of two quite different reactions involving the same nuclides but using different ion energies and if necessary different ion types. However, it is important that the validity of the approximations be checked by confirming that the parameter ξ is small, where ξ can be obtained from the Bethe-Bloch expression for stopping power (Ishii *et al.*, 1978b):

$$\xi = 1 + [\log(I_s/I_u)\log(m_{1i}\,\bar{E}_j/m_{1j}\,\bar{E}_i)]\,\log(4\bar{E}_i/\lambda_{1i}\,I_s)\,\log(4\bar{E}_j/\lambda_{1j}\,I_u)] \tag{12.20}$$

12.1.3 Thick Sample — Attenuation

Photons with energies below 10 keV can be significantly attenuated before they escape from the sample, Equation (1.19). For the geometry of Fig. 12.1a, l_3 is the path length in the sample:

$$l_3 = t/\sin\beta = (\sin\alpha/\sin\beta)\int_{E_1}^{E_1'}[1/S_m(E)]\,dE \tag{12.21}$$

where E_1' is the ion energy at the point of interaction and t is the depth normal to the surface. The observed yield, Y_3, after attenuation is obtained by combining equations (12.9), (1.19) and (12.21):

$$Y_3' = (c_i\,N_1e\,d\Omega/m_i)\int_{E_1}^{o}[d\sigma\,(E)/\varepsilon\,(E)]$$
$$\times \exp\left\{-\mu_m\,(\sin\alpha/\sin\beta)\int_{E_1}^{E_1'}[dE/S_m\,(E)]\right\}\,dE \tag{12.22}$$

where $\mu_m = \sum_i c_i\,\mu_i$ is the mass attenuation coefficient for a mixture or compound sample.

If additional absorbers are located between the sample and detector surface, Y_3' must be multiplied by additional attenuation factors, obtained from Equation (1.19), for each absorber. Approximations to the (cross-section/stopping power) integral can often be made using suitable analytical expressions. For example, the path length can be obtained from the difference between the range of the incident ion and the residual range for its energy at the point of interaction. This range difference can be expressed as a power law (Folkmann, 1976). It is important to include any significant concentrations of heavy elements in the calculation of the attenuation coefficient since they have a strong effect on X-ray transmission. Information on X-ray attenuation coefficients is given in Chapter 14.5 and the processing of X-ray data is discussed in Chapter 5.

12.2 DEPTH PROFILING

12.2.1 Principles

Ion energy loss has wide applications in depth profiling. Three separate factors are involved, viz. energy loss by the incident ion, the change in energy and possibly of radiation type as a result of a major interaction and energy loss on the outgoing path if the product is an ion. If $\mathscr{K}(E'_1, \theta)$ is the kinematic relation, Equation (1.21), then the energy of product ions as they leave the sample (Fig. 12.1a) is given by:

$$E''_3 = \mathscr{K}\left\{\left[E_1 - \left(\int_o^t S(E'_1)dt\right)/\sin\alpha\right], \theta\right\} - \left[\int_o^t S(E'_3)dt\right]/\sin\beta$$

$$(12.23)$$

For small depths and energy losses, average stopping powers ($\bar{S}_1$ and $\bar{S}_3$) can be used so that:

$$E''_3 = \mathscr{K}\{(E_1 - \bar{S}_1\, t/\sin\alpha), \theta\} - \bar{S}_3 t/\sin\beta \qquad (12.24)$$

Similar equations apply for the use of stopping cross-sections (where depths are given in atoms cm^{-2} rather than g cm^{-2}). Additional energy loss may occur if an absorber is placed between the sample and the detector. This reduces the energy of the product ion still further:

$$E_{obs} = E''_3 - \int_o^f S_f(E''_3)\, dy \qquad (12.25)$$

where f is the foil thickness and y is the depth of penetration in the foil.

In the case of elastic scattering, the kinematic factor is independent of energy and the incident and exit ions are identical. The energy loss relations are then much simpler, e.g. Equations (3.5) and (3.6). Energy loss by incident radiation alone is relevant to neutron or gamma-ray emission for which the last term in Equation (12.23) is zero. Their energies follow the energy of the incident ions as shown schematically in Fig. 12.1b. The shape of the observed spectrum depends on the product of the reaction

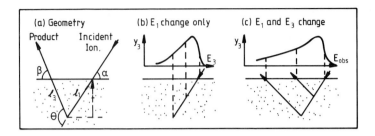

Fig. 12.1 Depth profiling from energy spectra: a. geometry; b. energy spectrum (top) showing the dependence on energy loss by incident ions only; c. energy spectrum (top) showing the effect of energy loss by incident and scattered (or product) ions.

cross-section and concentration profile through the equations given in Section 12.1. The energy spectrum of product ions is shown schematically in Fig. 12.1c. Energy loss by the incident ions is depicted by the lines sloping towards the left as they penetrate the sample. The slope of the exit lines is similar to that for incident ions in the case of scattering and different for a nuclear reaction.

a. The Depth Scale

The relation between energy and depth is provided by Equations (12.23) to (12.25) but the integrations require a knowledge of stopping power as a function of depth. Stopping powers are generally obtained from analytical expressions (applicable over a limited energy range) or tabulations as a function of ion type and energy (see Chapter 14.1). In the 'surface approximation', constant stopping powers are used so that the depth scale is:

$$t = (\sin \beta / \bar{S}_3) [\, \mathscr{K}(E'_1, \theta) - E''_3]$$
(12.26)

where E'_1 and hence $\mathscr{K}$ are functions of t. For scattering, the simple expression for the kinematic function ($\mathscr{K} = KE_1$) reduces Equation (12.26) to Equation (3.7).

For somewhat thicker layers, the 'mean energy approximation' can be used by calculating E'_1 and E'_3 and using stopping powers at $(E_1 + E'_1)/2$ and $(E'_3 + E''_3)/2$. For a layer thicker than 10 to 100 nm, it is necessary to consider this as consisting of a number of thin layers for each of which the changes in ion energy and stopping power are small and it is satisfactory to use constant stopping power values. For each sub-layer in turn, the incident energy and stopping power are determined and hence the corresponding product energy. Energy loss by a product ion from that layer is then followed step-by-step back through all the preceding layers until it reaches the surface. This procedure is readily implemented on a small computer and programs are available for use with various IBA techniques.

b. Depth Resolution

The depth resolution (dt) can be obtained from the system energy resolution (dE_{obs}) by differentiation of Equation (12.25). For constant stopping powers we obtain:

$$dt = -dE_{\text{obs}}/\{(dE'_3/dE'_1)\bar{S}_1/\sin\alpha + \bar{S}_3/\sin\beta\} \qquad (12.27)$$

For scattering, this simplifies to Equation (3.19).

The energy resolution depends on many factors (dE_i) each of which introduces a contribution (ΔE_i) to the energy resolution in the observed spectrum — depending on the experimental conditions, the nature of the sample and the interaction depth. The most important factors are:

- the energy spread of the incident beam (dE_1) is usually small (< 1 keV) compared to other contributions and if so can be neglected. If necessary its contribution (ΔE_1) to the energy resolution can be calculated from:

$$\Delta E_1 = (dE'_3/dE'_1)dE_1 \qquad (12.28)$$

- energy loss straggling along the incident ion trajectory (dE_{1s}) and, if applicable, along the outgoing trajectory of the product ion (dE_{3s}) gives contributions:

$$\Delta E_{1s} = (dE'_3/dE'_1)\,dE_{1s} \quad ; \quad \Delta E_{3s} = dE_{3s} \qquad (12.29)$$

  They are calculated using the appropriate Bohr formula or Vavilov distributions given in Chapter 14.2 with the ingoing and outgoing path lengths l_1 and l_3.

- geometrical contributions arising from finite beam size and angular divergence (ΔE_{1g}) and the detector acceptance angle (ΔE_{3g}) are given by:

$$\Delta E_{1g} = \{(dE'_3/d\theta) + (dE'_3/dE'_1)(E_1 - E'_1)\cot\alpha\}\,\Delta\alpha \qquad (12.30)$$

and

$$\Delta E_{3g} = \{(dE'_3/d\theta) + (E'_3 - E''_3)\cot\beta\}\,\Delta\gamma_{\text{det}} \qquad (12.31)$$

where the first term in each equation arises from changes in the kinematic relation for different product angles and the second term represents the effects of path length changes. $\Delta\alpha$ is the divergence angle of the incident beam (FWHM) and $\Delta\gamma_{\text{det}}$ is the effective detector acceptance angle which is approximated by:

$$\Delta\gamma_{\text{det}} = (1/D)\,(s^2 + d^2\sin^2\alpha/\sin^2\beta)^{1/2} \qquad (12.32)$$

where D is the detector-target distance, s is the detector entrance width and d is the incident beam width in the interaction plane. In most IBA situations, the incident beam collimation is sufficiently small ($< 0.1°$) so as to make ΔE_{1g} negligible in comparison to ΔE_{3g}.

- contributions due to multiple scattering (including angular path and path length variations) of the incident ions (ΔE_{1m}) and along the product ion trajectory (ΔE_{3m}) are given by:

$$\Delta E_{1m} = [\,(dE'_3/d\theta) + (dE'_3/dE'_1)\,\bar{S}_1\,(t/\sin\alpha)\,(\cot\alpha/\Gamma_1)]\,\Delta\theta_1 \qquad (12.33)$$

and

$$\Delta E_{3m} = [\,(dE'_3/d\theta) + \bar{S}_3\,t\cot\beta/\sin\beta\,\Gamma_3]\,\Delta\theta_3 \qquad (12.34)$$

where the first term in the square brackets is the additional angular (kinematic) effect which arises from the fact that products with a variety of interaction angles can reach the detector. The second term is the spatial (lateral spread) effect — resulting from path length differences due to multiple scattering. $\Delta\theta_1$ and $\Delta\theta_3$ are the respective ingoing and outgoing multiple scattering angles (FWHM) and Γ_1 and Γ_3 are the scaling functions (correlating angular and lateral spread) for the incident and exit path lengths. Data for multiple scattering angles and scaling functions are given in Chapter 14.2 as a function of reduced target thickness.

- if an absorber foil is used to remove scattered ions, energy loss straggling in the foil (ΔE_f) may well be the dominant factor. It can be calculated from data in Chapter 14.2 noting the information given for Mylar (the most common absorber material).

- contribution (ΔE_d) from detector energy resolution does not vary significantly with ion energy for solid state detectors. For electrostatic analysers and magnetic spectrometers, the attainable energy resolution has a complex dependence on experimental conditions and is a function of energy (E_3''). In such cases, it is desirable to measure the energy resolution at an ion energy close to that being detected (using a standard sample, radioactive source or a narrow reaction resonance).
- contribution due to surface roughness and other lateral inhomogeneities in the sample (ΔE_r).

Although not all of the contributions have a Gaussian distribution, they are usually added in quadrature:

$$\Delta E_{obs}^2 = \sum_i (\Delta E_i)^2 \qquad (12.35)$$

The contributions can all be measured or calculated except the effects of roughness and topography which are difficult to estimate.

If the time-of-flight method is used to measure energy, an important feature is that the flight time (and hence precision in energy determination) increases as the product mass increases and as the energy decreases. As a consequence, the equivalent energy resolution also decreases as the energy decreases, i.e. as depth increases. This tends to offset the worsening resolution caused by straggling and multiple scattering. Three factors then contribute to the detection resolution:

- The width of the incident beam pulse $(d\tau_b)$;
- The electronics timing resolution $(d\tau_e)$;
- The uncertainty in the flight path (d_D) arising from a finite detector thickness (d_d).

Once again, the contributions to detection resolution are not normally distributed but it is customary to combine them thus:

$$(\Delta E_d)^2 = (2E_3''/T)^2 \left[d\tau_b^2 + d\tau_e^2 + (m_3/5222\, E_3)d_D^2 \right] \qquad (12.36)$$

c. Optimisation of Depth Resolution

For calculations of depth resolution in elastic scattering, the terms ΔE_{1s}, ΔE_{3s}, ΔE_{3g}, ΔE_{1m}, ΔE_{3m}, ΔE_d and ΔE_r are the most significant. For ion–ion reactions, ΔE_f is likely to dominate and the terms corresponding

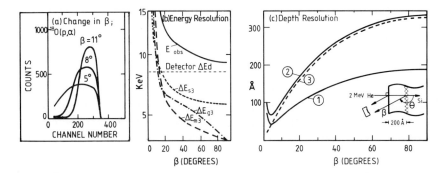

Fig. 12.2 Optimisation of depth resolution: a. energy spectra of α-particles from the ^{18}O (p, α) reaction for three values of the emergent angle; b. dependence on β of components of the energy resolution for 2 MeV He$^+$ RBS (see text); and c. dependence on β of the depth resolution calculated from the energy resolution for three different experimental conditions.

to incident and product ions having the smaller energy loss can often be neglected. For ion–gamma and ion–neutron reactions, only incident ion contributions are relevant. For near-surface analysis where energy losses are small, ΔE_r, ΔE_{3g}, ΔE_d and if applicable ΔE_f, are often dominant terms whereas, for analysis at large depths, ΔE_{1s}, ΔE_{3s}, ΔE_{1m} and ΔE_{3m} are dominant depending on the chosen geometric conditions. From Equations (12.27), the depth resolution can be improved either by improving the energy resolution (ΔE_{obs}) or by maximising the kinematic, stopping power and geometric factors which appear in the denominator. A particularly simple way of improving near-surface resolution is to decrease the angles α and/or β (Fig. 12.1a). This glancing angle method increases the incident and/or exit path lengths for interaction at a particular depth by the factors sin α and sin β respectively. This results in a corresponding magnification of the depth scale and improvement in depth resolution. Fig. 12.2a illustrates this effect for the case of oxygen profiling using the reaction ^{18}O (p, α)^{15}N. The alpha spectrum broadens as the angle β is decreased (i.e. the path length of the outgoing alpha particles is increased).

The geometric enhancement in depth resolution might be expected to scale roughly with sin α or sin β, depending on which term dominates in Equations (12.27). However, the increased path lengths in glancing angle geometries increase multiple scattering, energy straggling and solid angle effects and these all degrade the energy resolution. This is illustrated in Fig. 12.2b for the case of 2 MeV He$^+$ RBS at a depth of 200 Å in Si (Williams and Möller, 1978). In this case, the outgoing path length is increased by reducing β and hence the terms ΔE_{3s}, ΔE_{3g} and ΔE_{3m} (calculated using the expressions given in the previous section) begin to dominate over the detector resolution (assumed to be 8 keV) as β drops below about 15°.

The loss of energy resolution competes with the sine factors which improve the depth resolution and an optimum geometry for best depth resolution is observed at a particular depth (Fig. 12.2c). Curve (1) corresponds to the example in Fig. 12.2b and shows that there is an optimum exit angle of $\beta = 4°$. The maximum depth resolution at a depth of 200 Å is about 40 Å. In this case (c.f. Fig. 12.2b) multiple scattering is the dominant term at $\beta = 4°$. Curve (2) in Fig. 12.2c illustrates the effect of employing a detector with an inferior energy resolution (15 keV) as well as increasing the detector solid angle to improve the counting rate. In this case, the maximum resolution is approximately 65 Å, again at an angle of $\beta = 4°$ but now the geometry term dominates the depth resolution. Curve (3) in Fig. 12.2c shows that the depth resolution at the surface (corresponding to case (2)) continues to improve for values of β down to 2°. The path length effects are no longer significant and the detector resolution remains the dominant contribution at all angles unless surface roughness and topography (not included in Equation 12.35) act to degrade the resolution.

Further examples of improved and optimised depth resolution in energy loss profiling are given in Chapters 3, 4 and 6. Chapter 14.3 provides tabulations of depth resolution as a function of RBS scattering depth and geometry in Si substrates. Williams and Möller (1978) have shown that calculated depth resolutions, Equations (12.27), are in good agreement with measured resolutions for profiling to 1500 Å in smooth thin film structures.

d. Topography

Entrance and exit path length and energy loss, and hence the observed product energy, vary with the lateral position of the incident ion if the sample has a rough surface. This is illustrated in Fig. 12.3a for the simple case of a surface with parallel ribs. Contours can be drawn of points for which the combination of incident and product energy loss lead to the same observed energy. For normal incidence and exit, these contours are parallel to the sample surface and a depth profile can be derived which may contain useful information (Chapter 3.4.2). For other angles, the contours are no longer parallel to the surface and may enclose significant volumes within the sample which involve a specific value of energy loss. The consequence is unusual spectrum features such as the peaks in the RBS spectrum in Fig. 12.3b that arise because of sample topography rather than composition variations.

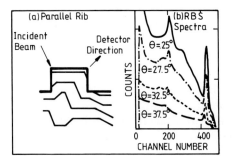

Fig. 12.3 Surface topography effects: a. contours for constant observed energy from a sample with parallel surface ribs; and b. RBS spectra for different scattering angles.

If it is known that a sample is uniform in composition, the spectrum features can be used to derive information on topography but, if the sample is non-uniform in composition, measurements with specific incident and exit angles are not sufficient to distinguish between the two effects. However, the spectral structure arising from regular topographic features has an unusual dependence on incident and exit angles. Measurements at a number of angles can, therefore, help distinguish between topography and composition gradients (Edge, 1981; Knudson, 1980; Bird, *et al.* 1983). Random roughness — a statistical distribution of widths, depths and spacings of peaks and valleys — does not produce prominent spectral features but the additional depth resolution term changes rapidly at glancing angles. Although measurements at several angles may be useful, high resolution depth profiling is only possible for smooth samples.

12.2.2 Energy Spectrum Method

a. Spectrum Shape

For a sample which is uniform laterally but has a concentration gradient ($f_2(t)$ or $c_2(t)$) at right angles to the surface, the yield from a thin layer dt at depth t is given by:

$$Y_3 \, dE_{obs} = N_1 \, e \, d\Omega \, f_i(t) \, d\sigma(E'_1) dE'_1 / \varepsilon_{1m} \, (E'_1)$$
$$= N_1 \, e \, d\Omega \, c_i(t) \, d\sigma(E'_1) dE'_1 / m_i \, S_{im} \, (E'_1) \qquad (12.37)$$

Both S_{1m} and ε_{1m} may be evaluated at the incident energy or at a lower energy which approximates the mean energy in the layer of interest. The methods described above can be used to calculate the depth scale, and the spectrum energy intervals (ΔE_{obs}) which correspond to equal increments in depth. For a constant cross-section and concentration, the height of the spectrum for a given concentration becomes less at lower energies because of deteriorating resolution since more ions are removed from a specific energy interval to lower energies than are brought in from higher energies. If the probability of a major interaction is sufficiently high, as is the case for RBS, multiple interactions can also introduce a major change in energy giving rise to a low energy tail which extends to zero energy even when the layer being investigated is relatively thin (Weber *et al.*, 1982).

Energy broadening can be introduced formally by the inclusion of integrals over the parameters affecting the energy resolution. One overall resolution function can be used (Equation 4.7). Alternatively it is possible to consider all factors affecting the incident ion energy combined to obtain the distribution function $P_1(E_1, \bar{E}_1, t)$ which describes the probability that the energy is E_1 at depth t where the average energy is $\bar{E}_1$. Likewise, factors affecting the product ion can be combined in the distribution function $P_3(E_3, \bar{E}_3, t)$. The number of events recorded in an interval in the observed energy spectrum which corresponds to a depth increment dt is given by:

$$Y_3\,(E_3'')\,dE_3'' = k \iiint \{c_i(t)\,[P_1\,d\sigma P_3]/S_m\}\,dE_1'\,dt\,dE_3 \quad (12.38)$$

where $k = N_1\,e\,d\Omega/m_2\,\sin\,\alpha$.

b. Spectrum Synthesis

Spectrum synthesis is the most common approach to deriving a depth distribution from measured spectra. An assumed concentration profile is convoluted with the energy spread and cross-section functions to obtain a predicted energy spectrum, Equation (12.38). A similar approach, ignoring energy spread, is to use Equation (4.6). The effects of attenuation must be included when relevant (for example in PIXE simulation). A comparison with the observed spectrum can then guide the modification of the assumed profile to obtain an improved fit. Numerous computer programs which have been written to use numerical or analytical information on energy loss, multiple scattering and cross-sections in spectrum synthesis are discussed in chapters on specific techniques.

c. Deconvolution

A channel-by-channel ratio of yields from standard (s) and unknown (u) samples can be used to obtain a depth profile:

$$f_u(t) = f_s(t') [Y_u/Y_s]_j [\varepsilon_s/\varepsilon_u]_{E_{ij}} \tag{12.39}$$

$$c_s(t) = c_u(t') [Y_u/Y_s]_j [S_s/S_u]_{E_{ij}} \tag{12.40}$$

where j is the channel number. It should be noted that the depths t and t' may be different in the two samples — being the depths at which the product ion energy is the same. The method will only be accurate if the incident ion energies are also equal at these depths (Schulte, 1976). A non-linear depth scale will be involved if the total thickness studied is sufficient to produce significant changes in stopping power. However, this is a very simple way to obtain useful information on depth distributions in relatively thin layers. A better method is to calculate observed energy intervals and mean incident energies for successive layers in a similar fashion to the use of Equation (4.10) for time-of-flight measurements. A third approach is to use the stopping power or cross-section data in tabulated form or as analytical expressions fitted to the region of interest — which may involve a quite limited energy range. Numerical tecniques can then be used to calculate the convolution integral in Equation (12.38) (see for example Möller *et al.*, 1977). Iterative procedures based on the use of Gaussian energy spread distributions are described by Lewis (1981). Deconvolution using matrix algebra is discussed in Section 12.2.4.

12.2.3 Resonance Scanning

Depth profiling by resonance scanning requires the presence of an isolated resonance in the interaction cross-section (Fig. 12.4a). For a given incident ion energy ($E_r < E'_1 < E'_r$), the observed yield depends on the resonance area (replacing $d\sigma$ in Equation 12.4) and the atomic concentration at a depth given by Equation 12.26 with $E'_1 = E_r$. These equations convert the axes of a yield curve (e.g. Fig. 12.4b) to a depth profile (f_2 vs. t). The depth resolution at the surface depends mainly on beam energy spread and the resonance width both of which can be much less than 1 kcV ($\sim$ 1 nm resolution). Broad resonances are sometimes used for depth profiling if poor resolution is acceptable. Below the surface, incident ion energy straggling and multiple scattering broaden the

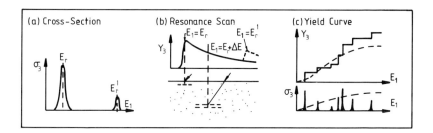

Fig. 12.4 Depth profiling by resonance scanning: a. nuclear reaction cross-section with two well separated resonances; b. yield curve as a function of energy; and c. yield curve for a reaction with many closely spaced resonances.

resolution (Section 12.2.2). Energy loss by product radiation, if it occurs, has no effect. For narrow resonances and depths up to $\sim 1\,\mu$m, the resolution function is a Vavilov distribution (Section 14.2) convoluted with the beam energy spread (usually Gaussian) and the resonance shape (usually Breit-Wigner). For depths greater than given by Equation (14.10) the energy loss distribution becomes Gaussian with a width given by Equation (14.19) which usually dominates over other contributions to the resolution. Spectrum synthesis or deconvolution techniques (Section 12.2.2) must be used to determine the depth profile but this is relatively straightforward ignoring product radiation terms.

The maximum depth which can be studied is dependent on the spacing between the resonance being used and the next higher energy resonance (E'_r, Fig. 12.4a). The dashed curve in Fig. 12.4b shows the rise in yield observed as the incident energy passes the energy of the second resonance. If this is relatively small a correction can be made by multiplying the early (low energy) profile by the ratio of the resonance areas (A_r, A'_r) and subtracting at the appropriate energy above the second resonance:

$$Y(E'_r+\Delta E) = Y_{obs} - A'_r\,Y(E_r+\Delta E)/A_r \qquad (12.41)$$

Several resonances can be handled in this way (Kregar *et al.*, 1977) provided that the correction remains small relative to the observed yield. Another important factor is the possible presence of any smooth cross-section component between resonances. This contributes to the observed count rate and requires a correction to the observed yield to obtain the resonance component only. The need for repeated measurement of the resonance yield increases the measurement time and the beam dose to the sample. A typical profile may take at least 1 hour to complete.

12.2.4 Yield Curve Unfolding

a. Energy Dependence

If there are many resonances in the cross-section, the thick sample yield as a function of incident ion energy shows a step for each resonance (Fig. 12.4c) and the integral relation, Equation (12.22), becomes a sum of terms, the number of which increases as the energy increases. The set of equations required to describe a yield curve can be expressed in matrix notation as:

$$(\mathbf{Y}) = k\,(\mathbf{C})\,(\mathbf{A}) \qquad (12.42)$$

The concentration profile is then given by:

$$(\mathbf{C}) = 1/k\,(\mathbf{Y})\,(\mathbf{A})^{-1} \qquad (12.43)$$

This approach also describes the measurement of a yield histogram for an interaction which has a cross-section which is a continuous function of the incident energy. In effect, any departure from the shape of the yield curve measured with a uniform standard sample contains information on the depth profile. However, small differences in yield are difficult to measure with the necessary precision and the depth resolution is usually poor. These are serious limitations and, as a consequence, this method has had only very limited application (e.g. Chittleborough *et al.*, 1978).

b. Angle Dependence

Instead of changing the incident energy, a number of measurements can be made with different angles of incidence or emergence or both (e.g Pabst, 1974 and 1975). These methods are suitable for the study of depths which are greater than can be achieved with resonance scanning although the accuracy and resolution at large depths is seriously compromised when compared with near-surface measurements.

12.3 STATISTICS

12.3.1 Experimental Uncertainties

When the probability of an ion interaction is small and constant (i.e. the interaction rate is not sufficient to remove significant numbers of target atoms), the counting of product radiation, whether all detected events or those within a defined energy, time of flight, or mass interval, follows a Poisson distribution:

$$P(N) = m^N e^{-m}/N! \approx (1/\sqrt{2\pi m})\,(m/N)^N\,e^{(N-m)} \qquad (12.44)$$

where P is the probability of observing N events with m incident ions. If N is large, the probability distribution becomes approximately normal (Gaussian):

$$dP(N) = (1/\sigma\sqrt{2\pi})\exp\left[-(N - \bar{N})^2/2\sigma^2\right]dN \qquad (12.45)$$

A set of n experiments gives values of the mean:

$$\bar{N} = \sum_1^n N/n \qquad (12.46)$$

and standard deviation:

$$\sigma_e = \left\{\sum_1^n (N - \bar{N})^2/(n - 1)\right\}^{1/2} \qquad (12.47)$$

which are approximate estimates of the results which would be obtained from an infinite number of measurements. An estimate of confidence limits for the true mean is given by:

$$\mu = \bar{N} \pm t\sigma_e/n^{1/2} \qquad (12.48)$$

where t is the obtained from the students-t distribution (Table 12.1) being the area in the tails of the normal distribution (beyond $\pm\, t\,\sigma$).

The expected standard deviation from purely statistical fluctuations in number of counts is ($\sigma = 100\,N^{-1/2}\%$) and this is usually observed for a number of successive measurements with stable equipment performance. In order to reduce the standard deviation by a factor f it is necessary to increase the count rate or counting time by f^2 which is an important limitation on high precision analyses. Random fluctuations in experimental parameters such as beam energy, dose measurement, detector gain, etc. broaden the observed distribution and the ratio of σ_e/σ is a measure of the contribution from such factors. When calculations such as background subtraction, peak fitting, solid angle and efficiency correc-

TABLE 12.1
Student's-*t* values for *n* measurements

n	DF*	Confidence Level (%)			
		90	95	98	99
4	3	2.353	3.182	4.541	5.841
6	5	2.015	2.571	3.365	4.032
8	7	1.895	2.365	2.998	3.499
10	9	1.833	2.262	2.821	3.250
15	14	1.761	2.145	2.624	2.977
20	19	1.729	2.093	2.539	2.861
25	24	1.711	2.064	2.492	2.797
30	29	1.699	2.045	2.462	2.756
∞	∞	1.645	1.960	2.326	2.576

* DF = Number of degrees of freedom = $n - 1$

tions, etc. are required to obtain element or isotope concentrations, the standard deviation of the final result requires error propagation using two rules:

i. $\qquad$ If $A = B \pm C$ then $\sigma_A^2 = \sigma_B^2 + \sigma_C^2$ $\qquad$ (12.49)

ii. $\qquad$ If $A = B^b C^c$ then $(\sigma_A/A)^2 = b\,(\sigma_B/B)^2 + c(\sigma_c/C)^2$ $\quad$ (12.50)

provided that each factor is subject to random errors. The standard deviations associated with various experimental and fitting factors can be determined by calibrations including the use of standard samples. These will also show whether non-random effects such as drift in beam energy or detector gain are present.

Ion beam analysis usually involves a small near-surface volume of sample and heterogeneities are of major significance. The use of a large beam area (> 1 cm) will help to minimise the effects of heterogeneities but the restricted depth usually involved in IBA can still cause problems. An alternative approach is to make measurements on a number of spots on the sample from which the level of heterogeneity can be estimated. The true mean (μ) will lie within $\pm t\sigma_n/n^{1/2}$ of the observed mean where t is the student's t-value for the required level of confidence. The same approach can be used to assess the variability of n samples taken from original bulk material. At least 4 or 5 measurements must be made for a useful estimate of heterogeneity to be obtained. If it is required to achieve a standard deviation σ_r, the minimum number of measurements required is given by:

$$n = (t\,\sigma_n/\sigma_r)^2 \qquad (12.51)$$

Because such a procedure affects the time and cost of a measurement program, a protocol should be established for material sampling and repeated runs to achieve the required precision as efficiently as possible (Taylor, 1986).

12.3.2 Cluster Analysis

The results of ion beam analysis are often a multi-element data set for collections of samples which may be the starting point for investigating grouping by statistical techniques. Such investigations involve two aspects:

- Cluster analysis or pattern recognition applied to establish the existence of groups which are distinguishable on the basis of measured composition; and
- Consideration of the probability with which a specific sample can be assigned to a particular group. For this purpose, the shape of the distribution of concentrations for each element amongst members of each group of samples must be determined.

Clustering can be studied by computerised counting of point densities in multi-dimensional space. It is also common to use some measure of multi-dimensional distance between points representing different samples and to search for discontinuities in the distribution of distance values which may indicate the boundary of a group. The distances and their distributions can then be used for sample/group assignments. Reference samples may be available to define sample groups and the shapes of the distributions for various elements in those groups. For satisfactory cluster analysis, the number of samples should be much larger than the number of parameters used (e.g. by a factor of 5). It is therefore advantageous to omit elements which are measured with poor precision or which show concentrations correlated with those of other elements. Even so, it is important not to overlook any data point which, if not in error, is significantly different from acceptable values for an adopted cluster. Computerised cluster analysis can be carried out using a variety of well-established methods (Clayton, 1982; Peisach *et al.*, 1982). Many standard packages (e.g. SPSS, BMDP) are available for this work.

a. Parameter Scaling

Element concentrations (c_{ij} for $i = 1$ to e elements and $j = 1$ to s samples) can differ by many orders of magnitude and trace elements may be more useful than major or minor elements for sample characterisation, provided that the measurements have been made with sufficient accuracy. Various methods are used to scale the data so as to improve cluster analysis:

    i.   The log-concentration ($\log c_{ij}$ or $\log (k + c_{ij})$) often show symmetric distributions with commensurate relative standard deviations for each element.

    ii.   The square root ($c_{ij}^{1/2}$) has similar properties but gives a range of values intermediate between those of c and $\log c$. Other powers between 0.3 and 0.6 can be used and $5 (c_{ij}^{0.2} - 1)$ can be useful for specific data sets (Leach and Manly, 1982).

    iii.   An auto-scaling function which is specific to a particular data set is $(c_j - c_{\min})_i / (c_{\max} - c_{\min})_i$. This expresses each measurement as a fraction of the observed maximum spread for each element. Every element is thereby placed on the same footing in relation to the intrinsic variation occurring within the data set.

    iv.   A more significant auto-scaling function is $(c_{ij} - c_i)/\sigma_i$ where σ_i is the observed standard deviation for element i. The ratio c_{ij}/c_i, where c_i is the mean value for element i, can also be used.

Weighting factors (w_i) can be included to take account of differences in precision imposed for different elements by the experimental techniques. The most common weighting factor is $1/v_i$ where v_i is the variance of element i.

It is also common to use element ratios to minimise the influence of systematic errors within the data set. Contributions to systematic errors include changes in equipment performance and calibration, the use of different standards or data processing algorithms and varying degrees of interference not adequately allowed for in data analysis. If measurements made by several techniques are included in the data set the systematic error component is usually increased and element ratios should be taken within each set and not from one set to another. The use of element ratios is only advantageous if they reduce the effects of correlations.

Missing data points are another problem occurring in instrumental analysis, either because the level of an element is below the detection limit or because of some malfunction in data collection or processing.

The use of zero values may introduce problems into cluster analysis in which case genuinely low values can be replaced by the estimated detection limit and other missing values can be replaced by estimated means.

b. Multi-dimensional Distances

A diversity of methods is available for constructing multi-dimensional distances. The Euclidean Distance (D_{st}) between points s and t in multi-dimensional space is usually written as:

$$D_{st} = \{w_i\,(c_{is} - c_{it})^2\}^{1/2} \tag{12.52}$$

with weighting factors (w_i) included if required. If data values are missing for some elements in some samples, the Mean Euclidean Distance can be used:

$$\bar{D}_{st} = \{D_{st}^2/e\}^{1/2} \tag{12.53}$$

Another useful measure is the Mean Character Difference:

$$MCD = \left\{\sum_i w_i\,|\,c_{is} - c_{it}\,|\,\right\}/e \tag{12.54}$$

c. Single Linkage Cluster Diagrams

Starting with a number of groups equal to the number of samples, interpoint distances are calculated between groups and the smallest value used to select two groups to be combined into a common group having a centroid from which interpoint distances are recalculated. Repeating this process produces a similarity dendrogram which when plotted shows the degree of clustering present within the data and the relative magnitude of the separation between clusters. The largest separation determines the scale of the dendrogram and omitting samples which are separated at this largest distance (which may be isolated outliers) allows a more detailed assessment of the remaining groups.

A different approach is the Minimal Spanning Tree which displays nearest neighbour links by lines joining points for each sample (Boulle and Peisach, 1979). There are no loops in this display and clusters are characterised by points in one region of the tree having shorter links between them than with other parts of the tree. Well separated groups appear as separate branches with outliers being at the end of long branches. The direction of branches is arbitrary so that the appearance is rather like a dendrogram folded in upon itself so that visually it can be more confusing than a dendrogram.

In both cases the length of the links between sample points is one criterion for establishing clusters and a decision must be made at what length (or level) to accept groups as well defined. Algorithms can be used to mechanise this decision and to provide numerical values describing the distinctions although such values must be used with considerable caution.

d. Principal Components Analysis

The distribution of points in e-dimensional space can be projected onto two- or three-dimensional plots in a variety of ways. Principal Components Analysis involves construction of new variables from linear combinations of the original e variables in such a way as to retain as much as possible of the total variance associated with the s points. The two or three largest eigenvalues of the covariance matrix are used and in the case of pseudo-three-dimensional plots, various rotations can be used to assess the separation between groups of points.

e. Non-linear Mapping

In non-linear mapping, the e dimensions are combined so as to minimise the difference between the representation of interpoint distances (d_{es}) in a two-dimensional plot from those in the original space (D_{es}). The difference function (F) is defined as:

$$F = \left\{ \sum_{e \neq s}^{s} (D_{es} - d_{es})^2 / D_{es} \right\} / \sum_{e \neq s}^{s} D_{es} \qquad (12.55)$$

and this is minimised by an iterative procedure which starts from a plot of the two elements exhibiting the greatest standard deviations. The axes in the final plot do not have a physical meaning in this presentation. Other non-linear plots can also be used — for example using distances to the origin and centroid of all data.

f. Correspondence Analysis

Correspondence Analysis is another mapping technique which superimposes plots based on group separations and on the significance of the contribution from various elements to these separations (Underhill and Peisach, 1985). The matrix of values of element concentrations are treated symmetrically to establish axes, the first of which has the most distinct sample groups at opposite ends. The elements which contribute most to this distinction are also plotted at the ends of the same axis. Groups with less distinct separations and less important elements are plotted along the second and subsequent axes. For each sample, a number of nearest neighbours are listed with their multi-dimensional distances which allow assessment of the clustering behaviour in more detail.

g. Sample Assignment Probabilities

Once groups have been postulated, or if they are defined by prior knowledge of reference materials, it is possible to assess the probability that a specific sample has a composition falling within the range of one group. This can be done by comparing the sample data with group distributions or by calculating the Mahalanobis distance:

$$D_M = \sum_i w_i (c_{is} - \bar{c}_i)/\sigma_i \qquad (12.56)$$

which is the distance of the sample point from the group centroid in e-dimensional space in units of the group variance in the direction of the sample point (Ward, 1974). This involves the assumption that the parameters are normally distributed, an assumption which can be far from the truth. Nevertheless, if the probability of assignment to the group is sufficiently high (e.g. greater than 90%) or low (e.g. less than 10%) then it may well be acceptable. Intermediate cases may require more evidence or more detailed consideration of distributions, correlations and errors as do cases in which overlapping groups occur and a significant probability is predicted for membership of more than one group (Leach and Manly, 1982).

h. Drawing Conclusions

Changing the choice of elements or ratios, using different scaling or omitting some samples all change the cluster patterns — sometimes dramatically. Features which remain constant through such changes have a strong basis in sample similarities and differences. However, changes in clustering can also provide useful information. For example, elements whose exclusion cause changes in clustering may have unacceptable measurement precision but they may also indicate sample heterogeneities which, if understood, can be used as a distinguishing feature in themselves. Great care must be taken not to confuse these with factors such as sample contamination, roughness and preparation problems. Although computerised cluster analysis may seem to be removed from human bias, the results are so subject to the choices made that it is very easy to run many versions of cluster analysis and then, consciously or unconsciously, select those which best display the result hoped for. The analyst must be an unbiased skeptic.

REFERENCES

Bird, J.R., Duerden, P., Cohen, D.D., Smith, G.B. and Hillery, P. (1983). *Nucl. Instrum. Methods* **218**, 53.

Boulle, B.J. and Peisach, M. (1979). *J. Radioanal. Chem.* **50**, 205.

Boulton, R.B. and Ewan, G.T. (1977). *Anal. Chem.* **49**, 1297.

Chittleborough, C.W., Chaudhri, M.A. and Rouse, J.L. (1978). *In* '2nd Australian Conf. on Nucl. Tech. of Analysis', Lucas Heights, 61.

Clayton, E. (1982). *In* 'Archaeometry: An Australian Perspective', (W. Ambrose and P. Duerden, eds.), ANU Press, Canberra, Australia, 90.

Deconninck, G.(1978). 'Introduction to Radioanalytical Physics', Elsevier, Amsterdam.

Edge, R.D. and Bill, U. (1979) *Nucl. Instrum. Methods* **168**, 157.

Engelmann, C. (1981). *Atomic Energy Rev.* **19**, 107.

Folkmann, F. (1976). *In* 'Ion Beam Surface Layer Analysis', (Meyer, O., Linker, G. and Kappeler, F., eds) Plenum Press, New York, Vol 2. 747.

Ishii, K., Valladon, M. and Debrun, J.-L. (1978a). *Nucl. Instrum. Methods* **150**, 213.

Ishii, K., Sastri, C.S., Valladon, M., Borderie, B., and Debrun, J.L. (1978b). *Nucl. Instrum. Methods* **153**, 507.

Kenny, M.J., Bird, J.R. and Clayton, E. (1980). *Nucl. Instrum. Methods* **168**, 115.

Knudson, A.R. (1980). *Nucl. Instrum. Methods* **168**, 163.

Kregar, M., Muller, J., Rupnik, P. and Spiler, F.(1977). *Nucl. Instrum. Methods* **142**, 495.

Leach, F. and Manly, B. (1982). *N.Z.J. Archaeology* **4**, 27.

Lewis, M.B.(1981). *Nucl. Instrum. Methods* **190**, 605.

Möller, W., Hufschmidt, M. and Kamke, D. (1977). *Nucl. Instrum. Methods* **140**, 157.

Pabst, W. (1974). *Nucl. Instrum. Methods* **120**, 543.

Pabst, W. (1975). *Nucl. Instrum. Methods* **124**, 143.

Peisach, M., Jacobson, L., Boulle, G.J., Gihwala, D. and Underhill, L.G. (1982). *J. Radioanal. Chem.* **69**, 47.

Ricci, E. and Hahn, R.L. (1965). *Anal. Chem.* **37**, 742.

Schulte, R.L. (1976). *Nucl. Instrum. Methods* **137**, 251.

Taylor, J.K. (1986). *In* 'Materials Characterisation', Metals Handbook, (ed. Whan, R.E.), *American Soc. for Metals, Ohio*, **10**, 12.

Underhill, L.G. and Peisach, M. (1985). *J. Trace and Microprobe Techniques* **3**, 41.

Ward, G.K. (1974). *Archaeometry*, **16**, 41.

Weber, A., Mommsen, H., Sarter, W. and Weller, A. (1982). *Nucl. Instrum. Methods* **198**, 527.

Williams, J.S. and Möller, W. (1978). *Nucl. Instrum. Methods* **157**, 213.

Section 3
Useful Data

13
Directory of Materials

L. WIELUNSKI
Division of Applied Physics, CSIRO
Menai, Australia

J.R. BIRD
ANSTO Lucas Heights Research Laboratories
Menai, Australia

J.S. WILLIAMS
Microelectronics and Materials Technology Centre, RMIT
Melbourne, Australia

13.1 THIN FILM STANDARDS

Thin film standards should preferably be uniform, amorphous and stable with time and during irradiation. Some useful materials are listed in Table 13.1. General methods for preparation are reviewed by Amsel and Davies (1983) and specific cases are summarised below.

ION BEAMS FOR
MATERIALS ANALYSIS
ISBN 0 12 099740 1

TABLE 13.1
Thin Layer Standards

Isotope	Standard	Comment
^{1}H, ^{2}D	H_2O Ice	Unstable
	TaH_x	Slightly unstable[a]
	TiH_x	Stable, requires calibration[b]
	H implanted Si	Stable, requires calibration
	Amorphous Si	Stable, requires calibration[a]
	Pyrolitic C	Stable, requires calibration[a]
	Mylar, Kapton	Unstable (beam sensitive)[a,c,d]
^{6,7}Li	$LiNbO_3$	Stable (insulator)
^{9}Be	Metal	Stable
	Be Window	Stable
^{10,11}B	B_2O_3 glass	Stable
	B implanted Si	Stable
C	Graphite	Stable
	C foil	Stable
^{14,15}N	Si_3N_4 on Si	Stable, requires calibration
	TiN	RF plasma deposited, requires calibration
	N implanted Si	Stable
^{16,17,18}O	Al_2O_3 on Al	Anodic or thermal oxidation
	SiO_2 on Si	Anodic or thermal oxidation
	Ta_2O_5 on Ta	Anodic or thermal oxidation
^{19}F	CaF_2 on Si	Epitaxial (insulator)
Bi	Bi implanted Si	Stable, RBS calibration

[a]Westerberg et al., (1985); [b]Kamykowski et al., (1979); [c]Rauhala and Raisanen (1985); [d]Venkatesan et al., (1983)

13.1.1 Compound Films

Thin oxide layers can be produced by anodisation in aqueous solutions (e.g. 1% NaCl or 3% ammonium citrate) (Siejka et al., 1971; Bradshaw et al., 1987). The relation between applied voltage and film thickness is linear provided that the metal substrate has been chemically polished. It is advisable to cool the electrolyte during anodisation. The oxidation rate is 1.64 nm V^{-1} for 10 to 200 nm of Ta_2O_5, 1.37 nm V^{-1} for 10 to 250 nm of Al_2O_3 and 0.62 nm V^{-1} for 5 to 50 nm of SiO_2.

Thin films can also be grown via solid state reactions by two main methods:

i. by heating in a reactive gas (e.g. O_2 or N_2) to obtain stoichiometric compounds such as SiO_2, Al_2O_3, Ta_2O_5, Si_3N_4 or TiN; or

ii. by deposition of a thin film which reacts with the substrate during later heating (e.g. Ni deposited on clean Si reacts to form Ni_2Si at 250°C, then NiSi at 400°C and finally $NiSi_2$ at ~750°C).

Table 13.2 gives the ratios of the compound film thickness to the initial film thickness (required to form the listed compound) for a number of commonly used binary compounds. Compounds formed from enriched isotopes (e.g. ^{15}N or ^{18}O) are very useful for calibration of NRA measurements. Thin films of commercial plastics can be used as H, C, N and O standards provided that beam intensities are sufficiently low to avoid serious radiation damage (see Sections 13.1.6 and 13.3). Beam induced charging may also be a serious problem, for example contributing to detector noise and current integration error. Aluminised mylar is useful for minimising these problems.

13.1.2 Vacuum Deposition

Commercial evaporators can be used for film deposition but a simple vacuum chamber with a heated filament on which the material to be deposited should be placed is often adequate for low melting point ($< 1000°C$) elements. The filament-sample distance should be large enough to ensure that thickness variations across the sample surface are sufficiently small. Quartz-crystal thickness monitors are normally required. Electron-beam evaporators are required for high melting point materials and they are subject to less contamination than filament evaporators. Sputter deposition is used for very high melting point materials such as SiO_2 or Al_2O_3 but such films can contain occluded gas from plasma-based deposition at 10^{-2} to 10^{-3} Torr.

Molecular Beam Epitaxy (MBE) can be used for preparation of thin films at less than 10^{-8} Torr. The substrate is usually cleaned using flash heating or ion beam irradiation and maintained at elevated temperature during deposition (Bean, 1981).

A useful calibration standard for RBS is a multilayer of 0.5 nm each of Au, Pd, Ni and Cr on 200 nm of SiO_2 on a thick Si substrate. This can be prepared by vacuum deposition and gives four well spaced peaks which can be used to determine the energy scale, linearity and resolution. Single layer standards (such as SiO_2 on Si or Al_2O_3 on Al) are also useful for quantitative yield calibrations.

Methods used for the measurement of film thickness include:

- weighing before and after deposition;

TABLE 13.2

Relative reacted and unreacted film thicknesses for compounds formed by solid state reactions

Compound (A_xB_y)	Density* $(g\ cm^{-3})$	Thickness B/A	Thickness $\dfrac{A_xB_y}{A}$
Al_2O_3	3.97	—	1.28
Co_2Si	7.46	0.91	1.48
$CoSi$	6.58	1.82	2.01
$CoSi_2$	4.95	3.64	3.51
$CrSi_2$	4.98	3.33	3.00
$FeSi$	6.16	1.70	1.92
$FeSi_2$	4.94	3.38	3.47
$HfSi$	10.27	0.88	1.48
$HfSi_2$	7.98	1.77	2.15
$IrSi$	13.10	1.41	1.97
Mg_2Si	1.94	0.44	1.41
$MnSi$	5.90	1.63	1.93
$MnSi_2$	5.24	3.26	2.85
$MoSi_2$	6.24	2.56	2.59
$NbSi_2$	5.66	2.18	2.38
Ni_2Si	7.44	0.91	1.52
$NiSi$	5.92	1.83	2.22
$NiSi_2$	4.83	3.66	3.53
Pd_2Si	9.59	0.68	1.65
$PdSi$	7.69	1.36	1.97
Pt_2Si	16.27	0.66	1.41
$PtSi$	12.39	1.32	1.97
SiO_2	2.27	—	2.20
Si_3N_4	3.44	—	1.13
Ta_2O_5	8.20	—	2.47
$TaSi_2$	9.08	2.21	2.40
$TiSi$	4.32	1.14	1.70
$TiSi_2$	4.04	2.27	2.39
VSi_2	4.82	2.90	2.76
WSi_2	9.80	2.53	2.71
$ZrSi_2$	4.86	1.72	2.16

*Density from Murarka (1983) and Jastrezebski (1987)

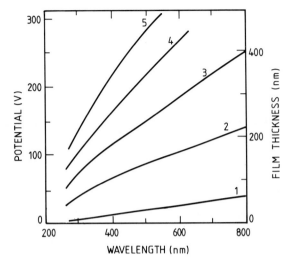

Fig. 13.1 The relation between formation voltage, film thickness and wavelength (λ) defining the colour of Ta_2O_5 layers formed by anodic oxidation (Bradshaw *et al.*, 1987). The numbers against each curve represent the sequence of appearance of each colour when the thickness reaches multiples of $\lambda/2$.

- interference colours (Fig. 13.1); reflection or transmission measurements in monochromatic light extend the range to at least 10 μm;
- optical interferometry (down to 40 nm);
- acoustical impedance monitoring using a quartz crystal oscillator (from 1 nm to > 10 μm);
- ellipsometry (useful down to 3 nm);
- Talystep measurements of step height (down to 2.5 nm); and
- RBS (down to 0.01 of a monolayer for heavy elements on a light substrate).

13.1.3 Metal Foils

Commercial metal foils (e.g. from entries 17 and 18 in Table 13.3) are useful as thin film standards but in the case of alloys the composition must be accurately known.

13.1.4 Ion Implantation

Implanted ions with accurately determined dose can be used for energy and yield calibrations in RBS, ERA and PNA. This is the only way in

TABLE 13.3
Suppliers of reference materials

1. International Atomic Energy Agency, Analytical Quality Control Service, Laboratory Siebersdorf, PO Box 590, A-1011 Vienna, Austria.
2. Central Bureau of Nuclear Measurements, Commission of the European Communities, Geel Establishment, Steenweg op Retie, B-2440 Geel, Belgium.
3. Community Bureau of Reference (BCR), Directorate General XII, CEE 200, Rue de la Loi, B-1049 Brussels, Belgium.
4. National Bureau of Standards, Office of Standard Reference Materials, B311, Chemistry Building, Washington DC 20234, USA.
5. National Physical Laboratory, Officer of Reference Materials, Teddington, Middlesex TW110LW, United Kingdom.
6. British Standards Institution, 10 Blackfriars Street, Manchester, M3 5DT, United Kingdom.
7. Commonwealth Scientific and Industrial Research Organisation, National Measurement Laboratory, PO Box 218, Lindfield, NSW, 2070, Australia.
8. Canadian Certified Reference Materials Project, c/- Mineral Science Laboratories, CAMNET, Canada Centre for Mineral and Energy Technology, 555 Booth Street, Ottawa, Ontario, K1A OG1, Canada.
9. Bureau National de Metrologie (BNM), 8-10, Rue Crillon, 75194, Paris Cedex 04, France.
10. Service des Materiaux de Reference (SMR), 1 Rue Gaston Boissier, 75015 Paris, France.
11. Bundesanstalt fur Materialprufung (BAM), Unter den Eichen 87, D-1000 Berlin 45, Germany, F.R.
12. National Office of Measures, PO Box 19, H-1531 Budapest, Hungary.
13. Standards Department, Agency of Industrial Science and Technology, Ministry of International Trade and Industry, 3-1, Kasumigaseki 1, Chiyodaku, Tokyo, Japan.
14. Division of Physico-Chemical Metrology, National Board for Quality Control and Measures, 2 Elektoralna Street, Warsaw, Poland.
15. South African Bureau of Standards, Private Bag X191, Pretoria, Transvaal 0001, Republic of South Africa.
16. U.S. Geological Survey, Dr F.J. Flanagan, USGS, Reston, Virginia 22092, USA.
17. Micromatter Co., 197-34th Avenue E., Seattle, Washington 98112, USA.
18. Goodfellow Metals Ltd., Milton Road, Cambridge CB4 4DJ, England.

which noble gases can be used for calibration but the substrate should be a material such as Si which minimises the migration of the implanted species. Calibration samples consisting of Bi implanted Si wafers have been prepared at Harwell with 4.77×10^{15} atom cm^{-2} (L'Ecuyer et al., 1979; Cohen et al., 1983; Davies et al., 1986). The ion dose, D ions cm^{-2}, is given by:

$$D = 6.2415 \times 10^{12} \, C/q \, a \qquad (13.1)$$

where C is the integrated charge in μC, q is the charge on each ion and a is the beam area (cm^2).

TABLE 13.4
Typical Erosion Rates of Frozen Gases[a]

Gas	Ion	E_1 (MeV)	Erosion Rate (molecules/ion)
H_2O, CO_2	H	1.0	0.5
H_2O, CO_2	He	0.75	10–20
N_2O	D	0.97	20
NO, NH	D	0.97	1

[a]Davies *et al.*, 1983

13.1.5 Frozen Gases

Thin stoichiometric layers of frozen gases are useful for comparing cross-sections, for example in light isotopes (Davies *et al.*, 1983). These can be prepared by using a clean metal sample (in good vacuum) mounted on a cooled backing which is heatable for temperature control. When the chosen gas is bled into the sample chamber a thin layer is frozen onto the surface of the metal. Ion irradiation causes erosion of the gas molecules and some typical rates are listed in Table 13.4 for sample temperatures at which thermal sublimation is negligible.

13.1.6 H Standards

Plastic films are very prone to ion beam damage which causes a loss of H. They can therefore only be used as standards for H analysis by ERA or NRA if a reliable extrapolation can be made to estimate the yield equivalent for zero dose (Rudolf *et al.*, 1986). H implanted Si is also useful as a reference material provided that there is no interference from a surface peak due to adsorbed H. Pyrolitic C or amorphous Si prepared in the presence of H and Ti or Ta hydrides are stable under ion irradiation and have adequate electrical conductivity. They can be used as secondary standards following independent calibration of their H content (Kamykowski *et al.*, 1979; Westerberg *et al.*, 1985).

13.1.7 Trace Elements

Trace quantities of a wide range elements on thin foils are available commercially (Heagney and Heagney, 1976). If necessary such materials can be calibrated by RBS (Mingay, 1983).

TABLE 13.5
Composition of typical geological materials (μg g^{-1})[a]

El.	Basalt[b] BHVO-1	Granite[b] MA-N	Rhyolite[b] RGM-1	Coal[a] SABS-SARM-18	Fly Ash[a] NBS-1633a	Soil[a] IAEA-Soil7	Sea Water[c]
Al	7.3E4	9.3E4	7.3E4	1.4E4	1.4E5	4.7E4	1.8E-3
As	1.5	1.3E1			1.5E2	1.3E1	4.E-3
Au	1.5E-3		3.3E-4				3.8E-5
B	2.3	1.7E1	2.9E1	3.0E1			4.4E1
Ba	1.4E2	4.2E1	8.2E2	7.8E1	1.5E3	1.6E2	2.0E-2
Be	0.9	2.8E2	2.5	4.1	1.2E1		
Br				3.		7.	6.7E1
Ca	8.2E4	4.2E3	8.1E3	1.3E3	1.1E4	1.6E5	4.1E2
Cl	9.2E1	1.4E2	4.9E2				1.9E4
Co	4.5E1	1.E1	2.0	6.7	4.6E1	8.9	
Cr	3.0E2	3.E1	2.5	1.6E1	2.0E2	6.0E1	2.9E-4
Cs	8.6E-2	6.4E2	9.9	1.	1.1E1	5.4	3.9E-4
Cu	1.4E2	1.4E2	1.1E1	5.9	1.2E2	1.1E1	4.8E-4
F	3.8E2	1.5E4	3.6E2			4.8E2	1.1
Fe	8.5E4	3.3E3	1.3E4	2.0E3	9.4E4	2.6E4	2.2E-3
Ga	2.2E1	5.9E1	1.4E1	8.	5.8E1	1.0E1	
Hf	4.2	4.5	6.1	1.7	7.6	5.1	
Hg	4.0E-3			4.E-2	1.6E-1	4.E-2	3.9E-5
K	4.6E3	2.6E4	3.6E4	1.2E3	1.9E4	1.2E4	4.0E2
La	1.7E1	1	2.5E1	1.0E1		2.8E1	
Li	4.5	4.9E3	5.1E1	1.1E1		3.1E1	1.8E-1
Mg	4.3E4	2.4E4	1.7E3	6.6E2	4.6E3	1.1E4	1.3E3
Mn	1.3E3	3.1E2	2.8E2	2.2E1	1.9E2	6.3E2	1.9E-4
Mo	1.0		2.3	1.	2.9E1	2.5	9.3E-3
Na	1.6E4	4.3E4	3.0E4	1.3E2	1.7E3	2.4E3	1.1E4
Nb	1.9E1	1.7E2	9	6.		1.2E1	
Ni	1.2E2	3	1.E1	1.1E1	1.3E2	2.6E1	1.7E-3
P	1.2E3	6.1E3	2.1E2	3.0E1		4.6E2	5.9E-2
Pb	4.0	2.9E1	2.2E1	5.	7.2E1	6.0E1	2.0E-5
Rb	1.0E1	3.6E3	1.6E2	8.1	1.3E2	5.1E1	1.2E-1
S	1.0E2	1.0E2	7.5E1	5.6E3			9.0E2
Sb	1.7E-1	1.9	1.3	0.3	7.	1.7	2.4E-4
Sc	3.0E1	0.2	5.0	4.3	4.0E1	8.3	
Si	2.3E5	3.1E5	3.4E5	2.9E4	2.3E5	1.8E5	2.0
Sn	2.2	1.1E3	3.9	1.			
Sr	4.4E2	8.4E1	1.1E2	4.4E1	8.3E2	1.1E2	7.8
Ta	1.1	3.1E2	1.0	0.3		0.8	
Th	1.1	1	1.6E1	3.4	2.5E1	8.2	
Ti	1.6E4	5.9E1	1.6E3	6.8E2	8.0E3	3.0E3	9.8E-4
U	0.4	1.2E1	5.8	1.5	1.0E1	2.6	3.1E-3
V	3.1E2	4.6	1.4E1	2.3E1	3.0E2	6.6E1	2.4E-3
W	0.3	7.0E1	1.6	2			9.0E-5
Y	2.8E1	1	2.7E1	1.2E1		2.1E1	
Zn	1.0E2	2.2E2	3.2E1	5.5	2.2E2	1.0E2	4.8E-3
Zr	1.8E2	2.7E1	2.1E2	6.7E1		1.9E2	2.7E-5

[a] Muramatsu and Parr (1985); [b] Gladney and Goode (1981); [c] Whitfield and Turner (1981)

TABLE 13.6
Composition of typical biological materials (μg g^{-1})[a]

El.	Citrus Leaves NBS-1572	Cotton Cellulose IAEA-V9	Pine Needles NBS-1575	Whole Blood[b]	Urine NBS-2670	Oyster Tissue NBS-1566	Milk Powder IAEA-A11	Bovine Liver NBS-1577a
Ag						8.9E-1		4.0E-2
Al	9.2E1	4.4E1	5.5E2	3.9E-1	1.8E-1		1.3	2.
As	3.1		2.1E-1	1.E-2	1.5E-2	1.3E1	4.8E-2	4.7E-2
B				0.1			1.7	
Ba	2.1E1	9.0		6.8E-2			2.5E-1	
Br	8.2	1.7E-1	9.	4.7		5.5E1	1.4E1	9.
Ca	3.2E4	2.4E2	4.1E3	6.1E1	1.1E2	1.5E3	1.3E4	1.2E2
Cd	3.0E-2	2.E-3	4.E-1	5.2E-3	4.0E-4	3.5	5.3E-1	4.4E-1
Cl	4.1E2	6.0E2		2.9E3	4.4E3	1.0E4	9.1E3	2.9E3
Co	2.0E-1		1.E-1	1.E-2		4.0E-1	5.E-3	2.1E-1
Cr	8.0E-1	1.1E-1	2.6	1.E-1	1.3E-2	6.9E-1	2.6E-1	
Cs	9.8E-2			3.8E-3			5.1E-2	
Cu	1.7E1	5.9E-1	3.0	1.0	1.3E-1	6.3E1	8.4E-1	1.6E2
F				0.5		5.2	2.6E-1	
Fe	9.0E1	1.1E1	2.0E2	4.5E2		2.0E2	3.7	1.9E2
Hg	8.0E-2	6.E-2	1.5E-1	7.8E-3	2.E-3	5.7E-2	2.5E-3	4.E-3
I	1.8			5.7E-2		2.8	1.5	
K	1.8E4		3.7E3	1.6E3	1.5E3	9.7E3	1.7E4	1.0E4
La	1.9E-1		2.E-1	4.E-3				
Mg	5.8E3	5.3E1		3.8E1	6.3E1	1.3E3	1.1E3	6.0E2
Mn	2.3E1	1.5E-1	6.8E2	1.E-2	3.E-2	1.8E1	3.8E-1	9.9
Mo	1.7E-1	3.4E-2					1.3	3.5
N	2.9E4		1.2E4					1.1E5
Na	1.6E2	5.6E1		2.0E3	2.6E3	5.1E3	4.4E3	2.4E3
Ni	6.0E-1	9.E-2	3.5	1.E-2	7.E-2	1.0	9.3E-1	
P	1.3E3		1.2E3	3.5E2		8.1E3	9.1E3	1.1E4
Pb	1.3E1	2.5E-1	1.1E1	2.1E-1	1.E-2	4.8E-1	2.7E-1	1.4E-1
Rb	4.8		1.2E1	2.5		4.5	3.1E1	1.3E1
S	4.1E3	5.4E1		1.8E3	4.3E2	7.6E3		7.8E3
Sb	4.0E-2		2.E-2	3.3E-3			1.4E-3	3.E-3
Sc	1.0E-2	9.E-3	3.E-2					
Se	2.5E-2	1.5E-2		1.7E-1	3.0E-2	2.1	3.4E-2	7.1E-1
Si				3.9			3.4E1	
Sm	5.2E-2	3.E-3						
Sn	2.4E-1	6.E-1					1.2E-1	
Sr	1.0E2	6.5E-1	4.8	3.1E-2		1.0E1	5.4	1.4E-1
Te	2.0E-2							
Ti					5.4E-2			
U		5.E-2	2.0E-2	5.E-4		1.2E-1		7.1E-4
V		9.E-2				2.3	1.0E-1	
Zn	2.9E1			7.0		8.5E2	3.9E1	1.2E2

[a]Muramatsu and Parr (1985); [b] Bowen (1979)

13.2 THICK SAMPLE STANDARDS

The composition of a few typical materials and standards of interest in geo-, bio- and museum science is given in Tables 13.5 to 13.7. More information can be obtained from the references used in the tables and in catalogs from organisations listed in Table 13.3. In the preparation of local standards, the expected composition may be achieved with varying degrees of success depending on factors such as mass loss during heating or contamination in the case of trace elements. Intercomparisons by different techniques or laboratories are essential to confirm actual compositions and accidental contamination must be avoided by careful handling (Moody, 1982; Keith *et al.*, 1983). Homogeneity is also vital for IBA since the volume analysed is often very small.

TABLE 13.7
Typical composition of archaeological and museum materials ($\mu g\ g^{-1}$)

El.	Ceramics	Cu Alloys	Ag Alloys	Glasses	Obsidian	Papers
Ag		1-5E4	matrix	1-1E4		
Al				0.2-7E4	5E4-1.2E5	50-1.7E4
As	1-2.5E2	1-1E5	0.1-1E2	5-5E3	1-2.5E2	<2
Au		<550	<5E4	<0.1		<60
B	25-3.5E2					
Ba	80-4E3	<500		<1E5	5-2.5E3	2-1.7E2
Bi		<5E4	<2E4			
Br	1-25			1-8	0.1-8	0.4-75
Ca	<2E5	20-5E4	1-1E4	<2E5	<6E4	5E2-8E4
Cd		<100	<5E4			
Ce	20-1.6E2			1-15	10-3.5E2	
Cl	45-3E3		1E2-2E4		50-9E3	
Co	1-1E2	<5E4	<5E2	<2E3	<2	<4
Cr	500-2E3	1-2E3		10-1.2E2	<10	<30
Cs	1-1.5E2			<1.5	<50	
Cu	1-6E2	matrix	1E2-2E5	1E2-6E4	1-30	2-1E2
Eu	1-5		<0.5	<4		
F					3E2-5E.	
Fe	<1.5E5	<5E4	<1E4	<1.2E5	<1E5	20-3.3E3
Ga	10-55		<0.15		5-70	
Ge	0.5-4		0.3-1			
Hf	1-20		0.2-1	1-50		
Hg		<5E2			27-45	
I	0.5-8		<0.4			
In		0.1-1E2	<25		4E-2	
Ir				<10		
K	<7E4	10-3E4		<2E5	<6E4	16-1.2E4
La	1-1E2			0.8-2	20-1.5E2	<6

TABLE 13.7 *cont.*

El.	Ceramics	Cu Alloys	Ag Alloys	Glasses	Obsidian	Papers
Mg	<5E4			<5E4	1-2.5E3	2E2-9E3
Mn	60-1E4	1-5E3	<1E2	<1.5E4	1E2-3E3	0.5-2E2
Mo		<500	<1			
Na	<4E4		<1E3	<2E5	<5E4	30-4E3
Nb	20-60				1-3E2	
Ni	10-6E2	<5E4	1-5E2	10-2E3	2-25	
P	1.5E2-6E3	<5E3		<5E4		
Pb	5-1.5E3	<3E5	1E2-5E4	<3E5	1-85	
Pt		<1000	<10			
Rb	1-1E3	<1E3		1-5E2	40-7E2	
S	1E2-2E3	10-3E4		10-5E3		
Sb	<2	<1E5	<100	<4E5	0.2-2	<5
Sc	1-1E2			0.3-1.5	0.1-10	<12
Se		1-2E3	<0.5			
Si	2.5E5-4E5	<5E3		2.5E5-4E5	2.5E5-4E5	
Sm	1-20			0.3	3-30	<1.5
Sn		<3E5	<5E4	<7E4		
Sr	10-2E3	<5E2		10-2E3	1-1.3E3	
Ta	1-10				1-20	
Te		<5E2	<1			
Th	5-50			0.3-2	1-1E2	
Ti	<1.5E4	1-1E3		1E3-5E3	<2E4	
U	2-5				1-5	
V	1-3E2	1-1E2	<0.5	10-2E2		<4
Y	20-1E2		<0.03		<2E2	
Zn	10-3.5E2	<3E5	<5E4	10-1.5E4	10-6E2	6-4.5E2
Zr	10-5E2	<1E2	<0.3	10-5E2	15-2.2E3	

13.3 PROPERTIES OF USEFUL MATERIALS

Some properties of useful materials are listed in Tables 13.8 to 13.9. The properties of foils used as external beam windows are listed in Table 13.10 and Table 13.11 provides information on detector and shielding materials. Photon attenuation can be calculated from:

$$I_t = I_0 \exp -[(\mu/\rho)\, t\, \rho] \tag{13.2}$$

where I_0 and I_t are the initial intensity and the intensity after passing through a thickness (t) of material having density ρ and mass attenuation coefficient (μ/ρ). Mass attenuation coefficients are independent of

TABLE 13.8

Properties of some alloys, compounds and composites

Material	Density (g cm^{-3})	Melting Point (°C)	Crystal Structure	Electrical Resistivity Ω cm	Thermal Conductivity Wcm^{-1}°C^{-1}
(a) *Compounds*					
AlN	3.26	Sublimes 2000	hcp	–	–
α-Al$_2$O$_3$	3.97	2872	hcp	10^{16}	0.2
BN	2.25	2730	hcp	2×10^{-3}	–
CaCl$_2$	2.15	782	fcc	–	0.04
CdS	4.82	Sublimes 980	fcc	–	0.2
GaAs	5.35	1338	fcc	370	0.54
Ga$_2$O$_3$	5.88	1795	rhombo	–	–
InP	4.79	1330	fcc	800	0.68
InSb	5.78	535	fcc	160	0.18
KCl	1.98	770	fcc	0.8	0.09
NaCl	2.17	801	fcc	0.2	0.07
NiSi$_2$	4.83	~1000	hcp	50	–
β-SiC	3.22	2700	fcc	~150	0.9
α-SiC	3.22	2700	hcp	~150	0.9
SiO$_2$	2.27	1723	a	10^{15}	0.01
α-SiO$_2$	2.6	1723	hcp	10^{15}	0.01
Si$_3$N$_4$	3.44	1900	a	10^{14}	–
Ta$_2$O$_5$	8.2	1872	rhombo	–	–
TiC	4.93	3140	fcc	2×10^{-4}	0.2
TiN	5.43	2930	bcc	2.2×10^{-5}	–
TiSi$_2$	4.04	1540	ortho	1.5×10^{-5}	–
WC	15.63	2870	hcp	7×10^{-5}	–
WSi$_2$	9.8	2050	tetra	3.3×10^{-5}	–
YBa$_2$Cu$_3$O$_7$	6.27	Decomp.>650	ortho	0(at<96K)	–
ZnSe	5.42	1100	fcc	–	0.2
ZrO$_2$	5.6	2715	monocl	~10^6	0.02
(b) *Metallic Alloys*					
Carbon Steel AISI-SAE 1020 (Fe, C, Mn, Si, P)	7.86	1515	fcc	10^{-5}	1.0
Type 304 Stainless Steel	8.02	1427	fcc	7×10^{-5}	0.3
Iconel X (Fe, Cr, Ni)	8.25	1399	fcc	1.2×10^{-4}	0.3
Aluminium Alloy 3003 (ASTM B221) (Al, Cu, Mg, Mn, Si)	2.73	649	fcc	4×10^{-6}	2.8
Aluminium bronze (ASTM B169, alloy A)	7.8	1038	fcc	1.2×10^{-5}	1.3

TABLE 13.8 *cont.*

Material	Density (g cm^{-3})	Melting Point (°C)	Crystal Structure	Electrical Resistivity Ω cm	Thermal Conductivity Wcm^{-1}°C^{-1}
Beryllium copper 25 (ASTM B194) (Cu, Be, Co)	8.25	927	fcc	–	0.2
Nickel silver 18% (ASTM B122, No. 2)	8.8	1110	fcc	3×10^{-5}	0.6
Cupronickel 30%	8.95	1227	fcc	3.5×10^{-5}	0.5
Cupronickel 55-45 (constantan)	8.9	1260	fcc	5×10^{-5}	0.4
Yellow brass (ASTM B36)	8.47	932	–	7×10^{-6}	2.2
Red brass (ASTM B30, 4A) (85Cu, 152n)	8.7	996	–	1×10^{-5}	1.3
Solder 50:50 (Pb, Sn)	8.9	216	–	1.5×10^{-5}	0.8
Magnesium Alloy (AZ31B) (Mg, Al, Zn, Mn)	1.77	627	hcp	9×10^{-6}	1.4
(c) *Non-Metallic Composites* Concrete (C, O, Si, Ca, H, Mg)	2.34	~2000	a	$>10^{10}$	$<5\times10^{-3}$
Glass (soda-lime) (O, Si, Na, Ca, Mg, Al)	2.6	800-1150	a	$>10^{13}$	10^{-2}
Glass (pyrex) (O, Si, B, Na, Al)	2.23	1200	a	$>10^{13}$	10^{-2}
PMMA resist (H, C, O)	0.95	135	a	$>10^{15}$	2×10^{-3}

TABLE 13.9
Properties of useful plastics[a]

Material	Composition	Density (g/cm³)	Resistivity (Ω-cm)	Manufacturer
Formvar	C, H, O (varies)	1.214–1.229		Monsanto
Havar	Be (0.3%) C (1%)[b] Cr (22.2%) Mn (1.7%) Fe (18.1%) Co (41.6%) Ni (12.8%) Mo (1.4%) W (0.9%)	8.3		Hamilton Watch Co.
Kapton	$(C_{22}H_{10}N_2O_4)_n$	1.42	10^{18}	Du Pont
Perspex (Lucite, Plexiglas)	$(C_5H_8O_2)_n$	1.18–1.19	$>10^{14}$	Du Pont; Rohm and Haas
Mica	$K_2O, 3Al_2O_3, 6SiO_2, 2H_2O$	2.76–3.00		
Mylar	$(C_{10}H_8O_4)_n$	1.38–1.395	10^{14}	Du Pont
Nylon	$(C_{12}H_{22}N_2O_2)_n$	1.08–1.14	10^{12}-10^{15}	Du Pont
Polyethylene (Polythene)	$(CH_2{:}CH_2)_n$	0.910–0.965	$>10^{16}$	
Polystyrene	$(C_6H_5CH{:}CH_2)_n$	0.98–1.10	$>10^{16}$	Dow Chemical
Teflon	$(CF_2)_n$	2.1–2.2	$>10^{18}$	Du Pont
VYNS	$CH_2 CHCl$ (90%) $CH_2CHO_2CCH_3$ (10%)	1.36		Union Carbide

[a]Marion and Young (1968); [b]Rauhala and Raisanen (1985)

density and the physical state of an absorber (Jaeger *et al.*, 1968). For a mixture, *m*:

$$(\mu/\rho)_m = \Sigma \, c_i \, (\mu/\rho)_i \qquad (13.3)$$

where c_i is the weight fraction of component *i*.

Mass attenuation coefficients for X-rays (E < 20 keV) can be calculated from information given in Table 14.24. Coefficients for gamma-rays ($0.1 < E_\gamma < 20$ MeV) are given in Table 13.11c for a number

TABLE 13.10
External beam windows

Material	Density g/cm^3	A Tensile Strength MPa	Thermal Conductivity W m^{-1} K^{-1}	B 2.7 MeV Proton Energy Loss keV/μm	Melting Point °C	C Typical Thickness μm	B/A Heat Build Up
Mylar[a]	1.39	172	0.15	20	250	3.5	133
Kapton[b]	1.42	172	0.15	20	–	8	133
Beryllium[c]	1.85	414	150	18	1285	25	0.12
Aluminium	2.70	82	222	24.5	660	5	0.11
Nickel	8.90	379	58	63	1435	1	1.09

[a]Mylar suffers radiation damage at doses greater than 5×10^8 rads (Koehler *et al.*, 1965; Rauhala and Raisanen, 1985)
[b]Kapton is >50 times more resistant to radiation damage than Mylar[a] (Rauhala and Raisanen, 1985).
[c]High background above 2 MeV from gamma-ray and neutron production (Raisanen and Antilla, 1982).

of materials. The thickness of concrete to reduce the radiation dose from MeV neutrons to one tenth of its initial value is 70 to 85 g cm^{-2} (NCRP, 1977). It should be noted that when considering radiation shielding, attenuation must be corrected for the build-up of secondary radiation (Price *et al.*, 1957).

13.4 PROPERTIES OF THE ELEMENTS

The periodic table of the elements with physical and chemical data are given in Tables 13.12 and 13.13 by permission of Sargent-Welch Co., Skokie, Illinois, USA.

13.5 TABLE OF ISOTOPES

Atomic masses (amu) (Wapstra and Bos, 1977) and isotopic abundances (Ab at%) (Holden *et al.*, 1984) are listed in Table 13.14. The atomic masses are for neutral atoms, including Z electrons. When calculating scattering or reaction kinematics, the nuclear masses ($M - Z m_e$) should be used but equivalent results are obtained by using atomic masses provided that they are also used for the incident ion. *DO NOT USE* ($M - q m_e$), where q is the ion charge. The lists of physical constants and units included on the endpapers are from Cohen (1976).

TABLE 13.11

Properties of detector and shielding materials

a. Scintillator Materials[a]

Type	Density g.cm^{-3}	R.I.	M.P. C	τ ns	λ nm	Application
NE102 Plastic	1.032	1.581	75	3	425	α, β, γ, n
NE213 Liquid	0.88	1.508	141	2.4	425	α, β, n
NE226 Liquid	1.61	1.38	80		425	γ
NE907 Glass	2.67	1.566	1200	5	395	n
Anthracene	1.25	1.62	217	32	447	
Stilbene	1.16	1.626	125	4	410	α, β, γ, n
NaI(Tl)	3.67	1.775	650	230	413	X, γ
LiI(Eu)	4.06		445	1200	475	n
CsI(Tl)	4.51	1.788	620	700	420	ions, γ
CsF	3.59	1.434	684	5	390	γ
CaI$_2$(Eu)	3.96		573	790	470	γ
ZnS(Ag)	4.09	2.356	1850	200	450	α, ions

τ = decay constant; γ = wavelength of emission

b. Semiconductor Materials[b]

	Si	Ge	GaAs	CdTe	HgI$_2$
Z	14	32	31/33	48/52	80/53
Band Gap, 300 K (eV)	1.12	0.66	1.43	1.47	2.13
Dielectric Constant	12	16	11	15	–
Intrinsic Carrier Density, 300 K (cm^{-3})	1.5 10^{10}	2.4 10^{13}	1.1 10^7		
Electron Mobility, 300 K (cm^2/Vs)	350	3900	8500	800	–
Hole Mobility, 300 K (cm^2/Vs)	80	1900	400		
Energy per e-h pair (eV)	3.62	2.90	4.2	4.43	6.5
Minority Carrier Lifetime (s)	2.5 10^{-3}	10^{-3}	10^{-8}		

c. Shielding Materials, Mass Attenuation Coefficients (μ/ρ cm^2 g^{-1}) for gamma-rays[c]

E$_\gamma$ (MeV)	0.1	0.2	0.5	1.0	2.0	5.0	10.0	20.0
H	.294	.243	.173	.126	.088	.051	.033	.022
C	.149	.122	.087	.064	.045	.027	.020	.016
Al	.162	.120	.084	.061	.043	.028	.023	.022
Fe	.342	.139	.083	.060	.043	.032	.030	.032
Mo	1.05	.228	.086	.058	.041	.034	.036	.043
Pb	5.62	.969	.154	.069	.045	.042	.049	.061
Concrete	.171	.125	.087	.064	.045	.029	.023	.021

[a]Nuclear Enterprises (n.d.); [b]Knoll (1979); Sze (1969); [c]Jaeger et al., (1968).

REFERENCES

Amsel, G. and Davies, J.A. (1983). *Nucl. Instrum. Methods* **218**, 177.

Bean, J.C. (1981). *In* 'Impurity Doping Processes in Silicon' (Wang, F.Y. ed.), North-Holland, Amsterdam.

Bowen, H.J.M. (1979). 'Environmental Chemistry of the Elements', Academic Press, London.

Bradshaw, S.D., Cohen, D., Katsaros, A., Tom, J. and Owen, F.J. (1987). *J. Appl. Physiol.* **161**, 1296.

Cohen, C., Davies, J.A., Drigo, A.V. and Jackman, T.E. (1983). *Nucl. Instrum. Methods* **218**, 147.

Cohen, E.R. (1976). *Atomic and Nuclear Data Tables* **188**, 587.

Davies, J.A., Jackman, T.E., Plattner, H. and Bubb, I. (1983). *Nucl. Instrum. Methods* **218**, 141.

Davies, J.A., Jackman, T.E., Eschbach, H.L., Domba, W., Watjen, U. and Chivers, D. (1986). *Nucl. Instrum. Methods* **B15**, 238.

Gladney, E.S. and Goode, W.E. (1981). *Geostandards Newsletter* **5**, 31.

Heagney, J.M. and Heagney, J.S. (1976). *Nucl. Instrum. Methods* **167**, 137.

Holden, N.E., Martin, R.L. and Barnes, I.C. (1984). *Pure and Applied Chem.* **56**, 675.

Jaeger, R.G., Blizard, E.P., Chilton, A.B., Grotenhuis, M., Honig, A., Jaeger, T.A. and Eisenlohr, H.H. (1968). 'Engineering Compendium on Radiation Shielding', Springer-Verlag, N.Y.

Jastrezebski, Z.D. (1987). 'The Nature and Properties of Engineering Materials (3rd Ed.)', John Wiley & Sons, New York.

Kamykowski, E.A., Kuehne, F.J., Schneid, E.J. and Schulte, R.L. (1979). *Nucl. Instrum. Methods* **165**, 573.

Keith, L.H., Crummett, W., Deegan Jr. J., Libby, R.A., Taylor, J.K. and Wentler, G. (1983). *Anal. Chem.* **55**, 2210.

Knoll, G.F. (1979). 'Radiation Detection and Measurement', Wiley, New York.

Koehler, A.M., Measday, D.F. and Morrill, D.H. (1965). *Nucl. Instrum. Methods* **33**, 341.

L'Ecuyer, J., Davies, J.A. and Matsunami, N. (1979). *Nucl. Instrum. Methods* **160**, 337.

Marion, J.B. and Young, F.C. (1968). 'Nuclear Reaction Analysis', North-Holland, Amsterdam, 136.

Mingay, D.W. (1983). *J. Radioanal. Chem.* **78**, 127.

Moody, J.R. (1982). *Anal. Chem.* **54**, 1358A.

Muramatsu, Y. and Parr, R.M. (1985). IAEA/RL/128.

Murarka, S.P. (1983). 'Silicides for VLSI Applications', Academic Press, New York.

NCRP (1977). 'Radiation Protection Guidelines for 0.1–100 MeV Particle Accelerator Facilities', Report No. 51, National Council for Radiation Protection and Measurements, Washington.

Nuclear Enterprises (n.d.). 'Table of Physical Constants', Nuclear Enterprises, UK.

Price, B.T., Horton, C.C. and Spinney, K.T. (1957). 'Radiation Shielding', Pergamon Press, London.

Rauhala, E. and Raisanen, J. (1985). *Nucl. Instrum. Methods* **B12**, 321.

Raisanen, J. and Antilla, A. (1982). *Nucl. Instrum. Methods* **196**, 489.

Rudolph, W., Bauer, C., Brankoff, K., Grotzschel, R., Heiser, C. and Herrman, F. (1986). *Nucl. Instrum. Methods* **B15**, 508.

Siejka, J., Nadai, J.P. and Amsel, G. (1971). *J. Electrochem. Soc.* **118**, 727.

Sze, S.M. (1969). 'Physics of Semiconductor Devies', Wiley, New York.

Venkatesan, T., Wolf, T., Allara, D., Wilkens, B.J. and Taylor, G.N. (1983). *Appl. Phys. Lett.* **43**, 934.

Wapstra, A.H. and Bos, K. (1977). *Atomic and Nuclear Data Tables* **18**, 587.

Westerberg, L., Svensson, L.E., Karlsson, E., Richardson, M.W. and Lundstrom, K. (1985). *Nucl. Instrum. Methods* **B9**, 49.

Whitfield, M. and Tuner, D.R. (1981). In 'Marine Electrochemistry', Whitfield, M. and Jagner, D. (Eds), Wiley, Chichester.

TABLE 13.12
Periodic table of the elements

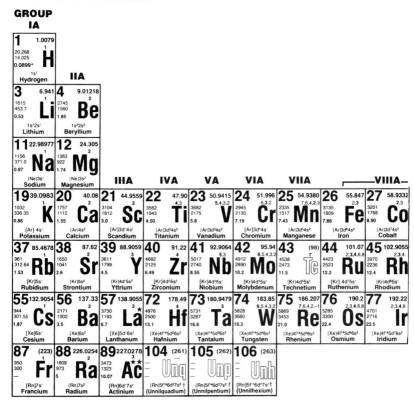

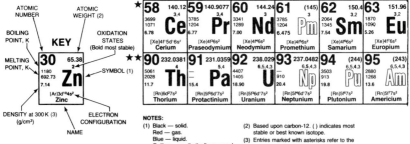

NOTES:

(1) Black — solid.
Red — gas.
Blue — liquid.
Outline — synthetically prepared.

(2) Based upon carbon-12. ( ) indicates most stable or best known isotope.

(3) Entries marked with asterisks refer to the gaseous state at 273 K and 1 atm and are given in units of g/l.

VIII

2	4.00260
4.215	
0.95 (at 26atm)	
0.1787*	**He**
1s²	
Helium	

IIIB	IVB	VB	VIB	VIIB

5 10.81	6 12.011	7 14.0067	8 15.9994	9 18.998403	10 20.179
3	±4,2	±3,5,4,2	-2	-1	
4275	4470*	77.35	90.18	84.95	27.096
2300 **B**	4100* **C**	63.14 **N**	50.35 **O**	53.48 **F**	24.553 **Ne**
2.34	2.62	1.251*	1.429*	1.696*	0.901*
1s²2s¹p¹	1s²2s²p²	1s²2s²p³	1s²2s²p⁴	1s²2s²p⁵	1s²2s²p⁶
Boron	Carbon	Nitrogen	Oxygen	Fluorine	Neon

13 26.98154	14 28.0855	15 30.97376	16 32.06	17 35.453	18 39.948
3	4	±3,5,4	±2,4,6	±1,3,5,7	
2793	3540	550	717.75	239.1	87.30
933.25 **Al**	1685 **Si**	317.30 **P**	388.36 **S**	172.16 **Cl**	83.81 **Ar**
2.70	2.33	1.82	2.07	3.17*	1.784*
[Ne]3s²p¹	[Ne]3s²p²	[Ne]3s²p³	[Ne]3s²p⁴	[Ne]3s²p⁵	[Ne]3s²p⁶
Aluminum	Silicon	Phosphorus	Sulfur	Chlorine	Argon

IB	IIB

28 58.70	29 63.546	30 65.38	31 69.72	32 72.59	33 74.9216	34 78.96	35 79.904	36 83.80
2,3	2,1	2	3	4	±3,5	-2,4,6	±1,5	
3187	2836	1180	2478	3107	876(subl.)	958	332.25	119.80
1726 **Ni**	1357.6 **Cu**	692.73 **Zn**	302.90 **Ga**	1210.4 **Ge**	1081(28atm) **As**	494 **Se**	265.90 **Br**	115.78 **Kr**
8.90	8.96	7.14	5.91	5.32	5.72	4.80	3.12	3.74*
[Ar]3d⁸4s²	[Ar]3d¹⁰4s¹	[Ar]3d¹⁰4s²	[Ar]3d¹⁰4s²p¹	[Ar]3d¹⁰4s²p²	[Ar]3d¹⁰4s²p³	[Ar]3d¹⁰4s²p⁴	[Ar]3d¹⁰4s²p⁵	[Ar]3d¹⁰4s²p⁶
Nickel	Copper	Zinc	Gallium	Germanium	Arsenic	Selenium	Bromine	Krypton

46 106.4	47 107.868	48 112.41	49 114.82	50 118.69	51 121.75	52 127.60	53 126.9045	54 131.30
2,4	1	2	3	2,4	±3,5	-2,4,6	±1,5,7	
3237	2436	1040	2346	2876	1860	1261	458.4	165.03
1825 **Pd**	1234 **Ag**	594.18 **Cd**	429.76 **In**	505.06 **Sn**	904 **Sb**	722.65 **Te**	386.7 **I**	161.36 **Xe**
12.0	10.5	8.65	7.31	7.30	6.68	6.24	4.92	5.89*
[Kr]4d¹⁰	[Kr]4d¹⁰5s¹	[Kr]4d¹⁰5s²	[Kr]4d¹⁰5s²p¹	[Kr]4d¹⁰5s²p²	[Kr]4d¹⁰5s²p³	[Kr]4d¹⁰5s²p⁴	[Kr]4d¹⁰5s²p⁵	[Kr]4d¹⁰5s²p⁶
Palladium	Silver	Cadmium	Indium	Tin	Antimony	Tellurium	Iodine	Xenon

78 195.09	79 196.9665	80 200.59	81 204.37	82 207.2	83 208.9804	84 (209)	85 (210)	86 (222)
2,4	3,1	2,1	3,1	4,2	3,5	4,2	±1,3,5,7	
4100	3130	630	1746	2023	1837	1235	610	211
2045 **Pt**	1337.58 **Au**	234.28 **Hg**	577 **Tl**	600.6 **Pb**	544.52 **Bi**	527 **Po**	575 **At**	202 **Rn**
21.4	19.3	13.53	11.85	11.4	9.8	9.4	—	9.91*
[Xe]4f¹⁴5d⁹6s¹	[Xe] 4f¹⁴5d¹⁰6s¹	[Xe]4f¹⁴5d¹⁰6s²	[Xe]4f¹⁴5d¹⁰6s²p¹	[Xe]4f¹⁴5d¹⁰6s²p²	[Xe]4f¹⁴5d¹⁰6s²p³	[Xe]4f¹⁴5d¹⁰6s²p⁴	[Xe]4f¹⁴5d¹⁰6s²p⁵	[Xe]4f¹⁴5d¹⁰6s²p⁶
Platinum	Gold	Mercury	Thallium	Lead	Bismuth	Polonium	Astatine	Radon

The A & B subgroup designations, applicable to elements in rows 4, 5, 6, and 7, are those recommended by the International Union of Pure and Applied Chemistry. It should be noted that some authors and organizations use the opposite convention in distinguishing these subgroups.

* Estimated Values

64 157.25	65 158.9254	66 162.50	67 164.9304	68 167.26	69 168.9342	70 173.04	71 174.967
3	3,4	3	3	3	3,2	3,2	3
3539	3496	2835	2968	3136	2220	1467	3668
1585 **Gd**	1630 **Tb**	1682 **Dy**	1743 **Ho**	1795 **Er**	1818 **Tm**	1097 **Yb**	1936 **Lu**
7.89	8.27	8.54	8.80	9.05	9.33	6.98	9.84
[Xe]4f⁷5d¹6s²	[Xe]4f⁹6s²	[Xe]4f¹⁰6s²	[Xe]4f¹¹6s²	[Xe]4f¹²6s²	[Xe]4f¹³6s²	[Xe]4f¹⁴6s²	[Xe] 4f¹⁴5d¹6s²
Gadolinium	Terbium	Dysprosium	Holmium	Erbium	Thulium	Ytterbium	Lutetium

96 (247)	97 (247)	98 (251)	99 (252)	100 (257)	101 (258)	102 (259)	103 (260)
3	4,3	3					
1340	900						
13.511 **Cm**	**Bk**	**Cf**	**Es**	**Fm**	**Md**	**No**	**Lr**
[Rn]5f⁷6d¹7s²	[Rn]5f⁸7s²	[Rn]5f¹⁰7s²	[Rn]5f¹¹7s²	[Rn]5f¹²7s²	[Rn]5f¹³7s²	[Rn]5f¹⁴7s²	[Rn]5f¹⁴6d¹7s²
Curium	Berkelium	Californium	Einsteinium	Fermium	Mendelevium	Nobelium	Lawrencium

TABLE 13.13

Table of periodic properties of the elements

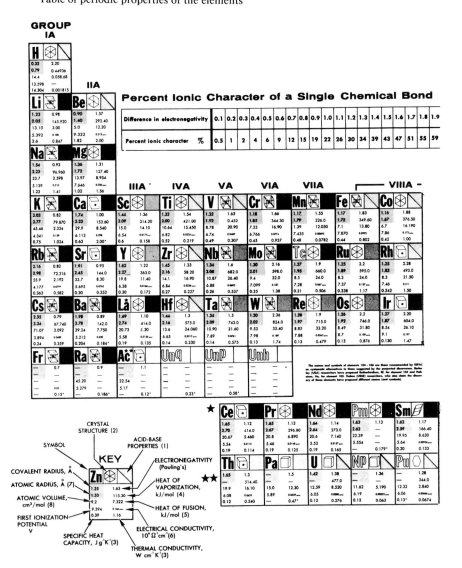

VIII

	IIIB	IVB	VB	VIB	VIIB	

He
- 0.93 —
- 0.49 0.0845
- 24.587 —
- 5.193 0.00152

Scale legend (oxide color / electronegativity): 2.0 2.1 2.2 2.3 2.4 2.5 2.6 2.7 2.8 2.9 3.0 3.1 3.2

Scale legend (atomic numbers): 63 67 70 74 76 79 82 84 86 88 89 91 92

Period 2 (B C N O F Ne):

- **B** — 0.83/2.04, 1.17/489.70, 4.6/50.20, 8.298/10⁻ⁿ, 102/0.270
- **C** — 0.77/2.55, 0.91/355.80, 4.58/—, 10⁻ⁿ, 0.71/1.29
- **N** — 0.75/3.04, 0.91/2.7928, 17.3/0.3604, 14.534/—, 1.04/0.0002598
- **O** — 0.73/3.44, 0.65/3.4099, 14.0/0.22259, 13.618/—, 0.92/0.0002674
- **F** — 0.72/3.98, 0.57/3.2698, 17.1/0.2552, 17.422/—, 0.82/0.000279
- **Ne** — 0.71/—, 0.51/1.7326, 16.7/0.3317, 21.564/—, 0.904/0.000493

Period 3 (Al Si P S Cl Ar):

| | IB | IIB | | | | |

- **Al** — 1.18/1.61, 1.82/293.40, 10.6/10.790, 5.986/8.3ⁿ, 0.90/2.37
- **Si** — 1.11/1.90, 1.46/384.220, 12.1/50.550, 8.151/2.3x10⁻ⁿ, 0.77/1.48
- **P** — 1.06/2.19, 1.23/12.129, 17.0/0.657, 10.486/10⁻ⁿ, 0.77/0.00235
- **S** — 1.02/2.58, 1.09/—, 15.5/1.7175, 10.360/8.3x10⁻ⁿ, 0.71/0.00269
- **Cl** — 0.99/3.16, 0.97/10.20, 22.7/3.203, 12.967/—, 0.48/0.000089
- **Ar** — 0.98/—, 0.88/6.447, 28.5/1.188, 15.759/—, 0.520/0.0001772

Period 4 (Ni Cu Zn Ga Ge As Se Br Kr):

- **Ni** — 1.35/1.91, 1.32/370.40, 6.59/17.470, 7.635/8.1ⁿ, 0.44/0.907
- **Cu** — 1.17/1.90, 1.57/300.30, 7.1/13.050, 7.726/8.9ⁿ, 0.38/4.01
- **Zn** — 1.25/1.65, 1.33/115.30, 9.2/7.322, 9.394/5.1ⁿ, 0.39/1.16
- **Ga** — 1.26/1.81, 1.83/258.70, 11.8/5.590, 5.999/8.8ⁿ, 0.37/0.406
- **Ge** — 1.22/2.01, 1.52/330.90, 13.6/36.940, 7.899/1.43x10⁻ⁿ, 0.32/0.599
- **As** — 1.30/2.18, 1.33/34.760, 13.1/—, 9.81/0.0343ⁿ, 0.33/0.500
- **Se** — 1.18/2.55, 1.22/37.70, 16.45/6.694, 9.752/10⁻ⁿ, 0.32/0.0204
- **Br** — 1.14/2.96, 1.12/15.438, 23.5/5.286, 11.814/—, 0.473/0.00122
- **Kr** — 1.12/—, 1.03/9.029, 38.9/1.638, 13.999/—, 0.248/0.0000949

Period 5 (Pd Ag Cd In Sn Sb Te I Xe):

- **Pd** — 1.28/2.20, 1.37/357.0, 8.9/17.60, 8.34/4.0ⁿ, 0.24/0.718
- **Ag** — 1.34/1.93, 1.75/250.580, 10.3/11.30, 7.576/6.4ⁿ, 0.235/4.29
- **Cd** — 1.48/1.69, 1.71/99.570, 13.1/6.192, 8.993/6.1ⁿ, 0.23/0.968
- **In** — 1.44/1.78, 2.00/231.50, 15.7/3.263, 5.786/8.1ⁿ, 0.23/0.816
- **Sn** — 1.41/1.96, 1.72/295.80, 16.3/7.210, 7.344/8.9ⁿ, 0.227/0.666
- **Sb** — 1.40/2.05, 1.53/77.140, 18.23/19.870, 8.641/2.8ⁿ, 0.21/0.243
- **Te** — 1.36/2.1, 1.42/52.550, 20.5/17.490, 9.009/2.x10⁻ⁿ, 0.20/0.0235
- **I** — 1.33/2.66, 1.32/20.752, 25.74/7.824, 10.451/4.1x10⁻ⁿ, 0.214/0.00449
- **Xe** — 1.31/—, 1.24/12.636, 37.3/2.297, 12.130/—, 0.158/0.0000569

Period 6 (Pt Au Hg Tl Pb Bi Po At Rn):

- **Pt** — 1.30/2.28, 1.35/334.40, 9.10/19.60, 9.0/8.9ⁿ, 0.13/0.716
- **Au** — 1.34/2.54, 1.79/59.229, 10.2/12.550, 9.225/4.1ⁿ, 0.128/3.17
- **Hg** — 1.49/2.00, 1.76/59.229, 14.82/2.295, 10.437/9.8x10⁻ⁿ, 0.139/0.0834
- **Tl** — 1.48/2.04, 2.04/164.10, 17.2/4.142, 6.108/5.0x10⁻ⁿ, 0.13/0.461
- **Pb** — 1.47/2.33, 1.81/177.70, 18.17/4.799, 7.416/0.04ⁿ, 0.13/0.353
- **Bi** — 1.46/2.02, 1.63/104.80, 21.3/11.30, 7.289/9.0x10⁻ⁿ, 0.12/0.0787
- **Po** — 1.46/2.0, 1.53/—, 22.23/—, 8.42/0.011ⁿ, —/0.20*
- **At** — (1.45)/2.2, 1.43/—, 25.74/—, —/—, —/0.017*
- **Rn** — 1.34/16.40, 50.5/2.890, 10.748/—, 0.09*/0.0000364*

Lanthanides (Eu Gd Tb Dy Ho Er Tm Yb Lu):

- **Eu** — 1.85/1.2, 2.56/143.50, 28.9/9.210, 5.67/0.0113ⁿ, 0.18/0.139*
- **Gd** — 1.61/1.20, 2.54/359.40, 19.9/10.050, 6.15/0.0078ⁿ, 0.23/0.106
- **Tb** — 1.59/1.2, 2.51/330.90, 19.2/10.80, 5.86/0.0089ⁿ, 0.18/0.111
- **Dy** — 1.59/1.22, 2.49/230.0, 19.0/11.060, 5.94/0.0105ⁿ, 0.17/0.107
- **Ho** — 1.58/1.23, 2.47/241.0, 18.7/12.20, 5.94/0.0113ⁿ, 0.16/0.162
- **Er** — 1.57/1.24, 2.45/261.0, 18.4/19.90, 6.101/0.0117ⁿ, 0.17/0.143
- **Tm** — 1.56/1.25, 2.42/191.0, 18.1/16.840, 6.184/0.0113ⁿ, 0.16/0.168
- **Yb** — 1.74/1.1, 2.40/128.90, 24.79/7.660, 6.254/0.0331, 0.15/0.349
- **Lu** — 1.56/1.27, 2.25/355.90, 17.78/18.60, 5.43/0.0183ⁿ, 0.15/0.164

Actinides (Am Cm Bk Cf Es Fm Md No Lr):

- **Am** — —/1.3, 17.86/14.40, 5.993/0.0025ⁿ, 0.11*/0.1*
- **Cm** — —/1.3, 18.28/15.0, 6.02/—, —/0.1*
- **Bk** — —/1.3, 6.23/—, —/0.1*
- **Cf** — —/1.3, 6.30/—, —/0.1*
- **Es** — —/1.3, 6.42/—, —/0.1*
- **Fm** — —/1.3, 6.50/—, —/0.1*
- **Md** — —/1.3, 6.58/—, —/0.1*
- **No** — —/1.3, 6.65/—, —/0.1*
- **Lr** — —/—, —/—, —/0.1*

NOTES: (1) For representative oxides (higher valence) of group. Oxide is acidic if color is red, basic if color is blue and amphoteric if both colors are shown. Intensity of color indicates relative strength.

(2) Cubic, face centered; cubic, body centered; cubic; hexagonal; rhombohedral; tetragonal; orthorhombic; monoclinic.

(3) At 300 K (27°C)
(4) At boiling point
(5) At melting point
(6) Generally at 293 K (20°C)
(7) Quantum mechanical value for free atom
(8) From density at 300 K (27°C) for liquid and solid elements; values for gaseous elements refer to liquid state at boiling point

TABLE 13.14
Table of isotopes

Z	El.	A	M (amu)	Ab (at%)
1	H	1	1.007825	99.99
		2	2.014102	0.015
		3	3.016049	–
2	He	3	3.016029	1.4E-4
		4	4.002603	100
3	Li	6	6.015123	7.5
		7	7.016004	92.5
4	Be	9	9.012182	100
5	B	10	10.012938	19.9
		11	11.009305	80.1
6	C	12	12.000000	98.90
		13	13.003355	1.10
7	N	14	14.003074	99.63
		15	15.000109	0.37
8	O	16	15.994915	99.76
		17	16.999131	0.04
		18	17.999159	0.20
9	F	19	18.998403	100
10	Ne	20	19.992439	90.51
		21	20.993845	0.27
		22	21.991384	9.22
11	Na	23	22.989770	100
12	Mg	24	23.985045	78.99
		25	24.985839	10.00
		26	25.982595	11.01
13	Al	27	26.981541	100
14	Si	28	27.976928	92.23
		29	28.976496	4.67
		30	29.973772	3.10
		83	82.914134	11.5
		84	83.911506	57.0
		86	85.910614	17.3
37	Rb	85	84.911800	72.17
		87	86.909184	27.83
38	Sr	84	83.913428	0.56
		86	85.909273	9.86
		87	86.908890	7.00
		88	87.905625	82.58
39	Y	89	88.905856	100
40	Zr	90	89.904708	51.45
		91	90.905644	11.22
		92	91.905039	17.15
		94	93.906319	17.38
		96	95.908272	2.80
41	Nb	93	92.906328	100
42	Mo	92	91.906809	14.84
		94	93.905086	9.25
		95	94.905838	15.92
		96	95.904676	16.68
		97	96.906018	9.55
		98	97.905405	24.13
		100	99.907473	9.63
43	Tc			–
44	Ru	96	95.907596	5.52
		98	97.905287	1.88
		99	98.905937	12.7
		100	99.904218	12.6
		101	100.905581	17.0
		144	143.910096	23.80
		145	144.912582	8.30
		146	145.913126	17.19
		148	147.916901	5.76
		150	149.920900	5.64
61	Pm			–
62	Sm	144	143.912009	3.1
		147	146.914907	15.0
		148	147.914832	11.3
		149	148.917193	13.8
		150	149.917285	7.4
		152	151.919741	26.7
		154	153.922218	22.7
63	Eu	151	150.919860	47.8
		153	152.921243	52.2
64	Gd	152	151.919803	0.20
		154	153.920876	2.18
		155	154.922629	14.80
		156	155.922130	20.47
		157	156.923967	15.65
		158	157.924111	24.84
		160	159.927061	21.86
65	Tb	159	158.925350	100
66	Dy	156	155.924287	0.06
		158	157.924412	0.10
		160	159.925203	2.34
		161	160.926939	18.9
		162	161.926805	25.5
		163	162.928737	24.9

TABLE 13.14 (*cont.*)
Table of isotopes

Z	El.	A	M (amu)	Ab (at%)
15	P	31	30.973763	100
16	S	32	31.972072	95.02
		33	32.971459	0.75
		34	33.967868	4.21
		36	35.969033	0.02
17	Cl	35	34.968853	75.77
		37	36.965903	24.23
18	Ar	36	35.967546	0.34
		38	37.962732	0.06
		40	39.962383	99.60
19	K	39	38.963708	93.26
		40	39.963999	0.01
		41	40.961825	6.73
20	Ca	40	39.962591	96.94
		42	41.958622	0.65
		43	42.958770	0.14
		44	43.955485	2.09
		46	45.953689	0.004
		48	47.952532	0.19
21	Sc	45	44.955914	100
22	Ti	46	45.952633	8.0
		47	46.951765	7.3
		48	47.947947	73.8
		49	48.947870	5.5
		50	49.944786	5.4
23	V	50	49.947161	0.25
		51	50.943963	99.75
24	Cr	50	49.946046	4.35
		52	51.940510	83.79
		53	52.940651	9.50

Z	El.	A	M (amu)	Ab (at%)
45	Rh	103	102.905503	100
46	Pd	102	101.904348	1.02
		104	103.904026	11.14
		105	104.905075	22.33
		106	105.903475	27.33
		108	107.903894	26.71
		110	109.905169	11.72
47	Ag	107	106.905095	51.84
		109	108.904754	48.16
48	Cd	106	105.906461	1.25
		108	107.904186	0.89
		110	109.903007	12.49
		111	110.904182	12.80
		112	111.902761	24.13
		113	112.904401	12.22
		114	113.903361	28.73
		116	115.904758	7.49
49	In	113	112.904056	4.3
		115	114.903875	95.7
50	Sn	112	111.904823	0.97
		114	113.902781	0.65
		115	114.903344	0.36
		116	115.901744	14.53
		117	116.902954	7.68
		118	117.901607	24.22
		119	118.903310	8.58
		120	119.902199	32.59
		122	121.903440	4.63

Z	El.	A	M (amu)	Ab (at%)
		164	163.929183	28.2
67	Ho	165	164.930332	100
68	Er	162	161.928787	0.14
		164	163.929211	1.61
		166	165.930305	33.6
		167	166.932061	22.95
		168	167.932383	26.8
		170	169.935176	14.9
69	Tm	169	168.934225	100
70	Yb	168	167.933908	0.13
		170	169.934774	3.05
		171	170.936338	14.3
		172	171.936393	21.9
		173	172.938222	16.12
		174	173.938873	31.8
		176	175.942576	12.7
71	Lu	175	174.940785	97.41
		176	175.942694	2.59
72	Hf	174	173.940065	0.16
		176	175.941120	5.21
		177	176.943233	18.61
		178	177.943710	27.30
		179	178.945827	13.63
		180	179.946561	35.10
73	Ta	180	179.947189	0.01
		181	180.948014	99.99
74	W	180	179.946727	0.13
		182	181.948225	26.3
		183	182.950245	14.3
		184	183.950953	30.67

TABLE 13.14 (*cont.*)
Table of isotopes

Z	El.	A	M (amu)	Ab (at%)
25	Mn	54	53.938882	2.37
		55	54.938046	100
26	Fe	54	53.939612	5.8
		56	55.934939	91.72
		57	56.935396	2.2
		58	57.933278	0.28
27	Co	59	58.933198	100
28	Ni	58	57.935347	68.27
		60	59.930789	26.10
		61	60.931059	1.13
		62	61.928346	3.59
		64	63.927968	0.91
29	Cu	63	62.929599	69.17
		65	64.927792	30.83
30	Zn	64	63.929145	48.6
		66	65.926035	27.9
		67	66.927129	4.1
		68	67.924846	18.8
		70	69.925325	0.6
31	Ga	69	68.925581	60.1
		71	70.924701	39.9
32	Ge	70	69.924250	20.5
		72	71.922080	27.4
		73	72.933464	7.8
		74	73.921179	36.5
		76	75.921403	7.8
33	As	75	74.921596	100
34	Se	74	73.922477	0.9
		76	75.919207	9.0

Z	El.	A	M (amu)	Ab (at%)
		124	123.903824	5.79
51	Sb	121	120.903824	57.3
		123	122.904222	42.7
52	Te	120	119.904021	0.10
		122	121.903056	2.60
		123	122.904278	0.91
		124	123.902825	4.82
		125	124.904435	7.14
		126	125.903310	18.95
		128	127.904464	31.96
		130	129.906229	34.80
53	I	127	126.904477	100
54	Xe	124	123.90612	0.10
		126	125.904281	0.09
		128	127.903531	1.91
		129	128.904780	26.4
		130	129.903510	4.1
		131	130.905076	21.2
		132	131.904148	26.9
		134	133.905395	10.4
		136	135.907219	8.9
55	Cs	133	132.905433	100
56	Ba	130	129.906277	0.11
		132	131.905042	0.10
		134	133.904490	2.42
		135	134.905668	6.59
		136	135.904556	7.85
		137	136.905816	11.23
		138	137.905236	71.70

Z	El.	A	M (amu)	Ab (at%)
		186	185.955437	28.6
75	Re	185	184.952977	37.40
		187	186.955765	62.60
76	Os	184	183.952514	0.02
		186	185.953852	1.58
		187	186.955762	1.6
		188	187.955850	13.3
		189	188.958156	16.1
		190	189.958455	26.4
		192	191.961487	41.0
77	Ir	191	190.960603	37.3
		193	192.962942	62.7
78	Pt	190	189.959937	0.01
		192	191.961049	0.79
		194	193.962679	32.9
		195	194.964785	33.8
		196	195.964947	25.3
		198	197.967879	7.2
79	Au	197	196.966560	100
80	Hg	196	195.965812	0.14
		198	197.966760	10.02
		199	198.968269	16.84
		200	199.968316	23.13
		201	200.970293	13.22
		202	201.970632	29.80
		204	203.973481	6.85
81	Tl	203	202.972336	29.52
		205	204.974410	70.48
82	Pb	204	203.973037	1.4

TABLE 13.14 (*cont.*)
Table of isotopes

Z	El.	A	M (amu)	Ab (at%)
		77	76.919908	7.6
		78	77.917304	23.6
		80	79.916521	49.7
		82	81.916709	9.2
35	Br	79	78.918336	50.69
		81	80.916290	49.31
36	Kr	78	77.920397	0.35
		80	79.916375	2.25
		82	81.913483	11.6

Z	El.	A	M (amu)	Ab (at%)
57	La	138	137.907114	0.09
		139	138.906355	99.91
58	Ce	136	135.907141	0.19
		138	137.905996	0.25
		140	139.905442	88.48
		142	141.909249	11.08
59	Pr	141	140.907657	100
60	Nd	142	141.907731	27.13
		143	142.909823	12.18

Z	El.	A	M (amu)	Ab (at%)
		206	205.974455	24.1
		207	206.975885	22.1
		208	207.976641	52.4
83	Bi	209	208.980388	100
88	Ra	226	226.025406	100
90	Th	232	232.038054	100
92	U	234	234.040947	0.006
		235	235.043925	0.72
		238	238.050786	99.27

14
Data Lists

J.R. BIRD
ANSTO Lucas Heights Research Laboratories
Menai, Australia

R.A. BROWN
School of Physics, University of Melbourne
Melbourne, Australia

D.D. COHEN
Australian Institute of Nuclear Science and Engineering
Menai, Australia

J.S. WILLIAMS
Microelectronics and Materials Technology Centre, RMIT
Melbourne, Australia

ION BEAMS FOR
MATERIALS ANALYSIS
ISBN 0 12 099740 1

14.1 STOPPING POWERS AND RANGES

The basic principles of energy loss and ion ranges are discussed in Chapter 1 with relevant equations in Table 1.3. In this Section we will present information on stopping power and range of various ions and target materials.

14.1.1 Stopping Powers

The stopping power of an ion (S) is defined as the rate of energy loss per unit distance:

$$S(E_1) = \frac{dE_1}{dx} \qquad (14.1)$$

where E_1 is the ion energy in MeV and x is the target thickness in g cm^{-2}. An alternative quantity sometimes used is the stopping cross-section (ε) which is measured in MeV cm^2 atom^{-1}. As discussed in Chapter 1, the stopping power of MeV ions, which are of interest for ion beam analysis, is made up of electronic and nuclear components. The former dominates if the ion velocity is comparable to or larger than the orbital electron velocity of target atoms; nuclear stopping dominates at the lower velocities.

The six volumes of Ziegler *et al.* (1977) give values of stopping power and range for all ions from H to U in all targets from H to U and for ion energies from 1 keV up to 100 MeV in some cases. These tables include experimental data where it exists. Polynomials obtained by fitting the data over various energy ranges are also given and will be used here as a convenient way of summarising the information available. The polynomial coefficients, taken from Ziegler *et al.* (1977) Vols III and IV, for H and He ions are listed in Table 14.1. Table 14.1a gives the polynomial function and its coefficients for proton stopping powers from 1 keV to 100 MeV, Table 14.1b does the same for He ions for energies from 1 keV to 10 MeV, and Table 14.1c gives the functional form and the coefficients for He ranges from 1 keV to 10 MeV (see Section 14.1.2). Fig. 14.1 shows typical stopping power curves for protons and He ions in Si for ion energies from 1 keV to 100 MeV. Table 14.2 gives selected values (in MeV cm^{2}g^{-1}) for C, Si, Ge and Pb. Vol. I (Ziegler *et al.*, 1985) contains an improved formulism for stopping power fits.

The position (E_{max}) and height (S_{max}) of the stopping power maximum increases with ion atomic number, Z_1, for a given target material, Z_2, moving to higher ion energies, for a given ion, with increasing Z_2. Table 14.3 gives the approximate stopping power maxima and their positions for various ions in C and Ge. The mean change in S_{max} per unit change in ion atomic number for C is (1.4 $\pm$ 0.1) MeV cm^{2}mg^{-1}Z$_1^{-1}$ for ions from B to U. Below $Z_1 = 5$ this reduces fairly rapidly. Estimates of stopping powers for ions heavier than protons and He ions in other target materials can be obtained from Table 14.1 or Vols III and IV of Ziegler *et al.* (1977), by the use of the scaling laws given below.

a. Electronic Stopping Power

The high energy (but non-relativistic) behaviour of the electronic stopping power of fully stripped ions is given by Kumakhov and Komarov (1981) as:

$$S_{el} = \frac{NZ_2 4\pi e^4 Z_i^2}{m_e v_1^2} L \tag{14.2}$$

where N is the number of stopping atoms cm^{-3}, m_e is the electronic mass and e its charge and L is a slowly varying log function of the ion energy and the stopping electron binding energy. Z_i is the effective charge of the ion travelling in the material with atomic number Z_2 and mass M_2, and M_1 is the mass of the ion. Equation (14.2) is called the Bethe formula and generally applies if the ion velocity (v_1) is much higher than that of the atomic electrons, (i.e. $v_1 \gg v_o Z_2^{2/3}$, where v_o is the Bohr velocity), and is the same as Equation 1.35.

a.

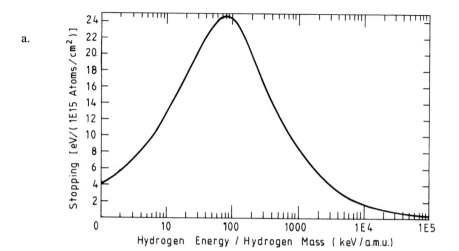

b.

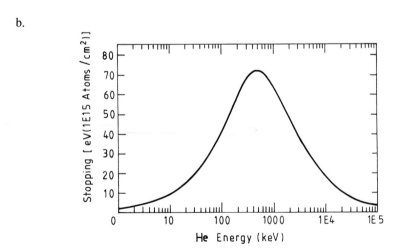

Fig. 14.1 Typical stopping power curves for a. protons and b. He ions in Si.

b. Nuclear Stopping Power

Wilson *et al.* (1977) derived the following expression for S_n based on the Thomas-Fermi screening potential:

$$S_n = \frac{0.5 ln\,(1 + \varepsilon)}{(\varepsilon + 0.10718\varepsilon^{0.37544})} \qquad (14.3a)$$

where ε is the reduced ion energy, and should not be confused with the stopping cross-section. The reduced ion energy is given by:

$$\varepsilon = \frac{32.53\,M_2 E_1}{Z_1 Z_2 (M_1 + M_2)(Z_1^{2/3} + Z_2^{2/3})^{1/2}} \qquad (14.3b)$$

and E_1 is in keV. S_n is in LSS (Lindhard Scharff Schiott) reduced stopping units and is converted to units of $(eV/(10^{15}$ atoms cm$^{-2}))$ by multiplying by $(8.462 Z_1 Z_2 M_1)/[(M_1 + M_2)(Z_1^{2/3} + Z_2^{2/3})^{1/2}]$ which in turn is converted to MeV cm^{2}g^{-1} by multiplying by $10 N/\rho$ where ρ is the density (g cm^{-3}). For protons and He ions this is not a significant term, compared with S_e for energies above 40 keV. For much heavier ions $S_n \ll S_e$ for energies above 200 keV/amu.

More recently Ziegler *et al.* (1985), Vol. I, defined a universal atomic potential and hence a universal nuclear stopping cross-section, namely:

$$S_n(E_1) = \frac{8.462 Z_1 Z_2 M_1 S_n(\varepsilon)}{(M_1 + M_2)(Z_1^{0.23} + Z_2^{0.23})} \qquad (14.4a)$$

where the reduced energy ε is defined as in Equation (14.3b) except $(Z_1^{2/3} + Z_2^{2/3})^{1/2}$ is replaced by $(Z_1^{0.23} + Z_2^{0.23})$ and $S_n(E_1)$ is in units of $(eV/(10^{15}$ atom cm$^{-2}))$. $S_n(\varepsilon)$ is given by:

$$S_n(\varepsilon) = \frac{ln\,(1 + 1.1383\varepsilon)}{2(\varepsilon + 0.01321\varepsilon^{0.21226} + 0.19593\varepsilon^{0.5})} \qquad (14.4b)$$

for $\varepsilon \le 30$ and:

$$S_n(\varepsilon) = ln(\varepsilon)/2\varepsilon \qquad (14.4c)$$

for $\varepsilon > 30$.

The semi-empirical approach of Equation (14.4) describes the available experimental nuclear stopping data for a wide variety of heavy ions, and is superior to the theoretical, Thomas-Fermi, Moliere, Lenz-Jensen or Bohr descriptions, for reduced energies $\varepsilon < 1$. Differences between Equations (14.3) and (14.4) for $\varepsilon > 0.1$ are negligible. However for $\varepsilon < 10^{-3}$ the Thomas-Fermi approach, Equation (14.3), over-estimates the nuclear stopping cross section by more than a factor of 2.

c. Stopping Power Scaling

If we define the stopping power of the proton at energy E_1 as $S_p(E_1)$, then the stopping power of a heavier ion S_i can be expressed as:

$$S_i(E_1) = (Z_i/Z_p)^2 S_p(E_1/M_1) \qquad (14.5)$$

where Z_i is the effective charge of the heavier ion and Z_p is the effective charge of the proton. It should be emphasised that Z_i and Z_p are generally not equal to Z_1, even for the lightest of ions, such as protons, unless the ion energies are sufficiently large (>1 MeV/amu). For deuterons ($^2_1$D) and $^3_2$He ions, Equation (14.5) implies a stopping power, relative to the proton stopping power, S_p, of $S_D(E_1) = S_p(E_1/2)$ and $S_{He}(E_1) = 4S_p(E_1/3)$ respectively. Equation (14.5) is a good approximation for the ion stopping power in terms of the proton stopping power if we ignore the slow logarithmic dependence of the function L defined in Equation (14.2) and for ions with energies >1 MeV/amu (Z_i/Z_p) $= Z_1$, since Z_p approaches unity.

The effective ion charge, $Z_i = Z_1$, for ion energies larger than some minimum energy E_{min}. Ziegler et al. (1977) give the following expressions for the effective ion charge relative to the effective proton charge:

$$\left[\frac{Z_i}{Z_p}\right] = 1 - \exp(-A)[1.034 - 0.1777\exp(-0.08114Z_1)] \quad (14.6a)$$

where:

$$A = B + 0.0378 \sin(\pi B/2) \qquad (14.6b)$$

and:

$$B = 0.886 (E_1/25M_1)^{1/2} Z_1^{-2/3} \qquad (14.6c)$$

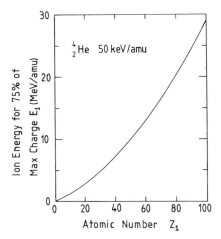

Fig. 14.2 The value of E_1 for which an ion of atomic number Z_1 reaches 75% of its maximum charge $Z_1 e$. The value for ^{4}He ions is 50 keV/amu.

where E_1 is in keV and mass in amu. Note that Z_i/Z_p contains only the ion parameters Z_1 and v_1 and is valid for all target materials. Equations (14.6a,b,c) are valid to about 5% for all ion energies above 200 keV/amu and for $6 \leq Z_1 \leq 92$ with $4 \leq Z_2 \leq 79$. The proton effective charge Z_p is 1 (to 5%) above 200 keV and the alpha particle charge is 2 (to 5%) for energies above 1.2 MeV. Once an accurate and reliable description of the proton stopping power is available, the corresponding values for heavier ions maybe obtained through Equation (14.5) and the effective charge, Equation (14.6). Table 14.4 shows the ion effective charge for protons, alphas, Li, N, Ar and Br ions for various ion energies between 0.2 and 20 MeV.

Fig. 14.2, shows the value of E_1 (MeV/amu) for which an ion reaches 75% of its maximum charge Z_1 versus the ion atomic number Z_1. For the heavy ions, E_1 for maximum charge is very large. For example, $^{36}_{17}$Cl ions have only reached (2/3) of their maximum charge by 55 MeV while, at the same energy, $^{236}_{92}$U ions have only 10% of their maximum charge. Hence the proper use of Equation (14.5), for most heavy ions with $E_1 < 1$ MeV/amu requires the calculations of Equation (14.6), as well.

Recently, Montenegro *et al.* (1982) obtained a simple analytical equation for S_i, valid for all ion-target combinations, in principle free of adjustable parameters:

$$S_i = Z_1^2 \left[\frac{1 - \exp(-\alpha\mu) - 1/6\alpha\mu\exp(-2\alpha\mu)}{1 - \exp(-\mu) - 1/6\mu\exp(-2\mu)} \right]^2 S_p(E_1/M_1) \quad (14.7)$$

where $\alpha = Z_1^{-2/3}$ and $u = v_1/v_o$ and Z_1 is the ion nuclear charge.

d. Compounds and Mixtures

The stopping powers of compounds and mixtures can be calculated from Bragg's rule which states that the stopping power of a compound or mixture, $S(E)$, is the sum of the individual stopping powers of each of its constituents, $S_j(E)$, weighted by the fraction, f_j, of each constituent j in the compound/mixture:

$$S(E) = \sum_{j=1}^{n} f_j S_j(E) \qquad (14.8)$$

where $\sum_{j=1}^{n} f_j = 1$. If $S_j(E)$ is in $MeVcm^2g^{-1}$ then f_j must be the fraction by weight of constituent j in the compound or mixture. If $S_j(E)$ is in $[eV/(10^{15} atoms/cm^2)]$ then f_j is the fraction of atoms in the compound/mixture. For light ions, such as protons and He ions, from 0.5 to 3 MeV on heavier materials, such as metals, alloys, oxides and semiconductors, Bragg's rule is generally good to within 2% (Baglin and Ziegler, 1974). However errors of more than 10% may occur when Bragg's rule is applied to compounds or mixtures containing a large proportion of lighter atoms (H, C).

14.1.2 Ion Ranges

As discussed in Chapter 1, the total pathlength of an ion in matter ($R_T(E_1)$) is obtained by integration of its stopping power (see Figure 1.8). It is generally more interesting to use the mean ion projected range $R_m(E_1)$ than the total pathlength. Fig. 14.3 gives the mean ion ranges (μm) of protons, deuterons and He ions in Si and mylar as well as the rate of energy loss in mylar. Below 100 keV, R_T becomes significantly greater than R_m and the final distribution of these ions in Si for lower energies must be taken into account. Table 14.5 shows the mean projected range for protons and He ions in C and Si for initial ion energies from 1 keV to 10 MeV; also given is the percentage range straggling, defined as one standard deviation of the range straggling divided by the mean projected ion range. For He ions below 10 keV the full width half maximum (FWHM) of the final ion distribution in materials heavier than C becomes greater than the mean projected ion range.

Substitution of Equation (14.5) into the definition of ion range gives the mean ion range $(R_m(E_1))$ as a function of the proton ion range R_p as:

$$R_m(E_1) = (Z_p^2 M_1 / Z_{ion}^2) R_p(E_1/M_1) \qquad (14.9)$$

where $R_p(E_1/M_1)$ is the proton range at the same ion velocity. Equation (14.9) is only an approximate expression and ignores the $E_1(v_1)$ dependence of the (Z_i/Z_p) term contained implicitly in Equations (14.6). However, it is still a very useful way of scaling ion ranges from proton ranges especially for high enough ion energies where (Z_i/Z_p) maybe near Z_1 (i.e. $E_1 > 1$ MeV/amu). For example, for $_1^2$D and $_2^3$He ions their ranges would be $2R_p(E_1/2)$ and $(3/4)R_p(E_1/3)$ respectively, for large enough E_1. Table 14.6 presents some results from a numerical calculation by Kumakhov and Komarov (1981) for the mean projected range of various slow heavy ions in Si.

References

Baglin, J.E., and Ziegler, J.F. (1974). *J. Appl. Phys.* **45**, 1413.

Kumakhov, M.A., and Komarov, F.F. (1981). "Energy loss and Ion Ranges in Solids", Gordon and Breach Science Publishers, NY.

Montenegro, E.C., Cruz, S.A., and Vargas-Aburto, C. (1982). *Phys. Lett.* **A92**, 195.

Wilson, W.D., Haggmark, L.G., and Biersack, J.P., (1977). *Phys. Rev.* **15B**, 2458.

Ziegler, J.F., Andersen, H.H., and Littmark, U. (1977). "Stopping Powers and Ranges in All Elements", Vols I to VI, Pergamon Press, NY.

Ziegler, J.F., Biersack, J.P. and Littmark, U. (1985). "Stopping Powers and Ranges in All Elements", Vol. I, Pergamon Press, N.Y.

14.2 ENERGY STRAGGLING, MULTIPLE SCATTERING AND LATERAL SPREAD

14.2.1 Introduction

In applications which require grazing (<15°) incidence or exit geometries, multiple scattering of the analysis beam in the target material is the principal factor limiting the available resolution. Otherwise, the major contribution (apart from detector resolution) is energy straggling.

14.2.2 Energy Straggling

The discrete nature of the energy-loss process results in a statistical uncertainty in the energy of particles penetrating any material. This phenomenon is known as energy straggling. The energy loss $\Delta(\equiv \Delta E)$ corresponding to a given path length x is distributed around a mean value

a.

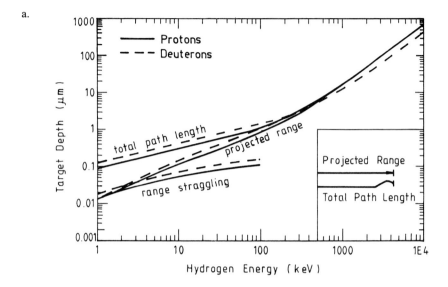

b.

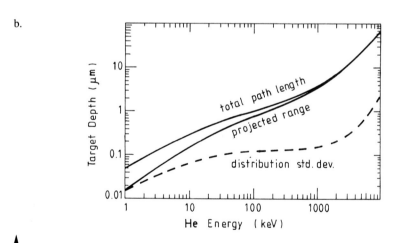

Fig. 14.3 Typical ranges for a. protons and deuterons, b. He ions in Si, c. protons and He ions in mylar and Si, and d. energy loss by protons and He ions in mylar.

c.

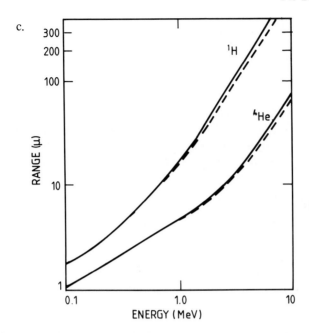

d.

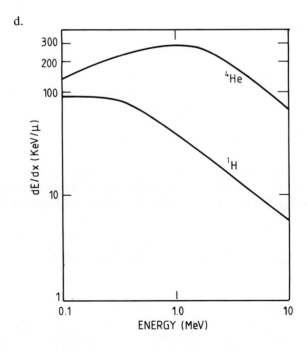

Δ_1, the corresponding distribution function being written $f(x, \Delta)$. There are four main theories describing this process, each applicable in a different regime of energy loss:

$\Delta_1/E_1 < 10\%$ **Vavilov's Theory.** For small path lengths and energy losses, the energy distribution is asymmetrical.

10–20%   **Bohr's Theory.** As the number of collisions becomes large, the distribution of particle energies becomes a simple Gaussian.

20–50%   **Symon's Theory.** This theory includes non-statistical broadening caused by the change in stopping power over the particle energy distribution, and predicts a full width at half maximum more than twice as large as that obtained with the Bohr theory (Kumakhov and Komarov, 1981).

50–90%   **Payne and Tschalär's Theory.** When the energy losses become very large, the distributions again become skewed, and Symon's theory overestimates the straggling width (Payne, 1969; Tschalär, 1968a, 1968b).

In materials analysis we are generally interested in energy losses less than 20%, for which we can use the Vavilov or Bohr theories. For path lengths smaller than x_G, where:

$$x_G = 3 \times 10^{-4} (M_2/Z_2)(E_1/Z_1 m_1)^2 \text{ g/cm}^2 \tag{14.10}$$

the Vavilov distribution should be used. However, the simplicity of the Bohr theory makes it a useful approximation unless accurate measurements are required at very small path lengths.

a. Vavilov's Theory of Straggling

Landau (in Fano, 1963 and Vavilov, 1957) calculated the energy distribution function assuming very small losses. These functions are asymmetric with a high-energy tail. The Vavilov distribution is defined as follows:

$$f(x, \Delta) = \frac{1}{\pi \zeta} \kappa \exp(\kappa(1 + \beta^2 C)) \int_0^\infty \exp(\kappa f_1) \cos(y\lambda_1 + \kappa f_2) \, dy \tag{14.11}$$

with

$$f_1 = \beta^2(\ln y - Ci(y)) - \cos y - ySi(y) \tag{14.12}$$

$$f_2 = y(\ln y - Ci(y)) + \sin y + \beta^2 Si(y) \tag{14.13}$$

where $C = 0.5772$ (the Euler constant), Si and Ci are the sine and cosine integral functions (Abramowitz and Stegun, 1965) and $\beta = (v_1/c)$. Also:

$$\xi = 0.3xm_e c^2 Z_1^2 Z_2/\beta^2 M_2$$

$$= 72m_1 Z_1^2 Z_2 x/E_1 M_2 \tag{14.14}$$

$$\lambda_1 = (\Delta - \Delta_1)/K_{max} - \kappa(\beta^2 - 0.423) \tag{14.15}$$

$$\Delta_1 = x \cdot S(E_1) \tag{14.16}$$

where x is the thickness in g/cm^2. The variable $\kappa \equiv \xi/K_{max}$ is a measure of the asymmetry: it compares the energy loss with $K_{max} = 4(m_e m_1/(m_e + m_1)^2)E_1$, the maximum energy loss in a single ion–electron collision. Table 14.7 gives values of the Vavilov distribution for $\beta^2 = 0$, valid for MeV ion interactions. The distributions are given (for several values of κ) as a function of the parameter λ:

$$\lambda = (\Delta - \Delta_1)/\xi - 0.423 - \ln \kappa \tag{14.17}$$

where:

$$\kappa = 3.3 \times 10^4 (m_1 Z_1/E_1)^2 (Z_2 x/M_2) \tag{14.18}$$

Average values for Z_2 and M_2 may be used for multielemental targets. Complete tables of the Vavilov distribution for different values of β^2 are given in Seltzer and Berger (1964).

EXAMPLE: **2 MeV He in 1000 Å of Si**

Input data: $Z_1 = 2$, $m_1 = 4.0026$, $E_1 = 2$ MeV, $Z_2 = 14$, $m_2 = 28.0$, $M_2 = 28$, $N = 4.978 \times 10^{22}$ cm^{-3}, $t = 1 \times 10^{-5}$ cm, $S(E_1) = 1056$ MeV g^{-1} cm^2.

The density of Si is $\rho = (m_2 N/N_A) = 2.32$ g cm^{-3} (N_A is Avogadro's number), so that $x = \rho \cdot t = 2.32 \times 10^{-5}$ g cm^{-2}. From Equations 14.14, 14.16 and 14.18 we have: $\xi = 6.69$ keV, $\Delta_1 = 24.5$ keV and $\kappa = 6.13$. Looking up Table 14.7 for $\kappa = 6$ we find the distribution given as a function of λ, where the corresponding energy units are (14.17): $6.69\lambda + 39.5$ keV.

b. Bohr Straggling

Bohr (1915) calculated the variance, σ^2, in the number of collisions experienced by an energetic particle passing through a solid of thickness, t, and atom density, N, to be:

$$\sigma^2 = 4\pi(Z_1e^2)^2N\,Z_2t \qquad (14.19)$$

This formula represents a simple Gaussian distribution with standard deviation σ and FWHM $= 2\sqrt{2\ln 2}\ \sigma = 2.354\sigma$. One of the assumptions of Bohr's theory is that the incident ion is fully ionised and that the target electrons are moving sufficiently slowly with respect to the velocity, v_1, of the incident ion that they may be considered to be stationary. Lindhard and Scharff (1953) extended Bohr's theory by including a correction term for energies where these assumptions may not be valid, with $\sigma^2 = \frac{1}{2}\sigma_B^2 L(\chi)$ for $\chi \le 3$ and $\sigma^2 = \sigma_B^2$ for $\chi > 3$, where $L(\chi)$ is the so-called stopping number and $\chi = v_1^2/Z_2v_0^2$ is a reduced energy variable. Here v_0 is the Bohr velocity (2.2×10^8 cm s^{-1}). Subsequent refinements to the Bohr theory have been made by Bonderup and Hvelplund (1971) and Chu (1976), principally by using a more realistic charge distribution for the target atoms. Tables of $(\sigma/\sigma_B)^2$ for He in various materials may be found in Mayer and Rimini (1977).

14.2.3 Multiple Scattering and Lateral Spread

Multiple scattering and lateral spread distributions are derived in Sigmund and Winterbon (1974) and Marwick and Sigmund (1975). Both are given as functions of the reduced thickness τ:

$$\tau = \pi\, a^2\, Nt = 6.895 \times 10^{-17} Nt/(Z_1^{1/2} + Z_2^{1/2})^{4/3} \qquad (14.20)$$

where a is the Thomas-Fermi screening radius, Equation 1.4. Fig. 14.4 shows $\tilde{\alpha}_{1/2}$, the reduced half-angle of the particle angular distribution, as a function of τ. We have fitted a simple functional form to the data in Sigmund and Winterbon (1974) and Anne *et al.* (1988):

$$\tilde{\alpha}_{1/2} = k\tau^p \qquad (14.21)$$

with the following parameters:

τ	k	p
3–20	0.3617	0.7237
30–200	0.4865	0.6285
100–2000	0.5899	0.5894
10^3–10^6	1.0	0.55

For most applications, the first set of parameters is probably adequate: the fit obtained is in error to less than 2%, 5% and 12% for τ values up to 20, 40 and 80, respectively ($\tau = 20$ corresponds to $\simeq 6000$ Å of Si). Having found the reduced half-angle, this must be translated into a 'real' half-angle $\alpha_{1/2}$ via the formula:

$$\begin{aligned} \alpha_{1/2} &= 2Z_1 Z_2 e^2 \tilde{\alpha}_{1/2}/aE_1 \\ &= 6.1464 \times 10^{-2} Z_1 Z_2 (Z_1^{1/2} + Z_2^{1/2})^{2/3} \tilde{\alpha}_{1/2}/E_1 \end{aligned} \tag{14.22}$$

with E_1 in keV.

The lateral spread is related to the angular spread via the relation:

$$\rho_{1/2} = x\,\alpha_{1/2}/\Gamma \tag{14.23}$$

where Γ is the scaling function given by Marwick and Sigmund (1975) and shown in Fig. 14.5. This function is slowly varying at large values of τ and may be approximated by $\Gamma = 0.8$. For more accurate calculations, we use the form:

$$\Gamma = 1.77 + 0.172\tau^{-0.335} \tag{14.24}$$

These values of $\alpha_{1/2} = \Delta\theta/2$ and Γ may then be used in the calculation of energy resolution (Equations 12.50 and 12.51).

EXAMPLE: **2MeV He in 2000 Å of Si**

The following data is used:

$$Z_1 = 2, Z_2 = 14, x = 2 \times 10^{-5} \text{ cm}, N = 4.978 \times 10^{22}\text{cm}^{-3},$$
$$S(E_1) = 245 \text{ keV}/\mu\text{m}.$$

From Equation 14.20 we have $\tau = 7.707$ and using Equation (14.21) for $\tilde{\alpha}_{1/2}$ (or reading from Fig. 14.4) we have:

$$\tilde{\alpha}_{1/2} = 0.3617 \times 7.707^{0.7237} = 1.586$$

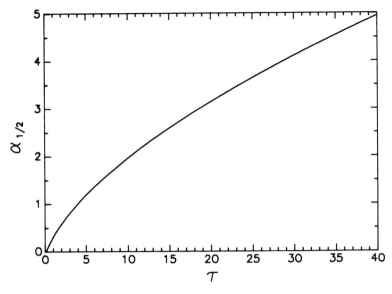

Fig. 14.4 Reduced multiple scattering half angle $\tilde{\alpha}_{1/2}$ plotted as a function of reduced thickness τ, Equation (14.21).

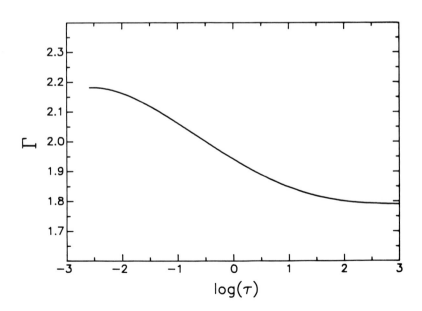

Fig. 14.5 The lateral spread scaling function Γ as a function of reduced thickness τ, Equation (14.24).

The energy lost by the beam is $\Delta E = 49.04$ keV, so the average energy may be taken as $E = 1975$ keV. Converting $\tilde{\alpha}_{1/2}$ to a non-reduced angle via Equation (14.22), we obtain the multiple scattering half-angle:

$$\alpha_{1/2} = 6.1464 \times 10^{-2} \frac{132.5}{1975} = 4.125 \times 10^{-3} \text{ radians} \simeq 4 \text{ mrad}$$

and with $\Gamma = 1.86$ (Equation 14.24), we have for lateral spread:

$$\rho_{1/2} = 4.44 \times 10^{-8} \text{ cm} \simeq 4\text{Å} .$$

References

Abramowitz, E.M. and Stegun, I. (1965) 'Handbook of Mathematical Functions', Dover Publ. Inc., New York

Anne, R., Herault, J., Bimbot, R., Gauvin, H., Bastin, G. and Hubert, F. (1988). *Nucl. Instrum. Methods* **B34**, 295.

Bohr, N., (1915). *Phil. Mag.* **30**, 581.

Bonderup, E. and Hvelplund, P., (1971). *Phys. Rev.* **A4**, 562.

Chu, W.K., (1976). *Phys. Rev.* **A13**, 2057.

Fano, F., (1963). *Ann. Rev. Nucl. Sci.* **13**, 1–66.

Kumakhov, M.A. and Komarov, F.F. (1981). 'Energy Loss and Ion Ranges in Solids', Gordon and Breach, New York.

Lindhard, J. and Scharff, E., (1953). *K. Dan. Vid. Selsk. Mat. Fys. Medd.* **27**, No. 15.

Marwick, M.A. and Sigmund, P., (1975). *Nucl. Instrum. Methods* **126**, 317.

Mayer, J.W. and Rimini, E. (1977) 'Ion Beam Handbook for Material Analysis', Academic Press, New York.

Payne, M.G. (1969). *Phys. Rev.* **185**, (2) 611.

Seltzer, S.M. and Berger, M.J. (1964). 'Studies in the Penetration of Charged Particles in Matter', N.A.S.-N.R.C. Publication 1133, Washington D.C., 187.

Sigmund, P. and Winterbon, K.B., (1974). *Nucl. Instrum. Methods* **119**, 541.

Tschalär, C., (1968a). *Nucl. Instrum. Methods* **61**, 141.

Tschalär, C., (1968b). *Nucl. Instrum. Methods* **64**, 237.

Vavilov, P.V., (1957). *Soviet Physics J.E.T.P.* **5**, 749.

14.3 ELASTIC COLLISION DATA

14.3.1 RBS Kinematic Factors, Cross-sections and Mass Scales

Table 14.8 lists kinematic factors, backscattered energies, channel numbers and RBS cross-sections for 2 MeV He$^+$ scattering from all isotopes at four geometries: $\theta = 98°$, 105°, 110° and 170°. The channel numbers assume an energy calibration of 4keV/channel and allow direct surface mass determination from RBS spectra for each scattering geometry.

14.3.2 RBS Depth Information for Si Targets

Table 14.9 lists backscattered energies, depth scales and depth resolution (as a function of depth for 2 MeV He in Si) for four typical RBS and ion channeling geometries (Chapters 3 and 6) as illustrated in Fig. 14.6. There are four listings for depth scale:

i.   In units of eV/Å using the stopping power factor S defined by Equation (3.6) and a density of 2.33 g cm^{-3} for Si.
ii. In Å/channel using 4 keV/channel energy calibration.
iii. In 10^{-15} eV cm^2 using the stopping cross-section factor ε defined by Equation (3.8).
iv. In 10^{15} cm^{-2}/channel using 4keV/channel energy calibration.

These depth scales have been calculated by computer using iterative techniques described in Chapter 12.2 and empirical stopping power formulae given in Chapter 14.1. The depth scales are *linear* to better than 10% indicating that constant stopping power values over the incident and exit path length are adequate to provide a simple analytical expression for depth as described in Chapters 3 and 12. The depth resolution is given for detector energy resolutions of 15, 7 and 2 keV, calculated at each depth using Equation (12.27) with the system energy resolution calculated from Equations (12.28) to (12.35).

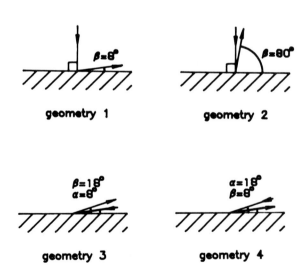

Fig. 14.6 Geometries employed in Tables 14.9, 14.10 and Fig. 14.7.

14.3.3 RBS Depth Scales in All Elements

Table 14.10 lists the near-surface depth scales (Å per channel) for 2 MeV He in elemental targets assuming the densities found in Section 13.4 and 4 keV/channel energy calibration. The four columns refer to the geometries in Fig. 14.6.

Table 14.10 only provides depth scales in elemental targets when 2 MeV He$^+$ scatters from the host atoms. Scaling factors, f_s, are needed to determine the near-surface depth scale for scattering from different mass impurities within a given host target. Figs 14.7 a) to d) give such scaling factors for five host targets (C, Si, Ge, Sn and Au) as a function of scattering mass for each of the four geometries of Fig. 14.6. Examples of the use of these figures in association with Table 14.10 are:

i) Depth scale for scattering from Sb in a Si host lattice for geometry 1. From Table 14.10 the depth scale for scattering from Si (in a Si host) is 18Å/channel. The appropriate scaling factor for Sb scattering in Si is found to be 0.925 (from Fig. 14.7a), giving the required depth scale as 19Å/channel (depth (Sb) = depth (Si)/f_s).

ii) Depth scale for scattering from Mo in a Ti host lattice for geometry 3. From Table 14.10 the depth scale for scattering from Ti in Ti is 14Å/channel. Ti falls between Si and Ge in atomic number and the scaling factors for scattering from Mo in Si and Ge are 1.17 and 1.03 respectively. The required scaling factor (1.11) can be approximated by linear interpolation (with respect to atomic number) between the values for Si and Ge. Thus, the depth scale is 15.5 Å/channel.

In general, the scaling factors alter the depth scales of Table 14.10 by less than 5% (particularly for large Z host lattices) and thus Table 14.10 can be employed as a good estimate of the near surface depth scale of all trace elements in elemental host targets.

14.3.4 Elastic Recoil and Forward Scattering

Tables 14.11 a) to d) provide the kinematic factors for elastic recoiling atoms and forward scattered ions in light element profiling using ERA ($M_1 > M_2$). Four forward-scattering geometries are shown with $\theta = 10°$, 20°, 30° and 40°. The upper part of each table gives the fraction of the incident ion energy which is imparted to the recoiling atom. The lower part of the tables gives the kinematic factors for ions scattered from five target masses. In most cases the forward scattered ions have energies close to their incident energy. Thus, in practice absorber foils must be chosen to

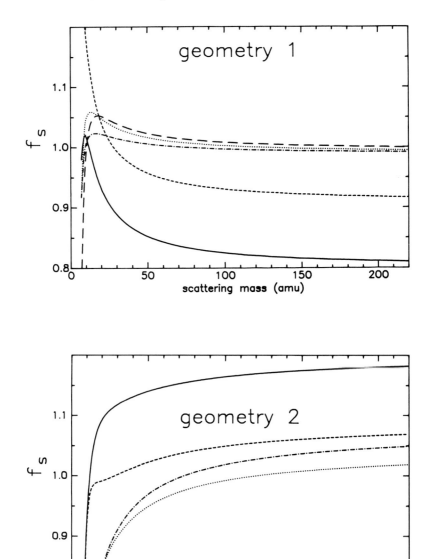

Fig. 14.7 Scaling factors for depth scale determination at four geometries as given in Fig. 14.6.

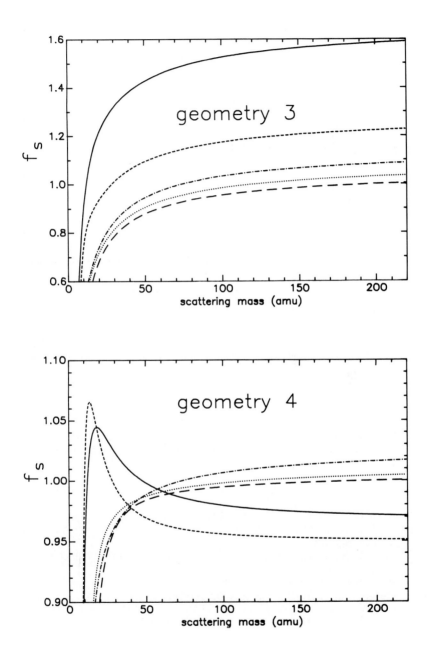

Fig. 14.7 (*cont.*)

stop incident ions but allow the lighter, lower energy recoiling atoms to be transmitted. As an example, for incident 2 MeV He^+ ions and $\theta = 10°$, a Mylar foil of 10 μm will stop the He but allow recoiling H atoms (energy 1.24 MeV) to pass through the foil with a loss of ~8% of this energy. The foil thickness to just stop scattered ions but allow recoiling atoms to pass can be calculated using the stopping power and range data of Chapter 14.1.

14.4 NUCLEAR REACTION DATA

14.4.1 Prompt Nuclear Analysis Catalog

Nuclear reactions which have been used for sample analysis are listed in Table 14.12 together with typical values of experimental parameters and performance characteristics. The list is not exhaustive since new applications are constantly being developed and the references given are illustrative only of the extensive literature on PNA. The experimental parameters and performance figures have been chosen to represent the capabilities of the technique and, in any one entry, they are not necessarily all obtained from the specific references given. Further information on resonances and gamma-ray production are given in Sections 14.4.2, 14.4.4 and 14.4.5. Reaction Q-values are listed in Table 4.2.

References for Table 14.12

1. Amsel, G. and David, D. (1969). *Rev. Phys. Appl.* **4**, 383.
2. Amsel, G., Nadai, J.P., D'Artemare, E., David, D., Girard, E. and Moulin, J. (1971). *Nucl. Instrum. Methods* **92**, 481.
3. Barrandon, J.-N. and Seltz, R. (1973). *Nucl. Instrum. Methods* **111**, 595.
4. Behrisch, R., Bottiger, J., Eckstein, W., Littmark, U., Roth, J. and Scherzer, B.M.U. (1975). *Appl. Phys. Lett.* **27**, 199.
5. Berzina, I.G., Gusev, E.B., Drushchits, A.V., Kulikauskas, V.S. and Tulinov, A.F. (1982). *Sov. Atom. Energy* **53**, 648.
6. Borderie, B. and Barrandon, J.-N. (1978). *Nucl. Instrum. Methods* **156**, 483.
6a. Borderie, B., Barrandon, J.-N., Delaunay, B. and Basutcu, M. (1979). *Nucl. Instrum. Methods* **163**, 441.
7. Borderie, B., Basutcu, M., Barrandon, J.-N. and Pinault, J.L. (1980). *J. Radioanal. Chem.* **56**, 185.
8. Bottiger, J. (1978). *J. Nucl. Mater.* **78**, 161.
9. Calvert, J.M., Derry, D.J. and Lees, D.G. (1974). *J. Phys. D.* **7**, 940.
10. Carnera, A., Della Mea, G., Drigo, A.V., Lo Russo, S. and Mazzoldi, P. (1977). *J. Non-Crystalline Solids* **23**, 123.
11. Caterini, M., Thompson, D.H., Wan, P.T. and Sawicki, J.A. (1986). *Nucl. Instrum. Methods* **B15**, 535.
12. Chen, N.S. and Fremlin, J.H. (1970). *Radiochem. Radioanal. Lett.* **4**, 365.
13. Chemin, J.-F., Roturier, J., Saboya, B. and Petit, G.Y. (1972). *J. Radioanal. Chem.* **12**, 221.

15. Clark, P.J., Neal, G.F. and Allen, R.O. (1975). *Anal. Chem.* **47**, 650.
16. Clark, G.J., Switkowski, Z.E., Petty, R.J. and Heggie, J.C.P. (1979). *Appl. Phys. Lett.* **50**, 4791.
17. Close, D.A., Malanify, J.J. and Umbarger, C.J. (1973). *Nucl. Instrum. Methods* **113**, 561.
18. Cohen, D.D., Katsaros, A. and Garton, D. (1986). *Nucl. Instrum. Methods* **B15**, 555.
19. Cowgill, D.F. (1977). *Nucl. Instrum. Methods* **145**, 507.
20. D'Agostino, M.D., Kamyowski, E.A., Kehna, F.J., Padawer, G.M., Schneid, E.J., Schulte, R.L., Stauber, M.C. and Swanson, F.R. (1978). *J. Radioanal. Chem.* **43**, 421.
21. Davis, J.C. and Anderson, J.D. (1975). *J. Vac. Sci. Technol.* **12**, 358.
22. Davis, J.C., Lefevre, H.W., Poppe, C.H., Drake, D.M. and Veeser, L.R. (1978). *Nucl. Instrum. Methods* **149**, 41.
23. Deconninck, G. (1972). *J. Radioanal. Chem.* **12**, 157.
24. Deconninck, G. and Demortier, G. (1973). *In* "Nuclear Techniques in the Basic Metal Industries", IAEA-SM-159/27, IAEA, Vienna, p. 573.
25. Deconninck, G. and Debras, G. (1975). *Radiochem. Radioanal. Lett.* **20**, 175.
26. Deconninck, G., Stroobants, J., Stone, W.E.E. and Charlier, H. (1976). *Radiochem. Radioanal. Lett.* **24**, 331.
27. Deconninck, G. and Van Oystaeyen, B. (1983). *Nucl. Instrum. Methods* **218**, 165.
28. Della Mea, G. (1986). *Nucl. Instrum. Methods* **B25**, 495.
29. Demortier, G. (1974). *Radiochem. Radioanal. Lett.* **16**, 329.
30. Didriksson, R., Gonczi, L. and Sundqvist, B. (1980). TLU-73-80.
31. Dieumegard, D., Dubreuil, D. and Amsel. G. (1979). *Nucl. Instrum. Methods* **166**, 431.
32. Dunning, K.L., Hubler, G.K., Comas, J., Lucke, W.H. and Hughes, H.L. (1973). *Thin Solid Films* **19**, 145.
33. Dunning, K.L. (1978). *Nucl. Instrum. Methods* **149**, 317.
34. Eckstein, W., Behrisch, R. and Roth, J. (1976). *In* "Ion Beam Surface Layer Analysis", (eds. Meyer, O., Linker, G. and Kappeler, F.), Plenum Press, NY, **2**, 821.
35. Falk, W.R., Abou-Zeid, O. and Roesch, L.P. (1976). *Nucl. Instrum. Methods* **137**, 261.
36. Gihwala, D., Giles, I.S., Olivier, C. and Peisach, M. (1978). *J. Radioanal. Chem.* **46**, 333.
37. Gihwala, D., Giles, I.S. and Peisach, M. (1978). *J. Radioanal. Chem.* **47**, 145.
38. Gihwala, D. and Peisach, M. (1979). *Radiochem. Radioanal. Lett.* **40**, 285.
39. Gihwala, D. and Peisach, M. (1980). *J. Radioanal. Chem.* **55**, 163.
40. Gihwala, D. and Peisach, M. (1982). *J. Radioanal. Chem.* **70**, 287.
41. Gippner, P., Bauer, C., Hohmuth, K., Mann, R. and Rudolph, W. (1981). *Nucl. Instrum. Methods* **191**, 341.
42. Gossett, C.R. (1980). *Nucl. Instrum. Methods* **168**, 217.
43. Gossett, C.R. (1981). *Nucl. Instrum. Methods* **191**, 335.
44. Gossett, C.R. (1983). *Nucl. Instrum. Methods* **218**, 149.
45. Hanninen, R., Raisanen, J. and Antilla, A. (1980). *Radiochem. Radioanal. Lett.* **44**, 201.
46. Hanson, A.L., Jones, K.W., Kraner, H.W., Osborne, J.W. and Nelson, G.V. (1982). BNL-31027.
47. Hanson, A.L., Kraner, H.W., Shroy, R.E. and Jones, K.W. (1984). *Nucl. Instrum. Methods* **B4**, 401.
48. Hayashi, S., Nagai, H., Aratani, M., Nozaki, T., Yanokura, M., Kohno, I., Kuboi, O. and Yatsurugi, Y. (1986). *Nucl. Instrum. Methods* **B16**, 377.
49. Heggie, J.C.P., Switkowski, Z.E. and Clark, G.J. (1980). *Nucl. Instrum. Methods* **168**, 125.

630   J. R. Bird, R. A. Brown, D. D. Cohen and J. S. Williams

50. Hubler, G.K., Comas, J. and Plew, L. (1978). *Nucl. Instrum. Methods* **149**, 635.
51. Hyvonen-Dabek, M. (1981). *J. Radioanal. Chem.* **63**, 367.
53. Johnson, P.B. (1974). *Nucl. Instrum. Methods* **114**, 467.
54. Kaim, R.E. and Palmer, D.W. (1979). *J. Radioanal. Chem.* **48**, 295.
55. Katsanos, A.A., Katselis, B. and Aravantinos, A. (1986). *Nucl. Instrum. Methods* **B15**, 647.
56. Kregar, M., Mueller, J., Rupnik, P., Ramsak, V. and Spiler, E. (1979). *In* "Nuclear Activation Techniques in the Life Sciences", IAEA, Vienna, IAEA-SM-227/83, 407.
57. L'Ecuyer, J., Brassard, C., Cardinal, C., Deschenes, L., Jutras, Y. and Labrie, J. (1977). *Nucl. Instrum. Methods* **140**, 305.
58. Lefevre, H.W., Davis, J.C. and Anderson, J.D. (1976). *In* "Scientific and Industrial Applications of Small Accelerators" (eds Duggan, J.L., Morgan, I.L. and Martin, J.), IEEE, NY, 225.
59. Lightowlers, E.C., North, J.C., Jordan, A.S., Derick, L. and Merz, J.L. (1979). *J. Appl. Phys.* **44**, 4758.
60. Ligeon, E., Bruel, M., Bontemps, A., Chambert, G. and Monnier, J. (1973). *J. Radioanal. Chem.* **16**, 537.
61. Lindh, U. and Tveit, A.B. (1980). *J. Radioanal. Chem.* **59**, 167.
62. Lorenzen, J. and Brune, D. (1972). AB Atomenergi Report AE-465.
63. Lorenzen, J. (1976). *Nucl. Instrum. Methods* **136**, 289.
64. Macias, E.S. and Barker, J.H. (1978). *J. Radioanal. Chem.* **45**, 387.
65. Macias, E.S., Radcliffe, C.D., Lewis, C.W. and Sawicki, C.R. (1978). *Anal. Chem.* **50**, 1120.
66. Maurel, B., Amsel, G. and Nadai, J.P. (1981). *Nucl. Instrum. Methods* **191**, 349.
67. McMurray, W.R., Peisach, M., Pretorius, R., Van der Merwe, P. and Van Heerden, I.J. (1968). *Anal. Chem.* **40**, 266.
68. McMillan, J.W. and Pummery, F.C.W. (1977). *J. Radioanal. Chem.* **38**, 51.
69. Mitchell, I.V. and Lennard, W.N. (1976). *In* "Ion Beam Surface Layer Analysis" (eds Meyer, O., Linker, G. and Kappeler, F.), Plenum, NY, 2, 925.
70. Moller, W., Hufschmidt, M. and Kamke, D. (1977). *Nucl. Instrum. Methods* **140**, 157.
71. Naramoto, H. and Ozawa, K. (1981). *Nucl. Instrum. Methods* **191**, 367.
72. Olivier, C. and Peisach, M. (1970a). *J. S. Afr. Chem. Inst.* **23**, 77.
73. Olivier, C. and Peisach, M. (1970b). *J. Radioanal. Chem.* **5**, 391.
74. Olivier, C. and Peisach, M. (1972). *J. Radioanal. Chem.* **11**, 105.
75. Olivier, C. and Peisach, M. (1974). *Radiochem. Radioanal Lett.* **19**, 227.
76. Olivier, C., McMillan, J.W. and Pierce, T.B. (1976). *J. Radioanal. Chem.* **31**, 515.
77. Olivier, C., Peisach, M. and Pierce, T.B. (1976). *J. Radioanal. Chem.* **32**, 71.
78. Olivier, C., Ras, H.A. and Peisach, M. (1982). *J. Radioanal. Chem.* **70**, 311.
79. Ollerhead, R.W., Almqvist, E. and Kuehner, J.A. (1966). *J. Appl. Phys.* **37**, 2440.
80. Overley, J.C. and Lefevre, H.W. (1976). *In* "Scientific and Industrial Applications of Small Accelerators" (eds Duggan, J.L., Morgan, I.L. and Martin. J.A.), IEEE, NY, 557.
81. Overley, J.C., Ebright, R.P. and Lefevre, H.W. (1979). IEEE *Trans. Nucl. Sci.* **NS-26**, 1624.
82. Peisach, M. (1966). *Chem. Communications* **18**, 632.
83. Peisach, M. (1972). *J. Radioanal. Chem.* **12**, 251.
84. Peisach, M. and Pretorius, R. (1973). *J. Radioanal. Chem.* **16**, 559.
85. Peisach, M. and Wilson, V.W. (1977). *J. Radioanal. Chem.* **38**, 67.
86. Pierce, T.B., Peck, P.F. and Cuff, D.R.A. (1967). *Analyst* **92**, 143.
87. Pollard, P.M., McMillan, J.W. and Malcolm-Lawes, D.J. (1982). *J. Radioanal. Chem.* **70**, 349.

88. Porte, L., Sandino, J.-P., Grea, J., Thomas, J.-P. and Tousset, J. (1972). *J. Radioanal. Chem.* **16**, 493.
89. Pronko, P.P. and Pronko, J.G. (1974). *Phys. Rev.* **B9**, 2870.
90. Pronko, P.P., Okamoto, P.T. and Wiedersich, H. (1978). *Nucl. Instrum. Methods* **149**, 77.
91. Pruppers, M.J.M., Zijderhand, F., Maessen, K.M.H., Bezemer, J., Habraken, F.H.P.M. and Van der Weg, W.F. (1986). *Nucl. Instrum. Methods* **B15**, 512.
92. Raisanen, J. (1984). *Nucl. Instrum. Methods* **B3**, 220.
93. Raisanen, J. and Lappalainen, R. (1986). *Nucl. Instrum. Methods* **B15**, 546.
94. Ricci, E. (1971). *Anal. Chem.* **43**, 1866.
95. Scanlon, P.J., Farrell, G., Ridgway, M.C. and Valizadeh, R. (1986). *Nucl. Instrum. Methods* **B16**, 479.
96. Schulte, R.L. (1976). *Nucl. Instrum. Methods* **137**, 251.
97. Schulte, R.L. (1978). *Nucl. Instrum. Methods* **149**, 65.
97a. Schulte, R.L., Papazian, J.M. and Adler, P.N. (1986). *Nucl. Instrum. Methods* **B15**, 550.
98. Shaanan, M., Kalish, R. and Richter, V. (1986). Israel Institute of Technology Report Technion-PH-85-34.
99. Shroy, R.E., Kraner, H.W., Jones, K.W., Jacobson, J.S. and Heller, L.I. (1978). *Nucl. Instrum. Methods* **149**, 313.
100. Sie, S.H., McKenzie, D.R., Smith, G.B. and Ryan, C.G. (1986). *Nucl. Instrum. Methods* **B15**, 525.
101. Simons, D.G., Land, D.J., Brennan, J.G. and Brown, M.D. (1976). *In* "Ion Beam Surface Layer Analysis" (eds Meyer, O., Linker, G. and Kappeler, F.), Plenum, NY, **2**, 863.
102. Sparks, R.J. (1977). NZ Institute of Nuclear Sciences Report INS-863.
103. Switkowski, Z.E., Overley, J.C., Shiu-Chin Wu, Barnes, C.A. and Roth, J. (1978). *J. Nucl. Materials* **78**, 64.
104. Switkowski, Z.E., Petty, R.J., Heggie, J.C.P. and Clark, G.J. (1979). *Nucl. Instrum. Methods* **159**, 407.
105. Thomas, J.-P., Engerran, J., Cachard, A. and Tardy, J. (1974). *Nucl. Instrum. Methods* **119**, 373.
106. Turos, A., Wielunski, L. and Barcz, A. (1973). *Nucl. Instrum. Methods* **111**, 605.
107. Ziegler, J.F. et al. (1978). *Nucl. Instrum. Methods* **149**, 19.
108. Zinke-Allmang, M., Kossler, V. and Kalbitzer, S. (1986). *Nucl. Instrum. Methods* **B15**, 563.

14.4.2 Resonance Parameters

Parameters are listed in Table 14.13 for resonances used in depth profiling and for neighbouring resonances which may need to be taken into account. Further information on performance is included in Table 14.12. The resonance energy (E_r) and total width (Γ) were obtained from the references in Table 4.1. The strengths listed are usually the resonance areas; in a few cases they are experimental count rates which give only the relative strengths.

14.4.3 Non-Rutherford Scattering

A list of useful backscattering resonances is given in Table 14.14. Many other such resonances occur and even relatively smooth backscattering cross-sections can be strongly affected by non-Rutherford scattering (see Chapter 3).

References for Table 14.14

1. Ajzenberg-Selove, F. (1986). *Nucl. Phys.* **A449**, 1 and references therein.
2. Jarjis, R.A. (1979). "Nuclear Reaction Cross-Section Data for Surface Analysis", Dept. Physics, University of Manchester.
3. Jones, C.M., Philips, G.C., Harris, R.W. and Beckner, E.H. (1962). *Nucl. Phys.* **37**, 1.
4. Mayer, J.W. and Rimini, E. (eds) (1977). "Ion Beam Handbook for Materials Analysis", Academic Press, New York.
5. Nikolic, N., Lidofsky, L.J. and Kruse, T.H. (1963). *Phys. Rev.* **132**, 2212.
6. Wang, Z.L., Westendorp, J.F.M. and Saris, F.W. (1983). *Nucl. Instrum. Methods* **211**, 193.

14.4.4 Nuclear Reaction Cross-sections

Publications containing references to measurements of ion–ion and ion–neutron cross-sections are listed in Table 4.1. However, of the many such measurements, relatively few are directly applicable to the conditions used in IBA (see the last three references in Table 4.1). Cross-sections for some of the most used (p, α) and deuteron induced reactions are plotted in Figs 14.8 and 14.9 respectively. Additional cross-section values at specific energies and analytical expressions are given in Table 14.15. Cross-sections at 0° for some ion–neutron reactions are plotted in Fig. 14.10 and total thick sample neutron yields are plotted in Fig. 14.11.

References for Table 14.15

Altstetter, C.J., Behrisch, R., Bottiger, J., Pohl, F. and Scherzer, B.M.U. (1978). *Nucl. Instrum. Methods* **149**, 59.
Amsel, G. and Samuel, D. (1967). *Anal. Chem.* **39**, 1689.
Annegarn, H.J., Mingay, D.W. and Sellschop, J.P.F. (1974). *Phys. Rev.* **C9**, 419.
Barschall, H.H. (1978). *Ann. Rev. Nucl. Particle Sci.* **28**, 207.
Burrill, E.A. (no date). "Neutron Production and Protection", HVEC, Burlington, Mass., USA.
Davies, J.A. and Norton, P.R. (1980). *Nucl. Instrum. Methods* **168**, 611.
Davies, J.A., Jackman, T.E., Plattner, H. and Bubb, I. (1983). *Nucl. Instrum. Methods* **218**, 141.

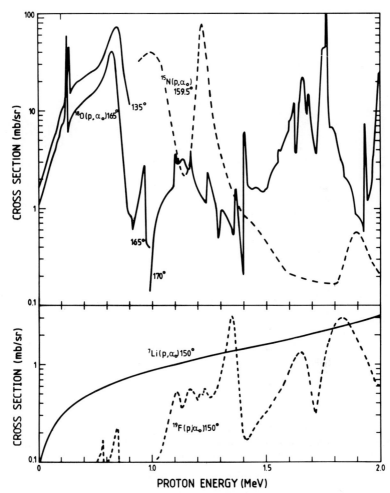

Fig. 14.8 Cross-sections of (p, α) reactions at the laboratory angles indicated (from Jarjis, 1979).

Debras, G. and Deconninck, G. (1977). *J. Radioanal. Chem.* **38**, 193.

Jarjis, R.A. (1979). "Nuclear Cross-Section Data for Surface Analysis", Dept. of Physics, University of Manchester, UK.

Liskien, H. and Paulsen, A. (1973). *Nucl. Data Tables* **11**, 569.

Liskien, H. and Paulsen, A. (1975). *Atomic and Nucl. Data Tables* **15**, 57.

Maurel, B., Amsel, G. and Dieumegard, D. (1981). *Nucl. Instrum. Methods* **191**, 349.

Marion, J.B. and Weber, G. (1956). *Phys. Rev.* **103**, 1408.

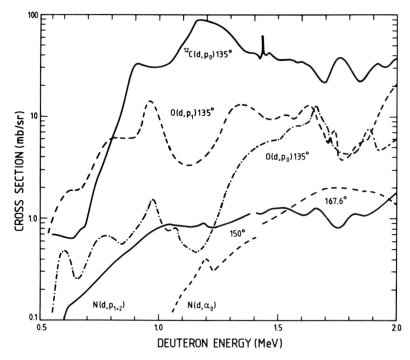

Fig. 14.9 Cross-sections of (d, p) and (d, α) reactions at the laboratory angles indicated (from Jarjis, 1979).

14.4.5 Gamma-ray Catalog

A number of systematic studies have been made of the energy and intensity of gamma-rays emitted during irradiation of thick samples by various ion types and energies (Table 14.16). Although only a few of the most intense gamma-rays are commonly used in sample analysis, detailed information is required to assess the possibility of inter-element interference and for interpretation of the observed spectra. The following tables can be used for this purpose:

- Table 14.17 — Gamma-ray Index
- Table 14.18 — Proton Induced Gamma-ray Yields
- Table 4.5a — Background Gamma-rays
- Table 4.5b — Neutron Induced Gamma-rays
- Table 14.19 — Gamma-rays from Ion Induced Activation

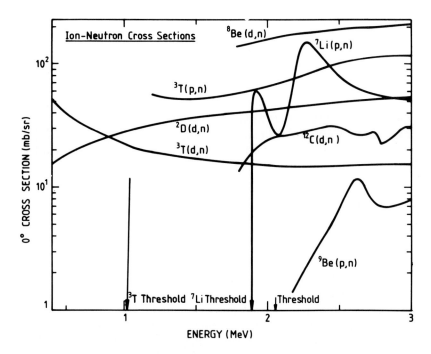

Fig. 14.10 0° cross-section for ^{2}D(d, n), ^{3}T(p, n), ^{3}T(d, n), ^{7}Li(p, n) and ^{12}C(d, n) reactions (Liskien and Paulsen, 1975) and ^{9}Be(p, n) (Marion, 1956).

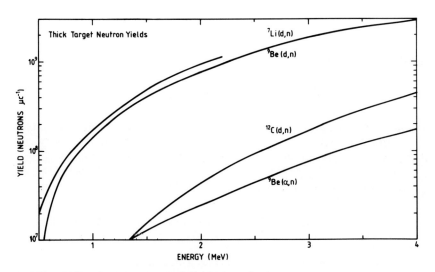

Fig. 14.11 Thick target neutron yields (Burrill, no date).

636   *J. R. Bird, R. A. Brown, D. D. Cohen and J. S. Williams*

a. Gamma-ray Index

Table 14.17 lists gamma-rays in order of increasing energy for proton, deuteron, triton, alpha irradiation and for Coulomb excitation by heavy ion beams (illustrated by results for ^{35}Cl irradiation). The source of the data is shown in the last column in Table 14.16. The proton list includes all gamma-rays observed for proton energies up to 2.5 MeV and the stronger gamma-rays only for higher energies. No systematic study has been made using deuterons and the gamma-rays listed are those reported in a small number of analytical papers.

b. PIGME Yields

Gamma-ray yields for triton, alpha and ^{35}Cl beams are included in the Gamma-ray Index (Table 14.17) for the experimental conditions listed in Table 14.16. The yields of the strongest gamma-rays from proton irradiation are summarised in Table 14.18 which includes data from several laboratories measured under different experimental conditions (see Table 14.16). They are not in good agreement where they overlap so that calibration with standard samples is necessary to achieve the accuracy intrinsically available from PIGME analyses. However, they show which gamma-rays are strongest and the nature of their dependence on proton energy. Detection limits, which depend on gamma-ray yields and background conditions, are plotted in Chapter 4 (Fig. 4.19).

References for Table 14.16

Antilla, A., Hanninen, R. and Raisanen, J. (1981). *J. Radioanal. Chem.* **62**, 293.
Bird, J.R., Scott, M.D., Russell, L.H. and Kenny, M.J. (1978). *Aust. J. Phys.* **31**, 209.
Bodart, F., Deconninck, G. and Demortier, G. (1977). *J. Radioanal. Chem.* **35**, 95.
Borderie, B. (1978). "Analyse par activation a l'aide de T, α and Heavy ions", Ph.D. Thesis, U. Paris.
Borderie, B. and Barrandon, J.-N. (1978). *Nucl. Instrum. Methods* **156**, 483.
Borderie, B., Barrandon, J.-N., Delaunay, B. and Basutcu, M. (1979). *Nucl. Instrum. Methods* **163**, 441.
Deconninck, G. (1972). *J. Radioanal. Chem.* **12**, 157.
Deconninck, G. and Demortier, G. (1972). *J. Radioanal. Chem.* **12**, 189.
Deconninck, G. and Debras, G. (1975). *Radiochem. Radioanal. Lett.* **20**, 175.
Deconninck, G. and Demortier, G. (1975). *J. Radioanal. Chem.* **24**, 437.
Demortier, G. and Bodart, F. (1972). *J. Radioanal. Chem.* **12**, 209.
Demortier, G. and Delsate, P. (1975). *Radiochem. Radioanal. Lett.* **21**, 219.
Demortier, G. (1978). *J. Radioanal. Chem.* **45**, 459.
Gihwala, D., Giles, I.S., Olivier, C. and Peisach, M. (1978a). *J. Radioanal. Chem.* **46**, 333.
Gihwala, D., Giles, I.S. and Peisach, M. (1978b). *J. Radioanal. Chem.* **47**, 145.
Gihwala, D. (1982). "Analytical Applications of Proton-Induced Prompt Photon Spectrometry", Ph.D. Thesis, U. Capetown.

Gihwala, D. and Peisach, M. (1982a). *Anal. Chim. Acta* **136**, 311.
Gihwala, D. and Peisach, M. (1982b). *Nucl. Instrum. Methods* **193**, 371.
Gihwala, D. and Peisach, M. (1982c). *J. Radioanal. Chem.* **70**, 287.
Giles, I.S. and Peisach, M. (1976). *J. Radioanal. Chem.* **32**, 105.
Giles, I.S. and Peisach, M. (1979). *J. Radioanal. Chem.* **50**, 307.
Golicheff, I. (1972). *J. Radioanal. Chem.* **16**, 503.
Kenny, M.J., Bird, J.R. and Clayton, E. (1980). *Nucl. Instrum. Methods* **168**, 115.
Kenny, M.J. (1981). *Aust. J. Phys.* **34**, 35.
Kiss, A.Z., Koltay, E., Nyako, B., Somorjai, E., Antilla, A. and Raisanen, J. (1985). *J. Radioanal. and Nucl. Chem.* **89**, 123.
Lappalainen, R., Antilla, A. and Raisanen, J. (1983). *Nucl. Instrum. Methods* **212**, 441.
Raisanen, J. and Hanninen, R. (1983). *Nucl. Instrum. Methods* **205**, 259.
Raisanen, J., Witting, T. and Keinonen, J. (1987). *Nucl. Instrum. Methods* **B28**, 199.
Stroobants, J., Bodart, F., Deconninck, G., Demortier, G. and Nicolas, G. (1976) 2nd Int. Conf. Ion Beam Surface Layer Analysis (eds. Meyer, O., Linker, G. and Kappeler, F.), Plenum, NY **2**, 933.

14.4.6 Ion Activation Catalog

A selection of results for PAA is given in Table 14.19. The first part of the table lists reactions used for light element analysis ($Z = 1$ to 13), a region for which PAA is especially useful (McGinley *et al.*, 1978; Schweikert, 1981). This is followed by results from surveys of 10 MeV proton activation (Debrun *et al.*, 1976; Borderie *et al.*, 1977) and 3.5 MeV triton activation (Barrandon *et al.*, 1976). The experimental conditions are as follows:

- Irradiation — 1 μA for 1 h
- Counting — 1.8 $\tau_{1/2}$ or 60 h (protons), 15 h (tritons)

Detection limits are plotted in Chapter 4 (Fig. 4.23).

References for Table 14.19
Barrandon, J.-N., Benaben, P. and Debrun, J.L. (1976). *Analytica Chimica Acta* **83**, 157.
Borderie, B., Barrandon, J.-N. and Debrun, J.L. (1977). *J. Radioanal. Chem.* **37**, 297.
Debrun, J.L., Barrandon, J.-N. and Benaben, P. (1976). *Anal. Chem.* **48**, 167.
McGinley, J.R., Stock, G.J., Schweikert, E.A., Cross, J.B., Zeissler, R. and Zikovsky, L. (1978). *J. Radioanal. Chem.* **43**, 559.
Schweikert, E.A. (1981). *J. Radioanal. Chem.* **64**, 195.

14.5 X-RAY DATA

Table 14.20 is a tabulation of the K shell line energies (in keV) ($K\alpha$, $K\beta_1$ and $K\beta_2$) the fluorescence yield of ω_K and the intensities of the $K\beta_1$ and $K\beta_2$ lines relative to the Kα ($= 100$) line, for selected elements from Na to Ba. The definition of these quantities is discussed in Sections 5.2.3 and 5.2.4.

Table 14.21 is a tabulation of the 15 commonly occurring L shell line energies (in keV) for selected elements from Zr to U.

Table 14.22 is a tabulation of the L subshell fluorescence yields ω_i and Coster-Kronig transition rates f_{ij} for selected elements for atomic numbers 28 to 96. The yield ω_i and rate f_{ij} are discussed in Section 5.2.4 and defined in Equations (5.5) and (5.6).

Table 14.23(a), (b) and (c) are tabulations of the L_1, L_2 and L_3 emission rates respectively, defined in Equation (5.8) and discussed in Section 5.2.4. These emission rates have to normalise to the major line within each subshell.

Table 14.24 gives the coefficients a, b and c of a fit to the mass attenuation coefficients (μ/ρ) for various photon energies (keV) and target atomic numbers. The functional form of the fit is $(\mu/\rho) = aE^{-b}Z^c$ for energies between the K, L_1, L_2 and L_3 and M absorption edges.

14.6 CRYSTALLOGRAPHY

14.6.1 Crystal Systems and Symmetry

The symmetry of a crystal, which is determined by the regular arrangement of atoms, is represented in 3 dimensions by the unit cell. The unit cell is the smallest volume containing atoms which, when translated by unit vectors representing its edges, reproduces the entire crystal lattice. There are seven basic crystal systems each specified by the shape of the unit cell. These are triclinic, monoclinic, orthorhombic, tetragonal, cubic, hexagonal and rhombohedral. Atoms can be arranged differently with respect to a similar unit cell. For example, there are three basic cubic arrangements:

- with atoms at the corners only, simple cubic (sc);
- with atoms at the corners and one at the centre, body centred cubic (bcc); or
- with atoms at the corners and also at face centres, face centred cubic (fcc).

Most common crystals are fcc, bcc or hexagonal (atoms at the corners of the unit cell and also at the hexagon face centres). Some crystals have interpenetrating lattices. For example, Si (and other semiconductors) are fcc but have atoms arranged on two interpenetrating fcc lattices, each translated 1/4, 1/4, 1/4 along the unit cube vectors with respect to each

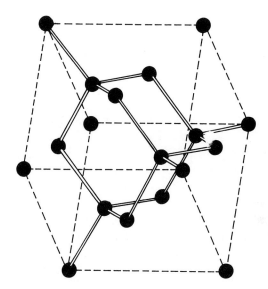

Fig. 14.12 The unit cell of a "diamond cubic" lattice (e.g. Si) consisting of interpenetrating fcc lattices.

other, as illustrated in Fig. 14.12. This atomic arrangement is called diamond cubic. A particular plan view of a Si lattice, showing the crystal symmetry, is illustrated in Fig. 1.17.

14.6.2 Miller Indices

In describing the interaction of ions with crystals we need to specify the atomic symmetry directions and crystal planes. This is achieved using a set of standard rules to determine directions with respect to the unit cell as illustrated for cubic symmetry in Fig. 14.13. Crystal directions (or axes) are described with respect to the unit cell (cube) vectors, $\hat{x}$, $\hat{y}$, $\hat{z}$ (Figs 14.13a,b). The origin of the unit cell is taken as the back, bottom-left corner and the cube edges have unit length. Consider the direction of a vector $\hat{r}$ which intersects the front face at a point 1/4 of a unit from each of the top and left edges, as shown. This vector can be written as $\hat{r} = 1$ $\hat{x} + 1/4\,\hat{y} + 3/4\,\hat{z}$. The direction of $\hat{z}$ is then denoted by the Miller indices [413] which are a triplet of smallest integer coefficients (with respect to the Cartesian vector equation) having similar ratios to the actual coefficients (i.e. multiplying the coefficients by 4 gives the Miller indices for the direction $\hat{r}$).

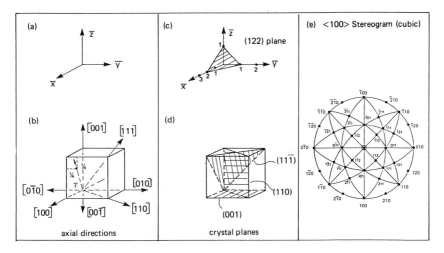

Fig. 14.13 Crystal structure nomenclature showing: a. Cartesian unit vectors, b. axial directions, c. the (122) crystal planes, d. common low index planes in a low index direction, and e. a (100) stereogram (cubic).

Common low index directions are those along the cube edge (e.g. [100]), along the face diagonals (e.g. [110]) and along the body diagonal (e.g. [111]) and these are shown in Fig. 14.13b. Specific directions are denoted by $[n_1n_2n_3]$ and all directions parallel to these have the same designation and symmetry. A bar may be used above integers (e.g. [0$\bar{1}$0]) to denote negative directions. The symmetry of the crystal can lead to several non-parallel directions being equivalent (e.g. the family of six cube edge directions [100], [$\bar{1}$00], [010], [0$\bar{1}$0], [001] and [00$\bar{1}$]). Such a family is denoted by ⟨100⟩ and hence a general family of non-parallel symmetry directions is given by ⟨$n_1n_2n_3$⟩. Directions with large indices (c.f. our example [4$\bar{1}$3]) have fewer atoms per unit length than do low index directions such as the ⟨111⟩ family.

Crystal planes (i.e. sheets of atoms) can also be designated by Miller indices, as indicated in Figs. 14.13c, d. In this case, the triplet of indices is given by the smallest integer set of inverse intercepts of the plane with respect to the unit cell vectors. For example, a plane which has intercepts 2, 1, 1 on the $\tilde{x}$, $\tilde{y}$, $\tilde{z}$ cubic axes (Fig. 14.13c) is denoted by the (122) plane. Here (*hkl*) denotes a specific plane and {*hkl*} refers to a family of non-parallel planes of similar symmetry. Some common low index planes for the cubic system are illustrated in Fig. 14.13d. An important property of cubic symmetry is that axes and planes which have identical indices are perpendicular (e.g. the [100] direction is perpendicular to the (100) plane). Low index planes contain more atoms than high index planes and are consequently separated by larger distances from adjacent planes. Directions and planes in hexagonal crystal structures are usually denoted by 4 Miller indices of the form [*hkil*] or {*hkil*} where $i = h + k$ is the negative intercept on the a (or x) axis and the c (or y) axis is perpendicular to the hexagonal basal plane in the unit cell; h and k are intercepts on the 120° unit cell directions in the basal plane.

14.6.3 Stereograms

It is often necessary to know the angular relationships between major symmetry directions (i.e. low index directions) in crystals. These relationships are conveniently depicted on a two dimensional plot (or stereogram) as illustrated in Fig. 14.13e for directions with respect to [100] in the cubic system. The stereogram denotes crystal directions as the intersection points of these directions with the surface of a unit sphere projected onto a horizontal plane. For the example shown, the [100] direction is perpendicular to the page and other low index direction points are shown relative to a sphere which projects as a circle. Directions lying around the circumference of the circle are hence perpendicular to the [100] axis. There is a four-fold rotational symmetry around the [100] axis (e.g. the family of four ⟨110⟩ directions which are 90° apart and perpendicular to [100]). Directions inside the circle are not perpendicular to [100] and their angular positions can be found using geometrical relationships or Wulff nets which are calibrated in angle and can be laid over the stereogram. The curved sectors within the circle denote projections of non-horizontal circles around the sphere surface and are useful in locating directions lying in the same plane. Stereograms of the more useful crystal systems for low index directions are given in Figs. 14.14 and 14.15 and Table 14.25 lists angles between major planes in the cubic system. The locations of various lattice sites within cubic crystals are illustrated in Fig. 14.16. Details of crystal structure can be found in a number of texts (e.g. Kittel, 1976).

14.6.4 Structural Data

Different crystal structures lead to differences in structural parameters along different crystal directions. Common structural parameters for crystals are illustrated in Table 14.26.

a.

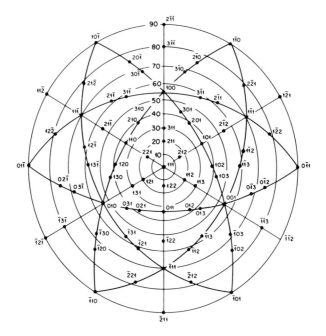

b.

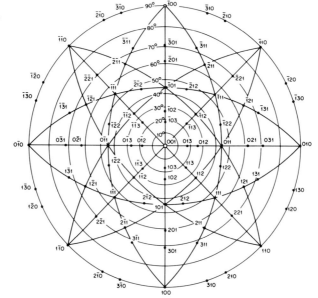

c.

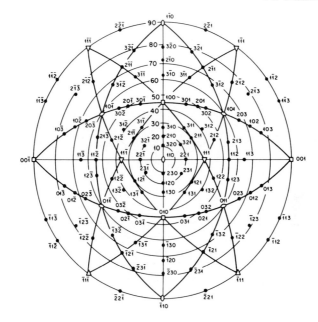

Fig. 14.14 Stereograms for a cubic crystal lattice (adapted from Mayer and Rimini, 1977), a. standard $\langle 111 \rangle$ projection for a cubic crystal, b. standard $\langle 100 \rangle$ projection for cubic crystal, and c. standard $\langle 110 \rangle$ projection for cubic crystal.

References

Gemmell, D.S. (1974). *Rev. Mod. Phys.* **46**, 129.

Kittel, C. (1976). "Solid State Physics", Wiley & Sons, N.Y.

Mayer, J.W. and Rimini, E. (eds). (1977). "Ion Beam Handbook for Materials Analysis", Academic Press, N.Y.

Swanson, M.L. (1982). *Rep. Progr. Phys.* **45**, 47.

14.7 LOW ENERGY ION DATA

14.7.1 Secondary Ion Yields

Relative positive and negative secondary ion yields from elemental or compound targets under O or Cs bombardment at normal incidence are plotted in Fig. 14.17 (Storms *et al.*, 1977).

14.7.2 Sputter and Etch Rates

Ion beam induced sputtering in UHV and a range of etch rates observed for typical laboratory vacuum conditions using 500 eV Ar ions at

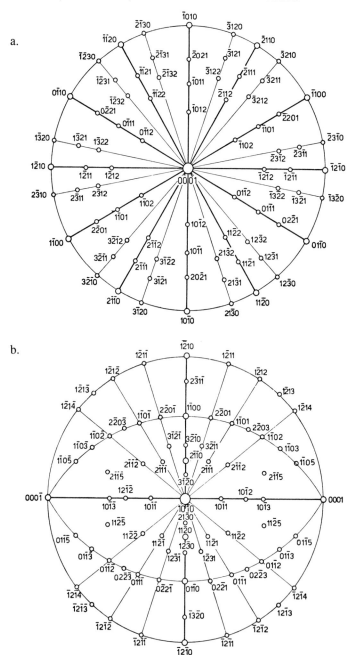

Fig. 14.15 Stereograms for a hexagonal close packed crystal lattice (adapted from Mayer and Rimini, 1977); a. standard ⟨0001⟩ projection for h.c.p. crystal, b. standard ⟨1010⟩ projection for h.c.p. crystal.

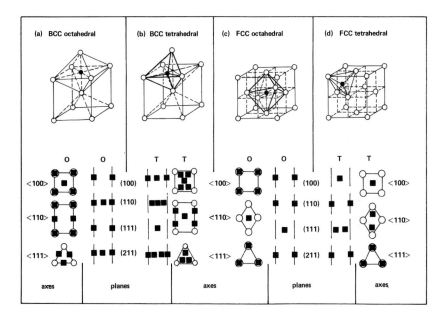

Fig. 14.16 Some lattice site positions in cubic crystals (adapted from Swanson, 1982).

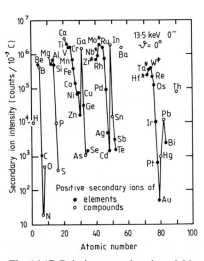

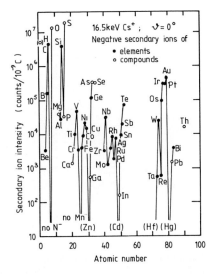

Fig. 14.17 Relative secondary ion yields.

1 mA cm^{-2} and normal to the sample surface are listed in Table 14.27 (Commonwealth Scientific Corporation, 1986).

14.7.3 SIPS Data

The yield of photons from 34 elements bombarded by 20 keV Ar$^+$ ions are listed in Table 14.28 together with wavelength (λ), upper level energy (E_u) and detection limit (MDL) (Tsong and Yusuf, 1978).

14.7.4 AES Data

Auger electron energies (measured in the derivative mode) are listed in Table 14.29. The concentration (c_i) of element i can be estimated approximately from:

$$c_i = (I_i/S_i)\bigg/ \sum_j (I_j/S_j) \tag{14.25}$$

where I is the measured Auger current and S is the relative sensitivity for each element as plotted in Fig. 14.18 for 5 keV incident electron energy (adapted from Perkin-Elmer, 1987).

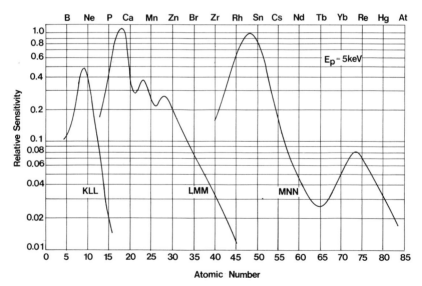

Fig. 14.18 Relative AES sensitivities in derivative mode.

References

Commonwealth Scientific Corporation (1986). "Data Sheets", CSC, Alexandra, Va 22314, USA.

Perkin-Elmer (1987). "Physical Electronic Data Sheets", Perkin-Elmer, Minneapolis, Mn 55344, USA.

Storms, H.A., Brown, K.F. and Stein, J.D. (1977). *Anal. Chem.* **49**, 2033.

Tsong, I.S.T. and Yusuf, N.A. (1978). *Appl. Phys. Lett.* **33**, 999.

ACKNOWLEDGMENTS

We gratefully acknowledge contributions from R. MacDonald, C.D. McKenzie and P. Paterson.

Tables 14.1 to 14.29 follow on pp. 648–712.

TABLE 14.1
Stopping power coefficients

a. Proton Stopping Powers

For $E_1/M_1 = 1$ to 10 keV/amu $S(E_1) = A_1 E_1^{1/2}$ eV/(10^{15} atoms cm^{-2})

For $E_1/M_1 = 10$ to 10^3 keV/amu $[S(E_1)]^{-1} = (S_L)^{-1} + (S_H)^{-1}$ eV/(10^{15} atoms cm^{-2})

where $S_L = A_2 E_1^{0.45}$, $S_H = (A_3/E_1)\ln[1 + (A_4/E_1) + A_5 E_1]$

For $E_1/M_1 = 10^3$ to 10^5 keV/amu $S(E_1) = (A_6/\beta^2)[f(E_1,\beta)]$ eV/(10^{15} atoms cm^{-2})

where $f(E_1,\beta) = \ln\left[A_7\beta^2/(1-\beta^2) - \beta^2 - \sum_{i=0}^{4} A_{i+8}(\ln E_1)^i\right]$

Target	A_1	A_2	A_3	A_4	A_5	A_6	A_7	A_8	A_9	A_{10}	A_{11}	A_{12}
H	1.262	1.44	242.6	1.2E4	0.1159	0.0005099	5.436E4	-5.052	2.049	-0.3044	0.01966	-0.0004659
He	1.229	1.397	484.5	5873	0.05225	0.00102	2.451E4	-2.158	0.8278	-0.1172	0.007259	-0.000166
Li	1.411	1.6	725.6	3013	0.04578	0.00153	2.147E4	-0.5831	0.562	-0.1183	0.009298	-0.0002498
Be	2.248	2.59	966	153.8	0.03475	0.002039	1.63E4	0.2779	0.1745	-0.05684	0.005155	-0.0001488
B	2.474	2.815	1206	1060	0.02855	0.002549	1.345E4	-2.445	1.283	-0.2205	0.0156	-0.000393
C	2.631	2.989	1445	957.2	0.02819	0.003059	1.322E4	-4.38	2.044	-0.3283	0.02221	-0.0005417
N	2.954	3.35	1683	1900	0.02513	0.003569	1.179E4	-5.054	2.325	-0.3713	0.02506	-0.0006109
O	2.652	3	1920	2000	0.0223	0.004079	1.046E4	-6.734	3.019	-0.4748	0.03171	-0.0007669
F	2.085	2.352	2157	2634	0.01816	0.004589	8517	-5.571	2.449	-0.3781	0.02483	-0.0005919
Ne	1.951	2.199	2393	2699	0.01568	0.005099	7353	-4.408	1.879	-0.2814	0.01796	-0.0004168
Na	2.542	2.869	2628	1854	0.01472	0.005609	6905	-4.959	2.073	-0.3054	0.01921	-0.0004403
Mg	3.792	4.293	2862	1009	0.01397	0.006118	6551	-5.51	2.266	-0.3295	0.02047	-0.0004637
Al	4.154	4.739	2766	164.5	0.02023	0.006628	6309	-6.061	2.46	-0.3535	0.02173	-0.0004871
Si	4.15	4.7	3329	550	0.01321	0.007138	6194	-6.294	2.538	-0.3628	0.0222	-0.0004956
P	3.232	3.647	3561	1560	0.01267	0.007648	5942	-6.527	2.616	-0.3721	0.02267	-0.000504
S	3.447	3.891	3792	1219	0.01211	0.008158	5678	-6.761	2.694	-0.3814	0.02314	-0.0005125
Cl	5.047	5.714	4023	878.6	0.01178	0.008668	5524	-6.994	2.773	-0.3907	0.02361	-0.0005209
Ar	5.731	6.5	4253	530	0.01123	0.009178	5268	-7.227	2.851	-0.4	0.02407	-0.0005294

TABLE 14.1 (*cont.*)
Stopping power coefficients

Target	A_1	A_2	A_3	A_4	A_5	A_6	A_7	A_8	A_9	A_{10}	A_{11}	A_{12}
K	5.151	5.833	4482	545.7	0.01129	0.009687	5295	-7.44	2.923	-0.4094	0.02462	-0.0005411
Ca	5.521	6.252	4710	553.3	0.01112	0.0102	5214	-7.653	2.995	-0.4187	0.02516	-0.0005529
Sc	5.201	5.884	4938	560.9	0.009995	0.01071	4688	-8.012	3.123	-0.435	0.02605	-0.0005707
Ti	4.862	5.496	5165	568.5	0.009474	0.01122	4443	-8.371	3.251	-0.4513	0.02694	-0.0005886
V	4.48	5.055	5391	952.3	0.009117	0.01173	4276	-8.731	3.379	-0.4676	0.02783	-0.0006064
Cr	3.983	4.489	5616	1336	0.008413	0.01224	3946	-9.09	3.507	-0.4838	0.02872	-0.0006243
Mn	3.469	3.907	5725	1461	0.008829	0.01275	3785	-9.449	3.635	-0.5001	0.02961	-0.0006421
Fe	3.519	3.963	6065	1243	0.007782	0.01326	3650	-9.809	3.763	-0.5164	0.0305	-0.00066
Co	3.14	3.535	6288	1372	0.007361	0.01377	3453	-10.17	3.891	-0.5327	0.03139	-0.0006779
Ni	3.553	4.004	6205	555.1	0.008763	0.01428	3297	-10.53	4.019	-0.549	0.03229	-0.0006957
Cu	3.696	4.175	4673	387.8	0.02188	0.01479	3174	-11.18	4.252	-0.5791	0.03399	-0.0007314
Zn	4.21	4.75	6953	295.2	0.006809	0.0153	3194	-11.57	4.394	-0.598	0.03506	-0.0007537
Ga	5.041	5.697	7173	202.6	0.006725	0.01581	3154	-11.95	4.537	-0.6169	0.03613	-0.0007759
Ge	5.554	6.3	6496	110	0.009689	0.01632	3097	-12.34	4.68	-0.6358	0.03721	-0.0007981
As	5.323	6.012	7611	292.5	0.006447	0.01683	3024	-12.72	4.823	-0.6547	0.03828	-0.0008203
Se	5.874	6.656	7395	117.5	0.007684	0.01734	3006	-13.11	4.965	-0.6735	0.03935	-0.0008425
Br	5.611	6.335	8046	365.2	0.006244	0.01785	2928	-13.4	5.083	-0.6906	0.04042	-0.0008675
Kr	6.411	7.25	8262	220	0.006087	0.01836	2855	-13.69	5.2	-0.7076	0.0415	-0.0008925
Rb	5.694	6.429	8478	292.9	0.006087	0.01886	2855	-13.92	5.266	-0.714	0.04173	-0.0008943
Sr	6.339	7.159	8693	330.3	0.006003	0.01937	2815	-14.14	5.331	-0.7205	0.04196	-0.0008962
Y	6.407	7.234	8907	367.8	0.005889	0.01988	2762	-14.36	5.397	-0.7269	0.04219	-0.000898
Zr	6.734	7.603	9120	405.2	0.005765	0.02039	2704	-14.59	5.463	-0.7333	0.04242	-0.0008998
Nb	6.902	7.791	9333	442.7	0.005587	0.0209	2621	-16.22	6.094	-0.8225	0.04791	-0.001024
Mo	6.425	7.248	9545	480.2	0.005367	0.02141	2517	-17.85	6.725	-0.9116	0.05339	-0.001148
Tc	6.799	7.671	9756	517.6	0.005315	0.02192	2493	-17.96	6.752	-0.9135	0.05341	-0.001147
Ru	6.108	6.887	9966	555.1	0.005151	0.02243	2416	-18.07	6.779	-0.9154	0.05342	-0.001145
Rh	5.924	6.677	1.018E4	592.5	0.004919	0.02294	2307	-18.18	6.806	-0.9173	0.05343	-0.001143
Pd	5.238	5.9	1.038E4	630	0.004758	0.02345	2231	-18.28	6.833	-0.9192	0.05345	-0.001142

TABLE 14.1 (*cont.*)
Stopping power coefficients

Target	A_1	A_2	A_3	A_4	A_5	A_6	A_7	A_8	A_9	A_{10}	A_{11}	A_{12}
Ag	5.623	6.354	7160	337.6	0.01394	0.02396	2193	−18.39	6.86	−0.9211	0.05346	−0.00114
Cd	5.814	6.554	1.08E4	355.5	0.004626	0.02447	2170	−18.62	6.915	−0.9243	0.0534	−0.001134
In	6.23	7.024	1.101E4	370.9	0.00454	0.02498	2129	−18.85	6.969	−0.9275	0.05335	−0.001127
Sn	6.41	7.227	1.121E4	386.4	0.004474	0.02549	2099	−19.07	7.024	−0.9308	0.05329	−0.001121
Sb	7.5	8.48	8608	348	0.009074	0.026	2069	−19.57	7.225	−0.9603	0.05518	−0.001165
Te	6.979	7.871	1.162E4	392.4	0.004402	0.02651	2065	−20.07	7.426	−0.9899	0.05707	−0.001209
I	7.725	8.716	1.183E4	394.8	0.004376	0.02702	2052	−20.56	7.627	−1.019	0.05896	−0.001254
Xe	8.231	9.289	1.203E4	397.3	0.004384	0.02753	2056	−21.06	7.828	−1.049	0.06085	−0.001298
Cs	7.287	8.218	1.223E4	399.7	0.004447	0.02804	2086	−20.4	7.54	−1.004	0.05782	−0.001224
Ba	7.899	8.911	1.243E4	402.1	0.004511	0.02855	2116	−19.74	7.252	−0.9588	0.05479	−0.001151
La	8.041	9.071	1.263E4	404.5	0.00454	0.02906	2129	−19.08	6.964	−0.9136	0.05176	−0.001077
Ce	7.489	8.444	1.283E4	406.9	0.00442	0.02957	2073	−18.43	6.677	−0.8684	0.04872	−0.001003
Pr	7.291	8.219	1.303E4	409.3	0.004298	0.03008	2016	−17.77	6.389	−0.8233	0.04569	−0.0009292
Nd	7.098	8	1.323E4	411.8	0.004182	0.03059	1962	−17.11	6.101	−0.7781	0.04266	−0.0008553
Pm	6.91	7.786	1.343E4	414.2	0.004058	0.0311	1903	−16.45	5.813	−0.733	0.03963	−0.0007815
Sm	6.728	7.58	1.362E4	416.6	0.003976	0.03161	1865	−15.79	5.526	−0.6878	0.0366	−0.0007077
Eu	6.551	7.38	1.382E4	419	0.003877	0.03212	1819	−15.13	5.238	−0.6426	0.03357	−0.0006339
Gd	6.739	7.592	1.402E4	421.4	0.003863	0.03263	1812	−14.47	4.95	−0.5975	0.03053	−0.0005601
Tb	6.212	6.996	1.421E4	423.9	0.003725	0.03314	1747	−14.56	4.984	−0.6022	0.03082	−0.0005668
Dy	5.517	6.21	1.44E4	426.3	0.003632	0.03365	1703	−14.65	5.018	−0.6069	0.03111	−0.0005734
Ho	5.219	5.874	1.46E4	428.7	0.003498	0.03416	1640	−14.74	5.051	−0.6117	0.03141	−0.0005801
Er	5.071	5.706	1.479E4	433	0.003405	0.03467	1597	−14.83	5.085	−0.6164	0.0317	−0.0005867
Tm	4.926	5.542	1.498E4	433.5	0.003342	0.03518	1567	−14.91	5.119	−0.6211	0.03199	−0.0005933
Yb	4.787	5.386	1.517E4	435.9	0.003292	0.03569	1544	−15	5.153	−0.6258	0.03228	−0.0006
Lu	4.893	5.505	1.536E4	438.4	0.003243	0.0362	1521	−15.09	5.186	−0.6305	0.03257	−0.0006066
Hf	5.028	5.657	1.555E4	440.8	0.003195	0.03671	1499	−15.18	5.22	−0.6353	0.03286	−0.0006133
Ta	4.738	5.329	1.574E4	443.2	0.003186	0.03722	1494	−15.27	5.254	−0.64	0.03315	−0.0006199
W	4.574	5.144	1.593E4	442.4	0.003144	0.03773	1475	−15.67	5.392	−0.6577	0.03418	−0.0006426

TABLE 14.1 (*cont.*)
Stopping power coefficients

Target	A_1	A_2	A_3	A_4	A_5	A_6	A_7	A_8	A_9	A_{10}	A_{11}	A_{12}
Re	5.2	5.851	1.612E4	441.6	0.003122	0.03824	1464	-16.07	5.529	-0.6755	0.03521	-0.0006654
Os	5.07	5.704	1.63E4	440.9	0.003082	0.03875	1446	-16.47	5.667	-0.6932	0.03624	-0.0006881
Ir	4.945	5.563	1.649E4	440.1	0.002965	0.03926	1390	-16.88	5.804	-0.711	0.03727	-0.0007109
Pt	4.476	5.034	1.667E4	439.3	0.002871	0.03977	1347	-17.28	5.942	-0.7287	0.0383	-0.0007336
Au	4.856	5.46	1.832E4	438.5	0.002542	0.04028	1354	-17.02	5.846	-0.7149	0.0374	-0.0007114
Hg	4.308	4.843	1.704E4	487.8	0.002882	0.04079	1352	-17.84	6.183	-0.7659	0.04076	-0.0007925
Tl	4.723	5.311	1.722E4	537	0.002913	0.0413	1366	-18.66	6.52	-0.8169	0.04411	-0.0008737
Pb	5.319	5.982	1.74E4	586.3	0.002871	0.04181	1347	-19.48	6.857	-0.8678	0.04747	-0.0009548
Bi	5.956	6.7	1.78E4	677	0.00266	0.04232	1336	-19.55	6.871	-0.8686	0.04748	-0.0009544
Po	6.158	6.928	1.777E4	586.3	0.002812	0.04283	1319	-19.62	6.884	-0.8694	0.04748	-0.000954
At	6.204	6.979	1.795E4	586.3	0.002776	0.04334	1302	-19.69	6.898	-0.8702	0.04749	-0.0009536
Rn	6.181	6.954	1.812E4	586.3	0.002748	0.04385	1289	-19.76	6.912	-0.871	0.04749	-0.0009532
Fr	6.949	7.82	1.83E4	586.3	0.002737	0.04436	1284	-19.83	6.926	-0.8718	0.0475	-0.0009528
Ra	7.506	8.448	1.848E4	586.3	0.002727	0.04487	1279	-19.9	6.94	-0.8726	0.04751	-0.0009524
Ac	7.649	8.609	1.866E4	586.3	0.002697	0.04538	1265	-19.97	6.953	-0.8733	0.04751	-0.000952
Th	7.71	8.679	1.883E4	586.3	0.002641	0.04589	1239	-20.04	6.967	-0.8741	0.04752	-0.0009516
Pa	7.407	8.336	1.901E4	586.3	0.002603	0.0464	1221	-20.11	6.981	-0.8749	0.04752	-0.0009512
U	7.29	8.204	1.918E4	586.3	0.002573	0.04691	1207	-20.18	6.995	-0.8757	0.04753	-0.0009508

b. Helium Stopping Powers for Solids

For $E_1 = 1$ to 10^4 keV $[S(E_1)]^{-1} = (S_L)^{-1} + (S_H)^{-1}$

where $S_L = A_1 E_1^{.42}$

$$S_H = (10^3 A_3/E_1) \ln[1 + (10^3 A_4/E_1) + (A_5 E/10^3)]$$

Target	A_1	A_2	A_3	A_4	A_5
H	0.9661	0.4126	6.92	8.831	2.582
He	2.027	0.2931	26.34	6.66	0.3409

Target	A_1	A_2	A_3	A_4	A_5
Ag	5.6	0.49	130	10	2.844
Cd	3.55	0.6068	124.7	1.112	3.119

TABLE 14.1 (*cont.*)
Stopping power coefficients

Target	A_1	A_2	A_3	A_4	A_5
Li	1.42	0.49	12.25	32	9.161
Be	2.206	0.51	15.32	0.25	8.995
B	3.691	0.4128	18.48	50.72	9
C	4.232	0.3877	22.99	35	7.993
N	2.51	0.4752	38.26	13.02	1.892
O	1.766	0.5261	37.11	15.24	2.804
F	1.533	0.531	40.44	18.41	2.718
Ne	1.183	0.55	39.83	17.49	4.001
Na	9.894	0.3081	23.65	0.384	92.93
Mg	4.3	0.47	34.3	3.3	12.74
Al	2.5	0.625	45.7	0.1	4.359
Si	2.1	0.65	49.34	1.788	4.133
P	1.729	0.6562	53.41	2.405	3.845
S	1.402	0.6791	58.98	3.528	3.211
Cl	1.117	0.7044	69.69	3.705	2.156
Ar	0.9172	0.724	79.44	3.648	1.646
K	8.554	0.3817	83.61	11.84	1.875
Ca	6.297	0.4622	65.39	10.14	5.036
Sc	5.307	0.4918	61.74	12.4	6.665
Ti	4.71	0.5087	65.28	8.806	5.948
V	6.151	0.4524	83	18.31	2.71
Cr	6.57	0.4322	84.76	15.53	2.779
Mn	5.738	0.4492	84.61	14.18	3.101
Fe	5.013	0.4707	85.58	16.55	3.211
Co	4.32	0.4947	76.14	10.85	5.441
Ni	4.652	0.4571	80.73	22	4.952
Cu	3.114	0.5236	76.67	7.62	6.385
Zn	3.114	0.5236	76.67	7.62	7.502
Ga	3.114	0.5236	76.67	7.62	8.514
Ge	5.746	0.4662	79.24	1.185	7.993

Target	A_1	A_2	A_3	A_4	A_5
In	3.6	0.62	105.8	0.1692	6.026
Sn	5.4	0.53	103.1	3.931	7.767
Sb	3.97	0.6459	131.8	0.2233	2.723
Te	3.65	0.64	126.8	0.6834	3.411
I	3.118	0.6519	164.9	1.208	1.51
Xe	2.031	0.7181	153.1	1.362	1.958
Cs	14.4	0.3923	152.5	8.354	2.597
Ba	10.99	0.4599	138.4	4.811	3.726
La	16.6	0.3773	224.1	6.28	0.9121
Ce	10.54	0.4533	159.3	4.832	2.529
Pr	10.33	0.4502	162	5.132	2.444
Nd	10.15	0.4471	165.6	5.378	2.328
Pm	9.976	0.4439	168	5.721	2.258
Sm	9.804	0.4408	176.2	5.675	1.997
Eu	14.22	0.363	228.4	7.024	1.016
Gd	9.952	0.4318	233.5	5.065	0.9244
Tb	9.272	0.4345	210	4.911	1.258
Dy	10.13	0.4146	225.7	5.525	1.055
Ho	8.949	0.4304	213.3	5.071	1.221
Er	11.94	0.3783	247.2	6.655	0.849
Tm	8.472	0.4405	195.5	4.051	1.604
Yb	8.301	0.4399	203.7	3.667	1.459
Lu	6.567	0.4858	193	2.65	1.66
Hf	5.951	0.5016	196.1	2.662	1.589
Ta	7.495	0.4523	251.4	3.433	0.8619
W	6.335	0.4825	255.1	2.834	0.8228
Re	4.314	0.5558	214.8	2.354	1.263
Os	4.02	0.5681	219.9	2.402	1.191
Ir	3.836	0.5765	210.2	2.742	1.305
Pt	4.68	0.5247	244.7	2.749	0.8962

TABLE 14.1 (*cont.*)
Stopping power coefficients

Target	A_1	A_2	A_3	A_4	A_5
As	2.792	0.6346	106.1	0.2986	2.331
Se	4.667	0.5095	124.3	2.102	1.667
Br	2.44	0.6346	105	0.83	2.851
Kr	1.491	0.7118	120.6	1.101	1.877
Rb	11.72	0.3826	102.8	9.231	4.371
Sr	7.126	0.4804	119.3	5.784	2.454
Y	11.61	0.3955	146.7	7.031	1.423
Zr	10.99	0.41	163.9	7.1	1.052
Nb	9.241	0.4275	163.1	7.954	1.102
Mo	9.276	0.418	157.1	8.038	1.29
Tc	3.999	0.6152	97.6	1.297	5.792
Ru	4.306	0.5658	97.99	5.514	5.754
Rh	3.615	0.6197	86.26	0.333	8.689
Pd	5.8	0.49	147.2	6.903	1.289

Target	A_1	A_2	A_3	A_4	A_5
Au	3.223	0.5883	232.7	2.954	1.05
Hg	2.892	0.6204	208.6	2.415	1.416
Tl	4.728	0.5522	217	3.091	1.386
Pb	6.18	0.52	170	4	3.224
Bi	9	0.47	198	3.8	2.032
Po	2.324	0.6997	216	1.599	1.399
At	1.961	0.7286	223	1.621	1.296
Rn	1.75	0.7427	350.1	0.9789	0.5507
Fr	10.31	0.4613	261.2	4.738	0.9899
Ra	7.962	0.519	235.7	4.347	1.313
Ac	6.227	0.5645	231.9	3.961	1.379
Th	5.246	0.5947	228.6	4.027	1.432
Pa	5.408	0.5811	235.7	3.961	1.358
U	5.218	0.5828	245	3.838	1.25

c. Helium Stopping Powers for Gases

Target	A_1	A_2	A_3	A_4	A_5
H	0.39	0.63	4.17	85.55	19.55
He	0.58	0.59	6.3	130	44.07
Li	15.23	0.1076	10.14	1232	31.24
C	3.47	0.4485	22.37	36.41	7.993
N	2	0.548	29.82	18.11	4.37
O	2.717	0.4858	32.88	25.88	4.336
F	2.616	0.4708	41.2	28.07	2.458
Ne	2.303	0.4861	37.01	37.96	5.092
Na	13.03	0.2685	35.65	44.18	9.175
S	3.116	0.5988	53.71	5.632	4.536

Target	A_1	A_2	A_3	A_4	A_5
Cl	3.365	0.571	63.67	6.182	2.969
Ar	2.291	0.6284	73.88	4.478	2.066
K	16.6	0.3095	99.1	10.98	1.092
Br	1.65	0.7	148.1	1.47	0.9686
Kr	1.413	0.7377	147.9	1.466	1.016
I	4.13	0.6177	152	2.516	1.938
Xe	3.949	0.6209	200.5	1.878	0.9126
Cs	25.84	0.3139	335.1	2.946	0.3347
Hg	8.15	0.4745	269.2	2.392	0.7467
Ru	4.822	0.605	418.3	0.8335	0.3865

TABLE 14.2
Stopping powers for protons and He ions in C, Si, Ge and Pb

Energy (MeV)	Ion Stopping Power S_i MeV cm^2 g^{-1}							
	C		Si		Ge		Pb	
	Proton	He	Proton	He	Proton	He	Proton	He
0.001	132	330	89	108	46	69	16	23
0.010	417	567	281	229	146	152	49	64
0.10	727	1254	519	881	247	403	121	197
1	229	1810	177	1357	114	670	63	401
2	141	1359	113	1027	76	600	45	349
3	104	1073	85	836	59	523	37	302
10	41	479	35	388	26	267	18	161
20	23	285	20	232	16	162	11	100

TABLE 14.3
Stopping power position (E_{max}) and maxima (S_{max}) for various ions in C, Si and Ge

Ion Z_1	C		Si		Ge	
	E_{max} MeV/amu	S_{max} MeV cm^2 mg^{-1}	E_{max} MeV/amu	S_{max} MeV cm^2 mg^{-1}	E_{max} MeV/amu	S_{max} MeV cm^2 mg^{-1}
H 1	0.070	0.75	0.078	0.54	0.066	0.26
He 2	0.15	1.93	0.12	1.57	0.22	0.68
B 5	0.27	5.8	0.17	4.4	0.55	2.30
Ne 10	0.45	13.6	0.52	9.7	0.90	6.1
Ca 20	0.80	28	1.0	22	1.4	14.2
Zn 30	1.5	42	1.8	34	2.3	22
Zr 40	1.8	56	2.3	45	2.9	31
Sn 50	2.3	69	2.8	56	3.7	39
Nd 60	2.75	84	3.6	68	4.5	48
Hg 80	3.75	110	4.5	92	6.0	65
U 92	4.5	125	6	105	7.0	76

TABLE 14.4
Ion effective charges

Energy (MeV)	Proton	He	Li	N	Ar	Br
0.200	0.974	1.494	1.603	1.575	1.158	0.149
0.600	1.000	1.648	1.753	2.029	1.608	0.551
1.000	1.000	1.866	2.045	2.758	2.412	1.328
2.000	1.000	1.979	2.413	3.445	3.268	2.182
5.000	1.000	2.000	2.903	4.480	4.828	3.811
10.00	1.000	2.000	2.998	5.268	6.381	5.550
20.00	1.000	2.000	3.000	5.962	8.258	7.840

TABLE 14.5
Mean ion projected ranges in C, density 2.26 g cm^{-3} and Si, density 2.322 g cm^{-3} for protons and He ions

Energy (MeV)	C				Si			
	Proton		He		Proton		He	
	Range mg cm^{-2}	Straggling % (1σ)	Range mg cm^{-2}	Straggling % (1σ)	Range mg cm^{-2}	Straggling % (1σ)	Range mg cm^{-2}	Straggling % (1σ)
0.001	0.0043	10	0.0023	50	0.0048	6.6	0.0037	95
0.010	0.030	5.6	0.019	24	0.034	5.0	0.037	38
0.10	0.16	3.6	0.11	7.4	0.208	3.8	0.183	16
1.0	2.75	2.2	0.626	2.4	3.85	2.6	0.813	4.9
2.0	8.44	1.9	1.27	2.1	11.3	2.2	1.70	2.9
5.0	33.4	1.6	4.1	2.5	50.7	1.9	5.57	2.5
10.0	136	1.4	13	3.3	186	1.6	16.0	3.3

TABLE 14.6
Calculated mean projected ranges (angstroms) of various heavy ions in Si

Ion	Energy (keV)						
	20	40	60	80	100	160	200
B	765	1566	2337	3081	3810	5800	7014
N	535	1089	1653	2199	2728	4259	5207
Al	288	563	846	1137	1430	2323	2907
P	253	488	729	974	1227	1988	2506
Ga	154	271	384	494	606	944	1175
As	150	262	368	473	577	894	1109
In	131	222	303	381	456	677	822
Sb	130	220	299	375	448	663	803

TABLE 14.7
Vavilov energy straggling distributions

λ \ κ	0.1	0.2	0.3	0.4	0.5	0.6	0.7	0.8	0.9	1.0
-4.0	.7435E-3		.8842E-5	.9824E-5	.1088E-4	.1213E-4	.1350E-4	.1488E-4	.1641E-4	.1821E-4
-3.5	.1065E-1	.8220E-3	.9085E-3	.1004E-2	.1110E-2	.1227E-2	.1356E-2	.1498E-2	.1656E-2	.1830E-2
-3.0	.4861E-1	.1177E-1	.1301E-1	.1438E-1	.1589E-1	.1756E-1	.1941E-1	.2145E-1	.2370E-1	.2620E-1
-2.5	.1111	.5372E-1	.5937E-1	.6562E-1	.7252E-1	.8015E-1	.8857E-1	.9788E-1	.1081	.1194
-2.0	.1673	.1228	.1357	.1500	.1658	.1832	.2024	.2232	.2456	.2693
-1.5	.1960	.1849	.2044	.2259	.2496	.2754	.3027	.3306	.3584	.3855
-1.0	.1977	.2166	.2394	.2645	.2917	.3189	.3444	.3669	.3860	.4012
-0.5	.1826	.2185	.2414	.2664	.2902	.3093	.3225	.3300	.3319	.3287
0.0	.1605	.2018	.2230	.2437	.2572	.2623	.2603	.2522	.2393	.2229
0.5	.1373	.1774	.1956	.2073	.2078	.2000	.1864	.1689	.1495	.1295
1.0	.1159	.1517	.1653	.1658	.1556	.1397	.1209	.1014	.8278E-1	.6600E-1
1.5	.9750E-1	.1281	.1349	.1257	.1094	.9057E-1	.7205E-1	.5539E-1	.4135E-1	.3008E-1
2.0	.8206E-1	.1077	.1063	.9105E-1	.7277E-1	.5509E-1	.3992E-1	.2789E-1	.1888E-1	.1243E-1
2.5	.6932E-1	.9006E-1	.8102E-1	.6438E-1	.4616E-1	.3167E-1	.2075E-1	.1307E-1	.7959E-2	.4706E-2
3.0	.5887E-1	.7448E-1	.6010E-1	.4277E-1	.2806E-1	.1732E-1	.1018E-1	.5740E-2	.3124E-2	.1649E-2
3.5	.5031E-1	.6043E-1	.4359E-1	.2794E-1	.1642E-1	.9056E-2	.4743E-2	.2379E-2	.1150E-2	.5377E-3
4.0	.4328E-1	.4801E-1	.3100E-1	.1774E-1	.9285E-2	.4545E-2	.2109E-2	.8356E-3	.3996E-3	.1651E-3
4.5	.3748E-1	.3745E-1	.2164E-1	.1099E-1	.5087E-2	.2197E-2	.8977E-3	.3499E-3	.1209E-3	
5.0	.3267E-1	.2880E-1	.1484E-1	.6645E-2	.2707E-2	.1027E-2	.3679E-3	.1257E-3		
5.5	.2865E-1	.2193E-1	.1000E-1	.3933E-2	.1402E-2	.4643E-3	.1449E-3			
6.0	.2527E-1	.1657E-1	.6644E-2	.2282E-2	.7090E-3	.2042E-3				
6.5	.2440E-1	.1245E-1	.4350E-2	.1299E-2	.3499E-3					
7.0	.1984E-1	.9290E-2	.2813E-2	.7269E-3	.1689E-3					
7.5	.1740E-1	.6883E-2	.1796E-2	.4001E-3						
8.0	.1499E-1	.5057E-2	.1134E-2	.2165E-3						
8.5	.1266E-1	.3682E-2	.7078E-3							
9.0	.1054E-1	.2659E-2	.4368E-3							
9.5	.8699E-2	.1907E-2	.2674E-3							
10.0	.7125E-2	.1359E-2	.1618E-3							
10.5	.5831E-2	.9643E-3								
11.0		.6805E-3								

TABLE 14.7 (cont.)
Vavilov energy straggling distributions

λ \ κ	1.5	2	3	4	5	6	7	8	9	10
-4.0										.8087E-4
-3.6	.3117E-4	.5165E-4	.1405E-3	.3752E-3	.2855E-3	.7208E-3	.1746E-2	.4019E-2	.8740E-2	.1796E-1
-3.5										
-3.4		.4964E-2	.1304E-1	.3188E-1	.2874E-2	.6863E-2	.1537E-1	.3224E-1	.6226E-1	.1162
-3.2					.1754E-1	.3889E-1	.7957E-1	.1502	.2615	.4210
-3.0	.3018E-2				.7110E-1	.1437	.2632	.4371	.6607	.9120
-2.8					.2030	.3660	.5865	.8390	.1077E+1	.1249E+1
-2.6	.4303E-1	.6941E-1	.1634	.3276	.4267	.6713	.9212	.1111E+1	.1187E+1	.1132E+1
-2.5										
-2.4					.6847	.9200	.1059E+1	.1055E+1	.9187	.7058
-2.2	.1914				.8646	.9714	.9181	.7404	.5156	.3133
-2.0		.2894	.5397	.7754	.8808	.8106	.6169	.3950	.2158	.1018
-1.8	.4004				.7392	.5463	.3285	.1639	.6895E-1	.2484E-1
-1.6		.5311	.6844	.6164	.5203	.3029	.1412	.5394E-1	.1720E-1	.4618E-2
-1.5										
-1.4					.3117	.1404	.4987E-1	.1433E-1	.3380E-2	.6897E-3
-1.2	.4957				.1611	.5514E-1	.1467E-1	.3106E-2	.5593E-3	.7183E-4
-1.0		.5383	.4193	.2091	.7262E-1	.1857E-1	.3632E-2	.5796E-3	.5907E-4	
-0.8					.2883E-1	.5415E-2	.7836E-3	.7405E-4		
-0.6	.4156				.1017E-1	.1394E-2				
-0.5										
-0.4		.3456	.1442	.3559E-1	.3220E-2	.3017E-3				
-0.2					.9147E-3					
0.0	.2568	.1541	.3087E-1	.3410E-2	.2329E-3					
0.5	.1238	.5081E-1	.4429E-2	.2025E-3						
1.0	.4853E-1	.1298E-1	.4505E-3	.8159E-5						
1.5	.1593E-1	.2660E-2	.3488E-4							
2.0	.4482E-2	.4419E-3	.2215E-5							
2.5	.1100E-2	.6429E-4								
3.0	.2387E-3									

TABLE 14.8 RBS Data

Element			kinematic factor				backscattered energy (keV)			
			98	105	110	170	98	105	110	170
3 Li		6	.1569	.1271	.1102	.0412	314	254	220	82
		7	.2255	.1913	.1708	.0761	451	383	342	152
4 Be		9	.3353	.2980	.2747	.1502	671	596	549	300
5 B		10	.3798	.3423	.3186	.1861	760	685	637	372
		11	.4187	.3815	.3577	.2203	837	763	715	441
6 C		12	.4529	.4162	.3926	.2523	906	832	785	505
		13	.4837	.4477	.4244	.2828	967	895	849	566
7 N		14	.5112	.4760	.4531	.3112	1022	952	906	622
		15	.5358	.5015	.4790	.3377	1072	1003	958	675
8 O		16	.5581	.5246	.5026	.3624	1116	1049	1005	725
		17	.5785	.5459	.5244	.3857	1157	1092	1049	771
		18	.5970	.5654	.5443	.4074	1194	1131	1089	815
9 F		19	.6140	.5832	.5627	.4278	1228	1166	1125	856
10 Ne		20	.6295	.5995	.5795	.4468	1259	1199	1159	894
		21	.6440	.6148	.5952	.4647	1288	1230	1190	929
		22	.6573	.6288	.6098	.4816	1315	1258	1220	963
11 Na		23	.6697	.6419	.6233	.4974	1339	1284	1247	995
12 Mg		24	.6811	.6541	.6360	.5124	1362	1308	1272	1025
		25	.6919	.6655	.6478	.5265	1384	1331	1296	1053
		26	.7019	.6762	.6589	.5399	1404	1352	1318	1080
13 Al		27	.7113	.6862	.6693	.5525	1423	1372	1339	1105
14 Si		28	.7201	.6956	.6791	.5645	1440	1391	1358	1129
		29	.7284	.7045	.6884	.5759	1457	1409	1377	1152
		30	.7363	.7129	.6971	.5867	1473	1426	1394	1173
15 P		31	.7437	.7208	.7054	.5970	1487	1442	1411	1194
16 S		32	.7506	.7283	.7132	.6068	1501	1457	1426	1214
		33	.7573	.7354	.7206	.6161	1515	1471	1441	1232
		34	.7635	.7422	.7277	.6250	1527	1484	1455	1250
		36	.7752	.7547	.7408	.6417	1550	1509	1482	1283
17 Cl		35	.7695	.7486	.7344	.6336	1539	1497	1469	1267
		37	.7806	.7605	.7468	.6495	1561	1521	1494	1299
18 Ar		36	.7752	.7547	.7408	.6417	1550	1509	1482	1283
		38	.7857	.7660	.7526	.6570	1571	1532	1505	1314
		40	.7953	.7764	.7635	.6710	1591	1553	1527	1342
19 K		39	.7906	.7713	.7582	.6642	1581	1543	1516	1328
		40	.7953	.7764	.7635	.6710	1591	1553	1527	1342
		41	.7998	.7812	.7686	.6776	1600	1562	1537	1355
20 Ca		40	.7953	.7764	.7635	.6710	1591	1553	1527	1342
		42	.8041	.7859	.7734	.6840	1608	1572	1547	1368

TABLE 14.8 *(cont.)*

channel number				Rutherford Cross-section (barns)			
98	105	110	170	98	105	110	170
78	64	55	21	.0242	.0180	.0147	.0037
113	96	85	38	.0277	.0213	.0179	.0054
168	149	137	75	.0553	.0438	.0376	.0136
190	171	159	93	.0891	.0711	.0613	.0233
209	191	179	110	.0910	.0730	.0632	.0249
226	208	196	126	.1331	.1071	.0930	.0375
242	224	212	141	.1347	.1087	.0946	.0389
256	238	227	156	.1851	.1497	.1305	.0545
268	251	240	169	.1865	.1511	.1319	.0557
279	262	251	181	.2450	.1988	.1737	.0741
289	273	262	193	.2463	.2000	.1749	.0752
299	283	272	204	.2473	.2010	.1759	.0762
307	292	281	214	.3141	.2555	.2237	.0974
315	300	290	223	.3889	.3166	.2773	.1214
322	307	298	232	.3899	.3176	.2783	.1223
329	314	305	241	.3907	.3184	.2791	.1231
335	321	312	249	.4737	.3862	.3386	.1498
341	327	318	256	.5646	.4605	.4039	.1792
346	333	324	263	.5654	.4613	.4048	.1800
351	338	329	270	.5662	.4621	.4055	.1807
356	343	335	276	.6652	.5430	.4766	.2128
360	348	340	282	.7723	.6306	.5536	.2476
364	352	344	288	.7730	.6313	.5543	.2482
368	356	349	293	.7736	.6319	.5549	.2489
372	360	353	298	.8888	.7261	.6377	.2863
375	364	357	303	1.0119	.8268	.7262	.3265
379	368	360	308	1.0125	.8274	.7269	.3271
382	371	364	313	1.0131	.8280	.7274	.3276
388	377	370	321	1.0141	.8290	.7284	.3286
385	374	367	317	1.1443	.9353	.8218	.3704
390	380	373	325	1.1453	.9364	.8228	.3715
388	377	370	321	1.2835	1.0492	.9219	.4159
393	383	376	328	1.2846	1.0503	.9230	.4170
398	388	382	336	1.2855	1.0512	.9239	.4179
395	386	379	332	1.4318	1.1708	1.0289	.4651
398	388	382	336	1.4323	1.1713	1.0294	.4656
400	391	384	339	1.4327	1.1717	1.0299	.4660
398	388	382	336	1.5870	1.2978	1.1406	.5159
402	393	387	342	1.5880	1.2988	1.1416	.5168

TABLE 14.8 *(cont.)*

Element		kinematic factor				backscattered energy (keV)			
		98	105	110	170	98	105	110	170
	43	.8082	.7903	.7781	.6901	1616	1581	1556	1380
	44	.8121	.7946	.7826	.6959	1624	1589	1565	1392
	48	.8264	.8100	.7988	.7174	1653	1620	1598	1435
21 Sc	45	.8159	.7987	.7869	.7016	1632	1597	1574	1403
22 Ti	46	.8196	.8026	.7910	.7071	1639	1605	1582	1414
	47	.8231	.8064	.7950	.7123	1646	1613	1590	1425
	48	.8264	.8100	.7988	.7174	1653	1620	1598	1435
	49	.8297	.8135	.8025	.7223	1659	1627	1605	1445
	50	.8328	.8169	.8060	.7270	1666	1634	1612	1454
23 Y	50	.8328	.8169	.8061	.7270	1666	1634	1612	1454
	51	.8358	.8202	.8095	.7316	1672	1640	1619	1463
24 Cr	50	.8328	.8169	.8061	.7270	1666	1634	1612	1454
	52	.8387	.8233	.8128	.7360	1677	1647	1626	1472
	53	.8415	.8263	.8160	.7403	1683	1653	1632	1481
	54	.8443	.8294	.8192	.7446	1689	1659	1638	1489
25 Mn	55	.8468	.8321	.8221	.7485	1694	1664	1644	1497
26 Fe	54	.8442	.8293	.8191	.7444	1688	1659	1638	1489
	56	.8493	.8349	.8250	.7524	1699	1670	1650	1505
	57	.8518	.8375	.8278	.7561	1704	1675	1656	1512
	58	.8541	.8401	.8305	.7598	1708	1680	1661	1520
27 Co	59	.8564	.8426	.8331	.7634	1713	1685	1666	1527
28 Ni	58	.8541	.8401	.8305	.7598	1708	1680	1661	1520
	60	.8586	.8450	.8356	.7668	1717	1690	1671	1534
	61	.8608	.8473	.8381	.7702	1722	1695	1676	1540
	62	.8629	.8496	.8405	.7734	1726	1699	1681	1547
	64	.8669	.8540	.8451	.7797	1734	1708	1690	1559
29 Cu	63	.8649	.8518	.8428	.7766	1730	1704	1686	1553
	65	.8688	.8561	.8473	.7827	1738	1712	1695	1565
30 Zn	64	.8669	.8540	.8451	.7797	1734	1708	1690	1559
	66	.8707	.8581	.8494	.7856	1741	1716	1699	1571
	67	.8725	.8601	.8515	.7884	1745	1720	1703	1577
	68	.8742	.8620	.8535	.7912	1748	1724	1707	1582
	70	.8776	.8656	.8574	.7965	1755	1731	1715	1593
31 Ga	69	.8759	.8638	.8555	.7939	1752	1728	1711	1588
	71	.8792	.8674	.8593	.7991	1758	1735	1719	1598
32 Ge	70	.8776	.8656	.8574	.7965	1755	1731	1715	1593
	72	.8808	.8691	.8611	.8016	1762	1738	1722	1603
	73	.8823	.8708	.8629	.8040	1765	1742	1726	1608
	74	.8838	.8724	.8646	.8064	1768	1745	1729	1613
	76	.8867	.8756	.8679	.8110	1773	1751	1736	1622

TABLE 14.8 *(cont.)*

channel number				Rutherford Cross-section (barns)			
98	105	110	170	98	105	110	170
404	395	389	345	1.5884	1.2992	1.1420	.5173
406	397	391	348	1.5888	1.2996	1.1424	.5177
413	405	399	359	1.5902	1.3010	1.1438	.5190
408	399	393	351	1.7521	1.4332	1.2599	.5711
410	401	396	354	1.9234	1.5734	1.3832	.6272
412	403	397	356	1.9238	1.5738	1.3836	.6276
413	405	399	359	1.9241	1.5742	1.3840	.6280
415	407	401	361	1.9245	1.5745	1.3844	.6284
416	408	403	364	1.9248	1.5749	1.3847	.6287
416	408	403	364	2.1038	1.7213	1.5134	.6871
418	410	405	366	2.1041	1.7216	1.5138	.6875
416	408	403	364	2.2907	1.8742	1.6479	.7482
419	412	406	368	2.2914	1.8749	1.6486	.7489
421	413	408	370	2.2918	1.8753	1.6489	.7492
422	415	410	372	2.2921	1.8756	1.6493	.7496
423	416	411	374	2.4874	2.0355	1.7899	.8136
422	415	410	372	2.6900	2.2012	1.9356	.8797
425	417	412	376	2.6907	2.2019	1.9363	.8804
426	419	414	378	2.6910	2.2022	1.9366	.8807
427	420	415	380	2.6913	2.2025	1.9369	.8810
428	421	417	382	2.9026	2.3755	2.0890	.9503
427	420	415	380	3.1213	2.5544	2.2463	1.0217
429	422	418	383	3.1219	2.5550	2.2469	1.0223
430	424	419	385	3.1222	2.5553	2.2472	1.0226
431	425	420	387	3.1225	2.5556	2.2475	1.0229
433	427	423	390	3.1230	2.5561	2.2480	1.0234
432	426	421	388	3.3498	2.7417	2.4112	1.0976
434	428	424	391	3.3503	2.7422	2.4118	1.0981
433	427	423	390	3.5851	2.9343	2.5807	1.1749
435	429	425	393	3.5856	2.9349	2.5812	1.1754
436	430	426	394	3.5859	2.9351	2.5815	1.1757
437	431	427	396	3.5861	2.9354	2.5817	1.1759
439	433	429	398	3.5866	2.9358	2.5822	1.1764
438	432	428	397	3.8294	3.1346	2.7569	1.2559
440	434	430	400	3.8299	3.1350	2.7574	1.2563
439	433	429	398	4.0807	3.3403	2.9379	1.3384
440	435	431	401	4.0812	3.3408	2.9384	1.3389
441	435	431	402	4.0814	3.3410	2.9386	1.3391
442	436	432	403	4.0817	3.3412	2.9389	1.3393
443	438	434	406	4.0821	3.3416	2.9393	1.3398

TABLE 14.8 *(cont.)*

Element			kinematic factor				backscattered energy (keV)			
			98	105	110	170	98	105	110	170
33 As	75		.8853	.8740	.8663	.8087	1771	1748	1733	1617
34 Se	74		.8838	.8724	.8646	.8064	1768	1745	1729	1613
	76		.8867	.8756	.8679	.8110	1773	1751	1736	1622
	77		.8881	.8771	.8695	.8132	1776	1754	1739	1626
	78		.8895	.8786	.8711	.8154	1779	1757	1742	1631
	80		.8921	.8814	.8741	.8196	1784	1763	1748	1639
	82		.8946	.8842	.8770	.8236	1789	1768	1754	1647
35 Br	79		.8908	.8800	.8726	.8175	1782	1760	1745	1635
	81		.8933	.8828	.8756	.8216	1787	1766	1751	1643
36 Kr	78		.8895	.8786	.8711	.8154	1779	1757	1742	1631
	80		.8921	.8814	.8741	.8196	1784	1763	1748	1639
	82		.8946	.8842	.8770	.8236	1789	1768	1754	1647
	83		.8958	.8855	.8784	.8255	1792	1771	1757	1651
	84		.8969	.8868	.8797	.8274	1794	1774	1759	1655
	86		.8992	.8892	.8824	.8310	1798	1778	1765	1662
37 Rb	85		.8981	.8880	.8811	.8292	1796	1776	1762	1658
	87		.9003	.8904	.8836	.8328	1801	1781	1767	1666
38 Sr	84		.8969	.8868	.8797	.8274	1794	1774	1759	1655
	86		.8992	.8892	.8824	.8310	1798	1778	1765	1662
	87		.9003	.8904	.8836	.8328	1801	1781	1767	1666
	88		.9014	.8916	.8849	.8345	1803	1783	1770	1669
39 Y	89		.9024	.8928	.8861	.8362	1805	1786	1772	1672
40 Zr	90		.9035	.8939	.8873	.8379	1807	1788	1775	1676
	91		.9045	.8950	.8885	.8395	1809	1790	1777	1679
	92		.9055	.8961	.8896	.8411	1811	1792	1779	1682
	94		.9074	.8982	.8918	.8443	1815	1796	1784	1689
	96		.9092	.9002	.8940	.8472	1818	1800	1788	1694
41 Nb	93		.9064	.8971	.8907	.8427	1813	1794	1781	1685
42 Mo	92		.9055	.8961	.8896	.8411	1811	1792	1779	1682
	94		.9074	.8982	.8918	.8443	1815	1796	1784	1689
	95		.9083	.8992	.8929	.8458	1817	1798	1786	1692
	96		.9092	.9002	.8940	.8472	1818	1800	1788	1694
	97		.9101	.9012	.8950	.8487	1820	1802	1790	1697
	98		.9110	.9021	.8960	.8501	1822	1804	1792	1700
	100		.9127	.9040	.8980	.8529	1825	1808	1796	1706
44 Ru	96		.9092	.9002	.8940	.8472	1818	1800	1788	1694
	98		.9110	.9021	.8960	.8501	1822	1804	1792	1700
	99		.9119	.9031	.8970	.8515	1824	1806	1794	1703
	100		.9127	.9040	.8980	.8529	1825	1808	1796	1706
	101		.9135	.9049	.8989	.8542	1827	1810	1798	1708

TABLE 14.8 *(cont.)*

channel number				Rutherford Cross-section (barns)			
98	105	110	170	98	105	110	170
443	437	433	404	4.3410	3.5535	3.1256	1.4246
442	436	432	403	4.6078	3.7719	3.3177	1.5120
443	438	434	406	4.6083	3.7724	3.3181	1.5125
444	439	435	407	4.6085	3.7726	3.3184	1.5127
445	439	436	408	4.6087	3.7728	3.3186	1.5129
446	441	437	410	4.6091	3.7732	3.3190	1.5133
447	442	438	412	4.6094	3.7736	3.3193	1.5136
445	440	436	409	4.8840	3.9982	3.5169	1.6034
447	441	438	411	4.8844	3.9986	3.5173	1.6038
445	439	436	408	5.1668	4.2297	3.7205	1.6961
446	441	437	410	5.1673	4.2302	3.7209	1.6965
447	442	438	412	5.1677	4.2306	3.7213	1.6969
448	443	439	413	5.1679	4.2308	3.7215	1.6971
448	443	440	414	5.1681	4.2310	3.7217	1.6973
450	445	441	416	5.1684	4.2313	3.7220	1.6977
449	444	441	415	5.4593	4.4695	3.9315	1.7931
450	445	442	416	5.4597	4.4698	3.9319	1.7935
448	443	440	414	5.7582	4.7141	4.1467	1.8911
450	445	441	416	5.7586	4.7145	4.1471	1.8915
450	445	442	416	5.7588	4.7147	4.1473	1.8917
451	446	442	417	5.7590	4.7149	4.1475	1.8919
451	446	443	418	6.0663	4.9665	4.3688	1.9930
452	447	444	419	6.3815	5.2246	4.5959	2.0966
452	448	444	420	6.3817	5.2248	4.5961	2.0968
453	448	445	421	6.3819	5.2250	4.5962	2.0970
454	449	446	422	6.3822	5.2253	4.5966	2.0973
455	450	447	424	6.3825	5.2256	4.5969	2.0977
453	449	445	421	6.7052	5.4897	4.8291	2.2033
453	448	445	421	7.0360	5.7605	5.0674	2.3119
454	449	446	422	7.0364	5.7609	5.0677	2.3123
454	450	446	423	7.0366	5.7611	5.0679	2.3125
455	450	447	424	7.0367	5.7612	5.0681	2.3127
455	451	448	424	7.0369	5.7614	5.0682	2.3128
456	451	448	425	7.0371	5.7616	5.0684	2.3130
456	452	449	426	7.0374	5.7619	5.0687	2.3133
455	450	447	424	7.7229	6.3230	5.5622	2.5382
456	451	448	425	7.7232	6.3234	5.5626	2.5385
456	452	449	426	7.7234	6.3235	5.5628	2.5387
456	452	449	426	7.7235	6.3237	5.5629	2.5388
457	452	449	427	7.7237	6.3238	5.5631	2.5390

TABLE 14.8 *(cont.)*

Element			kinematic factor				backscattered energy (keV)			
			98	105	110	170	98	105	110	170
		102	.9144	.9058	.8999	.8556	1829	1812	1800	1711
		104	.9159	.9075	.9017	.8581	1832	1815	1803	1716
45	Rh	103	.9152	.9067	.9008	.8569	1830	1813	1802	1714
46	Pd	102	.9144	.9058	.8999	.8556	1829	1812	1800	1711
		104	.9159	.9075	.9017	.8581	1832	1815	1803	1716
		105	.9167	.9084	.9026	.8594	1833	1817	1805	1719
		106	.9175	.9092	.9035	.8606	1835	1818	1807	1721
		108	.9189	.9108	.9052	.8630	1838	1822	1810	1726
		110	.9203	.9123	.9068	.8653	1841	1825	1814	1731
47	Ag	107	.9182	.9100	.9043	.8618	1836	1820	1809	1724
		109	.9196	.9116	.9060	.8642	1839	1823	1812	1728
48	Cd	106	.9175	.9092	.9035	.8606	1835	1818	1807	1721
		108	.9189	.9108	.9052	.8630	1838	1822	1810	1726
		110	.9203	.9123	.9068	.8653	1841	1825	1814	1731
		111	.9210	.9131	.9076	.8665	1842	1826	1815	1733
		112	.9217	.9138	.9084	.8676	1843	1828	1817	1735
		113	.9224	.9146	.9092	.8687	1845	1829	1818	1737
		114	.9230	.9153	.9100	.8697	1846	1831	1820	1739
		116	.9243	.9167	.9114	.8718	1849	1833	1823	1744
49	In	113	.9224	.9146	.9092	.8687	1845	1829	1818	1737
		115	.9237	.9160	.9107	.8708	1847	1832	1821	1742
50	Sn	112	.9217	.9138	.9084	.8676	1843	1828	1817	1735
		114	.9230	.9153	.9100	.8697	1846	1831	1820	1739
		115	.9237	.9160	.9107	.8708	1847	1832	1821	1742
		116	.9243	.9167	.9114	.8718	1849	1833	1823	1744
		117	.9249	.9174	.9122	.8729	1850	1835	1824	1746
		118	.9255	.9180	.9129	.8739	1851	1836	1826	1748
		119	.9261	.9187	.9136	.8749	1852	1837	1827	1750
		120	.9267	.9194	.9143	.8758	1853	1839	1829	1752
		122	.9279	.9206	.9156	.8778	1856	1841	1831	1756
		124	.9290	.9219	.9169	.8796	1858	1844	1834	1759
51	Sb	121	.9273	.9200	.9149	.8768	1855	1840	1830	1754
		123	.9285	.9213	.9163	.8787	1857	1843	1833	1757
52	Te	120	.9267	.9194	.9143	.8758	1853	1839	1829	1752
		122	.9279	.9206	.9156	.8778	1856	1841	1831	1756
		123	.9285	.9213	.9163	.8787	1857	1843	1833	1757
		124	.9290	.9219	.9169	.8796	1858	1844	1834	1759
		125	.9296	.9225	.9176	.8805	1859	1845	1835	1761
		126	.9301	.9231	.9182	.8814	1860	1846	1836	1763
		128	.9312	.9242	.9194	.8831	1862	1848	1839	1766

TABLE 14.8 *(cont.)*

channel number				Rutherford Cross-section (barns)			
98	105	110	170	98	105	110	170
457	453	450	428	7.7239	6.3240	5.5632	2.5392
458	454	451	429	7.7242	6.3243	5.5635	2.5394
458	453	450	428	8.0791	6.6149	5.8191	2.6561
457	453	450	428	8.4420	6.9120	6.0805	2.7752
458	454	451	429	8.4423	6.9123	6.0808	2.7755
458	454	451	430	8.4425	6.9124	6.0810	2.7757
459	455	452	430	8.4426	6.9126	6.0811	2.7758
459	455	453	432	8.4429	6.9129	6.0814	2.7762
460	456	453	433	8.4432	6.9132	6.0817	2.7764
459	455	452	431	8.8138	7.2166	6.3485	2.8980
460	456	453	432	8.8141	7.2169	6.3488	2.8983
459	455	452	430	9.1927	7.5268	6.6214	3.0225
459	455	453	432	9.1930	7.5271	6.6217	3.0228
460	456	453	433	9.1933	7.5274	6.6220	3.0231
461	457	454	433	9.1935	7.5275	6.6222	3.0232
461	457	454	434	9.1936	7.5276	6.6223	3.0234
461	457	455	434	9.1937	7.5278	6.6224	3.0235
462	458	455	435	9.1939	7.5279	6.6226	3.0237
462	458	456	436	9.1941	7.5282	6.6228	3.0239
461	457	455	434	9.5808	7.8447	6.9012	3.1508
462	458	455	435	9.5811	7.8450	6.9015	3.1511
461	457	454	434	9.9757	8.1680	7.1856	3.2806
462	458	455	435	9.9760	8.1683	7.1859	3.2809
462	458	455	435	9.9761	8.1685	7.1861	3.2810
462	458	456	436	9.9763	8.1686	7.1862	3.2811
462	459	456	436	9.9764	8.1687	7.1863	3.2813
463	459	456	437	9.9765	8.1689	7.1865	3.2814
463	459	457	437	9.9767	8.1690	7.1866	3.2815
463	460	457	438	9.9768	8.1691	7.1867	3.2816
464	460	458	439	9.9770	8.1693	7.1870	3.2819
465	461	458	440	9.9772	8.1696	7.1872	3.2821
464	460	457	438	10.3800	8.4992	7.4772	3.4144
464	461	458	439	10.3802	8.4995	7.4774	3.4146
463	460	457	438	10.7909	8.8357	7.7732	3.5494
464	460	458	439	10.7911	8.8359	7.7734	3.5497
464	461	458	439	10.7912	8.8361	7.7735	3.5498
465	461	458	440	10.7914	8.8362	7.7736	3.5499
465	461	459	440	10.7915	8.8363	7.7738	3.5500
465	462	459	441	10.7916	8.8364	7.7739	3.5501
466	462	460	442	10.7918	8.8366	7.7741	3.5504

TABLE 14.8 *(cont.)*

Element			kinematic factor				backscattered energy (keV)			
			98	105	110	170	98	105	110	170
		130	.9322	.9253	.9206	.8848	1864	1851	1841	1770
53	I	127	.9306	.9236	.9188	.8823	1861	1847	1838	1765
54	Xe	124	.9290	.9219	.9169	.8796	1858	1844	1834	1759
		126	.9301	.9231	.9182	.8814	1860	1846	1836	1763
		128	.9312	.9242	.9194	.8831	1862	1848	1839	1766
		129	.9317	.9248	.9200	.8840	1863	1850	1840	1768
		130	.9322	.9253	.9206	.8848	1864	1851	1841	1770
		131	.9327	.9259	.9212	.8857	1865	1852	1842	1771
		132	.9332	.9264	.9218	.8865	1866	1853	1844	1773
		134	.9341	.9275	.9229	.8881	1868	1855	1846	1776
		136	.9351	.9285	.9240	.8896	1870	1857	1848	1779
55	Cs	133	.9337	.9270	.9223	.8873	1867	1854	1845	1775
56	Ba	130	.9322	.9253	.9206	.8848	1864	1851	1841	1770
		132	.9332	.9264	.9218	.8865	1866	1853	1844	1773
		134	.9341	.9275	.9229	.8881	1868	1855	1846	1776
		135	.9346	.9280	.9234	.8889	1869	1856	1847	1778
		136	.9351	.9285	.9240	.8896	1870	1857	1848	1779
		137	.9355	.9290	.9245	.8904	1871	1858	1849	1781
		138	.9360	.9295	.9250	.8911	1872	1859	1850	1782
57	La	138	.9360	.9295	.9250	.8911	1872	1859	1850	1782
		139	.9364	.9300	.9256	.8919	1873	1860	1851	1784
58	Ce	136	.9351	.9285	.9240	.8896	1870	1857	1848	1779
		138	.9360	.9295	.9250	.8911	1872	1859	1850	1782
		140	.9369	.9305	.9261	.8926	1874	1861	1852	1785
		142	.9377	.9314	.9271	.8940	1875	1863	1854	1788
59	Pr	141	.9373	.9310	.9266	.8933	1875	1862	1853	1787
60	Nd	142	.9377	.9314	.9271	.8940	1875	1863	1854	1788
		143	.9382	.9319	.9276	.8948	1876	1864	1855	1790
		144	.9386	.9324	.9280	.8954	1877	1865	1856	1791
		145	.9390	.9328	.9285	.8961	1878	1866	1857	1792
		146	.9394	.9333	.9290	.8968	1879	1867	1858	1794
		148	.9402	.9341	.9299	.8981	1880	1868	1860	1796
		150	.9410	.9350	.9308	.8994	1882	1870	1862	1799
62	Sm	144	.9386	.9324	.9280	.8954	1877	1865	1856	1791
		147	.9398	.9337	.9295	.8975	1880	1867	1859	1795
		148	.9402	.9341	.9299	.8981	1880	1868	1860	1796
		149	.9406	.9346	.9304	.8988	1881	1869	1861	1798
		150	.9410	.9350	.9308	.8994	1882	1870	1862	1799
		152	.9417	.9358	.9317	.9007	1883	1872	1863	1801
		154	.9425	.9366	.9326	.9019	1885	1873	1865	1804

TABLE 14.8 *(cont.)*

channel number				Rutherford Cross-section (barns)			
98	105	110	170	98	105	110	170
466	463	460	442	10.7920	8.8368	7.7743	3.5506
465	462	459	441	11.2108	9.1797	8.0759	3.6881
465	461	458	440	11.6374	9.5290	8.3831	3.8282
465	462	459	441	11.6377	9.5292	8.3834	3.8285
466	462	460	442	11.6379	9.5294	8.3836	3.8287
466	462	460	442	11.6380	9.5296	8.3837	3.8288
466	463	460	442	11.6381	9.5297	8.3838	3.8290
466	463	461	443	11.6383	9.5298	8.3839	3.8291
467	463	461	443	11.6384	9.5299	8.3840	3.8292
467	464	461	444	11.6386	9.5301	8.3843	3.8294
468	464	462	445	11.6388	9.5303	8.3844	3.8296
467	463	461	444	12.0735	9.8862	8.6975	3.9724
466	463	460	442	12.5162	10.2486	9.0164	4.1178
467	463	461	443	12.5164	10.2489	9.0166	4.1181
467	464	461	444	12.5167	10.2491	9.0168	4.1183
467	464	462	444	12.5168	10.2492	9.0169	4.1184
468	464	462	445	12.5169	10.2493	9.0170	4.1185
468	465	462	445	12.5170	10.2494	9.0171	4.1186
468	465	463	446	12.5171	10.2495	9.0172	4.1187
468	465	463	446	12.9681	10.6188	9.3421	4.2671
468	465	463	446	12.9682	10.6189	9.3422	4.2672
468	464	462	445	13.4269	10.9945	9.6726	4.4179
468	465	463	446	13.4271	10.9947	9.6728	4.4181
468	465	463	446	13.4273	10.9949	9.6730	4.4184
469	466	464	447	13.4275	10.9951	9.6732	4.4186
469	465	463	447	13.8944	11.3774	10.0096	4.5721
469	466	464	447	14.3695	11.7665	10.3518	4.7285
469	466	464	447	14.3696	11.7666	10.3520	4.7287
469	466	464	448	14.3697	11.7667	10.3521	4.7287
469	466	464	448	14.3699	11.7668	10.3521	4.7288
470	467	464	448	14.3699	11.7669	10.3522	4.7290
470	467	465	449	14.3701	11.7671	10.3524	4.7291
470	467	465	450	14.3703	11.7672	10.3526	4.7293
469	466	464	448	15.3437	12.5642	11.0537	5.0492
470	467	465	449	15.3440	12.5645	11.0540	5.0495
470	467	465	449	15.3441	12.5646	11.0541	5.0497
470	467	465	449	15.3442	12.5647	11.0542	5.0498
470	467	465	450	15.3443	12.5648	11.0543	5.0499
471	468	466	450	15.3445	12.5650	11.0545	5.0500
471	468	466	451	15.3446	12.5652	11.0547	5.0502

TABLE 14.8 *(cont.)*

Element			kinematic factor				backscattered energy (keV)			
			98	105	110	170	98	105	110	170
63 Eu	151		.9413	.9354	.9313	.9000	1883	1871	1863	1800
	153		.9421	.9362	.9321	.9013	1884	1872	1864	1803
64 Gd	152		.9417	.9358	.9317	.9007	1883	1872	1863	1801
	154		.9425	.9366	.9326	.9019	1885	1873	1865	1804
	155		.9428	.9370	.9330	.9025	1886	1874	1866	1805
	156		.9432	.9374	.9334	.9031	1886	1875	1867	1806
	157		.9435	.9378	.9338	.9037	1887	1876	1868	1807
	158		.9439	.9382	.9342	.9043	1888	1876	1868	1809
	160		.9446	.9389	.9350	.9054	1889	1878	1870	1811
65 Tb	159		.9442	.9385	.9346	.9048	1888	1877	1869	1810
66 Dy	156		.9432	.9374	.9334	.9031	1886	1875	1867	1806
	158		.9439	.9382	.9342	.9043	1888	1876	1868	1809
	160		.9446	.9389	.9350	.9054	1889	1878	1870	1811
	161		.9449	.9393	.9354	.9060	1890	1879	1871	1812
	162		.9452	.9397	.9358	.9065	1890	1879	1872	1813
	163		.9456	.9400	.9362	.9071	1891	1880	1872	1814
	164		.9459	.9404	.9366	.9076	1892	1881	1873	1815
67 Ho	165		.9462	.9407	.9369	.9081	1892	1881	1874	1816
68 Er	162		.9452	.9397	.9358	.9065	1890	1879	1872	1813
	164		.9459	.9404	.9366	.9076	1892	1881	1873	1815
	166		.9465	.9411	.9373	.9087	1893	1882	1875	1817
	167		.9468	.9414	.9377	.9092	1894	1883	1875	1818
	168		.9471	.9417	.9380	.9097	1894	1883	1876	1819
	170		.9477	.9424	.9387	.9107	1895	1885	1877	1821
69 Tm	169		.9474	.9421	.9384	.9102	1895	1884	1877	1820
70 Yb	168		.9471	.9417	.9380	.9097	1894	1883	1876	1819
	170		.9477	.9424	.9387	.9107	1895	1885	1877	1821
	171		.9480	.9427	.9391	.9112	1896	1885	1878	1822
	172		.9483	.9431	.9394	.9117	1897	1886	1879	1823
	173		.9486	.9434	.9398	.9122	1897	1887	1880	1824
	174		.9489	.9437	.9401	.9127	1898	1887	1880	1825
	176		.9495	.9443	.9408	.9136	1899	1889	1882	1827
71 Lu	175		.9492	.9440	.9404	.9132	1898	1888	1881	1826
	176		.9495	.9443	.9408	.9136	1899	1889	1882	1827
72 Hf	174		.9489	.9437	.9401	.9127	1898	1887	1880	1825
	176		.9495	.9443	.9408	.9136	1899	1889	1882	1827
	177		.9498	.9446	.9411	.9141	1900	1889	1882	1828
	178		.9500	.9449	.9414	.9146	1900	1890	1883	1829
	179		.9503	.9452	.9417	.9150	1901	1890	1883	1830
	180		.9506	.9455	.9420	.9155	1901	1891	1884	1831

TABLE 14.8 *(cont.)*

channel number				Rutherford Cross-section (barns)			
98	105	110	170	98	105	110	170
471	468	466	450	15.8433	12.9735	11.4138	5.2141
471	468	466	451	15.8435	12.9737	11.4140	5.2143
471	468	466	450	16.3504	13.3887	11.7792	5.3811
471	468	466	451	16.3506	13.3889	11.7794	5.3813
471	469	466	451	16.3507	13.3890	11.7795	5.3814
472	469	467	452	16.3508	13.3891	11.7795	5.3815
472	469	467	452	16.3509	13.3892	11.7796	5.3816
472	469	467	452	16.3510	13.3892	11.7797	5.3817
472	469	468	453	16.3511	13.3894	11.7799	5.3818
472	469	467	452	16.8660	13.8110	12.1508	5.5512
472	469	467	452	17.3887	14.2390	12.5273	5.7231
472	469	467	452	17.3889	14.2392	12.5275	5.7233
472	469	468	453	17.3890	14.2393	12.5276	5.7235
472	470	468	453	17.3891	14.2394	12.5277	5.7235
473	470	468	453	17.3892	14.2395	12.5278	5.7236
473	470	468	454	17.3893	14.2396	12.5279	5.7237
473	470	468	454	17.3894	14.2397	12.5280	5.7238
473	470	468	454	17.9204	14.6745	12.9106	5.8986
473	470	468	453	18.4591	15.1156	13.2986	6.0758
473	470	468	454	18.4592	15.1158	13.2988	6.0759
473	471	469	454	18.4594	15.1159	13.2989	6.0761
473	471	469	455	18.4595	15.1160	13.2990	6.0762
474	471	469	455	18.4596	15.1161	13.2991	6.0762
474	471	469	455	18.4597	15.1163	13.2992	6.0764
474	471	469	455	19.0066	15.5640	13.6932	6.2564
474	471	469	455	19.5614	16.0183	14.0929	6.4389
474	471	469	455	19.5616	16.0185	14.0931	6.4391
474	471	470	456	19.5616	16.0186	14.0931	6.4392
474	472	470	456	19.5617	16.0187	14.0932	6.4393
474	472	470	456	19.5618	16.0188	14.0933	6.4393
474	472	470	456	19.5619	16.0188	14.0934	6.4394
475	472	470	457	19.5620	16.0190	14.0935	6.4396
475	472	470	457	20.1249	16.4799	14.4990	6.6248
475	472	470	457	20.1250	16.4800	14.4991	6.6249
474	472	470	456	20.6957	16.9473	14.9102	6.8127
475	472	470	457	20.6958	16.9474	14.9104	6.8128
475	472	471	457	20.6959	16.9475	14.9105	6.8129
475	472	471	457	20.6960	16.9476	14.9105	6.8130
475	473	471	458	20.6961	16.9477	14.9106	6.8131
475	473	471	458	20.6962	16.9478	14.9107	6.8131

TABLE 14.8 *(cont.)*

Element			kinematic factor				backscattered energy (keV)			
			98	105	110	170	98	105	110	170
73 Ta	180		.9506	.9455	.9420	.9155	1901	1891	1884	1831
	181		.9508	.9458	.9423	.9159	1902	1892	1885	1832
74 W	180		.9506	.9455	.9420	.9155	1901	1891	1884	1831
	182		.9511	.9461	.9427	.9164	1902	1892	1885	1833
	183		.9514	.9464	.9430	.9168	1903	1893	1886	1834
	184		.9516	.9467	.9433	.9172	1903	1893	1887	1834
	186		.9521	.9472	.9439	.9181	1904	1894	1888	1836
75 Re	185		.9519	.9470	.9436	.9177	1904	1894	1887	1835
	187		.9524	.9475	.9441	.9185	1905	1895	1888	1837
76 Cs	184		.9516	.9467	.9433	.9172	1903	1893	1887	1834
	186		.9521	.9472	.9439	.9181	1904	1894	1888	1836
	187		.9524	.9475	.9441	.9185	1905	1895	1888	1837
	188		.9526	.9478	.9444	.9189	1905	1896	1889	1838
	189		.9529	.9481	.9447	.9193	1906	1896	1889	1839
	190		.9531	.9483	.9450	.9197	1906	1897	1890	1839
	192		.9536	.9489	.9456	.9206	1907	1898	1891	1841
77 Ir	191		.9534	.9486	.9453	.9202	1907	1897	1891	1840
	193		.9538	.9491	.9458	.9209	1908	1898	1892	1842
78 Pt	190		.9531	.9483	.9450	.9197	1906	1897	1890	1839
	192		.9536	.9489	.9456	.9206	1907	1898	1891	1841
	194		.9541	.9494	.9461	.9213	1908	1899	1892	1843
	195		.9543	.9496	.9464	.9217	1909	1899	1893	1843
	196		.9545	.9499	.9466	.9221	1909	1900	1893	1844
	198		.9550	.9504	.9472	.9229	1910	1901	1894	1846
79 Au	197		.9548	.9501	.9469	.9225	1910	1900	1894	1845
80 Hg	196		.9545	.9499	.9466	.9221	1909	1900	1893	1844
	198		.9550	.9504	.9472	.9229	1910	1901	1894	1846
	199		.9552	.9506	.9474	.9232	1910	1901	1895	1846
	200		.9554	.9508	.9477	.9236	1911	1902	1895	1847
	201		.9556	.9511	.9479	.9240	1911	1902	1896	1848
	202		.9558	.9513	.9482	.9243	1912	1903	1896	1849
	204		.9563	.9518	.9487	.9251	1913	1904	1897	1850
81 Tl	203		.9561	.9516	.9484	.9247	1912	1903	1897	1849
	205		.9565	.9520	.9489	.9254	1913	1904	1898	1851
82 Pb	204		.9563	.9518	.9487	.9251	1913	1904	1897	1850
	206		.9567	.9522	.9492	.9258	1913	1904	1898	1852
	207		.9569	.9525	.9494	.9261	1914	1905	1899	1852
	208		.9571	.9527	.9496	.9264	1914	1905	1899	1853
83 Bi	209		.9573	.9529	.9499	.9268	1915	1906	1900	1854

TABLE 14.8 *(cont.)*

channel number				Rutherford Cross-section (barns)			
98	105	110	170	98	105	110	170
475	473	471	458	21.2751	17.4218	15.3278	7.0037
475	473	471	458	21.2751	17.4219	15.3278	7.0038
475	473	471	458	21.8619	17.9024	15.7506	7.1969
476	473	471	458	21.8621	17.9025	15.7507	7.1971
476	473	471	458	21.8622	17.9026	15.7508	7.1971
476	473	472	459	21.8622	17.9027	15.7509	7.1972
476	474	472	459	21.8624	17.9028	15.7510	7.1974
476	473	472	459	22.4572	18.3899	16.1795	7.3931
476	474	472	459	22.4573	18.3900	16.1797	7.3932
476	473	472	459	23.0599	18.8835	16.6138	7.5915
476	474	472	459	23.0601	18.8836	16.6139	7.5917
476	474	472	459	23.0602	18.8837	16.6140	7.5917
476	474	472	459	23.0602	18.8838	16.6141	7.5918
476	474	472	460	23.0603	18.8838	16.6141	7.5918
477	474	473	460	23.0603	18.8839	16.6142	7.5919
477	474	473	460	23.0605	18.8840	16.6143	7.5921
477	474	473	460	23.6713	19.3842	17.0544	7.7931
477	475	473	460	23.6714	19.3843	17.0545	7.7932
477	474	473	460	24.2901	19.8909	17.5002	7.9968
477	474	473	460	24.2902	19.8910	17.5003	7.9969
477	475	473	461	24.2904	19.8912	17.5004	7.9970
477	475	473	461	24.2904	19.8912	17.5005	7.9972
477	475	473	461	24.2905	19.8913	17.5006	7.9972
477	475	474	461	24.2906	19.8914	17.5007	7.9973
477	475	473	461	24.9174	20.4047	17.9523	8.2037
477	475	473	461	25.5521	20.9244	18.4095	8.4125
477	475	474	461	25.5522	20.9246	18.4097	8.4126
478	475	474	462	25.5523	20.9246	18.4097	8.4128
478	475	474	462	25.5524	20.9247	18.4098	8.4128
478	476	474	462	25.5524	20.9248	18.4099	8.4129
478	476	474	462	25.5525	20.9248	18.4099	8.4129
478	476	474	463	25.5526	20.9250	18.4101	8.4130
478	476	474	462	26.1954	21.4513	18.8732	8.6247
478	476	474	463	26.1955	21.4514	18.8733	8.6248
478	476	474	463	26.8462	21.9843	19.3421	8.8390
478	476	475	463	26.8464	21.9844	19.3422	8.8391
478	476	475	463	26.8464	21.9845	19.3423	8.8392
479	476	475	463	26.8465	21.9846	19.3423	8.8392
479	476	475	463	27.5053	22.5241	19.8171	9.0562

TABLE 14.9 RBS Depth Information

			geom 1	$\theta = 98°$	$\alpha = 90°$	$\beta = 8°$		
depth Å	energy (keV)	S eV/Å	4000/S Å/channel	ϵ $10^{-15} eV\ cm^2$	$4000/\epsilon$ $10^{15}\ cm^{-2}/ch$	\multicolumn{3}{c}{depth resolution (Å)}		
						15 keV	7 keV	2 keV
0	1441	215.3	18.6	432.6	9.2	75	42	29
200	1398	216.7	18.5	435.2	9.2	92	69	61
400	1354	218.0	18.3	438.0	9.1	116	99	95
600	1309	219.4	18.2	440.8	9.1	148	136	132
800	1264	220.9	18.1	443.7	9.0	185	176	173
1000	1219	222.3	18.0	446.6	9.0	227	220	218
1200	1172	223.9	17.9	449.7	8.9	274	268	267
1400	1125	225.4	17.7	452.9	8.8	325	320	319
1600	1078	227.1	17.6	456.1	8.8	381	377	376
1800	1029	228.7	17.5	459.5	8.7	440	437	436
2000	980	230.5	17.4	462.9	8.6	504	501	501
4000	442	249.7	16.0	501.5	8.0	1500	1499	1499

			geom 2	$\theta = 170°$	$\alpha = 90°$	$\beta = 80°$		
depth Å	energy (keV)	S eV/Å	4000/S Å/channel	ϵ $10^{-15} eV\ cm^2$	$4000/\epsilon$ $10^{15}\ cm^{-2}/ch$	15 keV	7 keV	2 keV
0	1131	44.2	90.5	88.8	45.0	340	161	53
200	1122	44.3	90.3	89.0	45.0	345	172	82
400	1113	44.3	90.2	89.1	44.9	350	183	103
600	1104	44.4	90.1	89.2	44.8	354	192	120
800	1096	44.5	89.9	89.3	44.8	359	202	135
1000	1087	44.5	89.8	89.5	44.7	363	210	148
1200	1078	44.6	89.7	89.6	44.6	367	218	160
1400	1069	44.7	89.5	89.7	44.6	371	226	171
1600	1059	44.7	89.4	89.9	44.5	375	234	181
1800	1050	44.8	89.3	90.0	44.4	379	241	190
2000	1041	44.9	89.1	90.1	44.4	382	248	199
10000	653	47.8	83.6	96.1	41.6	478	404	382

			geom 3	$\theta = 170°$	$\alpha = 8°$	$\beta = 18°$		
depth Å	energy (keV)	S eV/Å	4000/S Å/channel	ϵ $10^{-15} eV\ cm^2$	$4000/\epsilon$ $10^{15}\ cm^{-2}/ch$	15 keV	7 keV	2 keV
0	1130	195.0	20.5	391.7	10.2	78	37	15
200	1091	196.3	20.4	394.3	10.1	84	50	37
400	1051	197.6	20.2	397.0	10.1	91	62	53
600	1011	199.0	20.1	399.8	10.0	100	75	68
800	970	200.4	20.0	402.6	9.9	110	90	84
1000	928	201.9	19.8	405.6	9.9	123	106	101
1200	886	203.4	19.7	408.6	9.8	139	124	120
1400	843	204.9	19.5	411.6	9.7	157	144	141
1600	800	206.5	19.4	414.8	9.6	178	167	164
1800	756	208.1	19.2	418.0	9.6	201	192	189
2000	711	209.7	19.1	421.2	9.5	227	219	217
4000	245	221.3	18.1	444.6	9.0	847	844	844

			geom 4	$\theta = 170°$	$\alpha = 18°$	$\beta = 8°$		
depth Å	energy (keV)	S eV/Å	4000/S Å/channel	ϵ $10^{-15} eV\ cm^2$	$4000/\epsilon$ $10^{15}\ cm^{-2}/ch$	15 keV	7 keV	2 keV
0	1130	261.1	15.3	524.5	7.6	58	27	10
200	1077	263.3	15.2	528.9	7.6	67	45	37
400	1024	265.5	15.1	533.4	7.5	84	69	64
600	969	267.9	14.9	538.2	7.4	111	100	97
800	914	270.3	14.8	543.1	7.4	146	139	136
1000	857	272.8	14.7	548.1	7.3	189	184	182
1200	799	275.4	14.5	553.3	7.2	240	235	234
1400	741	278.1	14.4	558.6	7.2	298	294	294
1600	681	280.7	14.2	563.9	7.1	364	361	360
1800	620	283.4	14.1	569.3	7.0	439	437	436
2000	558	286.0	14.0	574.6	7.0	525	524	523

TABLE 14.10 RBS Depth Scales

Element	Depth Scale (Å/channel)			
	geom 1	geom 2	geom 3	geom 4
3 Li	51	497	146	75
4 Be	17	111	31	17
5 B	11	72	20	11
6 C	8	50	13	8
10 Ne	32	176	41	29
11 Na	44	225	52	38
12 Mg	25	126	29	21
13 Al	17	84	19	14
14 Si	18	90	20	15
18 Ar	31	146	31	25
19 K	53	250	54	43
20 Ca	31	145	31	25
21 Sc	16	75	16	13
22 Ti	12	57	12	10
23 Y	9	41	8	7
24 Cr	8	37	8	7
25 Mn	8	38	8	7
26 Fe	8	35	7	6
27 Co	8	35	7	6
28 Ni	8	34	7	6
29 Cu	9	38	8	7
30 Zn	11	48	10	9
31 Ga	14	61	12	11
32 Ge	15	67	14	12
33 As	14	62	13	11
34 Se	18	80	16	15
35 Br	26	116	23	21
36 Kr	28	125	25	23
37 Rb	49	216	43	39
38 Sr	30	131	26	24
39 Y	17	75	15	14
40 Zr	11	50	10	9
41 Nb	9	38	8	7
42 Mo	8	34	7	6
44 Ru	7	30	6	5
45 Rh	7	30	6	5
46 Pd	8	33	7	6
47 Ag	9	37	7	7
48 Cd	11	47	9	9

TABLE 14.10 *(cont.)*

Element	Depth Scale (Å/channel)			
	geom 1	geom 2	geom 3	geom 4
49 In	12	53	10	10
50 Sn	13	57	11	10
51 Sb	14	62	12	11
52 Te	16	67	13	12
53 I	20	88	17	16
54 Xe	28	121	24	22
55 Cs	47	205	40	38
56 Ba	26	112	22	21
57 La	15	63	12	11
58 Ce	14	62	12	11
59 Pr	14	62	12	11
60 Nd	14	62	12	12
62 Sm	14	61	12	11
63 Eu	21	91	18	17
64 Gd	14	59	12	11
65 Tb	14	59	11	11
66 Dy	14	59	12	11
67 Ho	14	60	12	11
68 Er	13	58	11	11
69 Tm	14	60	12	11
70 Yb	16	67	13	12
71 Lu	14	59	11	11
72 Hf	10	44	9	8
73 Ta	8	35	7	7
74 W	7	31	6	6
75 Re	7	28	5	5
76 Cs	6	26	5	5
77 Ir	6	26	5	5
78 Pt	7	30	6	6
79 Au	7	32	6	6
80 Hg	10	43	8	8
81 Tl	12	50	10	9
82 Pb	12	51	10	10
83 Bi	14	60	12	11

TABLE 14.11 Kinematic Factors for Elastic Recoil and Forward Scattering

M_2	M_1	^{4}He	^{7}Li	^{12}C	^{14}N	^{16}O	^{20}Ne	^{35}Cl	^{40}Ar
				Elastic Recoil: $K_4 = E_2/E_1$					
H	1	.6234	.4261	.2773	.2430	.2163	.1772	.1056	.0931
	2	.8639	.6723	.4774	.4265	.3853	.3226	.1998	.1772
	3	.9507	.8157	.6227	.5657	.5178	.4419	.2836	.2531
He	3	.9507	.8157	.6227	.5657	.5178	.4419	.2836	.2531
	4		.8973	.7276	.6707	.6211	.5392	.3575	.3210
Li	6		.9641	.8628	.8154	.7705	.6897	.4858	.4411
	7			.9032	.8627	.8222	.7460	.5399	.4928
Be	9			.9502	.9242	.8942	.8309	.6320	.5825
B	10			.9619	.9431	.9185	.8626	.6713	.6215
	11			.9680	.9560	.9368	.8884	.7065	.6569
C	12				.9641	.9501	.9093	.7379	.6890
	13				.9685	.9595	.9263	.7665	.7186
N	14					.9656	.9397	.7921	.7454
	15					.9688	.9501	.8150	.7698
O	16		5°	5°			.9579	.8354	.7919
	17		$\theta = 10°$				.9635	.8539	.8122
	18						.9672	.8703	.8306
M_2				Forward Scattering: $K_3 = E_3/E_1$					
10		.9939	.9894	.0926	.1704	.2364	.3438	.5891	.6436
20		.9970	.9947	.9909	.9894	.9879	.9848	.2798	.3435
50		.9988	.9979	.9964	.9958	.9951	.9939	.9894	.9879
100		.9994	.9989	.9982	.9979	.9976	.9970	.9947	.9939
200		.9997	.9995	.9991	.9989	.9988	.9985	.9973	.9970

M_2	M_1	^{4}He	^{7}Li	^{12}C	^{14}N	^{16}O	^{20}Ne	^{35}Cl	^{40}Ar
				Elastic Recoil: $K_4 = E_2/E_1$					
H	1	.5676	.3879	.2525	.2212	.1970	.1614	.0962	.0847
	2	.7866	.6121	.4347	.3883	.3508	.2937	.1819	.1613
	3	.8656	.7426	.5669	.5150	.4715	.4023	.2582	.2305
He	3	.8656	.7426	.5669	.5150	.4715	.4023	.2582	.2305
	4		.8170	.6625	.6106	.5655	.4909	.3255	.2923
Li	6		.8778	.7856	.7424	.7015	.6280	.4423	.4016
	7			.8224	.7854	.7486	.6792	.4916	.4487
Be	9			.8652	.8415	.8142	.7565	.5755	.5304
B	10			.8758	.8586	.8363	.7853	.6112	.5659
	11			.8814	.8704	.8529	.8089	.6432	.5981
C	12				.8778	.8650	.8279	.6719	.6273
	13				.8818	.8736	.8434	.6979	.6543
N	14					.8791	.8556	.7212	.6787
	15					.8821	.8650	.7420	.7009
O	16		10°	10°			.8721	.7606	.7210
	17		$\theta = 20°$				.8772	.7774	.7395
	18						.8806	.7924	.7562
M_2				Forward Scattering: $K_3 = E_3/E_1$					
10		.9760	.9580	.0981	.1825	.2562	.3831	.0000	.0000
20		.9880	.9790	.9641	.9581	.9520	.9397	.3062	.3828
50		.9952	.9916	.9856	.9832	.9808	.9761	.9581	.9520
100		.9976	.9958	.9928	.9916	.9904	.9880	.9791	.9761
200		.9988	.9979	.9964	.9958	.9952	.9940	.9895	.9880

TABLE 14.11 *(cont.)*

		Elastic Recoil: $K_4 = E_2/E_1$							
M_2 \ M_1		^{4}He	^{7}Li	^{12}C	^{14}N	^{16}O	^{20}Ne	^{35}Cl	^{40}Ar
H	1	.4821	.3295	.2144	.1879	.1673	.1371	.0817	.0720
	2	.6681	.5199	.3692	.3298	.2980	.2494	.1545	.1370
	3	.7352	.6308	.4815	.4374	.4004	.3417	.2193	.1958
He	3	.7352	.6308	.4815	.4374	.4004	.3417	.2193	.1958
	4		.6939	.5627	.5186	.4803	.4170	.2765	.2483
Li	6		.7456	.6672	.6306	.5958	.5334	.3757	.3411
	7			.6985	.6671	.6358	.5769	.4175	.3811
Be	9			.7348	.7147	.6915	.6425	.4888	.4505
B	10			.7439	.7293	.7103	.6670	.5191	.4806
	11			.7486	.7393	.7244	.6870	.5463	.5080
C	12				.7455	.7347	.7032	.5706	.5328
	13				.7490	.7420	.7163	.5928	.5557
N	14					.7467	.7267	.6125	.5765
	15					.7492	.7347	.6302	.5953
O	16	15°	15°				.7407	.6460	.6124
	17	$\theta = 30°$					.7451	.6603	.6281
	18						.7479	.6730	.6423
M_2				Forward Scattering: $K_3 = E_3/E_1$					
10		.9473	.9074	.1087	.2078	.3019	.5681	.0000	.0000
20		.9735	.9537	.9210	.9076	.8941	.8661	.3743	.5627
50		.9893	.9814	.9682	.9630	.9578	.9473	.9077	.8942
100		.9947	.9906	.9840	.9814	.9788	.9735	.9539	.9473
200		.9973	.9953	.9920	.9907	.9893	.9867	.9768	.9735

		Elastic Recoil: $K_4 = E_2/E_1$							
M_2 \ M_1		^{4}He	^{7}Li	^{12}C	^{14}N	^{16}O	^{20}Ne	^{35}Cl	^{40}Ar
H	1	.3772	.2578	.1678	.1470	.1309	.1072	.0639	.0563
	2	.5227	.4068	.2889	.2580	.2332	.1952	.1209	.1072
	3	.5752	.4935	.3768	.3423	.3133	.2674	.1716	.1532
He	3	.5752	.4935	.3768	.3423	.3133	.2674	.1716	.1532
	4		.5429	.4403	.4058	.3758	.3262	.2163	.1942
Li	6		.5834	.5221	.4934	.4662	.4173	.2939	.2669
	7			.5465	.5220	.4975	.4514	.3267	.2982
Be	9			.5750	.5592	.5411	.5027	.3824	.3525
B	10			.5820	.5706	.5558	.5219	.4062	.3761
	11			.5857	.5784	.5668	.5376	.4275	.3975
C	12				.5833	.5749	.5502	.4465	.4169
	13				.5860	.5806	.5605	.4638	.4348
N	14					.5842	.5686	.4793	.4510
	15					.5862	.5749	.4931	.4658
O	16	20°	20°				.5796	.5055	.4792
	17	$\theta = 40°$					.5830	.5167	.4915
	18						.5852	.5266	.5026
M_2				Forward Scattering: $K_3 = E_3/E_1$					
10		.9091	.8404	.1286	.2654	.0000	.0000	.0000	.0000
20		.9541	.9202	.8639	.8407	.8170	.7661	.0000	.0000
50		.9814	.9676	.9451	.9361	.9271	.9092	.8409	.8171
100		.9907	.9837	.9723	.9677	.9632	.9541	.9204	.9092
200		.9953	.9918	.9861	.9837	.9815	.9769	.9598	.9541

TABLE 14.12
Prompt nuclear analysis catalog

REACTION	ES	BEAM E_1 MeV	BEAM I_1 nA	D mm	PRODUCT E_3 MeV	PRODUCT θ	PRODUCT α	DETECTOR Type	DETECTOR $d\Omega$ msr	DETECTOR ΔE keV	FILTER t μm	FILTER MAT	FILTER CR	SENS. S μg.g^{-1}	SENS. M	DEPTH Res. nm	DEPTH Limit μm	REF.
^{1}H $(^{11}\text{B},\alpha)$	1	1.793R	20	1	3	90	45	Ann	740		3	My	5.4	40	R	40	0.4	107
$(^{7}\text{Li},\gamma)$	0, 1	3.070R	450	3.2	15, 18	90	90	NaI	3100				6E-3	11	R	100	10	107
$(^{15}\text{N},\alpha\gamma)$	0	6.385R	50	7	4.439	0	80	NaI	700				0.8	30	R	4	1.5	107
$(^{19}\text{F},\alpha\gamma)$	1–3	6.418R	2		6–7			NaI	1E4					30	R	11	>1	100
	1–3	16.586R	30	4	6–7	90	90	NaI	700				2.7	40	R	40		107
		26.5	100		9.6			NaI	1.3E4[y]						R			107
$(^{27}\text{Al},\gamma)$	0	4–5	1000		4	0	90	TOF	0.07	2ns				40	E	450	50	22
$(^{3}\text{H,n})$	0	0.2			2.5	178	90	Si	0.12	17	12	My			E	70	1	19
^{2}D (d,p)		0.55			2.45			Si	120		6	My	28					2
(d,p)	0	2.0	1000	1.5	4.5	40	25	Si500		15	ESA			40	E	50	3.5	70
(d,t)	0	0.2	2.5E3	3		60		Si	750	27	1.4	Ni						53
(d,n)	0	2.5	100			24	90	TOF	0.09	2ns				40	E	500	5	21
$(^{3}\text{He,p})$	0	0.700	70	2	12.817	165	20	Si	100	17	3	My	150		E	20	0.4	31
$(^{3}\text{He},\alpha)$		0.500			3.5	88	15	ESA	0.04	15					E	9		4
		0.750	30	1		50	30	Si	0.04				200		E	60		107
^{3}T $(^{15}\text{N,n}\gamma)$	1–3	13	50		6–7	90	90	BGO	1E4	2ns				1	R	10	20	48
(p,n)	0	2.5	100		1	24	90	TOF	0.09	2ns				10	E	250	1.8	21
(d,α)	0	0.500				40	30	Si		15	3	Al			E	70	30	11
^{3}He (d,p)	0	0.43R		2	13	145	15	Si	24		30	Al	400	10	R	600	30	89
(d,p)		2.0	1000	1.5	18	40	25	Si		15	ESA			26	E	100	4.5	68
^{4}He (d,α)	0	0.5		1	3.5	90	15	Si	39	15				30	E	1	1	34
$(^{7}\text{Li},\gamma)$		1.68R	120		4.4			NaI						300	R	30	1	97
$(^{10}\text{B,n})$		3.5			3									1500	R	60	1	8
^{6}Li (d,α)		2.0	1000	1.5	14	40	25	Si		20	ESA		35	0.1		150	50	70
^{7}Li (p,α)		2.3	1500		8.9	105	45	Si			51		30					93
(p,γ)		0.441R		0.075	14–18							My	9	4	R			20
(p,p'γ)	1	2.4	20	4.3	0.478	55		Ge(Li)	110[y]	2.1	1			0.3	R			51

TABLE 14.12 (cont.)
Prompt nuclear analysis catalog

REACTION	ES	BEAM E_1 MeV	BEAM I_1 nA	BEAM D mm	PRODUCT E_3 MeV	PRODUCT θ	PRODUCT α	DETECTOR Type	DETECTOR $d\Omega$ msr	DETECTOR ΔE keV	FILTER t μm	FILTER MAT	SENS. CR	SENS. S μg.g^{-1}	SENS. M	DEPTH Res. nm	DEPTH Limit μm	REF.
(p,n)		2	5000	1		160	90	TOF	0.6	2.5ns				2	E	1000	50	80
(^{3}He,p)		2.5						Ann			125	Ka		100				97a
(α,γ)		0.950R		2				NaI	1.3E4v						R			108
^{9}Be ($\alpha,\alpha'\gamma$)	1	3.5			3.7	90	45	Ge(Li)	60v	3.5				0.2				7
(p,d)		0.300	500		0.478	150	90	Si			2.5	My		150	R	60	1	90
(p,α)		0.330	100		0.557	165		Si		15	2.5	My			E	30	1.7	98
		2.25	40		1.198	45	22	Mag	1	3.2				3E-3[a]	E	6	1.2	50
(p,γ)	1	0.475	1E5		0.717	90		NaI	1200									86
($\alpha,n\gamma$)	1	3.5			4.439	90	45	Ge(Li)	60v	3.5				0.5				6
(^{6}Li,p)		6.5			14.5	165		Si		20	15	My		1000	E		10	57
^{10}B (p,$\alpha\gamma$)		2.4			0.429	55		Ge(Li)	110v	1.9				2				45
(d,p)	0	2.7	1000	3	11.7	30	70	Si		45	70	Au		10				74
(α,p)	0	2.67		0.016	4.714	150		Ann			13	My						76
($\alpha,p\gamma$)		3.5	500		0.169	135	45	Ge(Li)	60v	3.5				15				7
^{11}B (p,α)	0,1	0.165R	500		3.3	150	90	Ann	420		7	My		0.03	R	60	1	95
(p,γ)		0.685	400	1	4.439	55	45	Ge(Li)	126v	3	8	My	550	100	R	50	1	59
(p,p'γ)	1	0.165R			2.124	30	90	Ge(Li)		3				150				91
(p,n)		4.5	1400			0		TOF		1.5ns	13	My		3	E	300	50	78
(α,p)	0,1	3.45		0.016	2.035	150		Ann										81
^{12}C (p,γ)		2.67	100		2.366	0	90	Ge(Li)	75v	2.4	13	My		1000				17
(p,p'γ)		0.457R	20		4.439	90	90	Ge(Li)	60v	2.5	7mm	Pb		1[a]				65
(p,α)	0	7.0				40	80	PSD										35
		35			3.01	150	90	Si	120	16	19	My	278	20	E	500	4	2
(d,p)	0	1.0				150	90	Si	1.6		2.5	Ni		4	E	150	7	96
(d,pγ)		1.6	12	0.012	3.09	90	90	NaI	1600v				240					12
(d,n)		2.2	500		2.582	30	90	TOF	1.4	5ns			100			600	15	63
		3.0	500															
(t,p)	0	1.74	200	<1	5.06	150		Si500	2.4				5E-2[a]					3

TABLE 14.12 (cont.)
Prompt nuclear analysis catalog

Column groups: BEAM (ES, E_1, I_1, D); PRODUCT (E_3, θ, α); DETECTOR (Type, $d\Omega$, ΔE); FILTER (t, MAT); SENSITIVITY (CR, S, M); DEPTH (Res., Limit).

REACTION	ES	E_1 (MeV)	I_1 (nA)	D (mm)	E_3 (MeV)	θ	α	Type	$d\Omega$ (msr)	ΔE (keV)	t (μm)	MAT	CR	S ($\mu g \cdot g^{-1}$)	M	Res. (nm)	Limit (μm)	REF.
(^{3}He,p)	0	2.42			6.23	90	15	Si							240	E	11		44
(^{3}He,α)	0	2.8	500	2	0.378	135	90	Mag	3							E	30		43
(^{6}Li,p)	0	6.5	60	2	9.1	40	15	Si		20	15	My				E	20		57
^{13}C (p,γ)		0.551R	100	4	8.061	0	90	Ge(Li)	75^v	2.4					0.2%				17
(p,n)		4.5			1.38	30	90	TOF		4ns									82
(d,p)		1.5	2	0.1	3.373	135	0	Si	110	1.9					30				87
(d,nγ)						0		Ge(Li)	100^v						640				87
(^{3}He,p)	0	2.8	2	0.1		135	0	Si	110						630				87
(α,n)		2.8	2	0.1	4.936	0	90	Liq	100^v						100				87
^{14}N (p,γ)		1.061R			8.283	0		NaI	350^v							R			101
(p,p'γ)		4.2	100	4	2.313	0	0	Ge(Li)	80^v	1.8	5mm	Pb		100					47
(d,p)	0	1.2	180	1.7	9.647	45	55	Si	3		8	Au	2E-3	<20					77
(d,α)	0				12.289														77
(d,p)	4, 5	1.1	100	1	2.0	150		Si	120		19	My		5E-4[a]	E	700	15	1
(d,p)	0	2.5																	30
(d,pγ)	1	2.0	200	1	7–8	90		Ge(Li)	70^v				6.5	2E-3[a]					102
(d,α)		1.3			3.76	150		Si	120		19	My							2
(d,$\alpha\gamma$)		2.5			4.439	90		Ge(Li)											46
(d,n)	4	3.0	500		1.098	30	90	TOF	1.4	5ns				10	E	600	15	63
(α,pγ)		5.0	250		0.871	45	90	Ge(Li)	80^v	2.3				1%					36
^{15}N (p,α)		0.5		0.01	3.8	150	90	Si	800		7			900					5
(p,$\alpha\gamma$)		0.9	300		4.439	90	90	NaI	1600^v			Al							94
(p,n)		4.5			0.82	30	90	TOF		4ns									82
^{16}O (p,γ)		1.240	7000	4.3	0.493	90		Ge(Li)	68^v	2.1									16
(p,γ)		2.4	20			55		Ge(Li)	110^v					1%	E	20			52
(p,p'γ)		9.5	5		6–7	90	90	Ge(Li)	60^v	2.5	7mm	Pb							64

TABLE 14.12 (cont.)
Prompt nuclear analysis catalog

REACTION	ES	BEAM E_1 (MeV)	I_1 (nA)	D (mm)	PRODUCT E_3 (MeV)	θ	α	DETECTOR Type	$d\Omega$ (msr)	ΔE (keV)	FILTER t (μm)	MAT	CR	SENSITIVITY S ($\mu g.g^{-1}$)	M	DEPTH Res. (nm)	Limit (μm)	REF.
(d,p)	0	1.470	100	1		160	30	ΔE/E	120	40	19	My	22		E	33		105
(d,p)	1	0.830	100		1.52	150	90	Si	1.6	16	2.5	Ni		260	E	50	1	2
(d,pγ)		1.6	12	0.012	0.87	150	90	NaI	350[v]					10	E	150	7	96
(d,α)	0	3.0	500	1	2.63	145	60	Si	1.4	25					E	20	1.5	62
(d,α)		0.9	100			50	90	TOF		5ns					E	300	15	106
(d,n)	0	3.0	1000			140		Si500	2.4									63
(t,p)	0	1.870	200	8	5.01	90		Ge(Li)						100				3
(t,nγ)		1.9	50		0.940	150		Si	80[v]					5E-2[a]				83
(^{3}He,p)	0	2.9				55		Ge(Li)	37		12	My		0.1[a]				54
(^{3}He,pγ)	0	4.0	300		1	90		Si	60[v]					0.1%				49
(^{3}He,α)		2.42		2	5.40	90		TOF		4ns	12	My	60		E	9		44
(^{3}He,α)		4.5	250	3	0.86	150	15	Mag	5	43	15	My	90					82
^{17}O (p,n)		5.3	10	1		110	90	Si	120									79
(^{3}He,α)		0.730R		0.5	3.38	170	10	Ann										2
^{18}O (p,α)		0.846	500	2	3.7	30	90	TOF		4ns								18
(p,α)		1.765R			3.896	45	90	Ge(Li)										9
(p,α)		4.5		2	1.9	135	45	Ge(Li)										82
(p,n)		5	250		0.351	90	45	Ge(Li)	80	2.3				730	E	100	2.5	36
(p,nγ)		3.5	500		0.350	90	45	Ge(Li)	60	3.5				875	R	20		7
(α,nγ)	1	0.935R			0.110	90		Ge(Li)						2E-4[a]	E	10[a]		41
^{19}F (p,p'γ)		3.5	100		0.110	150	45	Ge(Li)	1.4[v]	1.2			12[b]	0.1	R	100	5	99
(p,p'γ)	2	2.0			0.197	90	45	Ge(Li)						0.1	R			26
(p,p'γ)		3.5	100		0.197	135	45	Ge(Li)					10[b]	0.1	E			99
(p,α)		1.340			6.92	150	90	Si	120	1.2	31	My	4.3					2
(p,αγ)	2–4	0.340R			6–7	90	45	NaI	820[v]					3E-3	R	12		27
(p,αγ)		0.872R			6–7	135	45	Ge(Li)					(60)	100	R	200	1.4	56
(p,αγ)		1.8	50		6–7	45		Ge(Li)										88
(p,αγ)		2.55			6–7			NaI	40[v]	3.5				25	Y		50	61

TABLE 14.12 (*cont.*)
Prompt nuclear analysis catalog

REACTION	ES	BEAM E₁ MeV	I₁ nA	D mm	PRODUCT E₃ MeV	θ	α	DETECTOR Type	dΩ msr	ΔE keV	FILTER t μm	MAT	SENS. CR	S μg·g⁻¹	M	DEPTH Res. nm	Limit μm	REF.
(d,α)		1.2				150		Si100							E			66
(α α'γ)	2	3.5	500		0.197	90	45	Ge(Li)	60^v	3.5				10				7
(α p'γ)		3.5	500		1.275	90	45	Ge(Li)	60^v	3.5				5				7
^{20}Ne (p,γ)		1.169R	2E4	3	3.545	55		Ge(Li)	97					2000	R	15	3	71,103
^{23}Na (p,γ)		1.417R	500	1	9–14			NaI						1	R	10		32
		1.458R	500	4.3	0.439	135	45	Ge(Li)	110^v	2.1				0.6				15
(p,p'γ)		2.4	20		0.439	55	45	Ge(Li)						3				51
(p,α)		0.592R	100			135	45	Si	100	14								10
(p,α'γ)		1.011R	100	1	1.632	0		IG			6	My	25			12	1	28
(d,p)		4.0	150		5.235	135		Si		2	21	Al					0.5	84
(α,a'γ)	1, 2	3.5	500		0.439	90	45	Ge(Li)	60^v	3.5				30				7
^{24}Mg (d,p)		1.8	10	0.02	5.5	135		Si	28					0.4%				75
^{25}Mg (p,p'γ)		2.4	20	4.3	0.390	55	90	Ge(Li)	110^v	2.1				50				51
		3			0.585	90		Ge(Li)						100				29
(α,α'γ)		5.0			0.585	45	90	Ge(Li)	80^v	2.3				0.2%				40
^{26}Mg (p,γ)		0.475			0.843	90		NaI	1180		140		2E-8					86
^{27}Al (p,γ)		0.992R	500		9–14			NaI	1600^v					1	R			32
(p,p'γ)		2.4	500	4.3	1.013	55	90	Ge(Li)	110^v	2.1				22				51
		4.0	20	3.5	1.013	45		Ge(Li)	80^v					15				40
(p,α)		1.365R			2.2	135	45	Mag Ann						7	R	1	0.3	33
		2.0	20			135		Si										69
^{28}Si (α,p)	0	4.0	150		5.235	135	45	Si	80^v					1000				84
(α,α'γ)		5.0			1.013	45	90	Ge(Li)			21	Al		970				40
(p,p'γ)		3.1	500	1	1.779	135	45	Ge(Li)	60^v	2.5	7mm	Pb		60				15
		9.5	5		1.779	90	90	Ge(Li)										64
(d,p)		1.9	10		7.50	135		Si	240		20	My		200				68

TABLE 14.12 (cont.)
Prompt nuclear analysis catalog

REACTION	ES	E₁ (MeV)	I₁ (nA)	D (mm)	E₃ (MeV)	θ	α	Type	dΩ (msr)	ΔE (keV)	t (μm)	MAT	CR	S (μg.g⁻¹)	M	Res. (nm)	Limit (μm)	REF.
^{29}Si (p,p'γ)		4.0		3.5	1.273	45	90	Ge(Li)	80[v]					240				40
^{30}Si (p,γ)		0.620R			7.90	55	45	Ge(Li)	126[v]	3					R	10	0.5	91
^{31}P (p,γ)		0.811R			7.42	55	45	Ge(Li)	126[v]	3					100	R	10	0.5	91
(p,p'γ)		2.5	100		1.266	90	45	Ge(Li)						100				26
^{32}S (p,α)		1.892R			3	145	90	ΔE/E					100	6E-3[a]	R		3	60
(p,p'γ)		3.430R	150		2.230		90	Ge(Li)	40[v]					700				13
(d,p)		7.0	20		4.744	90		Ge(Li)	60[v]	2.5	7mm	Pb		1[a]				65
^{35}Cl (p,p'γ)	7	2.0			1.220	135		Si	20		15	Au		10[a]				85
^{37}Cl (p,αγ)		2.9			2.128	90	45	Ge(Li)		3				1%				25
^{39}K (p,αγ)		2.9			2.168	90	45	Ge(Li)		3				1%				25
^{43}Ca (p,α'γ)		2.4			0.371	0	90	Ge(Li)	110[v]	1.8	5mm	Pb		2%				92
(p,p'γ)		4.0			1.477	45	90	Ge(Li)	80[v]					720				40
(p,n)		4.5	3000			0	90	TOF		4ns				1000				67
^{44}Ca (p,p'γ)	0.1	24	20	4.3	0.364	55		Ge(Li)	110[v]	2.1				0.8%				51
^{48}Ca (p,n)		4.0			1.157	45	90	Ge(Li)	80[v]	4ns				70				40
(p,nγ)		4.5	3000		1.64	0	90	TOF						1000				67
^{47}Ti (p,p'γ)		4.0			0.520	45	90	Ge(Li)	80[v]					0.9%				40
(p,nγ)		4.0			0.159	135	45	Ge(Li)						200				15
(α,α'γ)		3.5	500		0.159	90	45	Ge(Li)	60[v]	3.5				750				7
^{48}Ti (p,γ)		1.007R			7.582	135	45	Ge(Li)						0.8[a]	R		120[a]	42
(p,p'γ)		4.0			0.983	0	90	Ge(Li)						200				15
(p,n)	0, 1	2.02				0	90	TOF	4						E	1600		58
^{51}V (p,αγ)		5	200	3.5	0.320	45	90	Ge(Li)	80[v]					500				37
(p,nγ)		4			0.749	45	90	Ge(Li)	80[v]					270				40
^{52}Cr (p,γ)	1	1.005R	3000	8	0.378	0	90	Ge(Li)	68[v]						R	1	120[a]	104
(d,p)	0	3.5	500	3	9.217	60		Si700		45	30	Au						73
^{53}Cr (p,nγ)		4.5			0.378	45	90	Ge(Li)	80[v]					260				40

TABLE 14.12 (*cont.*)
Prompt nuclear analysis catalog

REACTION	ES	BEAM E_1 MeV	I_1 nA	D mm	PRODUCT E_3 MeV	θ	α	Type	DETECTOR $d\Omega$ msr	ΔE keV	FILTER t μm	MAT	SENSITIVITY CR	S μg.g^{-1}	M	DEPTH Res. nm	Limit μm	REF.
^{55}Mn (p,p'γ)		4.0			0.126	135	45	Ge(Li)						10				15
(p,n'γ)		4.0		3.5	0.412	45	90	Ge(Li)	80^y					190				40
					0.931	90	45	Ge(Li)		2.7				5				23
(α,α'γ)		3.5	500		0.126	90	45	IG	60^y	3.5				70				7
		5	100	3.5	0.126	45	90	IG	3^y									38
^{56}Fe (p,p'γ)		4.0			0.847	135	45	Ge(Li)						100				15
^{59}Co (p,n'γ)		4.0			0.343	45	90	IG	3^y					800				40
^{58}Ni (p,γ)		1.424R			0.492			Ge(Li)						1	R		100^a	42
(p,p'γ)		4.5	5		1.454	90		Ge(Li)		3				0.5%				55
(d,p)	0	3.5	7500	3	10.272	60	60	Si700		45	100	Au						72
^{68}Zn (p,p'γ)		3.7	150		1.078			Ge(Li)	40^y					300				13
^{75}As (p,p'γ)		4.0			0.199	135	45	Ge(Li)						2				15
^{77}Se (p,p'γ)		4.0			0.239	135	45	Ge(Li)						300				15
^{80}Se (p,p'γ)		4.0			0.666	135	45	Ge(Li)						200				15
^{93}Nb (p,n)	0, 1	2.8	4000					TOF	7.5	2.5ns								80
^{95}Mo (p,αγ)		5	200	3.5	0.204	45	90	IG	3^y	0.3				700				37
^{103}Rh (p,p'γ)		2.0			0.296	90	45	Ge(Li)		3				500				24
Ag (p,p'γ)		2.2			0.3	90	45	Ge(Li)		3				5000				24
(Cl,Cl'γ)		55			0.3	90	45	Ge(Li)		2				40				6a
^{181}Ta (p,p'γ)		4.0			0.301	135	45	Ge(Li)						10				15
^{182}W (p,αγ)		5.0			0.100	45	90	IG	3^y	0.3				700				39
^{194}Pt (p,p'γ)		2.0			0.328	90	45	Ge(Li)		3				1000				24
^{197}Au (p,p'γ)		4.0			0.279	135	45	Ge(Li)						200				15

ES: Sequence numbers of final states reached by observed radiation; Ann: Annular surface barrier detector; Liq: Liquid scintillator neutron detector; Mag: Magnetic analyser; S x: Si surface barrier detector with depletion depth x; TOF: Time of flight measurements; Ka: Kapton; MAT: Material; My: Mylar; CR: Count rate per μC for 100 msr solid angle and 10^{16} atoms cm^{-2} (or ^{b} μg.g^{-1}); S: Estimated detection limit (μg.g^{-1}) (or ^{a} μg.cm^{-2}); E: Energy Spectrum Method; Limit: Maximum depth that can be profiled (μm); R: Resonance scanning method; Res: Typical depth resolution at the surface (nm); W: Wedge scanning method; Y: Yield curve unfolding method.

TABLE 14.13
Resonance used in depth profiling

REACTION	E_r (MeV)	Γ (keV)	STRENGTH (Rel)
^{1}H (^{7}Li,γ)	3.070	70	6
(^{11}B,α)	1.793	66	100
(^{15}N,$\alpha\gamma$)	6.400	1.9	200
(^{19}F,$\alpha\gamma$)	6.418	45	60
	16.568	130	
(^{27}Al,γ)	26.5		
^{2}D (^{3}He,p)	0.645		70
^{3}He (^{2}D,p)	0.430		70
^{4}He (^{7}Li,γ)	0.701		0.04
	1.433		0.55
	1.677	20	3.5
(^{10}B,n)	3.77		
^{7}Li (p,γ)	0.441	12	6
(p,p' γ)	1.030	168	
(α,γ)	0.819	4	0.55
	0.958	7	3.5
^{9}Be (p,γ)	0.330	160	3
	0.991	89	2
	1.084	4	6
(p,α)	0.330	160	
	0.991	89	
^{11}B (p,γ)	0.163	6.5	0.157
(p,α)	0.163	6.5	
^{12}C (p,γ)	0.457	39	0.127
	1.698	72	0.035
^{13}C (p,γ)	0.551	30	9.2
	1.150	7	1.3
	1.748	0.07	15

TABLE 14.13 (*cont.*)
Resonance used in depth profiling

REACTION	E_r (MeV)	Γ (keV)	STRENGTH (Rel)
^{14}N (p,γ)	0.278	1.7	.014
	1.058	3.9	.95
	1.550	34	.16
^{15}N (p,$\alpha\gamma$)	0.429	0.9	200
	0.897	2.0	800
	1.210	22.5	600
^{18}O (p,$\alpha\gamma$)	0.633	2.1	
	0.846	47	
	1.766	3.6	
	1.928	0.2	
^{19}F (p,p'γ)	0.935	8.0	3
	1.137	3.7	0.04
(p,α)	0.778	10	0.08
	0.843	23	0.13
	1.358	54	1.1
	1.640	90	2.9
(p,$\alpha\gamma$)	0.227	1.0	
	0.340	2.4	12
	0.484	0.9	5
	0.672	6.0	8
	0.872	4.7	100
	0.935	8.1	39
	1.348	4.9	30
	1.373	12.4	77
	1.607	6.0	6
	1.694	35	17
^{20}Ne (p,γ)	1.172	0.02	375
	1.311	0.19	18
^{23}Na (p,γ)	1.318	1.9	48
	1.398	0.8	20
	1.417	0.09	38
	1.457	6.0	9

TABLE 14.13 (*cont.*)
Resonance used in depth profiling

REACTION	E_r (MeV)	Γ (keV)	STRENGTH (Rel)
(p,p'γ)	1.283	5.5	1000
	1.318	1.9	30
(p,α)	0.592	0.64	330
	0.744	<0.1	70
(p,$\alpha\gamma$)	1.011	<0.1	55
	1.091	5.0	27
^{24}Mg (p, γ)	0.224	<0.03	5
	0.419	<0.04	8
^{26}Mg (p, γ)	1.548	0.02	
^{27}Al (p, γ)	0.992	0.1	40
	1.002	0.001	1
(p, α)	1.365	1.4	1640
^{29}Si (p, γ)	0.416	<0.5	0.7
^{30}Si (p, γ)	0.620	0.07	3.1
	0.671		0.1
^{31}P (p, γ)	0.811	<0.4	2.2
	0.821		0.4
	1.121	<0.15	3.0
	1.151	<0.10	3.9
	1.251	1.4	11
	1.400		1.3
(p, α)	1.018	<0.3	0.1
	1.157		0.024
	1.892	23	7.0
^{32}S (p, p'γ)	3.379	1.0	600
	3.716	1.5	660
^{34}S (p, γ)	1.020	0.50	3.2
	1.055		0.4
^{48}Ti (p, γ)	1.007		
	1.013		
^{52}Cr (p, γ)	1.005	<0.1	0.89

TABLE 14.14
Useful elastic backscattering resonances (*cont.*)

Reaction	E_1 (MeV)	σ_0 mb/sr cm	Γ keV cm	Reference
^{12}C (p, p)	1.69	1100	60	1, 2, 4
	4.81	680	11	5
^{12}C (α, α)	4.26	1450	27	3
^{16}O (p, p)	2.66	80	19	1, 2, 4
	3.47	430	1.5	1, 2, 4
^{16}O (α, α)	3.04	1300	10	6
	3.38	250	8	1, 2, 4

TABLE 14.15
Selected cross-section values (*cont.*)

Target	Product	θ	$d\sigma/d\Omega$ mb sr^{-1}	Reference
Deuteron induced reactions (E_d = 972 keV)				
^2D	p_0	135	6	Jarmie and Seagrove (1957)
^3He	p_0	86	23	Altstetter *et al.* (1978)
^6Li	α_0	150	4.2	Maurel *et al.* (1981)
^9Be	p_0	150	1.6	Annegarn *et al.* (1974)
	p_1	140	0.75	Jarjis (1979)
	α_0	160cm	2.7	Jarjis (1979)
	α_1	160cm	0.8	Jarjis (1979)
^{10}B	p_0	150	0.44	Marion and Weber (1956)
	α_0	160	0.47	Debras and Deconninck (1977)
	α_1	160	0.26	Debras and Deconninck (1977)
^{12}C	p_0	150	25.5 ± 0.8	Davies and Norton (1980)
^{14}N	p_0	150	0.42 ± 0.01	Davies *et al.* (1983)
	p_1+p_2	150	1.03 ± 0.01	Davies *et al.* (1983)
	p_4	150	1.12 ± 0.03	Davies *et al.* (1983)
	p_5	150	6.0 ± 0.15	Davies *et al.* (1983)
	p_6	150	0.75 ± 0.05	Davies *et al.* (1983)
	α_0	150	0.11 ± 0.003	Davies *et al.* (1983)
	α_1	150	1.57 ± 0.03	Davies *et al.* (1983)
^{16}O	p_1	150	13.2 ± 0.3	Davies *et al.* (1983)
^3He Reactions (E_{3He} = 750 keV)				
^2D	p_0	150	54 ± 2	Davies and Norton (1980)

Cross-section relations

1) ^3He (d, p$_0$) from E_d = 0.1 to 2.5 MeV; θ = 86°, Altstetter *et al.* (1978):

$$d\sigma/d\Omega = 475E_1^3/(1 - 26.2E_1^{3.43} + 36.5E_1^{3.91}) \text{ mb sr}^{-1} \qquad (14.24)$$

Use also for $^2D(^3He, p)$ with $E_d = 0.667E_{3He}$.

2) $^6Li(d, \alpha_0)$ from E_d = 0.92 to 1.9 MeV; θ = 150°; Maurel *et al.* (1981):

$$d\sigma/d\Omega = 4.0 - 2.9 (E_1 - 1) \text{ mb sr}^{-1} \qquad (14.25)$$

3) $^{18}O(p, \alpha_0)$ from E_p = 0.4 to 0.6 MeV; θ = 165°; Amsel and Samuel (1967):

$$d\sigma/d\Omega = 3.38 \times 10^{-14} \exp (17.2E_1) \text{ mb sr}^{-1} \qquad (14.26)$$

TABLE 14.16
PIGME yield measurements

Z	θ	E₁ (MeV)	Reference	Table
a. Proton induced gamma-rays				
3–9	90°	0.4–1.5	Golicheff (1972)	
3–21	55°	1.0, 1.7, 2.4	Antilla et al. (1981)	
3–16	135°	2.0, 2.5	Bird et al. (1985)	
3–21	55°	2.4, 3.1, 3.8, 4.2	Kiss et al. (1985)	Table 14.18
3–82	55°	7, 9	Raisanen et al. (1987)	Table 14.18
3–92	45°	3.5, 4.0, 4.5, 5.0, 5.5, 6.0	Gihwala (1982)	Table 14.18
9	90°	0.3–2.4	Stroobants et al. (1976)	
9–79	135°	2.0, 2.514	Kenny et al. (1980)	
9	135°	1.5–2.75	Kenny (1981)	
9–27	45°	4.5	Gihwala and Peisach (1982c)	
11	90°	0.5–2.7	Bodart et al. (1977)	
11, 47	90°	1.4–2.7	Deconninck (1972)	
12	90°	1.2–3.0	Demortier and Delsate (1975)	
13	90°	0.5–2.5	Deconninck and Demortier (1972)	
15	90°	0.5–2.5	Demortier and Bodart (1972)	
17	90°	0.8–3.0	Deconninck and Debras (1975)	
22–30	90°	1.5–3.1	Demortier (1978)	Table 14.18
23	45°	2.0–5.3	Gihwala and Peisach (1982a)	
25	45°	4.5	Gihwala and Peisach (1982b)	
31–79	55°	1.7, 2.4	Raisanen and Hanninen (1983)	Table 14.18
35–37, 78, 79	90°	1.2–2.9	Deconninck and Demortier (1975)	

TABLE 14.16 (*cont.*)
PIGME yield measurements

Z	θ	E_1 (MeV)	Reference	Table
b. Triton induced gamma-rays				
3–12	90°	2.0	Borderie and Barrandon (1978)	Table 14.17
14–23	90°	3.0	Borderie and Barrandon (1978)	Table 14.17
22–79	90°	3.5	Borderie and Barrandon (1978)	Table 14.17
c. Alpha induced gamma-rays				
3–78	90°	3.5	Borderie (1978)	Table 14.17
3–14	55°	2.4	Lappalainen *et al.* (1983)	
3–80	45°	5	Giles and Peisach (1979)	
7	45°	3.5–5.0	Gihwala *et al.* (1978a)	
9	45°	3–5	Giles and Peisach (1976)	
23	45°	5	Gihwala *et al.* (1978b)	
d. ^{35}Cl induced gamma-rays				
3–90	90°	55	Borderie *et al.* (1979)	Table 14.17

TABLE 14.17 Gamma-ray Index

E_γ keV	T	E_γ keV	T	E_γ keV	T	E_γ keV	T	E_γ keV	T	E_γ keV	T	E_γ keV	T	E_γ keV	T
Proton Induced Gamma-Rays															
54	^{65}Cu	127	^{101}Ru	187	^{190}Os	273	^{117}Sn	361	^{72}Ge	466	^{67}Zn	595	^{48}Ti	743	^{128}Te
61	^{65}Cu	128	^{56}Fe	189	^{238}U	276	^{80}Se		^{73}Ge	469	^{72}Ge	598	^{69}Ga	744	^{92}Zr
80	^{158}Gd		^{57}Fe	190	^{81}Br		^{81}Br	364	^{44}Ca	475	^{60}Ni	602	^{69}Ga	745	^{51}V
81	^{162}Dy	129	^{104}Ru	192	^{197}Au	279	^{75}As		^{45}Sc	476	^{45}Sc	603	^{51}V		^{69}Ga
82	^{154}Sm	130	^{79}Br	193	^{153}Eu		^{197}Au	366	^{64}Ni	477	^{60}Ni		^{134}Ba	749	^{51}V
83	^{153}Eu		^{85}Rb	197	^{18}O	280	^{74}Ge	367	^{57}Fe		^{7}Li	605	^{51}V	754	^{139}La
88	^{46}Ti		^{104}Ru		^{19}F		^{75}As	368	^{200}Hg		^{55}Mn	608	^{51}V	755	^{29}Si
	^{47}Ti	131	^{48}Ca		^{59}Co		^{105}Pd	371	^{48}Ca	481	^{97}Mo		^{74}Ge		^{39}K
89	^{99}Ru		^{157}Gd		^{62}Ni	283	^{61}Ni	373	^{43}Ca		^{135}Ba	614	^{78}Se	770	^{64}Ni
	^{156}Gd	134	^{187}Re		^{71}Ga	286	^{75}As		^{61}Ni	485	^{87}Rb	617	^{79}Br		^{65}Cu
90	^{46}Ti	136	^{181}Ta	199	^{147}Sm		^{151}Eu	374	^{110}Pd		^{74}Ge		^{112}Cd	778	^{96}Mo
	^{48}Ti	137	^{57}Fe		^{151}Eu	287	^{69}Ga		^{127}I	491	^{58}Ni	621	^{111}Cd	779	^{48}Ca
	^{49}Ti	138	^{175}Lu		^{47}Ti		^{75}As	379	^{52}Cr	493	^{69}Ga	627	^{75}As	783	^{50}Cr
92	^{55}Mn	139	^{193}Ir		^{74}Ge	291	^{183}W		^{53}Cr	496	^{127}I	629	^{127}I	786	^{95}Mo
	^{92}Zr	141	^{75}As		^{75}As	293	^{75}As	382	^{123}Sb		^{16}O		^{95}Mo		^{98}Mo
93	^{67}Zn	143	^{59}Co	200	^{77}Ge		^{45}Sc	384	^{79}Br	497	^{115}In	633	^{133}Cs	789	^{69}Ga
	^{178}Hf		^{62}Ni	201	^{142}Ce	296	^{103}Rh		^{133}Cs	500	^{69}Ga	643	^{124}Sn	793	^{86}Sr
	^{180}Hf	144	^{79}Br	204	^{95}Mo	298	^{113}Cd	390	^{25}Mg	504	^{71}Ga	644	^{119}Sn	797	^{35}Cl
95	^{165}Ho	138	^{175}Lu		^{127}I		^{28}Si		^{71}Ga	508	^{70}Zn	649	^{51}V	807	^{63}Cu
98	^{48}Ti	145	^{127}I	205	^{78}Se	301	^{181}Ta	398	^{69}Ga		^{121}Sb	656	^{92}Zr	808	^{51}V
99	^{183}W		^{141}W		^{205}Tl	302	^{148}Nd	400	^{72}Ge	510	^{73}Ge	657	^{75}As	809	^{64}Zn
100	^{182}W	146	^{155}Gd	206	^{65}Cu	303	^{133}Cs		^{74}Ge	520	^{48}Ca	658	^{48}Ti		^{66}Zn
104	^{180}W		^{179}Hf		^{66}Zn	304	^{75}As	406	^{198}Pt	523	^{79}Br		^{49}Ti	811	^{32}S
105	^{91}Zr	147	^{70}Ge		^{67}Zn	306	^{79}Br	408	^{53}Cr	531	^{44}Ca	661	^{66}Zn		^{58}Fe
109	^{238}U	151	^{46}Ti	207	^{174}Hf	307	^{151}Eu		^{54}Cr		^{45}Sc	666	^{126}Te	822	^{75}As
110	^{18}O		^{48}Ti	208	^{199}Hg	308	^{47}Ti	412	^{198}Hg	536	^{100}Mo	667	^{80}Se	827	^{59}Co
	^{19}F		^{49}Ti	210	^{47}Ti	311	^{109}Ag	413	^{55}Mn	537	^{85}Rb	668	^{69}Ga		^{63}Cu
	^{151}Eu		^{85}Rb		^{165}Ho	313	^{41}K	414	^{9}Be	540	^{100}Ru	669	^{62}Ni	829	^{65}Cu
	^{169}Tm	152	^{65}Cu	211	^{195}Pt	316	^{51}V	415	^{109}Ag	542	^{123}Sb		^{63}Cu		^{65}Zn
111	^{79}Br	158	^{199}Hg	215	^{96}Mo		^{68}Zn	416	^{205}Tl	543	^{44}Ca	670	^{37}Cl	830	^{25}Mg
	^{151}Eu	160	^{47}Ti	216	^{80}Se	319	^{69}Ga	418	^{25}Mg		^{45}Sc	673	^{57}Fe	834	^{69}Ga
	^{184}W		^{123}Sb	217	^{79}Br	320	^{49}Ti		^{127}I	548	^{197}Au	677	^{29}Si		^{71}Ga
113	^{47}Ti	161	^{133}Cs	221	^{135}Ba		^{50}Ti	420	^{89}Y	550	^{85}Rb	681	^{59}Co	835	^{72}Ge
	^{75}As	165	^{181}Ta	226	^{47}Ti		^{51}V	423	^{58}Ni		^{148}Sm	684	^{49}Ti		^{54}Cr
	^{177}Hf	167	^{163}Dy		^{49}Ti	325	^{96}Mo		^{68}Zn	555	^{102}Pd	692	^{57}Fe	836	^{49}Ti
114	^{175}Lu		^{164}Dy		^{50}Ti		^{107}Ag	428	^{75}As		^{104}Pd	696	^{144}Nd	839	^{130}Te
115	^{65}Cu		^{66}Zn	230	^{57}Fe	327	^{71}Ga	429	^{7}Li	558	^{114}Cd	697	^{53}Cr	842	^{33}S
	^{67}Zn	170	^{26}Mg		^{85}Rb	328	^{194}Pt		^{10}B	559	^{75}As	700	^{119}Sn	843	^{26}Mg
	^{115}In		^{27}Al		^{87}Sr	334	^{150}Sm	431	^{45}Sc	560	^{76}Se	703	^{93}Nb		^{27}Al
	^{165}Ho	172	^{127}I	232	^{76}Ge	336	^{94}Mo	434	^{108}Pd	562	^{89}Y	709	^{29}Si	847	^{55}Mn
118	^{169}Tm	175	^{69}Ga	233	^{69}Ga	339	^{59}Co	439	^{23}Na	563	^{76}Ge	717	^{9}Be		^{56}Fe
119	^{103}Rh		^{71}Ga	239	^{77}Se		^{62}Ni		^{77}Se	565	^{53}Cr		^{10}B	848	^{51}V
121	^{147}Sm	175	^{67}Zn		^{195}Pt	342	^{111}Cd	440	^{123}Sb	568	^{37}Cl	719	^{117}Sn	850	^{93}Nb
	^{159}Tb		^{68}Zn	250	^{177}Hf	344	^{152}Gd	442	^{105}Pd	572	^{72}Ge	720	^{44}Ca	854	^{48}Ti
122	^{48}Ca	178	^{167}Er	251	^{175}Lu	353	^{57}Fe	443	^{208}Pb		^{73}Ge		^{45}Sc	865	^{65}Cu
	^{152}Sm	181	^{172}Yb	254	^{72}Ge	355	^{101}Ru	445	^{81}Br		^{74}Ge	724	^{115}Sn	871	^{17}O
	^{179}Hf		^{173}Yb		^{73}Ge	356	^{196}Pt	449	^{93}Nb	573	^{121}Sb	728	^{66}Zn	872	^{68}Zn
	^{186}W		^{179}Hf		^{74}Ge	357	^{103}Rh	454	^{146}Nd	574	^{68}Zn		^{62}Ni	875	^{62}Ni
123	^{57}Fe	183	^{79}Br	265	^{74}Ge	358	^{104}Rh	455	^{24}Mg		^{69}Ga	742	^{93}Nb	879	^{59}Co
	^{154}Gd	185	^{67}Zn		^{76}Ge	360	^{66}Zn			586	^{93}Nb				
125	^{185}Re		^{176}Lu	269	^{197}Au					588	^{61}Ni				
126	^{55}Mn			270	^{119}Sn						^{89}Y				
				271	^{115}Sn										

TABLE 14.17 *(cont.)*

Proton Induced Gamma-Rays

E_γ keV	T	E_γ keV	T	E_γ keV	T	E_γ keV	T	E_γ keV	T	E_γ keV	T	E_γ keV	T	E_γ keV	T	Yield $\mu C.sr^{-1}$
889	^{45}Sc	1040	^{69}Ga	1266	^{30}Si	1434	^{51}V	1930	^{60}Ni	2605	^{13}C	3685	^{13}C	125	^{185}Re	2.7E4
	^{46}Ti		^{70}Ge		^{31}P		^{52}Cr	1943	^{41}K	2646	^{35}Cl	3780	^{31}P	126	^{53}Cr	5.1E4
891	^{89}Y	1042	^{18}O		^{34}S	1454	^{58}Ni	1951	^{23}Na	2693	^{35}Cl	4238	^{23}Na		^{56}Fe	1.3E4
896	^{209}Bi	1049	^{45}Sc	1273	^{29}Si	1459	^{19}F	1965	^{25}Mg	2698	^{35}Cl	4343	^{28}Si	129	^{51}V	
899	^{41}K	1050	^{48}Ti	1290	^{52}Cr	1481	^{64}Ni	1970	^{35}Cl	2735	^{27}Al	4439	^{11}B	134	^{187}Re	4.3E4
909	^{88}Sr		^{49}Ti		^{53}Cr	1482	^{51}V	1972	^{35}Cl	2796	^{37}Cl		^{15}N	136	^{181}Ta	1.2E5
912	^{66}Zn	1065	^{48}Ti		^{62}Ni	1490	^{48}Ti	1982	^{18}O	2804	^{52}Cr	4492	^{27}Al	155	^{63}Cu	1.1E4
913	^{52}Cr		^{49}Ti	1293	^{59}Co	1507	^{88}Sr	2010	^{41}K	2839	^{27}Al	5270	^{15}N	157	^{52}Cr	3.7E5
	^{53}Cr	1071	^{52}Cr	1294	^{41}K	1511	^{53}Cr	2028	^{29}Si	2981	^{27}Al	5299	^{15}N	159	^{61}Ni	6.2E3
914	^{49}Ti	1077	^{85}Rb	1309	^{136}Ba	1513	^{48}Ti	2029	^{52}Cr	3003	^{35}Cl	6130	^{19}F		^{62}Ni	
	^{58}Ni	1078	^{68}Zn	1310	^{48}Ti		^{49}Ti	2035	^{32}S	3004	^{27}Al	6793	^{15}N		^{63}Cu	1.7E4
919	^{63}Cu	1081	^{18}O		^{60}Ni	1525	^{41}K	2081	^{46}Ti	3087	^{13}C	6917	^{19}F		^{185}Re	3.0E3
	^{94}Zr	1082	^{90}Zr	1313	^{88}Sr		^{42}Ca	2087	^{50}Ti	3163	^{35}Cl	7117	^{19}F	165	^{181}Ta	6.4E3
921	^{119}Sn		^{91}Zr	1316	^{55}Mn		^{68}Zn	2124	^{10}B	3180	^{13}C	7477	^{9}Be	167	^{187}Re	4.3E3
923	^{117}Sn	1092	^{47}Ti	1318	^{63}Cu	1530	^{93}Nb		^{11}B	3511	^{12}C	8062	^{13}C	175	^{66}Zn	2.0E4
928	^{50}Ti	1095	^{58}Fe	1327	^{63}Cu	1531	^{51}V	2127	^{34}S	3562	^{9}Be				^{69}Ga	3.3E4
	^{51}V		^{59}Co	1329	^{63}Cu	1547	^{62}Ni		^{37}Cl						^{70}Ge	1.2E4
931	^{55}Mn		^{62}Ni	1332	^{51}V	1552	^{47}Ti	2163	^{47}Ti					185	^{16}O	5.1E4
936	^{51}V	1107	^{68}Zn		^{59}Co	1570	^{48}Ti	2168	^{37}Cl					191	^{197}Au	1.5E3
940	^{18}O		^{69}Ga		^{60}Ni	1611	^{37}Cl		^{41}K					193	^{12}C	3.0E4
	^{69}Ga	1115	^{64}Ni	1335	^{68}Zn	1612	^{25}Mg	2176	^{59}Co					197	^{19}F	3.3E5
951	^{58}Ni	1117	^{52}Cr	1337	^{53}Cr	1614	^{50}Ti	2179	^{46}Ti					199	^{75}As	2.3E4
955	^{53}Cr		^{53}Cr		^{59}Co	1626	^{52}Cr	2186	^{89}Y					204	^{95}Mo	3.9E3
962	^{45}Sc	1121	^{45}Sc	1346	^{19}F	1634	^{23}Na	2209	^{35}Cl					205	^{63}Cu	1.4E4
	^{62}Ni	1131	^{61}Ni	1349	^{19}F	1635	^{13}C	2211	^{27}Al					211	^{54}Cr	1.0E5
	^{63}Cu	1139	^{69}Ga	1353	^{51}V	1640	^{60}Ni	2218	^{46}Ti						^{55}Mn	1.3E5
970	^{60}Ni	1140	^{48}Ti	1357	^{19}F	1642	^{37}Cl	2230	^{31}P						^{195}Pt	5.7E3
974	^{45}Sc		^{49}Ti	1361	^{48}Ti	1662	^{45}Sc		^{32}S					226	^{48}Ti	6.0E5
975	^{25}Mg	1154	^{48}Ti		^{49}Ti	1708	^{69}Ga		^{35}Cl					230	^{59}Co	3.1E4
980	^{41}K		^{49}Ti	1364	^{93}Nb	1717	^{52}Cr	2233	^{31}P					239	^{77}Se	1.3E4
981	^{232}Th	1157	^{44}Ca	1368	^{23}Na	1720	^{27}Al	2240	^{30}Si					244	^{60}Ni	2.0E4
986	^{48}Ti	1164	^{51}V		^{24}Mg	1727	^{37}Cl	2243	^{52}Cr					251	^{52}Cr	8.6E4
	^{49}Ti		^{59}Co		^{27}Al		^{51}V	2274	^{52}Cr					265	^{75}As	6.5E3
990	^{25}Mg	1168	^{62}Ni	1371	^{141}Pr	1745	^{88}Sr	2293	^{52}Cr					277	^{59}Co	1.2E4
991	^{127}I	1172	^{62}Ni	1379	^{56}Fe	1757	^{56}Fe	2313	^{13}C					278	^{63}Cu	1.6E4
992	^{63}Cu	1173	^{59}Co	1380	^{25}Mg		^{59}Co		^{14}N					279	^{75}As	3.8E4
	^{64}Zn	1174	^{49}Ti	1382	^{52}Cr	1763	^{34}S		^{33}S						^{197}Au	7.4E3
997	^{59}Co	1186	^{61}Ni	1384	^{28}Si		^{35}Cl	2319	^{89}Y					280	^{19}F	3.8E5
998	^{35}Cl	1190	^{59}Co	1398	^{58}Ni	1778	^{27}Al	2365	^{12}C						^{105}Pd	8.5E2
1012	^{49}Ti	1204	^{74}Ge	1400	^{14}N		^{28}Si	2391	^{23}Na					283	^{59}Co	8.9E3
1013	^{26}Mg	1206	^{68}Zn		^{53}Cr		^{31}P	2405	^{52}Cr					299	^{113}Cd	2.3E3
	^{27}Al	1208	^{90}Zr	1408	^{55}Mn	1792	^{59}Co		^{53}Cr					301	^{181}Ta	4.0E3
1022	^{48}Ti	1216	^{75}As		^{54}Fe	1798	^{47}Ti	2408	^{50}Ti					305	^{46}Ti	5.1E4
1023	^{9}Be	1220	^{34}S	1409	^{45}Sc	1799	^{63}Cu	2430	^{29}Si					311	^{109}Ag	2.7E4
1025	^{64}Ni		^{35}Cl	1410	^{37}Cl	1809	^{26}Mg	2470	^{55}Mn					325	^{107}Ag	2.4E4
1027	^{59}Co	1224	^{56}Fe		^{58}Ni	1854	^{85}Rb	2491	^{37}Cl					328	^{194}Pt	7.7E3
	^{69}Ga	1229	^{87}Rb	1412	^{62}Ni	1855	^{59}Co	2497	^{52}Cr					340	^{58}Ni	2.1E4
1029	^{232}Th	1236	^{19}F		^{63}Cu	1865	^{58}Ni	2522	^{39}K					342	^{48}Ti	7.0E4
1032	^{59}Co	1237	^{45}Sc	1416	^{52}Cr	1896	^{52}Cr	2530	^{47}Ti						^{49}Ti	
1035	^{55}Mn	1238	^{55}Mn	1422	^{49}Ti	1897	^{56}Fe	2576	^{41}K						^{111}Cd	1.3E3
1039	^{65}Cu		^{59}Co	1429	^{93}Nb	1911	^{52}Cr	2578	^{55}Mn					351	^{19}F	1.8E6
	^{66}Zn	1244	^{51}V	1432	^{59}Co	1920	^{56}Fe	2604	^{41}K					352	^{56}Fe	7.2E4

^{2}D Induced γ-Rays

E_γ keV	T	E_γ keV	T
495	^{16}O	4439	^{11}B
717	^{9}Be		^{12}C
871	^{16}O		^{14}N
1273	^{28}Si	4617	^{27}Al
1634	^{14}N	4934	^{28}Si
1779	^{27}Al	5270	^{14}N
1885	^{14}N	5299	^{14}N
2297	^{14}N	6264	^{14}N
2878	^{24}Mg	6324	^{14}N
3088	^{12}C	7286	^{14}N
	^{15}N	7301	^{14}N
3371	^{24}Mg	8313	^{14}N
3488	^{19}F	9050	^{14}N
3601	^{12}C		

^{3}T Induced γ-Rays

E_γ keV	T	Yield $\mu C.sr^{-1}$
090	^{23}Na	3.6E5
093	^{66}Zn	1.3E4
094	^{48}Ti	3.5E5
098	^{195}Pt	2.7E3
100	^{182}W	3.9E4
104	^{55}Mn	3.8E4
110	^{19}F	6.1E4
111	^{184}W	4.7E4
115	^{63}Cu	2.1E4
	^{64}Zn	1.1E4
116	^{56}Fe	9.4E4
122	^{56}Fe	2.6E4
	^{186}W	4.6E4
124	^{51}V	6.3E4
125	^{51}V	

TABLE 14.17 *(cont.)*

Triton Induced Gamma-Rays

E_γ keV	T	Yield µC.sr⁻¹
356	^{196}Pt	3.3E3
367	^{57}Fe	1.1E5
374	^{110}Pd	1.3E4
415	^{109}Ag	1.3E4
417	^{24}Mg	2.6E5
423	^{107}Ag	1.2E4
434	^{108}Pd	1.7E4
440	^{24}Mg	1.7E5
	^{77}Se	3.8E3
441	^{105}Pd	4.0E3
460	^{58}Ni	3.1E4
461	^{32}S	4.4E5
472	^{23}Na	1.4E5
478	^{7}Li	3.6E6
534	^{100}Mo	3.3E3
556	^{59}Co	3.3E4
	^{104}Pd	2.1E3
	^{114}Cd	3.3E3
559	^{76}Se	5.7E3
564	^{52}Cr	1.1E5
596	^{72}Ge	2.8E4
	^{73}Ge	
614	^{78}Se	1.1E4
617	^{112}Cd	1.6E3
656	^{19}F	1.6E5
659	^{16}O	3.9E5
665	^{32}S	1.9E5
666	^{80}Se	8.6E3
677	^{28}Si	4.0E5
709	^{28}Si	4.1E5
811	^{56}Fe	1.0E5
	^{57}Fe	
826	^{58}Ni	2.2E4
830	^{24}Mg	2.6E5
831	^{69}Ga	3.4E4
835	^{52}Cr	2.0E5
	^{53}Cr	
	^{55}Mn	8.0E5
840	^{32}S	4.5E5
877	^{23}Na	3.8E5
937	^{16}O	1.1E6
945	^{59}Co	6.0E4
1042	^{16}O	8.5E5
1081	^{16}O	4.4E5
1164	^{16}O	
1173	^{58}Ni	2.2E4
1275	^{23}Na	4.2E5
1332	^{58}Ni	8.5E4
1381	^{48}Ti	3.8E5
1381	^{49}Ti	
1554	^{48}Ti	1.8E4
	^{49}Ti	
1632	^{12}C	2.5E6
1982	^{16}O	1.1E6
2104	^{16}O	
2313	^{16}O	
2525	^{16}O	
5271	^{16}O	
5299	^{16}O	
6324	^{16}O	

α Induced γ-rays / Alpha Induced Gamma-Rays

E_γ keV	T	Yield µC.sr⁻¹
54	^{51}V	
67	^{73}Ge	2.5E3
73	^{193}Ir	
74	^{19}F	
77	^{197}Au	
78	^{29}Si	
89	^{99}Ru	6.4E2
93	^{67}Zn	
	^{178}Hf	1.2E4
93	^{180}Hf	
98	^{107}Ag	
99	^{183}W	
	^{195}Pt	1.9E2
100	^{183}W	5.7E3
103	^{109}Ag	
107	^{193}Ir	
110	^{19}F	5.1E4
111	^{184}W	5.0E3
113	^{177}Hf	1.7E3
122	^{57}Fe	4.4E2
	^{186}W	4.2E3
123	^{179}Hf	8.6E2
125	^{185}Re	2.2E3
126	^{55}Mn	2.8E4
127	^{101}Ru	5.4E2
129	^{51}V	
130	^{195}Pt	
134	^{187}Re	2.7E3
136	^{111}Cd	8.1E3
137	^{57}Fe	
139	^{193}Ir	
140	^{195}Pt	
151	^{85}Rb	
157	^{51}V	
158	^{199}Hg	
159	^{47}Ti	2.2E3
	^{185}Re	
165	^{181}Ta	
167	^{187}Re	
170	^{10}B	6.7E4
178	^{167}Er	
179	^{191}Ir	
180	^{193}Ir	
182	^{170}Er	
184	^{168}Er	
	^{166}Er	
185	^{67}Zn	
192	^{197}Au	
197	^{19}F	1.2E5
199	^{75}As	2.4E3
204	^{95}Mo	3.2E2
207	^{51}V	
208	^{199}Hg	
211	^{195}Pt	1.0E2
214	^{191}Ir	
217	^{79}Br	3.8E3
219	^{193}Ir	
230	^{57}Fe	
239	^{77}Se	1.3E3
	^{195}Pt	
261	^{79}Br	2.7E2
265	^{75}As	3.4E2
269	^{197}Au	
276	^{81}Br	2.6E3
279	^{75}As	3.1E3
	^{197}Au	
281	^{105}Pd	
285	^{185}Re	
295	^{103}Rh	2.8E3
299	^{113}Cd	
301	^{181}Ta	
	^{187}Re	
306	^{79}Br	6.0E2
312	^{109}Ag	7.1E2
313	^{39}K	
319	^{69}Ga	2.5E2
317	^{192}Pt	
320	^{51}V	2.7E3
325	^{107}Ag	5.4E2
329	^{194}Pt	
342	^{111}Cd	
343	^{191}Ir	
351	^{18}O	1.3E3
353	^{57}Fe	
356	^{196}Pt	
358	^{103}Rh	
	^{104}Ru	3.6E2
358	^{193}Ir	
362	^{103}Rh	8.8E2
362	^{193}Ir	
364	^{45}Sc	
367	^{57}Fe	
368	^{200}Hg	
374	^{110}Pd	2.2E2
397	^{79}Br	
407	^{198}Pt	
412	^{198}Hg	
415	^{109}Ag	1.2E2
417	^{23}Na	
423	^{107}Ag	93
434	^{108}Pd	
439	^{202}Hg	
440	^{23}Na	3.3E4
	^{77}Se	1.1E2
	^{110}Pd	
475	^{102}Ru	
478	^{7}Li	6.5E6
481	^{97}Mo	
523	^{79}Br	
536	^{100}Mo	
538	^{81}Br	
540	^{100}Ru	
548	^{197}Au	
556	^{104}Pd	
557	^{102}Pd	
558	^{114}Cd	
564	^{53}Cr	
583	^{19}F	2.5E4
584	^{113}Cd	
585	^{25}Mg	
598	^{10}B	
606	^{79}Br	
609	^{51}V	
617	^{112}Cd	
637	^{19}F	
658	^{110}Cd	
670	^{63}Cu	
677	^{27}Al	
709	^{27}Al	
718	^{10}B	3.1E4
749	^{48}Ti	
755	^{26}Mg	
768	^{10}B	
771	^{65}Cu	
778	^{96}Mo	
783	^{50}Cr	
786	^{95}Mo	
787	^{98}Mo	
811	^{58}Fe	
830	^{23}Na	
835	^{54}Cr	
844	^{24}Mg	
847	^{27}Al	
858	^{55}Mn	
871	^{14}N	
	^{94}Mo	
889	^{46}Ti	
891	^{19}F	
899	^{39}K	
929	^{51}V	
962	^{63}Cu	
983	^{48}Ti	
992	^{64}Zn	
1003	^{23}Na	
1015	^{24}Mg	
	^{27}Al	
1039	^{66}Zn	
1077	^{68}Zn	
1116	^{65}Cu	
1130	^{23}Na	9.5E3
1176	^{31}P	
1210	^{35}Cl	
1236	^{19}F	
1263	^{27}Al	
1266	^{28}Si	
	^{31}P	
1273	^{26}Mg	
	^{29}Si	
1275	^{19}F	2.5E5
1280	^{19}F	
1311	^{27}Al	
1332	^{27}Al	
1349	^{19}F	
1357	^{19}F	
1369	^{19}F	
1395	^{18}O	
1400	^{19}F	
1412	^{23}Na	
1454	^{27}Al	
1459	^{19}F	
1461	^{37}Cl	
1524	^{39}K	
1528	^{19}F	
1534	^{27}Al	
1552	^{27}Al	
1594	^{96}Zr	
1634	^{17}O	
1643	^{35}Cl	
1779	^{23}Na	
	^{25}Mg	
1779	^{28}Si	
1794	^{26}Mg	
1809	^{23}Na	5.1E4
1897	^{27}Al	
2028	^{26}Mg	
2081	^{19}F	

TABLE 14.17 *(cont.)*

Alpha Induced Gamma-Rays

E_γ keV	T	Yield μC·sr⁻¹	E_γ keV	T	Yield μC·sr⁻¹
2127	^{31}P		2595	^{27}Al	
2132	^{23}Na		2614	^{17}O	
2168	^{35}Cl		2839	^{25}Mg	
2236	^{27}Al		2938	^{23}Na	
	^{28}Si		3086	^{10}B	
	^{30}Si		3092	^{23}Na	
2313	^{11}B		3182	^{19}F	
2426	^{26}Mg		3304	^{31}P	
2438	^{18}O		3498	^{27}Al	
2476	^{27}Al		3684	^{10}B	
2511	^{23}Na		3770	^{27}Al	
2524	^{23}Na		3854	^{10}B	
2541	^{23}Na		4439	^{12}C	
2574	^{27}Al		6131	^{16}O	

Coulomb Excitation Gamma-Rays

E_γ keV	T	Yield μC·sr⁻¹	E_γ keV	T	Yield μC·sr⁻¹	E_γ keV	T	Yield μC·sr⁻¹	E_γ keV	T	Yield μC·sr⁻¹	E_γ keV	T	Yield μC·sr⁻¹
67	^{73}Ge	1.4E4	100	^{167}Er	8.4E3	146	^{155}Gd	6.5E3	245	^{111}Cd	4.8E2	523	^{79}Br	1.3E4
	^{171}Yb	3.4E4		^{182}W	8.9E4	159	^{47}Ti	1.4E4		^{152}Sm	1.2E3	535	^{100}Mo	7.5E4
68	^{10}B	2.1E3	101	^{173}Yb	8.5E3		^{185}Re	7.3E3	247	^{163}Dy	2.5E2	538	^{100}Ru	4.5E3
73	^{163}Dy	1.3E5	104	^{109}Ag	1.7E3	160	^{123}Sb	2.1E3	250	^{77}Se	9.0E2	544	^{197}Au	2.0E3
	^{159}Dy			^{159}Tb	3.4E3	161	^{179}Hf	1.4E3		^{165}Ho	3.9E2	547	^{101}Ru	1.9E3
75	^{157}Gd	9.5E4	108	^{153}Eu	2.3E4	162	^{77}Se	2.8E2		^{177}Hf	9.2E3	548	^{154}Sm	.6.3E2
	^{160}Gd		110	^{19}F	5.4E3	164	^{171}Yb	7.4E2	261	^{79}Br	3.5E3	556	^{104}Pd	4.0E3
76	^{171}Yb	1.9E5	111	^{184}W	1.1E5	165	^{181}Ta	2.0E4	265	^{75}As	6.6E3	559	^{76}Se	7.1E3
	^{174}Yb		113	^{177}Hf	5.5E4	167	^{163}Dy	4.2E4	268	^{179}Hf	8.9E2	560	^{120}Te	7.2E3
79	^{158}Gd	1.0E5		^{232}Th	2.6E3		^{164}Dy		276	^{81}Br	3.9E4	563	^{76}Ge	7.3E3
	^{159}Tb	1.1E5	115	^{165}Ho	4.7E4		^{187}Re	1.1E4	279	^{75}As	9.6E4	564	^{53}Cr	
	^{172}Yb		122	^{57}Fe	3.7E4	168	^{11}B	1.1E3		^{197}Au	1.9E4	572	^{75}As	2.4E4
	^{173}Yb			^{147}Sm	1.4E5	171	^{173}Yb	1.1E3	280	^{105}Pd	1.3E3	582	^{113}Cd	1.4E3
80	^{166}Er	3.5E5		^{152}Sm		172	^{127}I	3.2E3	286	^{185}Re	1.1E3	596	^{74}Ge	3.0E4
	^{167}Er			^{159}Tb	1.1E3	173	^{160}Gd	4.7E3	295	^{101}Ru	1.0E3	614	^{78}Se	1.1E4
	^{168}Er			^{171}Yb	6.4E4	175	^{167}Er	3.0E3	297	^{103}Rh	1.3E5	616	^{101}Ru	5.2E2
	^{170}Er			^{186}W	1.2E5	176	^{174}Yb	1.1E4	299	^{111}Cd	4.7E3	617	^{113}Cd	3.5E3
81	^{162}Dy	9.9E4	123	^{154}Gd	1.7E4	179	^{173}Yb		301	^{181}Ta	1.1E4	628	^{127}I	2.5E3
82	^{154}Sm	6.7E4		^{179}Hf	3.5E4	182	^{158}Gd	4.3E3		^{187}Re	1.6E3	657	^{76}Se	1.6E3
	^{176}Sm	6.9E4	125	^{185}Re	6.5E4		^{170}Er	1.6E3	304	^{151}Eu	2.9E4	658	^{110}Cd	1.1E3
83	^{153}Eu	1.7E5	126	^{55}Mn	1.7E5		^{172}Yb		307	^{79}Br	1.2E4	666	^{80}Se	1.3E4
84	^{170}Yb		127	^{101}Ru	9.3E3	183	^{159}Tb	1.2E3		^{101}Ru	7.2E2		^{126}Te	1.0E3
86	^{155}Gd	1.7E4	130	^{195}Pt	7.6E2	184	^{101}Ru	1.5E3	311	^{109}Ag	5.3E4	669	^{63}Cu	2.8E3
87	^{160}Dy	9.4E3	131	^{150}Nd	6.5E4		^{166}Er	8.9E3	319	^{69}Ga	6.6E3	717	^{10}B	
88	^{77}Se	4.3E2		^{157}Gd	3.9E3	185	^{154}Sm	2.3E3		^{105}Pd	8.5E2	720	^{101}Ru	5.6E2
	^{176}Hf	1.8E4	134	^{187}Re	1.1E5		^{162}Dy	4.5E3	320	^{51}V	3.6E4	743	^{128}Te	6.9E2
89	^{156}Gd	9.7E4	135	^{165}Ho	2.8E3	188	^{176}Yb	1.5E3	325	^{107}Ag	4.5E4	744	^{93}Nb	8.6E2
90	^{99}Ru	6.3E3	136	^{181}Ta	2.8E5	191	^{153}Eu	6.3E4	328	^{149}Sm	3.2E2	778	^{96}Mo	9.2E2
91	^{164}Er	1.3E4	137	^{159}Tb	1.4E4		^{197}Au	3.4E3		^{194}Pt	2.1E4	787	^{98}Mo	1.3E3
93	^{163}Dy	5.5E3		^{177}Hf	2.5E3	194	^{151}Eu	1.6E4	334	^{150}Sm	5.4E3	809	^{93}Nb	4.5E2
	^{180}Hf	2.2E5	138	^{156}Dy	1.9E3	196	^{165}Er	2.9E2	342	^{111}Cd	3.4E3	822	^{75}As	4.2E3
95	^{165}Ho	2.9E5	140	^{195}Pt	3.1E3	197	^{19}F	1.3E5	350	^{11}B	3.4E3	835	^{72}Ge	2.7E3
98	^{107}Ag	1.6E3	145	^{179}Hf	3.0E3		^{147}Sm	1.0E3	352	^{57}Fe	1.1E3	844	^{27}Al	2.3E3
99	^{195}Pt	6.9E3	146	^{11}B	5.5E2		^{160}Dy	5.5E2	356	^{196}Pt	1.1E4	871	^{94}Mo	2.1E2
100	^{158}Dy	1.2E3		^{127}I	1.8E3	198	^{101}Ru	4.1E2	358	^{104}Ru	4.3E4	889	^{46}Ti	1.4E3
						199	^{75}As	3.2E4	359	^{103}Rh	9.1E4	892	^{7}Li	4.1E4
							^{156}Gd	3.0E3	374	^{110}Pd	2.6E4	983	^{48}Ti	6.4E3
							^{195}Pt	5.5E2	375	^{127}I	1.3E3	1014	^{27}Al	1.6E3
						200	^{77}Se	3.8E3	397	^{79}Br	1.1E3	1369	^{24}Mg	1.5E3
						202	^{176}Hf	5.8E2	407	^{176}Hf	1.5E3	1584	^{7}Li	7.1E4
						203	^{127}I	2.5E4	415	^{109}Ag	2.7E4			
						204	^{95}Mo	7.5E3	418	^{127}I	1.3E3			
						208	^{163}Dy	1.4E3	422	^{101}Ru	1.9E2			
						210	^{165}Ho	5.7E3	423	^{107}Ag	2.4E4			
							^{164}Er	2.0E2	434	^{108}Pd	3.2E4			
						211	^{195}Pt	1.4E4	440	^{23}Na	1.8E5			
						215	^{167}Er	2.5E2		^{77}Se	4.5E3			
						217	^{79}Br	3.8E4	441	^{105}Pd	7.6E3			
						225	^{159}Tb	5.6E2	469	^{75}As	1.4E3			
						236	^{11}B	1.6E3		^{102}Ru	2.5E4			
						239	^{77}Se	2.2E4	478	^{7}Li				
							^{195}Pt	4.3E3	481	^{97}Mo	4.0E2			

TABLE 14.18 Proton Induced Gamma-ray Yields

REACTION	E_γ (keV)	YIELD (γs μC^{-1} sr^{-1})											
		1.0	1.7	2.4	3.1	3.8	4.2	3.5	4.0	4.5	5.0	6.0	9.0
^{7}Li n1,0	429				9.2E6	2.6E7	3.3E7	1.1E8	1.5E8	1.5E8	1.6E8	2.0E8	2.2E8
p1,0	477	6.5E5	8.6E6	2.6E7	5.6E7	8.9E7	1.1E8	4.5E8	4.9E8	5.2E8	5.7E8	6.2E8	2.1E9
^{9}Be γ3,2	414	4E2	6E2	6E2				9.3E3	1.0E4	1.4E4	1.8E4	3.6E4	
γ1,0	717	1.2E3	3.5E3	5.3E3				2.1E3	3.6E3	5.3E3	7.8E3	1.7E4	
γ2,1	1023	4E2	8E2	1.3E3				1.3E3	1.9E3	2.1E3	4.6E3	1.2E4	
γ1,0	3562		1E2	2.5E4	2.5E6	5.1E6	6.2E6	7.8E3	9.4E3	1.1E4	1.4E4	2.2E4	1.9E8
γr,0	7477	3.3E3	3.0E3										
^{10}B α1,0	429	1.4E4	9.1E5	3.5E6	7.2E6	8.3E6	8.3E6	2.1E7	2.2E7	2.7E7	3.3E7	3.7E7	
p1,0	717		4E3	1.2E5	1.3E6	2.6E6	4.7E6	6.8E6	9.0E6	1.5E6	1.9E7	2.2E7	1.5E8
^{11}B p1,0	2125				4.8E6	1.3E7	1.5E7	3.5E7	4.5E7	4.8E7	7.4E7	1.4E8	4.3E8
^{12}C γ1,0	2365	2.6E2	2.2E2	2.3E2	2.7E2					2.9E3			
2,0	3511		2.3E2	2.1E2				5.8E4	6.2E4	7.5E4	8.3E4	1.3E5	
p1,0	4439												2.9E9
^{13}C γ2,1	1635	6.0	9.0	26	35					1.5E4			
γ1,0	2313	17	26	30	1.4E2			1.5E4	2.0E4	2.4E4	3.2E4	5.4E4	
p1,0	3089					6.2E3	4.1E4	2.6E4	3.3E4	5.4E4	6.5E4	9.5E4	
pr,0	8062	25	25										
^{14}N p1,0	2313					5.5E4		3.4E5	4.6E5	5.2E5	6.0E5	2.4E6	4.2E7
^{15}N γ2,0	429							5.5E5	6.3E5	7.1E5	1.1E6	1.3E6	
α1,0	4439	1.2E2	7.7E2	5.0E3	5.0E4	6.0E4							
^{16}O γ1,0	496	25	2.5E2	9.5E2	7.3E2	1.1E3	2.2E3	1.6E5	2.2E5	2.7E5	3.1E5	5.8E5	
pl,0	6129												1.8E9
^{17}O / ^{18}O p1,0	871			35	1.3E3	3.8E3	7.1E3	1.7E5	2.7E5	3.8E5	6.1E5	1.1E6	
n1,0	937							2.2E4	2.1E5	9.0E5	1.5E6	3.5E6	
n2,0	1042							1.6E4	8.8E4	1.2E6	3.1E6	9.7E6	
n3,0	1081							2.6E4	1.1E5	2.7E5	6.1E5	1.4E6	
p1,0	1982				2.1E3	1.5E4	2.7E4	5.0E5	2.0E6	3.0E6	5.1E6	9.6E6	
^{19}F p1,0	110	2E4	1.2E5	3.5E5	7.2E6	1.1E7		2.1E7	4.4E7	8.5E7	9.7E7	2.0E8	
p2,0	197	4E4	2.3E5	2.9E6	2.9E6	3.7E7		1.5E8	3.0E8	4.8E8	5.7E8	7.7E8	
p3,1	1236			1.5E5	3.0E6	5.4E6		3.1E7	4.9E7	5.5E7	6.6E7	9.3E7	3.0E8
p4,1	1349				1.3E6	2.1E6		1.6E6	2.0E6	2.4E6	2.9E6	3.6E6	7.7E8
p5,2	1357			1.0E5	1.4E6	4.2E6							
p4,0	1459				9.4E5	3.9E6							
α2,0	6129	5.3E5	2.6E6		6.0E6	6.7E7	9.5E7	5.6E5	7.1E5	8.8E5	1.3E6	1.7E6	9.4E8
α3,0	6917				1.8E4	2.3E5	2.7E5						
α4,0	7114			5.4E3	4.4E4	4.6E4							
^{23}Na p1,0	439	2.2E3	8.3E5	3.4E6	9.6E6	1.6E7	1.5E7	3.7E7	5.0E7	9.3E7		2.2E8	6.7E8
γ1,0	1368	5E2	4.6E3										
α2,1	1634	4E2	2.3E5	1.5E6	9.9E6	1.9E7	1.9E7	1.1E7	4.9E7	8.1E7	1.2E8	1.9E8	4.6E8
^{24}Mg γ1,0	417	1E2	3E2										
p1,0	1368			1.5E5	9.3E5	5.1E6	8.5E6	2.2E6	9.6E6	2.1E7	3.3E7	8.8E7	9.3E8
^{25}Mg p2,1	390			2.4E4	6.2E4	2.5E5	2.8E5	5.4E5	8.6E5	1.3E6	1.8E6	3.7E6	
γ1,0	452	2E2	3E2										
p1,0	586		5.1E3	6.7E4	2.2E5	8.6E5	9.9E5	1.8E6	2.4E6	3.2E6	4.9E6	1.1E7	
p2,0	976			2.2E4	1.2E5	4.6E5	5.2E5	5.1E5	8.4E5	1.4E6	2.0E6	3.8E6	
p3,0	1612				7.6E4	9.0E5	1.2E6	8.1E5	1.9E6	2.8E6	3.7E6	6.5E7	
^{26}Mg pl,0	1809												8.6E7
^{27}Al p1,0	843		8.4E3	1.5E5	2.3E6	5.5E6	5.1E6	4.0E6	6.5E6	8.1E6	1.5E7	3.1E7	3.8E8
p2,0	1013		1.1E3	3.3E5	4.6E6	1.1E7	1.2E7	1.1E7	1.6E7	2.3E7	3.4E7	7.4E7	8.9E8
α1,0	1368		2.0E3	3.5E4	5.7E5	2.4E6	3.7E6	3.3E6	6.2E6	1.0E7	1.5E7	2.8E7	7.6E8
p4,2	1720												2.4E8

TABLE 14.18 *(cont.)*

REACTION		E_γ (keV)	YIELD (γ's μC⁻¹ sr⁻¹)											
			1.0	1.7	2.4	3.1	3.8	4.2	3.5	4.0	4.5	5.0	6.0	9.0
^{27}Al	γ1,0	1778	1.1E3	5.0E3	8.4E3	1.0E4					4.5E6			
	p3,0	2211				1.1E4	8.1E5	2.5E6	1.6E6	4.5E6	8.8E6	1.2E7	2.1E7	4.2E8
	p6,0	3004												5.4E8
^{28}Si	p1,0	1778			2.3E2	1.2E6		7.2E6	1.7E7	2.7E7	3.2E7	6.4E7	8.1E7	4.7E8
^{29}Si	p2,1	755				1.0E5	6.3E5		2.8E5	4.2E5	7.2E5	1.1E6	1.6E6	
	p1,0	1273				1.8E3	1.2E5	7.6E5		1.6E6	2.4E6	3.6E6	5.8E6	
	p2,0	2028				2.8E3	5.3E4		9.6E4	2.6E6	6.9E5	1.6E6	2.2E6	1.1E7
^{30}Si	p1,0	2235			1.3E2	4.7E3	3.4E4		6.6E4	1.9E5	5.7E5	6.1E5	7.9E5	1.2E7
^{31}P	pl,0	1266		7.2E2	3.8E4	1.6E6	5.2E6	2.3E7	2.6E7	3.0E7	3.9E7	5.2E7	1.0E8	1.7E8
	α1,0	1778			2.0E3	2.1E5	6.5E5	1.6E6	2.6E6	3.2E6	7.7E6	1.3E7	2.6E7	8.1E7
	γ1,0	2230	1.3E2	1.7E3	3.5E3	1.2E4	4.0E5	9.5E5	2.4E6	4.3E6	7.6E6	1.8E7	3.7E7	1.3E8
^{32}S	p1,0	2230				5.3E3	1.5E5	8.9E5	3.3E6	1.1E7	1.6E7	3.6E7	4.9E7	1.5E8
^{33}S	p1,0	841		9	1.4E2	8.7E3	1.7E4	2.9E4	2.8E5	9.6E5	1.4E6	3.3E6	9.3E6	
^{34}S	γ1,0	1220	10	37	65	4.3E3	4.6E4	2.8E5						
	p1,0	2127					2.1E4	4.8E4	2.7E4	3.2E5	8.3E5	1.4E6	2.0E6	
^{35}Cl	p1,0	1220			3.5E3	2.2E5	1.5E6		1.9E6	5.1E6	1.1E7	1.2E7	2.4E7	8.6E8
	p2,0	1763				5.2E4	6.8E5		7.1E5	2.6E6	7.4E6	1.2E7	1.8E7	1.6E8
	γ1,0	1970	2.2E2	4.5E2	1.1E3	3.3E3								
^{37}Cl	n1,0	1410					2.1E5		3.5E4	4.2E5	2.0E6	3.3E6	9.8E6	
	n2,0	1611					1.9E5		7.0E4	4.2E5	2.7E6	4.5E6	1.0E7	
	γ2,1	1642	30	4.9E2	1.0E3									
	α1,0	2127			1.2E3	3.1E4	1.8E5				1.8E6			
	γ1,0	2168	1.6E2	1.6E3	2.9E3	3.9E3	5.3E3							
^{39}K	α1,0	1970							2.1E4	6.6E4	1.1E5	1.9E5	8.1E5	3.1E7
	p2,0	2814												1.0E8
^{41}K	p1,0	980			5.2E2	1.1E4	5.5E4	9.3E4	8.3E3	7.4E4	1.4E5	2.6E5	4.2E5	
	n1,0	1943					8.0E4	1.4E5						
	α1,0	2168			2.0E2	2.6E3	2.5E4	9.5E4	1.6E5					
^{40}Ca	p2,0	3736												2.7E8
	p3,0	3904												2.8E8
^{42}Ca	p1,0	1525					5.9E2	7.7E3	3.9E4	6.2E4	1.4E5	2.4E5	5.3E5	
^{43}Ca	p1,0	373		32	8.7E2				6.4E4	1.2E5	1.8E5	2.6E5	4.8E5	
^{44}Ca	p1,0	1157							1.3E5	2.8E5	5.6E5	1.4E6	1.9E6	
^{48}Ca	n2,1	122							9.5E4	1.1E5	2.9E5	3.4E5	7.9E5	
	n1,0	131							1.3E5	2.5E5	3.1E5	6.1E5	1.1E6	
	n2,1	371					3.7E3	1.6E4						

REACTION		E_γ (keV)	YIELD (γ's μC⁻¹ sr⁻¹)					REACTION		E_γ (keV)	YIELD (γ's μC⁻¹ sr⁻¹)				
			2.4	3.5	4.5	6.0	9.0				2.4	3.5	4.5	6.0	9.0
^{45}Sc	n3,1	293		8.3E4	4.0E6	1.7E7		Ti		226		2.2E5	1.6E6	8.8E6	
	p2,1	364	3.6E4	1.0E6	3.2E6	1.3E7				308	1.0E3		1.8E4		1.5E8
	p3,1	531	2.5E4	8.7E5	3.9E6	6.5E6				320	1.4E3		6.4E4		
	p3,0	543		5.7E5	2.7E6	5.8E6				658	1.6E3		1.2E5		
	p4,0	720	2.5E4	1.2E6	3.8E6	7.5E6				889	1.4E3	5.6E5	2.0E6	1.3E7	2.6E7
	γ1,0	889	6.4E4	6.0E5	2.1E6	8.2E6				986	8.5E3	1.3E6	1.8E7	8.5E7	1.3E8
	p6,0	974		5.4E5	2.5E6	5.2E6				1140	1.0E3				
	p8,0	1237		5.3E5	3.2E6	9.9E6		^{51}V	n5,3	316			2.3E6	3.3E6	8.1E6
	p10,0	1409		2.3E5	1.5E6	5.2E6			p1,0	320	1.8E5	2.8E6	4.1E6	1.1E7	2.5E8
Ti		90		1.3E5	2.9E6	9.7E6			n4,1	603		1.0E6	3.6E6	9.8E6	
		160		4.1E5	2.1E6	7.4E6			p2,1	609		3.9E5	3.6E6	7.9E6	

TABLE 14.18 (cont.)

REACTION	E_γ (kev)	YIELD (γ's μC^{-1} sr^{-1}) 2.4	3.5	4.5	6.0	9.0
^{51}V n1,0	749	1.4E3	1.6E5	3.4E6	4.0E6	5.4E8
n6,1	808		3.3E6	5.2E6	1.4E7	2.2E8
γ2,1	936	3.5E3		3.2E5		
n3,0	1164		8.4E5	1.4E6	3.2E6	4.9E8
γ3,1	1332	3.4E3		3.5E4		
n4,0	1353		9.5E4	1.1E5	7.0E6	
γ1,0	1434	1.2E4		5.0E5		
n5,0	1480		4.1E5	5.0E5	1.1E7	1.7E8
^{50}Cr p1,0	783	8.4E2	2.6E5	3.1E6	6.2E6	
^{52}Cr γ1,0	379	1.9E4	1.0E6	8.1E6	2.7E7	
γ2,1	913	1.0E3	3.6E5	1.8E6	6.2E6	
γ2,0	1292	1.0E3	7.8E5	2.3E6	6.2E6	
p1,0	1434		1.1E7	3.7E7	7.9E7	3.3E8
^{53}Cr p1,0	565	7.5E2		7.0E5		
^{55}Mn p1,0	126	1.4E4	2.0E6	6.0E6	1.4E7	
n1,0	413	1.3E4	2.2E6	7.6E6	1.6E7	1.3E8
n4,2	477		1.7E7	6.0E6	2.1E7	
γ1,0	847	3.3E3		6.7E5		
n2,0	933	3.5E4	2.5E7	3.4E7	4.2E7	4.7E8
n3,0	1316		4.8E6	1.3E7	1.7E7	3.1E8
n4,0	1408		1.9E6	5.4E6	1.4E7	9.9E7
^{56}Fe p1,0	847	3.9E3	2.0E6	2.2E7	1.0E8	7.6E8
γ2,0	1378	2.1E3	8.0E4	5.9E5	2.2E6	
^{59}Co n1,0	340	1.7E4	1.2E6	1.5E7	3.0E7	1.9E8
n3,0	879		3.1E5	2.2E6	8.3E6	1.9E7
n5,1	997		3.8E5	4.5E6	1.2E7	4.8E7
p2,0	1190		5.1E5	4.9E6	1.7E7	
γ1,0	1332	1.9E4		7.5E5		3.4E7
n6,1	1337		4.4E5	3.7E6	8.7E6	
p4,0	1432		3.7E5	3.4E6	8.4E6	3.8E7
^{58}Ni p1,0	1454		4.1E5	4.4E6	3.8E7	8.8E8
^{60}Ni p1,0	1332	15	9.8E4	2.1E6	2.0E7	3.3E8
^{62}Ni p1,0	1172		4.0E4	3.2E5	2.0E6	
^{63}Cu p1,0	669	5.9E2		1.2E7		
γ2,1	807	2.8E3		1.0E6		
α2,1	827	2.2E3		5.3E5		
p2,0	962	4.6E2	9.1E5	1.9E7	5.0E7	3.4E7
γ1,0	992	8.8E3	8.4E5	4.4E6	2.0E7	
	1318	1.5E3		9.1E5		
p3,0	1327		1.2E6	9.9E6	3.2E7	1.5E7
α1,0	1329		1.6E6	5.6E6	1.2E7	1.4E7
p4,0	1412		6.3E5	4.3E6	9.9E6	
γ2,0	1799	1.2E3		4.3E5		
^{65}Cu n1,0	115		1.1E6	6.4E6	2.1E7	
n3,1	152		2.3E6	4.8E6	9.6E6	
n4,0	770		1.3E6	4.5E6	1.7E7	
n5,0	865		9.4E5	4.3E6	2.6E7	
γ1,0	1039	4.9E3		6.5E5		
^{64}Zn p1,0	992	1.6E2	2.1E6	1.0E7	2.1E7	7.6E8
^{66}Zn p1,0	1039		7.8E5	5.2E6	1.3E7	2.1E8
^{67}Zn γ1,0	175		3.8E4	1.2E6	2.3E6	
n2,0	360	1.1E3	3.3E4	1.5E6	2.0E6	

REACTION	E_γ (kev)	YIELD (γ's μC^{-1} sr^{-1}) 2.4	3.5	4.5	6.0	9.0
^{68}Zn γ1,0	320	7.3E2	2.4E4	9.3E4	1.5E6	
p5,0	874	5.5E2		3.0E5		
^{70}Zn n1,0	508		4.5E4	1.3E6	2.4E6	
^{69}Ga γ2,1	175	1.2E4	7.0E5	2.3E6	1.7E7	
n2,0	233		1.3E5	7.9E5	8.7E6	
n3,2	287		5.7E4	9.7E5	1.1E7	
p1,0	319	1.5E3		2.5E5		
p4,0	398		2.0E5	5.4E5	1.2E6	
p7,4	500	4.3E3	4.3E4	9.5E5	1.7E6	
γ3,1	668	3.0E3		1.0E5		
p4,0	1027		4.1E5	1.5E6	4.3E6	
γ1,0	1040	1.2E4	1.4E5	6.0E5	1.1E6	
^{71}Ga n	327	3.0E3				
p1,0	390	1.7E3	1.5E5	5.6E5	2.1E6	
^{70}Ge p1,0	1040					2.1E8
^{73}Ge n5,0	510		1.2E5	1.5E6	2.7E6	
^{74}Ge p	595	5.6E3				
^{75}As n1,0	113	4.8E3	2.9E5	3.2E5	1.5E7	
n5,3	141		2.2E4	1.4E6	3.2E6	
p1,0	199	4.3E3		1.9E5		
p3,0	280	8.8E3	5.4E3	3.3E5	1.6E6	
n3,0	286	7.4E3	4.7E5	4.9E6	2.0E7	
n5,0	428		2.2E4	8.3E5	2.0E6	
^{77}Se p	239	4.1E3				
^{78}Se p1,0	614	2.4E3	2.1E5	1.2E6	1.8E6	
^{80}Se α2,0	216		5.3E4	1.0E6	2.7E6	
^{79}Br n4,2	144		8.5E4	7.4E5	3.0E6	
n3,0	183		1.0E5	1.0E6	4.2E6	
p2,0	217	6.2E3		2.1E5		
n5,0	384		3.5E4	4.9E5	3.7E6	
γ1,0	617	3.9E3		1.2E5		
^{81}Br n1,0	190		9.5E4	1.0E6	3.8E6	
p1,0	276	6.2E3		1.5E5		
n8,3	445		5.0E4	9.0E5	4.2E6	
^{85}Rb n1,0	232	5.0E3	2.4E5	1.3E6	6.0E6	
n3,2	504		5.5E4	4.4E5	2.0E6	
n5,1	537		5.6E4	3.2E5	1.4E6	
n9,5	550		4.1E4	3.6E5	3.3E6	
^{87}Rb γ4,2	485	3.5E2	5.7E4	3.4E5	1.6E6	
^{87}Sr γ1,0	232		5.0E4	3.0E5	6.1E5	
^{88}Sr p	1507	2.2E2				
^{89}Y n1,0	588		4.3E4	2.6E5	8.7E6	3.3E8
p	2186	6.5E2				
^{90}Zr p	1208	75				
^{90}Zr p2,0	2186					4.2E7
^{91}Zr n1,0	105		3.5E4	2.0E5	5.7E5	
^{93}Nb p1,0	686					1.0E8
γ1,0	871	2.7E2		9.0E4		
n2,0	1364		2.2E5	1.3E6	2.8E6	9.3E7
γ5,1	1429		2.2E5	1.4E6	3.0E6	
n3,0	1477					2.8E8
γ6,1	1530		2.1E5	1.4E6	2.6E6	

TABLE 14.18 *(cont.)*

REACTION		E_γ (keV)	YIELD (γs μC^{-1} sr^{-1}) 2.4	3.5	4.5	6.0	9.0
^{95}Mo	p1,0	204	9.4E2		8.0E4		
	n1,0	336					5.3E7
	n3,0	630		3.1E4	1.8E5	1.0E6	3.3E7
^{100}Mo	p1,0	536	9.3E2		1.5E4		
^{102}Ru	p1,0	475	3.8E3	1.3E5	3.3E5	1.2E6	
^{104}Ru		358	7.4E3	1.6E5	5.0E5	1.4E6	4.0E7
^{103}Rh	n1,0	119		1.4E5	6.2E5	2.4E6	
	p3,0	296	1.6E4	1.2E5	5.9E5	7.8E5	
	p4,0	358	1.2E4	2.1E5	6.9E5	1.0E6	
^{108}Pd	p1,0	434	4.8E3	3.1E5	2.7E6	6.4E6	9.0E6
^{110}Pd	p1,0	374	4.0E3	2.4E5	1.3E6	2.3E6	
^{107}Ag	p2,0	325	6.0E3	9.3E4	2.9E5	5.6E5	
	n3,0	365					9.0E7
	p3,0	425	3.6E3	1.1E5	2.8E5	4.9E5	
^{109}Ag	p3,0	309	7.3E3	1.1E5	2.3E5	5.8E5	
	p4,0	414	4.2E3	1.1E5	2.6E5	6.7E5	
^{111}Cd	p4,0	621		4.3E5	2.7E5	6.3E5	
^{112}Cd	p1,0	617	4.5E2	2.3E4	9.2E5	3.9E5	
^{113}Cd	p2,0	298	7.7E2		1.8E4		
^{114}Cd	n	210					3.1E8
	p1,0	558	9.8E2	2.5E4	1.2E5	4.2E5	
^{115}In	n2,1	115	60	2.5E4	1.5E5	6.2E5	
	n1,0	497	30	1.0E5	8.7E5	1.6E6	4.3E8
Sn		270		1E4	1.5E5	1.9E6	
		643		8E4	1.9E5	1.6E6	
		700		3E4	2.4E5	1.9E6	5.4E7
		719		5E3	1.0E5	1.7E6	
	p1,0	1294					1.1E8
^{121}Sb	n1,0	212					5.8E8
^{123}Sb	p1,0+	160	3.2E2		2.2E5		
	n3,0+	440	2.9E5	1.5E6	3.6E6		
^{126}Te	p1,0	666	5.7E2	3.2E4	3.9E5	6.2E5	
^{128}Te	p1,0	743		3.7E4	4.0E5	8.9E5	
^{130}Te	p1,0	839		1.6E4	2.3E5	7.8E5	
^{127}I	p2,0	204	2.2E3		2.1E5		
	pl,0	375					1.2E8
	p10,0	991		3.6E5	2.0E6	1.2E7	
^{133}Cs	p2,0	161	6.2E2	1.0E4	6.5E4	2.1E5	
	p3,1	303	4.1E2	2.5E4	4.4E4	2.5E5	
	p6,0	633		5.4E4	1.4E5	3.2E5	
^{136}Ba	p8,1	1309		1.7E4	2.6E5	8.4E5	
^{137}Ba	p1,0	279	75		2.3E4		
^{139}La	n3,0	754		3.2E3	1.8E4	6.2E4	
^{138}Ce	n2,0	200		8.8E2	2.1E4	4.1E4	
^{141}Pr	p1,0	145	55		2.6E3		
	n5,1	1371		4.0E4	2.4E5	7.4E5	
^{148}Nd	p1,0	302	1.0E3	3.3E4	7.8E4	2.1E5	
^{150}Nd	p1,0	130	4.4E3	4.8E4	1.1E5	2.8E5	
^{150}Sm	p1,0	334	9.9E2	1.7E5	5.0E5	1.3E6	

REACTION		E_γ (keV)	YIELD (γs μC^{-1} sr^{-1}) 2.4	3.5	4.5	6.0	9.0
^{152}Sm	p1,0	122	3.4E4	2.0E5	3.8E5	2.6E6	
^{154}Sm	p	82	1.5E4				
^{153}Eu	p1,0	83	3.0E4	2.0E5	4.8E5	2.1E6	
^{154}Gd	p1,0	123	2.2E3	4E3	1.6E5	3.8E4	
^{156}Gd	p1,0	89	1.7E4				
	p1,0	297					3.4E7
^{158}Gd	p	80	9.0E3				
^{159}Tb	n3,1	121		8E2	1.9E5	9.4E5	
^{163}Dy	p2,0	167	3.3E3	8.7E4	3.3E5	6.1E5	
^{165}Ho	p1,0	95	1.2E5	8.1E4	1.5E5	3.2E5	
	p2,1	115	1.6E4	7.5E4	2.2E5	3.6E6	
^{167}Er	p2,0	178		4E3	3.2E4	5.7E4	
^{169}Tm	p2,1	110	1.8E5	2.5E5	5.2E5	1.1E6	
^{175}Lu	p1,0	114	1.6E4	3.1E5	7.3E5	2.0E6	
	p2,1	138	2.4E3	1.0E5	2.9E5	1.4E6	
	p2,0	251	1.6E3	2.0E5	5.6E5	1.5E6	
^{174}Hf	p2,1	207		3.4E5	3.4E6	1.1E7	
^{177}Hf	p	113	9.0E3				
^{178}Hf	p1,0	93	4.2E4		4.2E5		
	γ3,0	181		9.3E4	2.1E6	1.0E7	
	n3,0	365					3.1E7
^{183}W	p5,0	292					6.0E6
^{179}Hf	p1,0	123	4.9E3		5.1E5		
^{181}Ta	p2,0	136	1.6E5	7.2E5	8.1E5	1.7E6	
^{184}W	p1,0	111	2.1E4	1.4E5	3.0E5	1.7E6	
^{186}W	p1,0	122	2.2E4	7.4E4	4.8E5	1.3E6	
^{185}R	p1,0	125		4.2E5	1.0E6	2.7E6	
^{187}Re	p1,0	134		3.1E5	1.2E6	2.5E6	
^{190}Os	p1,0	187		5E4	1.6E5	9.0E5	
^{192}Os	p1,0	206		1.1E5	2.3E5	9.1E5	
^{193}Ir	p3,0	139		4.6E4	1.2E5	3.1E5	
^{194}Pt	p1,0	328	1.9E3	6.5E5	1.4E6	3.3E6	3.4E7
^{195}Pt	p4,0	211	1.2E3	1.9E5	2.9E5	9.0E6	
	n3,0	262					4.7E7
^{196}Pt	p1,0	356	9.3E2		9.6E5	2.3E6	
^{197}Au	p3,0	279	1.3E3	6.0E4	2.4E5	3.8E5	
	p6,0	548					4.0E6
^{199}Hg	p1,0	158	4.6E2		2.0E4		
^{202}Hg	p1,0	440		4.5E4	2.4E5	7.6E5	
^{203}Tl	p1,0	279	2.3E2	2.3E4	1.3E5	2.7E5	
^{205}Tl	p	204	6.7E2				
^{207}Pb	p3,1	571					6.6E6
^{208}Pb	n1,0	443		7.3E3	2.0E4	3.7E4	
^{209}Bi	p1,0	896		7.3E3	8.2E4	2.7E5	
^{232}Th	p19,2	981		2.2E6	3.7E6	5.7E6	
	p15,1	1029		2.0E6	3.8E6	7.1E6	
^{235}U	n7,4	109		4.0E4	1.5E5	2.7E5	
	p	189		3.5E4	7.0E5	2.1E5	

TABLE 14.19
Ion activation catalog

Target	REACTION	E_T / MeV	$\tau_{1/2}$	E_γ / keV	T	RI	$\tau_{1/2}$	E_γ / keV	I_γ *	T	RI	$\tau_{1/2}$	E_γ / keV	I_γ *
Light element activation														
^{1}H	$(^7\text{Li},n)^7\text{Be}$	13.2	53d	478	V	^{51}Cr	28d	320	3.9E2	I	^{127}Xe	36d	203	3.3E2
	$(^{10}\text{B},\alpha)^7\text{Be}$	12.7	53d	478	Cr	^{52}Mn	21m	1434	6.8E5	Ba	^{135}La	20h	481	12
	$(^{10}\text{B},n)^{10}\text{C}$		19s	720	Fe	^{56}Co	77d	847	4.3E2	W	^{182}Re	2.7d	192	10
	$(^{11}\text{B},n)^{11}\text{C}$		20m	511	Ni	^{60}Cu	23m	826	1.9E4	Re	^{185}Os	94d	646	27
	$(^{16}\text{O},\alpha)^{13}\text{N}$		10m	511				1333	9.5E4	Ir	^{191}Pt	2.9d	539	32
	$(^{19}\text{F},n)^{19}\text{Ne}$		18s	511				1792	5.7E4	Pt		6.2d	356	72
^{2}D	$(^7\text{Li},p)^8\text{Li}$	4.4	0.8s		Cu	^{63}Zn	38m	670	6.0E4	Au	^{196}Au	24h	134	1.1E2
	$(^{11}\text{B},p)^{12}\text{B}$	7.4	20ms	4400				962	5.2E4	Hg	^{197}Hg	5.3h	412	3.3E2
	$(^{19}\text{F},p)^{20}\text{F}$		11s	1630	Zn	^{66}Ga	9.4h	834	1.4E3		^{198}Tl	1.1d	368	2.3E2
	$(^{22}\text{Ne},p)^{23}\text{Ne}$		38s	440				1039	1.1E4	Tl	^{200}Tl	2.2d	279	1.5E2
^{4}He	$(^{10}\text{B},n)^{13}\text{N}$		10m	511	Ga	^{68}Ga	1.1h	1077	7.2E3	Pb	^{203}Pb	6.2d	803	48
^{6}Li	$(d,n)^7\text{Be}$		53d	478		^{69}Ge	1.6d	574	2.8E3		^{206}Bi		881	37
	$(^{14}\text{N},d)^{18}\text{F}$		110m	511				872	3.0E3					
^{7}Li	$(p,n)^7\text{Be}$	1.9	53d	478	Ge	^{70}As	52m	1107	8.3E3					
	$(d,2n)^7\text{Be}$	1.9	53d	478				745	1.0E4					
	$(^{12}\text{C},n)^{18}\text{F}$		110m	511				1040	4.5E4					
	$(^{14}\text{N},t)^{18}\text{F}$		110m	511				1115	9.5E3					
^{9}Be	$(t,n)^{11}\text{Be}$		14s	2125	As	^{72}As	1.1d	834	1.0E4					
	$(\alpha,2n)^{11}\text{C}$	18.8	20m	511		^{76}As	1.1d	559	2.4E3					
	$(^{14}\text{N},\alpha n)^{18}\text{F}$		110m	511	Se	^{76}Se	120d	136	3.5E2					
	$(^3\text{He},n)^{11}\text{C}$		20m	511				264	3.3E2					
^{10}B	$(p,\alpha)^7\text{Be}$	–	53d	478	Br	^{80}Br	18m	666	1.8E5					
	$(d,n)^{11}\text{C}$	–	20m	511		^{79}Kr	1.5d	261	2.7E3					
	$(\alpha,n)^{13}\text{N}$	–	10m	511				398	2.0E3					
	$(^9\text{Be},n)^{18}\text{F}$		110m	511				606	1.9E3					
^{11}B	$(p,n)^{11}\text{C}$	3	20m	511	Rb	^{85}Sr	1.2h	151	2.1E4					
	$(d,2n)^{11}\text{C}$	5.9	20m	511		^{87}Sr	2.8h	232	1.4E5					
	$(^9\text{Be},2n)^{18}\text{F}$		110m	511				389	3.5E4					

3.5 MeV triton activation

T	RI	$\tau_{1/2}$	E_γ / keV	I_γ *
Be	^{11}Be	14s	2125	4.1E3
B	^{11}C	20m	511	1.7E4
N	^{15}O	2.0m	511	3.3E4
O	^{18}F	110m	511	6.4E5
F	^{20}F	11s	1633	6.8E5
Na	^{25}Na	60s	391	1.5E4
			586	1.5E4
			975	2.7E4
Mg	^{24}Na	15h	1612	1.6E4
	^{28}Al	2.3m	1369	4.4E3
	^{27}Mg	9.5m	1779	4.8E4
			844	1.5E4
Al	^{28}Al	2.3m	1014	5.9E3
	^{29}Al		1779	7.7E4
		6.6m	1273	1.3E5

TABLE 14.19 (cont.)
Ion activation catalog

REACTION	E_T /MeV	$\tau_{1/2}$	E_γ /keV	T	RI	$\tau_{1/2}$	E_γ /keV	I_γ *	T	RI	$\tau_{1/2}$	E_γ /keV	I_γ *
^{12}C $(p,\gamma)^{13}$N	–	10m	511	Sr	^{86}Y	15h	1077	3.5E3	Si	^{30}P	2.5m	511	1.3E5
$(d,n)^{13}$N	0.3	10m	511		^{87}Y	3.3d	485	4.6E3		^{28}Al	2.3m	1779	2.2E3
$(^{3}$He$,\alpha)^{11}$C	–	20m	511	Y	^{89}Zr	3.3d	909	1.8E4		^{29}Al	6.6m	1273	1.1E3
$(^{6}$Li$,\alpha n)^{13}$N		10m	511	Zr	^{90}Nb	15h	141	1.1E4	S	^{34}Cl	32m	146	6.7E4
^{13}C $(p,n)^{13}$N	3.2	10m	511				1129	1.5E4				2129	2.8E3
^{14}N $(p,n)^{14}$O	6.4	1.2m	511	Nb	^{93}Mo	7h	685	1.6E3	Cl	^{39}Cl	56m	250	1.0E3
$(p,\alpha)^{11}$C	3.1	20m	511				1477	1.8E3				1267	1.1E3
$(d,n)^{15}$O	–	2m	511	Mo	^{94}Tc	5h	703	1.4E3				1517	1.2E3
$(^{3}$He$,\alpha)^{13}$N	–	10m	511				871	1.6E3	K	^{43}K	22h	372	28
$(^{9}$Be$,\alpha n)^{18}$F		110m	511		^{95}Tc	20h	766	6.3E3				616	26
^{16}O $(p,\alpha)^{13}$N	5.5	10m	511		^{96}Tc	4.4d	778	1.5E3	Ca	^{42}Sc	61s	438	7.5E2
$(d,n)^{17}$F	1.8	66s	511	Ru	^{96}Rh	9.3m	631	4.8E3				1220	7.8E2
$(t,n)^{18}$F	–	110m	511				685	7.3E3				1520	7.6E2
$(^{3}$He$,p)^{18}$F	–	110m	511		^{98}Rh	9m	652	1.1E4	Tl	^{51}Ti	5.8m	320	7.7E2
$(^{3}$He$,n)^{18}$Ne	3.8	1.5s	511		^{99}Rh	4.7h	341	8.5E3	V	^{53}V	1.6m	1006	3.3E3
^{18}O $(p,n)^{18}$F	2.6	110m	511	Rh	^{100}Rh	20h	540	3.4E3		^{52}V	3.8m	1434	9.6E3
^{19}F $(\alpha,n)^{22}$Na	2.4	2.6y	1275	Pd	^{103}Pd	17d	357	0.55	Cr	^{52}Mn	21m	1434	3.8E2
^{23}Na $(\alpha,n)^{26}$Al	3.5	6.4s	511		^{104}Ag	1.1h	556	2.1E4	Mn	^{57}Mn	1.5m	122	4.4E2
^{25}Mg $(\alpha,p)^{28}$Al	–	2.3m	1779				767	1.3E4		^{56}Mn	2.6h	847	1.5E3
^{27}Al $(\alpha,n)^{30}$P	3.1	2.5m	2240	Ag	^{107}Cd	6.5h	93	2.0E3	Fe	^{58}Co	71d	811	2.3
				Cd	^{113}In	1.7h	392	1.0E4	Co	^{60}Co	10m	1333	4.5
				In	^{113}Sn	115d	392	5.8	Ni	^{60}Cu	23m	1333	2.6E2
				Sn	^{114}Sb	3.5m	933	1.6E4	Cu	^{66}Cu	5.1m	1039	15
					^{117}Sb	2.8h	159	6.3E3	Zn	^{66}Ga	9.4h	1039	17
				Sb	^{121}Te	17d	573	4.5E2	Ga	^{73}Ga	4.8h	298	3
					^{128}I	25m	443	2.5E4	Ge	^{75}Ge	46s	140	7
				Te	^{130}I	12h	536	6.5E3		^{78}As	1.5h	614	7
							669	6.6E3	As	^{77}Se	18s	162	65
							740	5.4E3					

10 MeV proton activation

T	RI	$\tau_{1/2}$	E_γ /keV	I_γ *
Li	^{7}Be	53d	478	8.8E2
B	^{7}Be	53d	478	1.2E2
S	^{34}Cl	32m	146	4.3E3
Ca	^{44}Sc	3.9h	1157	1.1E4
Ti	^{48}V	16d	984	3.3E3
			1312	3.3E3

RI: radionuclide emitting the gamma-ray; T: target element; * (ppm$\cdot$ks)$^{-1}$

TABLE 14.20
K shell line energies, fluorescence yields and emission rates

Elt	Z	K_{abs}	$K\alpha_1$	$K\beta_1$	$K\beta_2$	$\omega\kappa$	$K\beta_1$	$K\beta_2$
			Energy (keV)				Relative Intensity $K\alpha = 100$	
Na	11	1.072	1.041			0.023		
Mg	12	1.305	1.253			0.030		
Al	13	1.559	1.486			0.039		
Si	14	1.838	1.736	1.839		0.050	2.7	
P	15	2.142	2.013	2.139		0.063	4.3	
S	16	2.472	2.307	2.464		0.078	5.9	
Cl	17	2.822	2.621	2.816		0.097	8.2	
Ar	18	3.202	2.955	3.190		0.118	10.5	
K	19	3.607	3.312	3.589		0.140	11.7	
Ca	20	4.038	3.690	4.012		0.163	12.8	
Sc	21	4.496	4.088	4.460		0.188	13.1	
Ti	22	4.965	4.508	4.931		0.214	13.4	
V	23	5.465	4.949	5.426		0.243	13.5	
Cr	24	5.989	5.411	5.946		0.275	13.5	
Mn	25	6.540	5.894	6.489		0.308	13.5	
Fe	26	7.112	6.398	7.057		0.340	13.5	
Co	27	7.709	6.924	7.648		0.373	13.5	
Ni	28	8.333	7.471	8.263		0.406	13.5	
Cu	29	8.979	8.040	8.904		0.440	13.7	
Zn	30	9.659	8.630	9.570		0.474	13.8	
Ga	31	10.368	9.241	10.263		0.507	14.3	
Ge	32	11.104	9.874	10.980		0.535	14.7	
As	33	11.868	10.530	11.724		0.562	15.2	
Se	34	12.658	11.207	12.494		0.589	15.7	
Br	35	13.474	11.907	13.289	13.467	0.618	15.8	1.1
Rb	37	15.201	13.373	14.959	15.183	0.667	16.0	1.7
Sr	38	16.105	14.140	15.833	16.082	0.690	16.0	2.0
Y	39	17.037	14.931	16.735	17.013	0.710	16.3	2.2
Zr	40	17.998	15.744	17.665	17.967	0.730	16.6	2.4
Nb	41	18.986	16.581	18.619	18.949	0.747	16.8	2.6
Mo	42	20.002	17.441	19.610	19.962	0.765	17.0	2.7
Tc	43	21.054	18.325	20.615	21.002	0.780	17.3	2.8
Ru	44	22.118	19.233	21.653	22.070	0.794	17.5	2.9
Rh	45	23.224	20.165	22.720	23.169	0.808	17.7	3.0
Pd	46	24.350	21.121	23.815	24.295	0.820	17.9	3.1
Ag	47	25.514	22.101	24.938	25.452	0.831	17.9	3.3
Cd	48	26.711	23.106	26.091	26.639	0.843	17.8	3.5
In	49	27.940	24.136	27.271	27.856	0.853	18.1	3.6
Sn	50	29.200	25.191	28.481	29.104	0.862	18.5	3.7
Sb	51	30.491	26.271	29.721	30.388	0.870	18.5	3.7
I	53	33.169	28.607	32.289	33.036	0.884	18.9	4.0
Ba	56	37.441	32.062	36.372	37.251	0.902	19.2	4.5

TABLE 14.21
L shell line energies (keV)

Elt	Z	$L\alpha_1$	$L\ell$	$L\eta$	$L\beta_1$	$L\beta_2$	$L\beta_3$	$L\beta_4$	$L\beta_5$	$L\beta_6$	$L\gamma_1$	$L\gamma_2$	$L\gamma_3$	$L\gamma_4$	$L\gamma_5$	$L\gamma_6$
ZrL	40	2.040	1.792	1.876	2.124	2.219	2.201	2.187		2.171	2.302	2.502			2.255	
NbL	41	2.166	1.902	1.996	2.257	2.367	2.334	2.319		2.312	2.461	2.663			2.406	
MoL	42	2.293	2.015	2.120	2.394	2.518	2.473	2.455		2.455	2.623	2.830			2.563	
RuL	44	2.558	2.252	2.382	2.683	2.835	2.763	2.771		2.763	2.964	3.180			2.891	
RhL	45	2.696	2.376	2.519	2.834	3.001	2.915	2.890		2.922	3.143	3.363			3.064	
PdL	46	2.838	2.503	2.660	2.990	3.171	3.072	3.045		3.087	3.328	3.553			3.243	
AgL	47	2.984	2.633	2.806	3.150	3.347	3.234	3.203		3.255	3.519	3.743			3.428	
CdL	48	3.133	2.767	2.956	3.316	3.528	3.401	3.367		3.429	3.716	3.951			3.619	
InL	49	3.286	2.904	3.112	3.487	3.713	3.572	3.535		3.608	3.920	4.160		4.236	3.815	
SnL	50	3.443	3.044	3.272	3.662	3.904	3.750	3.708		3.792	4.130	4.376		4.463	4.018	
SbL	51	3.604	3.188	3.436	3.843	4.100	3.932	3.886		3.979	4.347	4.599		4.696	4.228	
TeL	52	3.796	3.335	3.605	4.029	4.301	4.120	4.069		4.173	4.570	4.828		4.936	4.443	
IL	53	3.937	3.484	3.780	4.221	4.507	4.313	4.257		4.370	4.800	5.065		5.184	4.665	
CsL	55	4.286	3.794	4.141	4.619	4.935	4.716	4.649		4.780	5.279	5.541	5.552	5.702	5.128	
BaL	56	4.465	3.953	4.330	4.827	5.156	4.926	4.851		4.993	5.530	5.796	5.808	5.972	5.370	
LaL	57	4.650	4.124	4.524	5.041	5.383	5.143	5.061		5.211	5.788	6.059	6.073	6.251	5.620	
CeL	58	4.839	4.287	4.731	5.261	5.612	5.364	5.276		5.433	6.051	6.324	6.340	6.527	5.874	
PrL	59	5.033	4.452	4.935	5.488	5.849	5.591	5.497		5.659	6.321	6.597	6.615	6.814	6.135	
NdL	60	5.229	4.632	5.145	5.721	6.088	5.828	5.721		5.892	6.601	6.882	6.900	7.106	6.405	
SmL	62	5.635	4.994	5.588	6.204	6.586	6.317	6.195	6.711	6.369	7.177	7.465	7.485	7.712	6.967	7.306
EuL	63	5.845	5.176	5.816	6.455	6.842	6.570	6.438	6.975	6.616	7.479	7.766	7.795	8.029	7.255	7.613
GdL	64	6.056	5.361	6.049	6.712	7.102	6.830	6.686	7.236	6.866	7.784	8.086	8.104	8.354	7.553	7.924
TbL	65	6.272	5.546	6.283	6.977	7.365	7.095	6.939	7.508	7.115	8.100	8.396	8.422	8.683	7.852	8.245

TABLE 14.21 (cont.)
L shell line energies (keV)

Elt	Z	Lα₁	Lℓ	Lη	Lβ₁	Lβ₂	Lβ₃	Lβ₄	Lβ₅	Lβ₆	Lγ₁	Lγ₂	Lγ₃	Lγ₄	Lγ₅	Lγ₆
DyL	66	6.494	5.742	6.533	7.246	7.634	7.369	7.203	7.804	7.369	8.417	8.713	8.752	9.018	8.165	8.574
HdL	67	6.719	5.942	6.787	7.524	7.910	7.650	7.470	8.061	7.634	8.746	9.049	9.086	9.373	8.480	8.903
ErL	68	6.947	6.152	7.057	7.809	8.188	7.938	7.744	8.349	7.908	9.087	9.384	9.429	9.721	8.812	9.253
TmL	69	7.179	6.341	7.308	8.100	8.467	8.229	8.024	8.639	8.176	9.424	9.728	9.778	10.083	9.143	9.606
YbL	70	7.414	6.544	7.579	8.400	8.757	8.535	8.312	8.938	8.455	9.788	10.088	10.141	10.458	9.484	9.975
LuL	71	7.654	6.752	7.856	8.708	9.047	8.845	8.605	9.239	8.736	10.142	10.458	10.509	10.840	9.841	10.341
HfL	72	7.898	6.958	8.138	9.021	9.346	9.162	8.904	9.553	9.021	10.514	10.832	10.889	11.238	10.199	10.731
TaL	73	8.145	7.172	8.427	9.342	9.650	9.486	9.211	9.873	9.314	10.893	11.215	11.276	11.643	10.569	11.129
WL	74	8.396	7.386	8.723	9.671	9.960	9.817	9.524	10.199	9.610	11.284	11.606	11.672	12.061	10.947	11.537
ReL	75	8.651	7.602	9.026	10.008	10.274	10.158	9.845	10.530	9.909	11.683	12.008	12.080	12.490	11.332	11.954
OsL	76	8.910	7.821	9.335	10.354	10.597	10.509	10.174	10.869	10.215	12.093	12.420	12.498	12.921	11.728	12.383
IrL	77	9.174	8.040	9.649	10.706	10.919	10.866	10.509	11.209	10.523	12.510	12.840	12.922	13.366	12.132	12.818
PtL	78	9.441	8.267	9.973	11.069	11.249	11.233	10.852	11.559	10.840	12.940	13.268	13.359	13.826	12.550	13.269
AuL	79	9.712	8.493	10.307	11.440	11.583	11.608	11.203	11.914	11.158	13.379	13.707	13.807	14.297	12.972	13.728
HgL	80	9.987	8.720	10.649	11.821	11.922	11.993	11.561	12.275	11.480	13.828	14.160	14.262	14.776	13.408	14.196
TlL	81	10.267	8.952	10.992	12.211	12.270	12.388	11.929	12.641	11.810	14.289	14.623	14.734	15.269	13.850	14.684
PbL	82	10.554	9.183	11.347	12.612	12.621	12.791	13.013	12.304	12.141	13.370	14.762	15.200	15.765	14.305	15.176
BiL	83	10.837	9.419	11.710	13.021	12.978	13.208	12.689	13.393	12.479	15.245	15.580	15.708	16.292	14.771	15.683
RaL	88	12.338	10.620	13.661	15.233	14.839	15.442	14.745	15.375	14.234	17.845	18.176	18.354	19.081	17.271	18.411
ThL	90	12.967	11.117	14.507	16.199	15.621	16.423	15.640	16.211	14.973	18.979	19.302	19.503	20.289	18.361	19.596
PaL	91	13.288	11.364	14.944	16.699	16.022	16.927	16.101	16.634	15.343	19.565	19.869	20.094	20.879	18.925	20.212
UL	92	13.612	11.616	15.397	17.217	16.425	17.452	16.573	17.067	15.723	20.164	20.481	20.709	21.559	19.504	20.839

TABLE 14.22
L subshell fluorescence yields and Coster-Kronig transition rates

Z	ω_1	ω_2	ω_3	f_{12}	f_{13}	f_{23}	f'_{13}	v_1	v_2	v_3
28	0.0014	0.0086	0.0093	0.300	0.55	0.028		0.0092	0.0089	0.0093
30	0.0018	0.011	0.012	0.290	0.54	0.026		0.0116	0.0113	0.012
32	0.0024	0.013	0.015	0.280	0.53	0.050		0.0142	0.0138	0.015
34	0.0032	0.016	0.018	0.280	0.52	0.076		0.0174	0.0174	0.018
36	0.0041	0.020	0.022	0.270	0.52	0.100		0.0215	0.0222	0.022
38	0.0051	0.024	0.026	0.270	0.52	0.117		0.0259	0.0270	0.026
40	0.0068	0.028	0.031	0.260	0.52	0.132	0.0001	0.0313	0.0321	0.031
42	0.010	0.034	0.037	0.100	0.61	0.141	0.0001	0.0365	0.0392	0.037
44	0.012	0.040	0.043	0.100	0.61	0.148	0.0001	0.0429	0.0464	0.043
46	0.014	0.047	0.049	0.100	0.60	0.151	0.0001	0.0488	0.0544	0.049
48	0.018	0.056	0.056	0.100	0.59	0.155	0.0001	0.0575	0.0647	0.056
50	0.037	0.065	0.064	0.170	0.27	0.157	0.0003	0.0671	0.0750	0.064
52	0.041	0.074	0.074	0.180	0.28	0.155	0.0003	0.0771	0.0855	0.074
54	0.046	0.083	0.085	0.190	0.28	0.154	0.0004	0.0881	0.0961	0.085
56	0.052	0.096	0.097	0.190	0.28	0.153	0.0005	0.1003	0.1108	0.097
58	0.058	0.110	0.111	0.190	0.29	0.153	0.0006	0.1144	0.1270	0.111
60	0.064	0.124	0.125	0.190	0.30	0.152	0.0007	0.1288	0.1430	0.125
62	0.071	0.140	0.139	0.190	0.30	0.150	0.0008	0.1434	0.1609	0.139
64	0.079	0.158	0.155	0.190	0.30	0.147	0.0010	0.1600	0.1808	0.155
66	0.089	0.178	0.174	0.190	0.30	0.143	0.0012	0.1800	0.2029	0.174
68	0.100	0.200	0.192	0.190	0.30	0.140	0.0014	0.2010	0.2269	0.192
70	0.112	0.222	0.210	0.190	0.29	0.138	0.0018	0.2210	0.2510	0.210
72	0.128	0.246	0.231	0.180	0.28	0.135	0.0023	0.2431	0.2772	0.231
74	0.147	0.270	0.255	0.170	0.28	0.133	0.0028	0.2708	0.3039	0.255
76	0.130	0.295	0.281	0.160	0.39	0.128	0.0029	0.2934	0.3310	0.281
78	0.114	0.321	0.306	0.140	0.50	0.124	0.0028	0.3181	0.3589	0.306
80	0.107	0.347	0.333	0.130	0.56	0.120	0.0030	0.3448	0.3870	0.333
82	0.112	0.373	0.360	0.120	0.58	0.116	0.0035	0.3718	0.4148	0.360
84	0.122	0.401	0.386	0.110	0.58	0.111	0.0042	0.3963	0.4438	0.386
86	0.134	0.429	0.411	0.100	0.58	0.110	0.0052	0.4219	0.4742	0.411
88	0.146	0.456	0.437	0.090	0.58	0.108	0.0064	0.4475	0.5032	0.437
90	0.161	0.479	0.463	0.090	0.57	0.108	0.0078	0.4761	0.5290	0.463
92	0.176	0.467	0.489	0.080	0.57	0.167	0.0097	0.5034	0.5487	0.489
94	0.205	0.464	0.514	0.050	0.56	0.198	0.0130	0.5278	0.5658	0.514
96	0.228	0.479	0.539	0.040	0.55	0.200	0.0160	0.5565	0.5868	0.539

TABLE 14.23
Emission rates

a. L_1 Emission Rates (normalised to $L\beta_3 = 100$)

Z	β_3	β_4	γ_2	γ_3	$\gamma_{44'}$	$\beta_{9,10}$	Total
28	100.0	54.0				0.30	154.30
30	100.0	54.6				0.45	155.05
32	100.0	55.1	0.95	17.3		0.60	173.95
34	100.0	55.6	2.80	18.0		0.70	177.10
36	100.0	56.2	5.30	18.2		0.80	180.50
38	100.0	56.8	7.20	18.8		1.00	183.80
40	100.0	57.4	8.30	19.0		1.10	185.80
42	100.0	70.6	9.10	19.6		1.20	200.50
44	100.0	67.8	10.20	20.2		1.40	199.60
46	100.0	65.5	10.90	20.6		1.60	198.60
48	100.0	63.5	11.60	21.3		1.75	198.15
50	100.0	62.1	12.50	22.0	0.70	1.90	199.20
52	100.0	60.7	13.30	22.6	2.10	2.10	200.80
54	100.0	59.8	14.00	23.3	4.00	2.40	203.50
56	100.0	59.5	14.90	24.0	5.50	2.60	206.50
58	100.0	59.2	15.40	24.6	5.60	2.85	207.65
60	100.0	59.4	15.90	25.4	5.60	3.10	209.40
62	100.0	60.0	16.40	26.3	5.60	3.40	211.70
64	100.0	60.8	19.20	27.0	6.00	3.70	216.70
66	100.0	61.8	19.50	28.0	5.56	4.10	218.96
68	100.0	63.5	19.80	29.0	5.50	4.50	222.30
70	100.0	65.5	20.70	29.8	5.50	4.90	226.40
72	100.0	67.8	21.20	30.7	6.20	5.30	231.20
74	100.0	70.5	21.80	31.8	6.90	5.80	236.80
76	100.0	73.2	23.00	32.8	7.70	6.30	243.00
78	100.0	76.5	24.50	33.8	8.30	6.90	250.00
80	100.0	80.3	26.30	35.0	9.10	7.60	258.30
82	100.0	84.2	28.60	36.0	10.00	8.30	267.10
84	100.0	88.5	31.30	37.2	11.00	9.10	277.10
86	100.0	93.4	34.20	38.2	12.00	9.90	287.70
88	100.0	98.9	37.50	39.6	13.10	10.90	300.00
90	100.0	104.5	41.20	41.0	14.20	12.00	312.90
92	100.0	110.2	45.00	42.6	15.20	13.20	326.20
94	100.0	116.2	49.50	44.0	16.20	14.70	340.60
96	100.0	123.0	55.70	45.7	17.40	16.50	358.30

b. L_1 Emission Rates (normalised to $L\beta_1 = 100$)

Z	β_3	β_4	γ_2	γ_3	$\gamma_{44'}$	$\beta_{9,10}$	Total
28	100.0	7.60				0.43	108.03
30	100.0	6.80				0.30	107.10
32	100.0	6.28				0.23	106.51
34	100.0	5.80				0.43	106.23
36	100.0	5.35				0.46	105.81
38	100.0	4.93				0.51	105.44

TABLE 14.23 (*cont.*)
Emission rates

b. L_2 Emission Rates (normalised to $L\beta_1 = 100$)

Z	β_1	η	γ_1	γ_6	γ_5	Total
40	100.0	4.60	3.30		0.53	108.43
42	100.0	4.30	5.50		0.55	110.35
44	100.0	4.00	7.33		0.56	111.89
46	100.0	3.75	10.67		0.56	114.98
48	100.0	3.55	10.60		0.58	114.73
50	100.0	3.35	11.80		0.59	115.74
52	100.0	3.20	12.70		0.61	116.51
54	100.0	3.00	14.00		0.62	117.62
56	100.0	2.85	14.50		0.63	117.98
58	100.0	2.70	15.30	0.13	0.64	118.77
60	100.0	2.60	16.00	0.13	0.64	119.37
62	100.0	2.45	16.50	0.12	0.64	119.71
64	100.0	2.35	17.00	0.11	0.65	120.11
66	100.0	2.25	17.40	0.11	0.65	120.40
68	100.0	2.16	17.80	0.10	0.65	120.70
70	100.0	2.10	18.17	0.08	0.65	121.00
72	100.0	2.08	18.43	0.19	0.66	121.36
74	100.0	2.10	18.80	0.72	0.66	122.28
76	100.0	2.12	19.34	1.65	0.67	123.78
78	100.0	2.18	19.73	2.40	0.68	124.99
80	100.0	2.25	20.35	3.10	0.69	126.39
82	100.0	2.30	20.93	3.65	0.70	127.58
84	100.0	2.40	21.54	4.15	0.71	128.80
86	100.0	2.46	22.20	4.55	0.73	129.94
88	100.0	2.50	22.87	4.87	0.74	130.98
90	100.0	2.60	23.43	5.02	0.75	131.80
92	100.0	2.65	24.10	5.12	0.76	132.63
94	100.0	2.70	24.40	5.16	0.78	133.04
96	100.0	2.75	25.07	5.20	0.79	133.81

c. L_3 Emission Rates (normalised to $L\alpha_1 = 100$)

Z	α_1	$\beta_{2,15}$	α_2	β_5	β_6	$1001/\alpha$*	l	Total
28	100.0		11.10		0.51	8.05	8.94	120.56
30	100.0		11.10		0.37	6.60	7.33	118.80
32	100.0		11.10		0.49	5.80	6.44	118.03
34	100.0		11.10		0.54	5.18	5.75	117.39
36	100.0		11.10		0.58	4.75	5.28	116.95
38	100.0		11.10		0.65	4.43	4.92	116.67
40	100.0	0.70	11.10		0.70	4.20	4.67	117.16
42	100.0	5.17	11.10		0.72	4.00	4.44	121.44
44	100.0	9.30	11.12		0.75	3.85	4.28	125.45
46	100.0	11.80	11.12		0.77	3.70	4.11	127.80
48	100.0	14.33	11.12		0.81	3.66	4.07	130.32

* $\alpha = \alpha_1 + \alpha_2$

TABLE 14.23 (*cont.*)
Emission rates

c. L_3 Emission Rates (normalised to $L\alpha_1 = 100$)

Z	α_1	$\beta_{2,15}$	α_2	β_5	β_6	$100 1/\alpha*$	l	Total
50	100.0	16.00	11.13		0.84	3.60	4.00	131.97
52	100.0	18.00	11.13		0.88	3.60	4.00	134.01
54	100.0	19.40	11.13		0.92	3.60	4.00	135.45
56	100.0	20.67	11.13	0.14	0.97	3.62	4.02	136.93
58	100.0	21.00	11.14	0.14	0.99	3.68	4.09	137.36
60	100.0	21.33	11.14	0.13	0.88	3.72	4.13	137.61
62	100.0	21.07	11.14	0.12	0.93	3.74	4.16	137.41
64	100.0	20.83	11.14	0.11	0.99	3.78	4.20	137.27
66	100.0	20.50	11.14	0.13	1.05	3.83	4.26	137.08
68	100.0	20.04	11.15	0.15	1.12	3.90	4.33	136.80
70	100.0	19.40	11.15	0.17	1.17	4.02	4.47	136.36
72	100.0	21.33	11.15	0.30	1.21	4.13	4.59	138.58
74	100.0	22.74	11.16	0.50	1.25	4.28	4.76	140.41
76	100.0	23.40	11.16	1.32	1.37	4.45	4.95	142.20
78	100.0	24.00	11.17	1.98	1.43	4.62	5.14	143.72
80	100.0	24.50	11.17	2.62	1.50	4.83	5.37	145.16
82	100.0	24.83	11.17	3.21	1.56	5.02	5.58	146.35
84	100.0	25.13	11.17	3.73	1.62	5.22	5.80	147.45
86	100.0	25.60	11.18	4.25	1.68	5.40	6.00	148.71
88	100.0	25.92	11.18	4.73	1.76	5.63	6.26	149.85
90	100.0	26.17	11.18	5.18	1.82	5.88	6.54	150.89
92	100.0	26.40	11.18	5.58	1.89	6.11	6.79	151.84
94	100.0	26.67	11.18	5.92	1.95	6.30	7.00	152.72
96	100.0	26.93	11.18	6.26	2.01	6.60	7.34	153.72

$* \alpha = \alpha_1 + \alpha_2$

TABLE 14.24
Mass Attenuation Coefficients (μ/ρ cm^2 g^{-1}) E in keV ($E \leq 20$ keV)

		$E > E_K$	$E_K > E > E_{L1}$	$E_{L2} > E > E_{L3}$	$E_{L3} > E > E_M$
$3 < Z \leq 10$	a	8.41			
	b	2.92			
	c	3.07			
	%	2			
$11 \leq Z \leq 18$	a	15.50	0.528		
	b	2.79	2.74		
	c	2.73	3.03		
	%	2	4		
$19 \leq Z \leq 36$	a	25.26	0.859		$1.27.10^{-2}$
	b	2.66	2.70		2.44
	c	2.47	2.90		3.47
	%	2	1		3
$37 \leq Z \leq 54$	a		0.992		
	b		2.70		
	c		2.88		
	%		1		
$55 \leq Z \leq 71$	a		1.11	0.588	$8.55.10^{-2}$
	b		2.70	2.62	2.50
	c		2.83	2.82	2.98
	%		1	6	3
$72 \leq Z \leq 86$	a		$5.58.10^{-2}$		$5.76.10^{-2}$
	b		2.50		2.55
	c		3.38		3.09
	%		4		1
$87 \leq Z \leq 94$	a				$4.32.10^{-4}$
	b				2.63
	c				4.26
	%				3

$(\mu/\rho) = a\, E^{-b}\, Z^c$
% is the three standard deviation percentage error on the fit

TABLE 14.25
Angles between planes in cubic crystals (Mayer and Rimini, 1977).

HKL	hkl						
100	100	0.00	90.00				
	110	45.00	90.00				
	111	54.75					
	210	26.56	63.43	90.00			
	211	35.26	65.90				
	221	48.19	70.53				
	310	18.43	71.56	90.00			
	311	25.24	72.45				
	320	33.69	56.31	90.00			
	321	36.70	57.69	74.50			
	322	43.31	60.98				
	331	46.51	76.74				
	332	50.24	64.76				
	410	14.04	75.96	90.00			
	411	19.47	76.37				
110	110	0.00	60.00	90.00			
	111	35.26	90.00				
	210	18.43	50.77	71.56			
	211	30.00	54.74	73.22	90.00		
	221	19.47	45.00	76.37	90.00		
	310	26.56	47.87	63.43	77.08		
	311	31.48	64.76	90.00			
	320	11.31	53.96	66.91	78.69		
	321	19.11	40.89	55.46	67.79	79.11	
	322	30.96	46.69	80.12	90.00		
	331	13.26	49.54	71.07	90.00		
	332	25.24	41.08	81.33	90.00		
	410	80.96	46.69	59.04	80.12		
	411	33.56	60.00	70.53	90.00		
111	111	0.00	70.53				
	210	39.23	75.04				
	211	19.47	61.87	90.00			
	221	15.79	54.74	78.90			
	310	43.09	68.58				
	311	29.50	58.52	79.98			
	320	36.81	80.78				
	321	22.21	51.89	72.02	90.00		
	322	11.42	65.16	81.95			
	331	22.00	48.53	82.39			
	332	10.02	60.50	75.75			
	410	45.56	65.16				
	411	35.26	57.02	74.21			
210	210	0.00	36.87	53.13	66.42	78.46	90.00
	211	24.09	43.09	56.79	79.48	90.00	
	221	26.56	41.81	53.40	63.43	72.65	90.00
	310	8.13	31.95	45.00	64.90	73.57	81.87
	311	19.29	47.61	66.14	82.25		
	320	7.12	29.74	41.91	60.25	68.15	75.64	82.87
	321	17.02	33.21	53.30	61.44	68.99	83.14	90.00

TABLE 14.26

Structural data for some elemental crystals (adapted from Gemmell, 1974); θ and μ_1^* are the Debye temperature and thermal vibration amplitude (at room temperature).

Z_2	M_2	Name	Structure		$0.4685Z_2^{-1/3}$ (Å)	θ (°K)	μ_1^* (Å)	Lattice Constant (Å)
6	12.01	C	f.c.c.	(dia)	0.258	2000	0.04	3.567
13	26.98	Al	f.c.c.		0.199	390	0.105	4.050
14	28.09	Si	f.c.c.	(dia)	0.194	543	0.075	5.431
23	50.94	V	b.c.c.		0.165	360	0.082	3.024
24	52.00	Cr	b.c.c.		0.162	485	0.061	2.884
26	55.85	Fe	b.c.c.		0.158	420	0.068	2.867
28	58.71	Ni	f.c.c.		0.154	425	0.065	3.524
29	63.54	Cu	f.c.c.		0.152	315	0.084	3.615
32	72.59	Ge	f.c.c.	(dia)	0.148	290	0.085	5.657
41	92.91	Nb	b.c.c.		0.136	275	0.079	3.300
42	95.94	Mo	b.c.c.		0.135	380	0.057	3.147
45	102.91	Rh	f.c.c.		0.132	340	0.061	3.803
46	106.4	Pd	f.c.c.		0.131	275	0.074	3.890
47	107.87	Ag	f.c.c.		0.130	215	0.074	4.086
73	180.95	Ta	b.c.c.		0.112	245	0.064	3.306
74	183.85	W	b.c.c.		0.112	310	0.050	3.165
78	195.09	Pt	f.c.c.		0.110	225	0.066	3.923
79	196.97	Au	f.c.c.		0.109	170	0.087	4.078
82	207.19	Pb	f.c.c.		0.108	88	0.164	4.951

* (293° K)

TABLE 14.27
Ion beam sputter and etch rates (500 V Ar ions, 90°, 1 mA cm^{-2})

Elements	SPR	ER	Compounds	SPR	ER
Be	190	170–200	Al_2O_3	129	80–130
B		290	Al_2O_3TiC		813–825
C	40	44–80	* AZ1350	231	200–250
Mg	225		* AZ1450		140
Al	540	730	* AZ1470		180
Si	340	200–380	$Bi_{12}Ge_{20}O$	1290	1250–1450
Ti	336	150–380	BiTe	10	
V	337	370–550	BN	65	
Cr	530	160–580	CdS	2100	2050–2300
Mn	874	850–900	CdTe	560	2055
Fe	490	200–530	CoCr	143–300	
Co	510	500–550	Cr_2O_3		50
Ni	570	550–660	FeO		450–490
Cu	880	450–1100	GaAs	1500	630–1600
Ge	920	900–1000	GaGd		260–290
Rb	4000	3090–4020	GaP	1400	1400–1600
Y	840	820–960	GaSb	1700	1600–1900
Zr	570	220–620	Glass		200–210
Nb	390	200–440	HgCdTe		1300–1500
Mo	470	220–540	$In_2O_3SnO_2$	157	80–200
Ru	580	560–610	InSb	1300	1200–1500
Rh	650	620–740	* KTFR		260–290
Pd	1150	1000–1300	$LiNbO_3$	400	390–420
Ag	1900	1200–2200	MgF	28	
Sn	1217	1200–1800	Mo_2C	210	
Sb	3238	3100–3300	NiCr		180–200
Te		2000	NiFe	350	330–660
Sm	1000	950–1100	PbTe	3600	2900–3800
Gd	1027	990–1100	* PMMA		550–580
Dy	1040	1100–1300	* PSG		162–300
Er	881	890–980	* RISTON		240–260
Hf	590	580–660	SiC	320	310–350
Ta	380	180–420	Si_3N_4	140	250
W	340	190–380	SiO		240
Re	470	450–520	SiO_2	400	260–400
Os	440	440–510	S.Steel304		240–260
Ir	540	530–600	TaC	110	150
Pt	792	760–880	TiAu		340
Au	1553	1000–1700	WTi		125
Pb	3100	2800–3200	YO		75
Bi	8800	8500–8900			
Th	740	720–820			
U	660	540–740			

* Photoresists
SPR = Sputter Rate (Å/min); ER = Etch Rate (Å/min)

TABLE 14.28
Sputter induced photon yields

Element	λ Å	E_μ eV	I_γ 10^{-3}	MDL $\mu g \cdot g^{-1}$
Ag	3281	3.77	10.14	200
Al	3962	3.14	2.55	50
B	2497	4.96	0.26	100
Ba	5535	2.24	1.44	800
Be	2348	5.27	0.22	50
Ca	4227	2.93	3.00	50
Cd	2288	5.41	0.21	200
Ce	4187	3.51	0.095	2500
Co	3453	4.01	0.41	500
Cr	4254	2.91	1.73	300
Cs	4555	2.72	4.58	100
Cu	3247	3.81	3.83	150
F	6902	14.52	0.007	3000
Fe	3581	4.31	0.75	300
Ge	3039	4.95	0.75	500
H	6563	12.09	0.35	15
In	4511	3.02	2.18	500
K	7665	1.61	2.69	20
Li	6708	185	8.37	1
Mg	2852	4.34	6.99	40
Mn	4034	3.07	2.47	200
Mo	3798	3.26	5.33	150
Na	5890	2.10	7.41	1
Ni	3415	3.65	0.71	400
P	2536	7.20	0.003	7500
Pb	4058	4.37	0.082	5000
Re	3460	3.58	0.75	400
Si	2882	5.07	1.28	250
Sr	4607	2.69	4.20	200
Ta	3311	4.43	0.083	3500
Ti	3653	3.44	0.18	700
Tl	3519	4.48	4.65	500
Zn	2138	5.79	0.009	6300
Zr	3601	3.59	0.11	1300

TABLE 14.29
Selected Auger electron energies (eV)

El.	KLL			
Li	30	36	42	
Be		98		
B			172	
C			263	
N		358	375	
O	466	482	503	
F	603	620	645	
Na	837	915	948	979
Mg	1105	1139	1173	1184
Al	1340	1360	1375	1390
Si	1557	1577	1597	1614
	LMM			
K		214	243	265
Ca		249	284	330
Sc		290	331	364
Ti	362	381	416	449
V	408	428	469	509
Cr	476	486	525	568
Mn	509	528	582	631
Fe	589	645	700	712
Co	646	708	771	784
Ni	707	772	844	860
Cu	764	836	916	932
Zn	825	904	990	1012
Ga	888	971	1066	1093
Ge	952	1041	1143	1174
As	1016	1112	1222	1259
Se	1084	1188	1310	1350
Br	1167	1268	1390	1437

El.	LMM				
Y	1442	1578	1730	1808	
Zr	1527	1665	1836	1921	
Nb	1630	1762	1936	2032	
Mo	1704	1872	2036	2142	
	MNN				
Y	75	99	107	126	
Zr		90	115	146	173
Nb		104	131	164	196
Mo		119	159	184	219
Rh		219	251	298	347
Pd	184	236	268	321	
Ag		257	291	349	354
Cd		276	310	375	381
In		299	331	402	409
Sn		314	351	427	435
Sb		332	369	452	460
Te		350	390	479	488
Gd		750	867	1014	1165
Hf		1213	1406	1614	1671
Ta		1252	1453	1674	1731
W		1294	1500	1729	1792
Re		1334	1548	1782	1852
Ir		1420	1650	1899	1974
Pt		1460	1698	1960	2039
Au		1499	1744	2015	2101
Bi		1927	2035	2226	2336
	NVV				
Th	53	148	221	235	
U	72	88	176	279	

Index

9 0 1 2 3 4 5 6 7 8
A B C D E F G H I J